NATO ASI Series

Advanced Science Institutes Series

A series presenting the results of activities sponsored by the NATO Science Committee, which aims at the dissemination of advanced scientific and technological knowledge, with a view to strengthening links between scientific communities.

The Series is published by an international board of publishers in conjunction with the NATO Scientific Affairs Division

A Life Sciences B Physics	Plenum Publishing Corporation London and New York
C Mathematical and Physical Sciences D Behavioural and Social Sciences E Applied Sciences	Kluwer Academic Publishers Dordrecht, Boston and London
F Computer and Systems Sciences G Ecological Sciences H Cell Biology I Global Environmental Change	Springer-Verlag Berlin Heidelberg New York London Paris Tokyo Hong Kong Barcelona Budapest

PARTNERSHIP SUB-SERIES

1. Disarmament Technologies	Kluwer Academic Publishers
2. Environment	Springer-Verlag
3. High Technology	Kluwer Academic Publishers
4. Science and Technology Policy	Kluwer Academic Publishers
5. Computer Networking	Kluwer Academic Publishers

The Partnership Sub-Series incorporates activities undertaken in collaboration with NATO's Cooperation Partners, the countries of the CIS and Central and Eastern Europe, in Priority Areas of concern to those countries.

NATO-PCO DATABASE

The electronic index to the NATO ASI Series provides full bibliographical references (with keywords and/or abstracts) to about 50000 contributions from international scientists published in all sections of the NATO ASI Series. Access to the NATO-PCO DATABASE compiled by the NATO Publication Coordination Office is possible in two ways:

- via online FILE 128 (NATO-PCO DATABASE) hosted by ESRIN, Via Galileo Galilei, I-00044 Frascati, Italy.
- via CD-ROM "NATO Science & Technology Disk" with user-friendly retrieval software in English, French and German (© WTV GmbH and DATAWARE Technologies Inc. 1992).

The CD-ROM can be ordered through any member of the Board of Publishers or through NATO-PCO, Overijse, Belgium.

Series H: Cell Biology, Vol. 94

Springer
Berlin
Heidelberg
New York
Barcelona
Budapest
Hong Kong
London
Milan
Paris
Santa Clara
Singapore
Tokyo

Gene Technology

Stem Cell and Leukemia Research

Edited by

Axel R. Zander

Universitätskrankenhaus Eppendorf
Martinistraße 52
D-20246 Hamburg, Germany

Wolfram Ostertag

Heinrich-Pette-Institut für Experimentelle Virologie
und Immunologie der Universität Hamburg
Martinistraße 52
D-20246 Hamburg, Germany

Boris V. Afanasiev

City Center for Advanced Medical Technologies
Bone Marrow Transplantation Center
Pr. Dynamo 3, 197042 St. Petersburg, Russia

Frank Grosveld

Laboratory for Gene Structure Expression
Medical Research Council Institute
The Ridgeway, Mill Hill, London NW7 1AA, UK

Springer

Published in cooperation with NATO Scientific Affairs Div sion

Proceedings of the NATO Advanced Study Institute "Gene Technology in Analysis and Treatment of Malignant and Inherited Human Diseases Related to Development", held at Wilsede, Germany, and St. Petersburg, Russia, June 19–26, 1994

Cataloging-in-Publication Data applied for

Die Deutsche Bibliothek - CIP-Einheitsaufnahme

Gene technology : stem cell and leukemia research ; [proceedings of the NATO Advanced Study Institute "Gene Technology in Analysis and Treatment of Malignant and Inherited Human Diseases Related to Development" held at Wilsede, Germany, and St. Petersburg, Russia, June 19 - 26, 1994] / ed. by Axel R. Zander. Publ. in cooperation with NATO Scientific Affairs Division. - Berlin ; Heidelberg ; New York ; Barcelona ; Budapest ; Hong Kong ; London ; Milan ; Paris ; Santa Clara ; Singapore ; Tokyo : Springer, 1996
(NATO ASI series : Ser. H, Cell biology ; Vol. 94)
ISBN-13: 978-3-642-64704-8
NE: Zander, Axel R. [Hrsg.]; Advanced Study Institute Gene Technology in Analysis and Treatment of Malignant and Inherited Human Diseases Related to Development <1994, Wilsede; Sankt-Peterburg>; NATO: NATO ASI series / H

ISBN-13: 978-3-642-64704-8 e-ISBN-13: 978-3-642-6112-3
DOI: 10.1007/978-3-642-61122-3

Softcover reprint of the hardcover 1st edition 1996

Typesetting: Camera ready by authors/editors
Printed on acid-free paper
SPIN 10476627 31/3137 - 5 4 3 2 1 0

PREFACE

The 11th meeting in "Modern Trends in Human Leukemia" took place from June 19 to 21, 1994 in Wilsede in the middle of the Lüneburger Heide, South of Hamburg. Interwoven with the Leukemia program was the Nato-sponsored Symposium of the ASI-Series "Gene Technology in Analysis of Malignant and Inherited Human Diseases Related to Development".

The Wilsede meeting was continued on a ship of the Neva leading through lake Ladoga and lake Onega.

The topics of both meetings included discussion on recent progress isolation and development of hematopoietic stem cells, genes crucial for development and diseases, methods of gene transfer, application of gene transfer; oncogenes and anti-oncogenes as targets for gene therapy; receptors and their ligands in normal development and diseases, immunology and immunotherapy, radiation biology, clinical leukemias and bone marrow transplantation.

The Nato workshop concentrated not only on analysis of cell systems useful for somatic gene therapy, but also on actual themes directly related to correction of human diseases. The latter aspects emphasized themes related to biotechnology, the first part was by nature more general. We also included a few contributions that discussed perspectives for the future of gene therapy and possible relationships to evolution.

Both somatic gene therapy of inherited human diseases as well as of cancer, with the emphasis on malignancies of the hemopoietic system, were considered. Let us first outline some of the main features related to analysis and correction of human inherited diseases. Some of the highlights were two talks by F. Grosveld on hemoglobinoparthies emphasizing the need for a thorough knowledge of the normal control and the molecular basis of hemoglobin synthesis as well as its abnormal regulation leading to the human disease entity. Meaningful ways can only be developed to correct this most frequent family of human inherited diseases if the basic features are known, and we are now very close to understand the molecular basis of most hemoglobinopathies.

Possible vectors for correcting are now being constructed and true prospects are visible that somatic gene therapy can be used in the next decades for most persons suffering from these diseases.

One important new aspect of somatic gene therapy was introduced by Dr. H.G. Kräusslich summarizing data on potential correction of AIDS as a consequence of HIV infection. Correction of AIDS by somatic gene therapy is still very far off, but the methods required are related to use of hemopoietic stem cells as targets for gene transfer. This hopefully results in correction in the not too far future.

One major emphasis in somatic gene therapy is treatment of inherited and malignant diseases by manipulation of hematopoietic stem cells involving procedures of bone marrow transplantation. One aspect of improving stem cell transplantation is improvement of the knowledge of the stem cell and thus better use of stem cell biology for transplantation as emphasized by J. van Rood (The Netherlands), I. Lemischka (Princeton, U.S.A.), Dr. S. Watt (Oxford, U.K.), W. Knapp (Vienna, Austria), A.R. Migliaccio (New York, U.S.A.), J. Hatzfeld (Paris, France), P. Mannoni (Marseille,France), T. Hara (Palo Alto, U.S.A.) and W. Ostertag (Hamburg, Germany). Basic knowledge on mechanism how leukemias and other malignant diseases causally arise will also provide a handle to initiate somatic gene therapy. The increasing complexity inherent in malignant disease progression will make such therapy especially difficult.

The basic and normal features of cellular controls which contribute to malignant diseases were dicussed by Joe Bolen, Princeton, U.S.A. (receptors, signal transfer), R. Bernards, Amsterdam, The Netherlands (transcription factors, cell cycle control). The role of such cellular controls in malignant diseases were outlined in examples by C. Stocking (Hamburg, Germany, P.H. Roméo (Créteil, France), D. Afar (Los Angeles, U.S.A.) and others. Actual ways how to treat malignant diseases by application of somatic gene therapy were also discussed by M. Collins (London, U.K.), C. Baum (Hamburg, Germany), P. Mannoni (Marseille, France), M. Cavazzana (Paris, France) and other speakers. It was evident that direct replacement of deficient genes in malignancies, unlike in the case of genetic diseases, is still far off. Protection of normal bone marrow by somatic gene therapy during treatment with carcinostatic agents was thus discussed as one immediate prospect at length. Similarly strengthening of the immune response of the patient may at present be obtained using somatic gene therapy.

The addendum of the book reports on the clinical Neva-Meeting and summarizes the state of the art of leukemia treatment and bone marrow transplantation in Russia.

Hamburg - St. Petersburg
B. Afanasiev
F. Grosveld
W. Ostertag
A.R. Zander

CONTENTS

Part 8 Clinical Leukemias

Part 9 Bone Marrow Transplantation

Part 1

Stem Cells

Hematopoietic Differentiation of Embryonal Stem Cells *in vitro.*

Nicholas Hole, Roland Leung, Lobat Doostdar, Ursula Menzel, Kay Samuel, Janice Murray, Helen Taylor, Gerry Graham[1] & John Ansell

ICAPB
University of Edinburgh
West Mains Road
Edinburgh EH9 3JT
Scotland

Introduction

The hematopoietic system can be considered as a spectrum of differentiation and self renewal with at one extreme terminally differentiated cells such as erythrocytes or macrophages, and at the other, pluripotential hematopoietic stem cells (HSC) , which can give rise to all the different lineages of the hematopoietic system, and due to their capacity for self renewal, persist throughout adult life (Spangrude et al. 1991). Introduction of limiting numbers of these cells into recipients whose own hematopoietic systems have been ablated (by, for example, irradiation) results in long-term reconstitution by the donor cells. These properties of HSC have make them tempting targets for gene manipulation, both for investigating the molecular and cellular biology of genes that are expressed in different lineages of the hematopoietic system and in somatic gene therapy. In the former case, gene manipulation of HSC would have particular benefit where the mutations, if transmitted through the germ line, would result in embryonic lethality. Gene manipulation in HSC has proved difficult, at least in part due to an inadequate understanding of the biology of these cells; addition of genes (transgenesis) is dependent on retroviral infection (Fairbairn and Spooncer 1993). Gene targeting by homologous recombination in HSC has not yet been shown to be practical. In contrast, there is considerable experience in specific and facile manipulation of the mouse genome using embryonal stem (ES) cells, which as totipotential cells, can give rise to all the different cellular lineages in the organism (Evans and Kaufman 1981, Robertson 1986) . In particular, ES cells have proved to be amenable to

[1]CRC Beatson Laboratories, Garscube Estate, Bearsden, Glasgow G61 1BD, Scotland

NATO ASI Series, Vol. H 94
Gene Technology
Edited by A. R. Zander, W. Ostertag,
B. V. Afanasiev, F. Grosveld

gene targeting by homologous recombination. Early work demonstrated the appearance of erythrocytes following differentiation of ES cell under specific conditions *in vitro* (Doetschman et al. 1985). Under conventional circumstances, these cells could only have arisen from differentiation of pluripotent hematopoietic precursors, themselves derived from the ES cells. ES cells are competent to produce HSC; using HSC isolated from fetal livers of mice derived in their entirety from ES cells are capable of rescuing and reconstituting secondary recipient mice (Forrester et al. 1991). The question of whether HSC can be derived from ES cell differentiation *in vitro* is currently an active area of research (Hole et al. 1994, Muller and Dzierzak 1993). However, most mature hematopoietic lineages can be detected following *in vitro* differentiation of ES cells (Chen et al. 1993, Gutierrez-Ramos and Palacios 1992, Keller et al. 1993, Nakano et al. 1994, Wiles 1993). As part of our study of the hematopoietic potential of ES cells, we have examined the lymphoid commitment of ES cells differentiated *in vitro*. The ability to isolate T lymphocytes from ES cells *in vitro* may offer an alternative strategy in the study of thymocyte development and positive and negative T cell selection.

Materials and Methods

ES cell differentiation.

This was carried out essentially as described (Hole and Smith 1994). In brief, ES cells were cultured as hanging drops in the presence of LIF for 48 hours prior to harvesting (day 0 of differentiation) and suspension culture. During this suspension culture, the ES cell aggregates differentiate into cystic structures known as embryoid bodies (Doetschman, et al. 1985).

Flow cytometry

Embryoid bodies of varying periods of differentiation were disaggregated into single cell suspension by digestion with Dispase (1u/ml). Primary rat monoclonal antibody (Pharmingen) binding was detected with either goat anti-rat FITC conjugated antibody (TAGO tissue culture services) or streptavidin-FITC/PE (Sigma) as appropriate. Fluorescence was determined on FACScan flow cytometer (Becton Dickenson).

RT-PCR.

Embryoid bodies from various time points were harvested and washed. mRNA was isolated using RNAzol (Biotecs Lab) and cDNA reverse transcribed by AMV reverse transcriptase (Promega) according to the respective manufacturers instructions. PCR reactions were as described (Schmidt et al. 1991). In brief, denaturation (95°C), anneal and extension (72°C) times were 20, 60 and 60 seconds respectively for thirty cycles. Anneal temperatures and primers were as noted below.

Gene	5' Primer	3' Primer	Product size	Anneal Temp. (°C)
FLK2	GCTGCTGCTTGTTGTTTTGTCA	TTCATACTCTTCTTGCGAGCTG	1060	62
FLK2 ligand	CTGTTACTTCAGCCACAGTCCC	CTTCTTGCCCTTTGCCAGCGAA	560	62
RAG1	GATGGGAAGTCAAGCGACAAAG	GCAGTTGGAGATCTTCTTCAGG	520	62
Ly-6E	CCCCTACCCTGATGGAGTCTGT	GGATTAGAGCACCTACCTACCC	450	62
CD2	GAGACAATGAGACCATCTGGGG	GAATCTGTGTGCCTGATGGAGC	850	62
β-globin	AACCCTCATCAATGGCCTGTGG	TCAGTGGTACTTGTGGGACAGC	415	62
CD5	CACAGGAGTGAACCAGAACA	TATAGTGGAGGTCCTTGTAAGTAC CAC	640	50
Thy-1	ACAGCCTGCCTGGTGAACCAAA	GGCCCAACCAGTCACAGAGAAA	423	68
Actin	GTGACGAGGCCCAGAGCAAGAG	AGGGGCCGGACTCATCGTACTC	934	62
CD45	GATATGAGTCACTTCTCTTCCA	GATGAAAAGCAGCTGATGGA	520	50

PCR products from both reverse transcribed and mock-transcribed reactions were run on agarose gels and detected by ethidium bromide staining.

Results

Flow cytometry

The expression of surface antigens by single cells liberated from the differentiating embryoid bodies is shown in figure 1a-c. HSC from mice are believed to have a characteristic phenotype, namely Ly-6E^{+ve}, Thy-1^{+ve} (Spangrude, et al. 1991). Expression of these antigens was assessed individually (fig.1a and b) and together (data not shown). Thy-1 was found on undifferentiated ES cells, the numbers of cells expressing the antigen decreasing over the first 5 days of differentiation and apparently increasing thereafter to approximately 30% of embryoid body cells over the time course studied. The numbers of cells expressing both antigens were less than 0.1% of total cells at any time point examined. Thy-1 and Ly6E are not HSC-restricted, and are expressed on other cell types, including T cells. In order to examine the total hematopoietic commitment and T-cell commitment, expression of CD45, CD3 and αβTCR were examined (Figs 1b and 1c). CD45+ve hematopoietic cells increase in number over the time course, from about 1% of cells at day 7 to approximately 7% of cells after 20 days of differentiation. However, a relatively small number of these cells are likely to be T cells; CD3 or αβ TCR +ve cells are first detected around day 10, again increasing over the time course to about 10% of the numbers of CD45+ve hematopoietic population. The small numbers of TCR +ve cells recovered precluded two-colour flow cytometry. However, CD4+ve and CD2+ve cells were also found around the time point when these T cell markers could be detected.

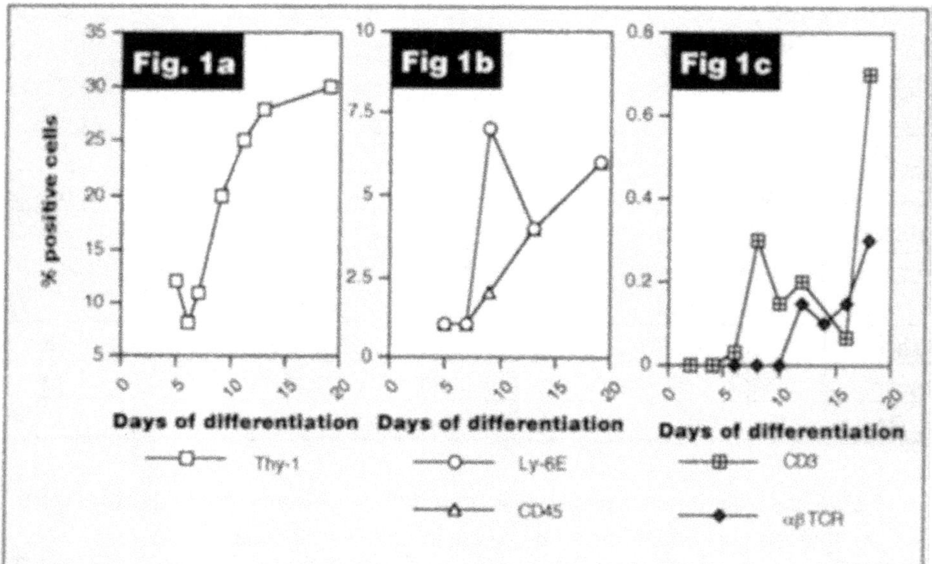

Figure 1. Surface antigen expression of ES derived cells after differentiation as embryoid bodies in suspension culture. Representative figures of 2-3 independent experiments shown

These data suggested that in common with several other studies, we were able to derive lymphoid cells from ES cells by *in vitro* differentiation (in a separate study, B cell development was indicated by detection of IgM+ve, IgD+ve and/or B220+ve cells in differentiation at around 12 days (data not shown). To confirm these results and examine the transcription of developmentally regulated genes, we carried out a selected series of RT-PCR reactions.

RT-PCR

A schematic of the results obtained is shown in table 1. The integrity of the cDNA for PCR was confirmed by a positive result for actin transcripts in all of the samples.

In brief, the results support the observation that lymphoblastoid cells are appearing later in the differentiation pathway. RAG-1 expression was detected as early as day 6 after initiation of differentiation, coincident with detection of CD2 transcripts. CD45 was detected by day 8. Thy-1 and Ly-6E transcripts were detected early in differentiation; in ES cells or day 2 differentiated cells respectively. Of interest was the relatively early appearance of a marker of another hematopoietic lineage, β-globin. However, no red blood cells were observed in embryoid bodies during this series of experiments. Detection of FLK2 ligand transcripts in differentiating ES cells was in contrast to its receptor, whose transcripts were not detected at the level of analysis adopted.

	Days of differentiation						
Transcript	**0**	**2**	**4**	**6**	**8**	**10**	**16**
Rag-1	-	-	-	+	+	+	+
CD2	-	-	-	+	+	-	+
CD5	-	-	-	-	-	-	+
Thy-1	+	+	+	+	+	+	+
CD45	-	-	-	-	+	-	±
β-globin	-	-	-	++	++	++	++
FLK2 ligand	-	+	-	+	-	+	+
FLK2	-	-	-	-	-	-	-
Ly-6E	-	+	+	+	+	+	+
Actin	++	++	++	++	++	++	++

Table 1. RT-PCR detection of mRNA transcripts in differentiating ES cells Results expressed as either absence of detectable PCR product (-) or a subjective assessment of the intensity of the appropriately sized PCR product (+, +, ++).

Discussion

This works suggests that lymphoid commitment of ES cells can be demonstrated *in vitro*. Detectable numbers of cells bearing lymphoblastoid characteristic markers can be detected at or around 10 days of differentiation *in vitro*, albeit at a low level. The presence of markers for CD5, CD2, CD3 all indicate that T cell development is detected. Productive expression of αβTCR is dependent on the presence of the product of RAG-1 gene (Wayne et al. 1994). It is not surprising therefore to find that RAG-1 expression in these differentiating cells precedes the appearance of cells bearing αβTCR (fig1). The numbers of cells expressing the T cell markers Thy-1 and Ly6E are dramatically higher than those expressing the markers listed above. This is indicative of one of the problems in this system; embryoid bodies contain cells of many different lineages, not just hematopoietic. Indeed undifferentiated ES cells contain already contain transcripts for Thy-1. Embryonal stem cells are clearly competent to produce the entire range of hematopoietic lineages; ES-derived fetal livers contain pluripotent repopulating HSC capable of repopulating the entire hematopoietic system of recipient mice (Forrester, et al. 1991). However, inducing ES cells to progress down all hematopoietic lineages *in vitro* has proved more elusive. Myeloid lineage progenitors can be detected following *in vitro* differentiation of ES cells (Wiles 1993), yet reconstitution of the myeloid lineages in lethally irradiated mice by adoptive transfer of ES-derived progeny has proved more elusive. In contrast, in at least one report the lymphoid lineages of mice have been reconstituted with the *in vitro* differentiated progeny of ES cells (Muller and Dzierzak 1993). Part of the problem with this system may the relatively low numbers of hematopoietic cells arising as a result of *in vitro* differentiation. In this study, although percentages varied between experiments, the numbers of αβTCR+ve cells were typically less than 5% at best. One approach to enhance the numbers produced may be to exploit cytokines to enhance hematopoietic development. One cytokine that is thought to play a role in early hematopoietic development is FLK2 ligand (Zeigler et al. 1994). At least in this system, transcripts for FLK2 ligand can be detected within the developing embryoid bodies, allowing the possibly of paracrine regulation of hematopoiesis. However, the failure to detect its receptor at this level of analysis suggests that the numbers of cells that may express the receptor may be very low. An alternative, less well defined system would be to use stromal cell lines to support hematopoietic development. This approach has been used by others (Gutierrez-Ramos and Palacios 1992, Nakano, et al. 1994). The presence of certain cytokines may interfere with hematopoietic differentiation *in vitro*, as has been suggested for M-CSF (Nakano, et al. 1994). Of interest is that transcripts for M-CSF cannot be detected in the early differentiating embryoid bodies in our system (data not shown). The approaches of cytokine enhancement and stromal cell support are currently under study.

The detection of cells expressing productive αβTCR suggests that these cells may be capable of positive and negative selection. The relatively facile nature of transgenesis of ES cells may allow us the opportunity to address issues not just of the control of primitive hematopoietic commitment, but of T cell development. The development of these transgenic cells endogenously in embryoid bodies, *in vitro* in fetal thymus organ culture and *in vivo* in rescued and heamtopoietically reconstituted mice is currently under study.

Acknowledgements

This work was supported by grants from The Wellcome Trust, The Leukaemia Research Fund and The Royal Society.

References

Chen, J., Gorman, J. R., Stewart, V., Williams, B., Jacks, T. and Alt, F. W. (1993) Generation of normal lymphocyte populations by Rb-deficient embryonic stem cells Curr. Biol. 3: 405-413

Doetschman, T., Eisetter, H., Katz, M., Schmidt, W. and Kemler, R. (1985) The in vitro development of blastocyst-derived embryonic stem cell lines: formation of visceral yolk sac, blood islands and myocardium. J.Embryol.exp.morphol. 87: 27-45

Evans, M. J. and Kaufman, M. (1981) Establishment in culture of pluripotential cells from mouse embryos Nature 292: 154-156

Fairbairn, L. J. and Spooncer, E. (1993) Retroviral gene transfer into haemopoietic cells In: Haemopoiesis N. G. Testa and I. Molyneux (eds.) Oxford University Press Oxford

Forrester, L. M., Bernstein, A., Rossant, J. and Nagy, A. (1991) Long-term reconstitution of the mouse hematopoietic system by embryonic stem cell-derived fetal liver Proc.Natl.Acad.Sci.(USA) 89: 7514-7517

Gutierrez-Ramos, J. C. and Palacios, R. (1992) In vitro differentiation of embryonic stem cells into lymphocyte precursors able to generate T and B lymphocytes in vivo Proc.Natl.Acad.Sci.(USA) 89: 9171-9175

Hole, N., Graham, G. and Ansell, J. D. (1994) Self renewing and repopulating hematopoietic progenitors can be derived from embryonal stem cells in vitro J.Cell.Biochem. 18B: 181

Hole, N. and Smith, A. G. (1994) Embryonic stem cells and hematopoiesis In: Culture of hematopoietic cells R. I. Freshney, I. B. Pragnell and M. G. Freshney (eds.) Wiley-Liss New York

Keller, G., Kennedy, M., Papayannopoulou, T. and Wiles, M. V. (1993) Hematopoietic commitment during embryonic stem cell differentiation in culture Mol.Cell.Biol. 13: 473-486

Muller, A. M. and Dzierzak, E. A. (1993) ES cells have only a limited ymphopoietic potential after adoptive transfer into mouse recipients Development 118: 1343-1351

Nakano, T., Kodama, H. and Honjo, T. (1994) Generation of lymphohematopoietic cells from embryonic cells in culture Science 265: 1098-1101

Robertson, E. J. (1986) Pluripotential stem cell lines as a route into the mouse germ line Trends. Genet. 2: 9-13

Schmidt, R. M., Bruyns, E. and Snodgrass, H. R. (1991) Hematopoietic development of embryonic stem cells in vitro: cytokine and receptor expression Genes Develop. 5: 728-740

Spangrude, G. J., Smith, L., Uchida, N., Ikuta, K., Heimfiled, S., Friedman, J. and Weissman, I. L. (1991) Mouse hematopoietic stem cells Blood 78: 1395-1402

Wayne, J., Suh, H., Misulovin, Z., Sokoi, K. A., Inaba, K. and Nussenzwig, M. C. (1994) A regulatory role for recombinase activating genes RAG-1 and RAG-2 in T cell development Immunity 1: 95-107

Wiles, M. V. (1993) Embryonic stem cell differentiation in vitro Meth.Enzymol 225: 900-918

Zeigler, F. C., Bennett, B. D., Jordan, C. T., Spencer, S. D., Baumhueter, S., Carroll, K. J., Hooley, J., Bauer, K. and Matthews, W. (1994) Cellular and molecular characterisation of the role of the FLK-2/FLT-3 receptor tyrosine kinase in hematopoietic stem cells Blood 84: 2422-2430

Post-irradiated canine serum (PICS-J) stimulates megakaryocyte (MK) progenitors and thrombopoiesis in-vitro

Varda R. Deutsch [1,2], Amiram Eldor[2], Tom Olson[3], Shimon Slavin[1] and Arnon Nagler[1]
[1]Departments of Hematology and Bone Marrow Transplantation, Hadassah University Hospital, Jerusalem, [2]Hematology Institute, Tel Aviv Medical Center, Tel Aviv, Israel and [3]Children's Hospital, Columbus, Ohio, USA

Protracted thrombocytopenia and bleeding is a serious complication in bone marrow transplantation (BMT). While engraftment of the erythroid and myeloid progenitors occurs promptly (within two to three weeks), there is often an unexplained period of profound thrombocytopenia. In contrast to the myeloid and erythroid lineages, lineage specific megakaryocytic-colony stimulating factors are not yet available for clinical use. Sera from thrombocytopenic subjects and animals have been previously shown to contain megakaryocyte (MK) colony stimulating activity (CSA). We therefore used sera from irradiated (850 cGy) beagle dogs that developed severe thrombocytopenia (<2000 platelets/mm^3) (PICS-J) and a patient who was accidentally exposed to high dose radiation (approx. 2000 cGy) (ML) as a source of MK-CSF. We evaluated the effect of PICS-J on the proliferation and maturation of MK progenitors in human bone marrow (BM), peripheral blood (PB), and cord blood (CB) using plasma clot and liquid culture techniques. Colony forming units (CFU)-MK and burst forming units (BFU)-MK were identified by immunofluorescence with anti-platelet GPIIb/IIIa. In liquid cultures, the percentage of cells expressing platelet GPIIb/IIIa was detected by the use of anti-CD41 McAb. The PICS-J was found to be a very potent stimulator of CFU-MK and BFU-MK growth for human BM (153.2±27 CFU-MK and 9.8±3.4BFU-MK per 2x10^5 BM cells). Similarly, PICS-J was the most potent stimulator of CFU-MK and BFU-MK proliferation from human PB and CB samples. PICS-J was a stronger stimulator of MK growth than optimal doses of other known thrombopoietic cytokines (IL-3, IL-6, SCF, GM-CSF, FGF, ECGF and PDGF). Dose response curves disclosed maximal MK-CSF activity at 7.5-10% PICS-J. PICS-J

NATO ASI Series, Vol. H 94
Gene Technology
Edited by A. R. Zander, W. Ostertag,
B. V. Afanasiev, F. Grosveld

contained significantly higher MK-CSF than sera collected from patients with severe aplastic anemia (SAA), sera from BMT patients that received total body irradiation (TBI) as part of conditioning prior to BMT, and sera from thrombocytopenic carboplatin treated ovarian cancer patients. PICS-J did not contain CSA for erythroid or myeloid progenitors. 10% PICS-J increased in the percentage of cells expressing GPIIb/IIIa in liquid cultures of human BM, PB, and CB samples. Moreover, the combination of SCF and PICS-J was found to be a very potent stimulator of MK progenitors in liquid culture assays. PICS-J also increased MK size and ploidy distributions. In parallel, plasma clot assays revealed that aplastic canine serum increased the cellularity of CFU-MK and the MK maturation manifested as the frequency of fragmented megakaryocytes within CFU-MK. No IL-6, GM-CSF, IL-1β, Epo, or SCF could be detected in the post irradiated sera and only 30% of the MK-CSF activity could be attributed to IL-3. We have demonstrated maximal stimulation of secondary MK precursor growth by PICS-J following pre-incubation of marrow allografts with combinations of IL-3 and GM-CSF. The ex-vivo expansion of the MK precursors resulted in enhanced platelet recovery post BMT. We conclude that post irradiation sera contains factors which stimulate megakarycytopoiesis and thrombopoiesis. This may be of benefit for ex-vivo expansion of MK to support accelerated platelet engraftment post BMT. In addition, post radiation sera may be a unique source for the characterization of novel cytokine(s) which enhance megakaryocytopoiesis.

INTRODUCTION

Protracted thrombocytopenia and bleeding is a serious complication in bone marrow transplantation (BMT). While engraftment of the erythroid and myeloid progenitors occurs promptly (within two to three weeks), there is often an unexplained period of profound thrombocytopenia. In contrast to the myeloid and erythroid lineages, lineage specific megakaryocytic-colony stimulating factors are not yet available for clinical use. The production of platelets is thought to be regulated by lineage specific growth factors which stimulate megakaryocyte (MK) proliferation and induce MK maturation (Hoffman 1989).

Plasma from thrombocytopenic patients with aplastic anemia (Mazur et al. 1981; Messner et al. 1982; Hoffman et al. 1981) or following bone marrow transplantation (BMT) (deAlarcon and Schnieder 1986; de Alarcon et al. 1988, Olson et al. 1992) or high dose chemotherapy (Mazur et al. 1984) are known to contain MK-CSF.

Sera from animals that have been rendered aplastic by total body irradiation have been shown to be a good source of both MK-CSF (Miura et al. 1984; Mazur and South 1985; Mazur et al. 1990). A protein that seems to have this activity has been identified in the plasma of irradiated pigs (de Sauvage et al. 1994) and thrombocytopenic rabbits (Kuter 1994). Recently the Mpl ligand - the putative thrombopoietin was purified by affinity chromatography from plasma obtained from irradiated pigs (de Sauvage 1994).

We now demonstrate that serum from a human subject following accidental lethal total body irradiation and post irradiated aplastic canine serum (PICS-J) contain highly potent and specific MK-CSF. These sera stimulate megakaryocytopoiesis in vitro far better than sera from patients with aplastic anemia or following BMT. This cytokine is distinct from known cytokines and can be enhanced by stem cell factor.

MATERIALS AND METHODS

Human bone marrow and blood specimens.

BM cells were obtained from healthy transplant donors. Peripheral blood (PB) mononuclear cells and umbilical cord blood (CB) cells were obtained from healthy subjects and normal full term deliveries.

Preparation of post irradiation aplastic canine serum.

MK-CSF was raised in beagle dogs that were irradiated over a period of several hours with a total irradiation dose of 850 cGY. The animal then received full medical support until severe aplasia developed (less than 2000 platelets/mm^3). Dogs were bled through the femoral artery and serum was separated, inactivated at 56°C for 30 minutes, filtered through a 0.45 μm filter and stored frozen in aliquots at -70°C until use.

Serum from a patient who underwent accidental lethal irradiation.

We obtained serum on days 14 and 24 post lethal irradiation from a patient who was accidentally exposed to high dose radiation (approx. 2000 cGy).

Sera from patients following total body irradiation.

Sera were collected from AML, ALL, and Lymphoma patients who received TBI prior to allogeneic BMT. Sera were collected at day -10 and then on days 0, 10, 20, 30 and 40 post BMT. Sera were also collected at the same time intervals from aplastic anemia patients undergoing BMT and from ovarian cancer patients that developed thrombocytopenia following carboplatin chemotherapy.

Megakaryocyte cultures.

Progenitor derived MK colonies were grown in plasma clots. Nonadherent mononuclear cells (NAMC) were seeded in triplicate 0.33 ml cultures containing 2×10^5 cells per well as previously described (Deutsch et al. 1993, Deutsch et al 1995).

Identification of megakaryocyte colonies.

MK colonies were identified by indirect immunofluorescence using a highly specific rabbit antibody to human platelet glycoprotein complex IIb/IIIa and FITC conjugated goat anti-rabbit IgG antibody. Fluorescent colonies were counted on day 12 as clusters of three or more megakaryocytes. BFU-MK were identified as clusters that contained at least three separate growth foci and consisting of at least 40 cells.

Megakaryocyte suspension cultures.

NAMC were grown at a concentration of $1\text{-}2 \times 10^6$ cells/ml in complete (Iscove Modified Dulbecco's Medium) IMDM to which growth supplements were added. Cultures were maintained for periods of up to 10 days. At the termination of culture viability was estimated by trypan blue dye exclusion and viability was at least 85%. Cultured cells were transferred to new flasks every 48 hours to remove contaminating adherent macrophages. Small aliquots from these cultures were removed daily to monitor the development and maturation of megakaryocytes. At the termination of suspension cultures the non adherent cells were collected.

FACS analysis.

Cells were stained with monoclonal antibodies to CD41, platelet glycoprotein complex IIb/IIIa and anti-CD34. Cells were washed and stained with FITC-labeled goat anti-mouse immunoglobulins. Quantitation of fluorescence was performed using a FACScan.

Detection of cytokines.

IL-6 and IL-1b and GM-CSF levels were tested by Elisa, R & D Systems, Minneapolis, Minn. Neutralizing antibodies to IL-3, GM-CSF and erythropoietin were purchased from Genzyme Corp., Cambridge, Mass. SCF was assessed by immunoblots.

Statistical analysis.

Results are reported as the mean number of colonies counted ± standard error of the mean of duplicate or triplicate cultures. Statistical significance was determined using a Students t test.

RESULTS

Highly active MK-CSF in serum following lethal irradiation.

The post-irradiation sera contained highly active MK-CSF which induced the proliferation of individual MK colonies, CFU-MK and burst forming units BFU-MK. The stimulatory effect of post lethal irradiation aplastic canine or human serum on the cloning efficiency of MK progenitors derived from PB, CB and BM is summarized in Table 1.

Table 1. The effect of post lethal irradiation aplastic serum on MK colony formation.

	Aplastic canine serum		Aplastic human serum	
	CFU-MK	BFU-MK	CFU-MK	BFU-MK
PB	64.5±24.2 (n=8)	0	24.1±4.3 (n=3)	1±0.5
CB	101±12.1 (n=11)	9.3±2.7	53.5±4.6 (n=3)	2.6±0.9
BM	153.2±27 (n=9)	9.8±3.4	160.5±23.4 (n=3)	14.5±3.2

Adherent cell depleted MC were cultured the presence of 15% AB plasma and 7% aplastic serum at a concentration of 2x10^5 cells per well in triplicate cultures. Results represent the average number of colonies counted ± SD, n= number of experiments.

The response of MK to MK-CSF in suspension cultures.

NAMC were grown in liquid culture for up to 10 days in the presence of normal AB serum alone (15%) or with PICS-J (10%) and AB serum (5%). The megakaryocytes were identified as the platelet glycoprotein complex IIb/IIIa positive population and their percentage assessed by FACS analysis. At the initiation of cultures the percentage of immunodetectable MK was <0.5%. By day six a significant increase in the frequency of gpIIb/IIIa positive cells was observed. In the presence of PICS-J, 4-6% of the cells displayed platelet proteins. Maximal MK growth was obtained from all three hematopoietic sources by day 10 (fig. 1).

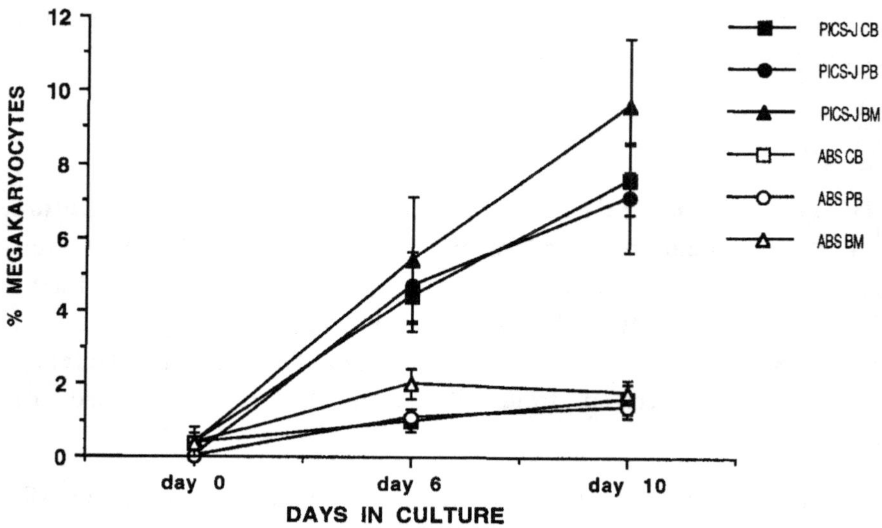

Figure 1. Megakaryocytes in suspension cultures. Non-adherent mononuclear cells ($1\text{-}2 \times 10^6$/ml) were grown in IMDM supplemented with 15% AB serum (ABS) or 10% PICS-J + 5% ABS. The percentage of gpIIb/IIIa positive cells was assessed by FACS analysis.

Comparison of colony MK-CSA in post-irradiation aplastic serum to colony stimulation by cytokines or aplastic anemia serum.

MK growth factor activity in the post-irradiation sera were compared to that of cytokines. The post lethal irradiation sera both from the human patient (ML) and from the dogs (PICS-J) were far more active at stimulating megakaryocyte colony formation than any of the cytokines tested (fig. 2).

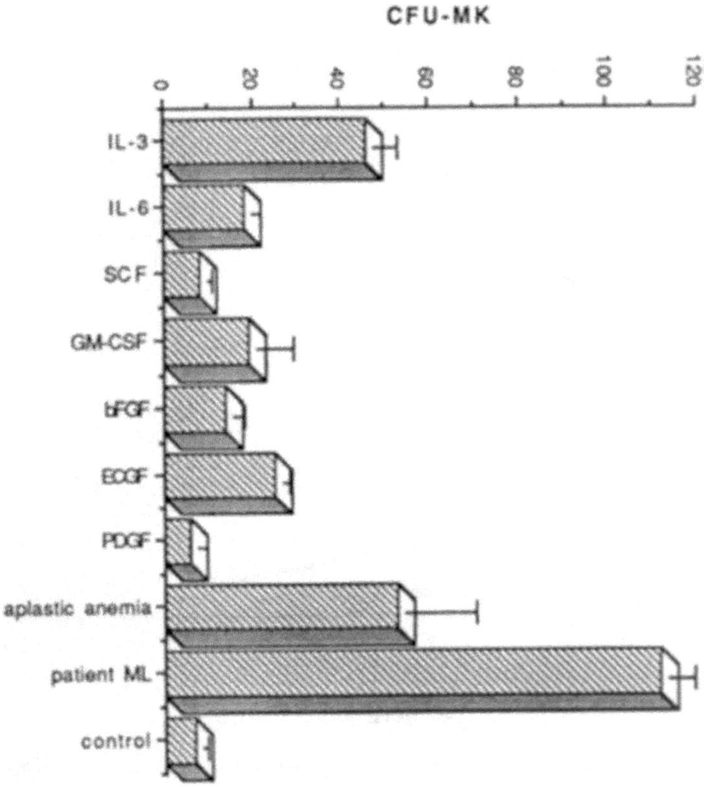

Figure 2. Colony formation by BM-MK progenitors in response to cytokines, aplastic anemia serum, and post lethal irradiation serum. BM-MC cells (1×10^5/well) were cultured in plasma clots supplemented with the indicated growth factors. Results represent the average number of MK colonies±SEM counted after 12 days. Human recombinant cytokine concentrations/ml were IL-3 10 ng, IL-6 100 U, GM-SCF 10 ng, b-FGF 50 ng, ECGF 10 ng, PDGF 10 ng, SCF 50 ng, aplastic anemia serum 10%, post-irradiation serum 10%. Each cytokine was tested on at least three separate normal bone marrow samples, each cultured in triplicate.

The response of MK progenitors to varying concentrations of MK-CSF in post-irradiation serum.

In order to determine the optimal concentration for culturing MK the dose response of progenitors to different concentrations of post-irradiation aplastic serum was tested. Peak MK-CSA activity was found to be at 7%-10% serum in both the canine and human sera (figure 3). In comparison to aplastic anemia serum, which was previously screened and known to contain MK-CSF, the post-irradiation sera were far more active at all concentrations tested.

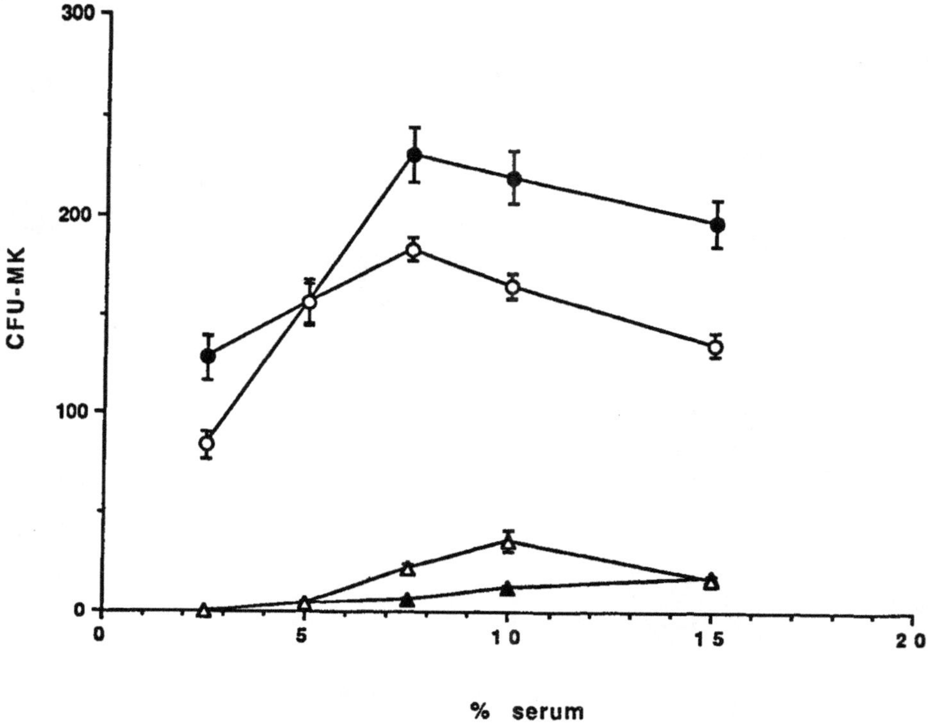

Figure 3. Dose response of bone marrow MK progenitor cells to MK-CSF in post-irradiation sera. Adherent cell depleted BM-MC cells (2×10^5) were cultured in plasma clots as described in the presence of varying concentrations of post-irradiation aplastic human (o) or canine serum (o), aplastic anemia serum (Δ), or normal AB serum (Δ). Numbers represent the average CFU-MK±SEM derived from three separate BM samples each cultured in triplicate.

Normal human AB serum and normal canine serum did not stimulate megakaryocytopoiesis at any of the concentrations tested.

MK-CSF activity in patients following total body irradiation or chemotherapy.

The activity of MK-CSF in post-lethal irradiation serum was compared to the average activity found in: seven aplastic anemia patients ten days prior to BMT and 20 days after BMT, 11 thrombocytopenic patients (platelets <40,000) who had received 3-5 cycles of Carboplatin treatment for ovarian cancer, 20 leukemia patients in remission prior to BMT, and 12 leukemia patients 20 days post BMT. Serum following lethal irradiation contains much higher levels of MK-CSF.

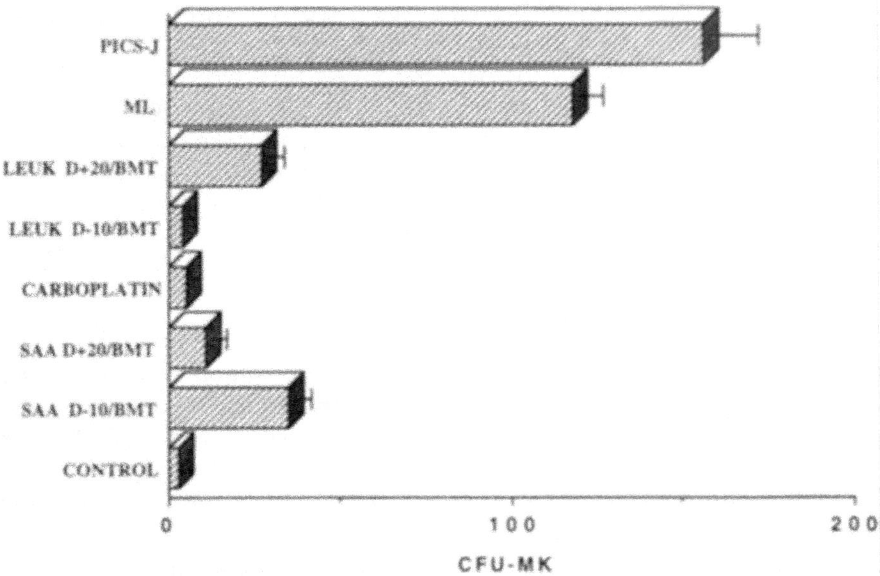

Figure 4. MK-CSF activity in serum following irradiation or chemotherapy. BM-MC cells ($2x10^5$/well) were cultured in plasma clots supplemented with the indicated sera (10%). Results represent the average number of MK colonies±SEM counted after 12 days. Sera were obtained from Severe Aplastic Anemia (SAA) 10 days prior to and 20 days after BMT, thrombocytopenic carbolplatin treated patients, leukemia patients 10 days prior to BMT and 20 days after BMT, serum from lethally irradiated patient ML, and post lethally irradiated canine serum (PICS-J).

The effect of post-irradiation aplastic serum on MK size, maturation and cellularity of colonies.

The thrombopoietin activity in PICS-J increased MK size and MK maturation manifested by the frequency of fragmented megakaryocytes (proplatelets) within CFU-MK. In addition the cellularity of CFU-MK was also significantly increased (Deutsch et al. 1995).

Table 2. Effect of PICS-J on megakaryocyte proliferation and maturation.

Cell Source	CB		BM	
SERUM	PICS-J	ABS	PICS-J	ABS
Colony Cellularity[a]	14.3±2.1	5.7±1.3	15.8±0.8	3.9±0.6
MK size[b]	30.4±3.1	24.5±1.0	35.6±4.2	53.8±5.2
% fragmentation[c]	36.4±9.4	0.0	54.0±2.6	0.0

Megakaryocyte cultures were grown from cord blood (CB) or bone marrow (BM) adherent cell depleted mononuclear cells in the presence of 10% post-irradiation aplastic canine serum (PICS-J)or normal AB serum (ABS).

[a] Data represents the mean number of cells per colony±SEM in four separate experiments.

[b] Cytospin preparations of cells that were grown in suspension cultures for seven days were stained by APAAP using an anti-GPIIb/IIIa monoclonal antibody. Cell diameters were determined in over 100 megakaryocytes in three separate samples using an Olympus microscope and the Cue-2 image analysis software. Results represent the mean diameters in uM ±SD.

[c] Data represent the average percentage±SD of colonies that contained fully mature fragmented megkaryocytes. The total number of colonies assessed ranged from n=20 in controls to n=100 in PICS-J stimulated cultures in four separate experiments.

FACS analysis of megakaryocytes in liquid cultures stimulated with post-irradiated canine serum.

BMMC were grown in suspension cultures for 3 days in the presence of 15% AB serum (control), 100 U/ml SCF, 10% post-irradiated canine serum (PICS-J) or combination of SCF and PICS-J. The percentage of CD34 and CD41 (GP IIb/IIIa) positive cells were determined by FACScan. In the presence of PICS-J alone 3.0% of the cells were positive for CD34 and 6.7% were positive for CD41 while in the presence of SCF 0.57% and 1.17% of the cells expressed CD34 and CD41, respectively. The combination of SCF and PICS-J was found to be a very potent stimulator of both CD34+ (13.4%) and CD41+ (32.9%) cells.

Table 3. FACS analysis of CD41+ and CD34+ progenitors in liquid cultures stimulated with post lethal irradiation aplastic serum.

	CD41	CD34
CONTROL	0.17	0.64
KL	1.07	0.57
PICS-J	3.00	6.70
PICS-J + KL	32.92	13.40

Bone marrow mononuclear cells were grown in suspension cultures for three days in the presence of 15% normal ABS (control), AB serum with 100 U/ml stem cell factor (Kit-ligand, KL), 10% post-irradiated canine serum (PICS-J) or the combination of PICS-J and KL. Numbers represent the percentage of positive cells detected by FACS analysis of 10,000 cells per sample.

Identification of cytokines contained in post lethal irradiation serum.

We evaluated the levels of the cytokines that are known to affect megakaryocytopoiesis in human and canine post lethal irradiation sera as well as in the sera from the SAA patients. Normal sera and sera from patients that received granulocyte monocyte colony stimulating factor (GM-CSF) served as controls. Elevated levels of Interleukin-6 (IL-6), GM-CSF or Interleukin-1β (IL-1β) could not be detected by Elisa (Table 4). No stem cell factor (SCF) was detected by western blot analysis.

The effect of neutralizing antibodies on MK-CSF activity in post irradiated aplastic canine and human sera..

The effect of neutralizing antibodies to Interleukin-3 (IL-3), Erythropoietin (EPO) and GM-CSF on the megakaryocyte colony stimulating activity was tested. Antibodies to EPO and GM-CSF had no effect on the activity of serum from the irradiated human and canine sera. Anti IL-3 reduced the activity of these sera by 30% but was unable to eliminate the activity even at very high antibody concentrations (Table 5).

Table 4. Quantitative analysis of cytokine levels in aplastic anemia and post-irradiation sera.

Cytokine	GM-CSF (pg/ml)	IL-6 (pg/ml)	IL-1β (pg/ml)
ELISA control			
[range]	[50-200]	[50-100]	[200-400]
Control serum 1	199	46	480
Control serum 2	160	96	430
SAA1	143	50	580
SAA 2	126	42	450
SAA 3	156	45	520
SAA 4	ND	43	ND
SAA 5	ND	46	ND
PICS-J	150	46	280
Pat-GM 1	1068	45	600
Pat-GM 2	216	14	730

Sera samples were tested from patients with aplastic anemia (SAA) and post-irradiation aplastic canine serum (PICS-J). Pat-GM represent patients who received GM-CSF (250 $\mu g/m^3$/day) within 48 hours of serum collection (1) and 72 hours prior to serum collection (2). ND - not done.

Stimulation of megakaryocyte precursors by preincubation with the combination of IL-3 and GM-CSF followed by PICS-J.

Bone marrow cells were grown in liquid cultures for 3 days in the presence or absence of IL-3 and GM-CSF. The number of MK progenitors was assessed by their colony forming capacity in plasma clots stimulated PICS-J or no cytokine (secondary cultures). The cloning efficiency of these cells was also compared to that of fresh bone marrow, grown with the same cytokines without preincubation.

Maximal MK colony growth was stimulated in cultures grown with post-irradiated aplastic canine serum (PICS-J). Prior to incubation with cytokines, BM cells gave rise to 122.7±53 CFU-MK ($p<0.008$) and 15.9±3.8 BFU-MK per $2 \cdot 10^5$ BM cells. Following preincubation with IL-3 and GM-CSF a significant expansion of CFU-

Table 5. The effect of neutralizing antibodies on MK-CSF activity in post-irradiated aplastic canine and human serum

		Canine serum		Human serum	
Exp. number		1	2	1	2
Neutralizing antibody activity/ml					
Anti-IL-3	100U	70	80	59	63
	200U	78	ND	53	63
	400U	77	ND	52	63
Anti-EPO	50U	92	97	91	93
	100U	93	ND	81	92
	200U	102	ND	96	91
Anti-GM-CSF	2ng	91	ND		
	5ng	91	ND		
	10ng	97	90		
Anti-EPO +	50U				
Anti-IL-3	100U	73	71	63	71

Results are expressed as the % of MK colony formation as compared to control.

MK (175.6±72, $p<0.008$) and a small expansion of BFU-MK (17.3±6.2 BFU-MK per $2 \cdot 10^5$ BM cells, $p<0.05$) was observed. Preincubation of BM cells without the addition of cytokines dramatically reduced the number of progenitors below control levels ($p<0.001$) (fig. 5).

The effect of IL-3 and GM-CSF preincubation of platelets recovery post-allogeneic bone marrow transplantation.

The effect of IL-3 and GM-CSF preincubation on platelet recovery post-allogeneic BMT was assessed. Twenty consecutive patients with leukemia received T-cell-depleted allografts using Campath-1G. Two thirds of the marrow was infused on the scheduled day of transplant and one third of the marrow following preincubation with IL-3 and GM-CSF (0.1 μg/ml each) on day 4. A group of 40 patients served as non-randomized, nonpaired but nearly contemporaneous controls: 20 of them were transplanted just before

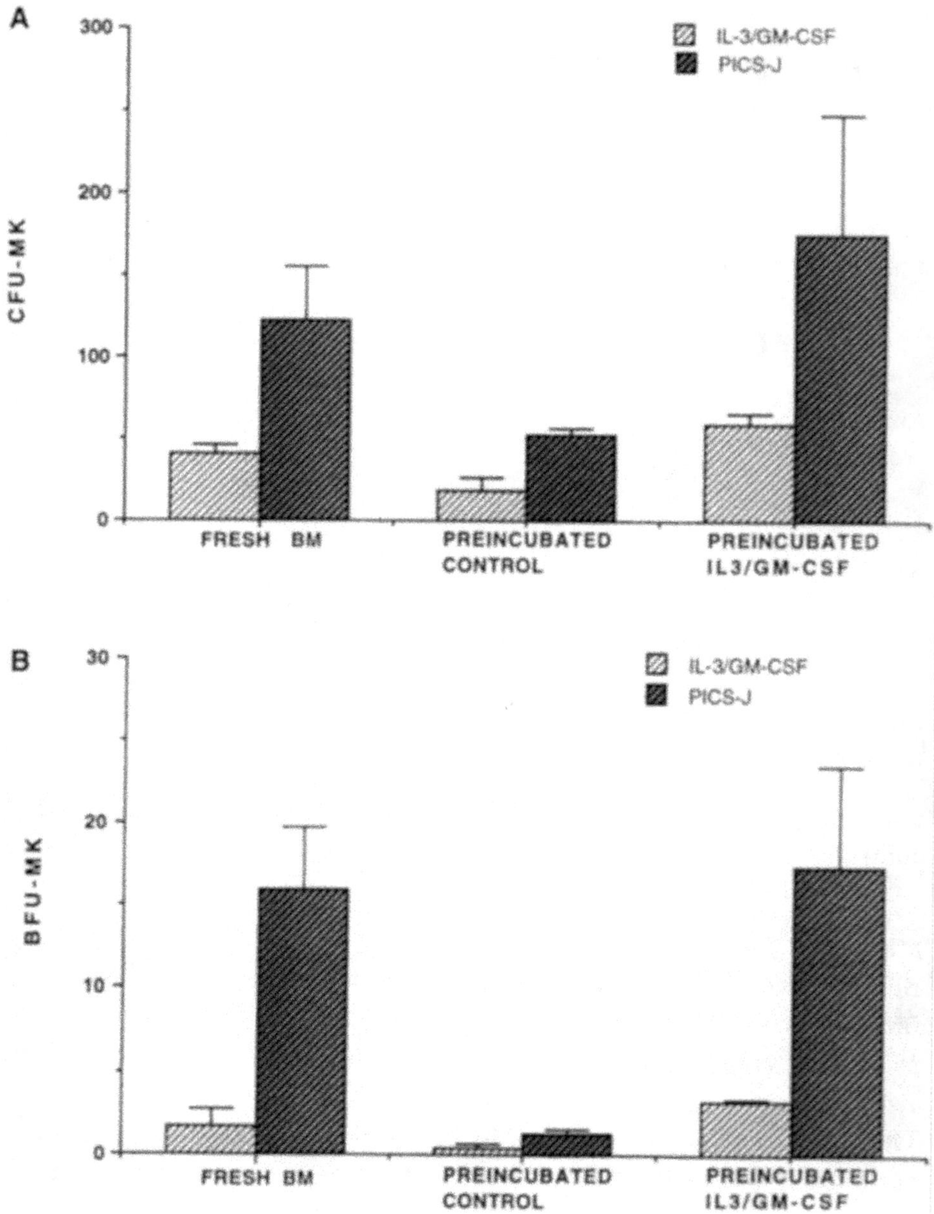

Figure 5. The effect of PICS-J on BM megakaryocyte colony formation. Prior to colony assays BM cells were preincubated for 3 days in the presence or absence of IL-3 and GM-CSF (0.1 μg/ml each). MK colonies were grown in plasma clots in the presence of PICS-J or IL-3 and GM-CSF. Results represent the average number of CFU-MK (a) or BFU-MK (b) ±SD per 2×10^5 BMMC of at least 5 experiments done in duplicate or triplicate

the group under study. IL-3 and GM-CSF preincubation resulted in acceleration of platelet recovery post-BMT. The median day to reach unsupported platelet count of $25 \cdot 10^9$/L was 17 days for patients in the study group and 23 days for the control (p<0.017).

Table 6. Platelets engraftment post-IL-3 and GM-CSF preincubation.

Days to engraftment	Study group	Control group
Median	17 (11-34)	23 (11-48)
Mean	17.8±6*	22.7±8.4[a]

[a]Three patients did not have platelets engraftment and are not included in the statistics.
*p<0.017.

SUMMARY AND FUTURE DIRECTIONS

Post irradiated sera is known to contain MK-CSF which supports the development of MK colony forming units. We therefore studied the effects of ionizing irradiation on the elaboration of factors which stimulate megakaryocytopoiesis both in canine and humans. We demonstrated post-irradiative canine and human serum to be a very potent stimulator of early and late MK progenitors both in colony assays and in liquid cultures. This serum does not contain elevated levels of any of the cytokines that are known to affect megakaryocytopoiesis. The nature of the active cytokine (s) in this serum is not yet resolved but may be c-Mpl. The post-irradiation sera elicited significantly higher MK-CSF than sera collected from patients with severe aplastic anemia (SAA), sera from BMT patients that received total body irradiation (TBI) as part of their conditioning prior to BMT, and sera from carboplatin treated patients.

PICS-J was also a potent stimulator of cell proliferation within the MK colonies as well as having a thrombopoietin activity which supports MK maturation. This sera increased both the MK size and enhanced fragmentation of MK into platelet like particles.

Preincubation of marrow allografts with the combination of IL-3 and GM-CSF expanded MK progenitors in vitro and facilitated platelet engraftment post-BMT. The expansion of MK progenitors that is affected by preincubation with IL-3 and GM-CSF was further enhanced in vitro by PICS-J. Maximal stimulation of MK progenitor expansion by post lethal irradiation sera (and possibly Mpl ligand) could be of vast clinical importance. Further experimental and clinical data will be necessary to evaluate the full potential of the MK-CSF contained within post lethal irradiation sera for clinical use in transplantation.

References

Gordon MS, Hoffman R (1992) Growth factors affecting human thrombopoiesis: Potential agents for the treatment of thrombocytopenia. Blood 80: 302-307

deAlarcon PA, Schnieder JA (1986) Megakaryocyte colony stimulating activity (MK-CSA) in serum from patients undergoing bone marrow transplantation. Megakaryocyte development and function. Liss, Alan R, 335-340.

deAlarcon PA, Schnieder JA, Gingrich R, Klugman MP (1988) Pattern of response of megakaryocyte colony stimulating activity in the serum of patients undergoing bone marrow transplantation. Exp Hematol 16: 316

de Sauvage FJ, Hass PE, Spencer SD, Malloy BE, Gurney AL, Spencer SA, Darbonne WC, Henzel WJ, Wong SC, Kuang W-J, Oles KJ, Hultgren B, Solberg Jr LA, Goeddel DV, Eaton DL. (1994) Stimulation of megakaryocytopoiesis and thrombopoiesis by the c-Mpl ligand. Nature 369: 533-538

Deutsch VR, Nagler A, Slavin S, Condiotti R, Levine RF, Eldor A (1993) Effect of bone marrow T lymphocytes treated with CAMPATH 1G on megakaryocyte colony formation. Exper Hematol 21: 1427-1435

Deutsch VR, Olson TA, Nagler A, Slavin S, Levine RF (1995) The response of cord blood megakaryocyte progenitors to IL-3, IL-6 and aplastic canine serum varies with gestational age. 89: 791

Hoffman R, Mazur E, Bruno E, Floyd V (1981) An assay of an activity in the serum of patients with disorders of thrombopoiesis that stimulate the formation of megakaryocyte colonies New Eng J Med 305: 533-538

Hoffman R (1989) Regulation of megakaryocytopoiesis. Blood 74: 1996-1212

Kuter DJ, Rosenberg RD (1994) Appearance of a megakaryocyte growth-promoting activity, megapoietin, during acute thrombocytopenia in the rabbit. Blood 84: 1464-1472

Mazur EM, Hoffman R, Chasis J, Marchesi S, Bruno E (1981) Immunofluorescent identification of human megakaryocyte colonies using an antiplatelet glycoprotein antiserum. Blood 57: 277-286

Mazur EM, deAlarcon P, South K, Miceli L (1984) Human serum megakaryocyte colony stimulating activity increases in response to intensive cytotoxic chemotherapy. Exp Hematol 12: 624-629

Mazur EM, South K (1985) Human megakaryocyte colony stimulating factor in sera from aplastic dogs: Partial purification, characterization and determination of hematopoietic cell lineage specificity. Exp Hematol 13: 1164-1171

Messner HA, Jamal N, Izaguirre C (1982) The growth of large megakaryocyte colonies from human bone marrow. J Cell Physiol (suppl 1):45-51

Miura M, Jackson CW, Steward SA (1984) Increase in circulating growth promoting activity (Meg. GPA) following sublethal irradiation is not related to decreased platelets. Exp Hematol 16: 139-142

The Adhesion Molecules, PECAM-1, ICAM-3 and CD33, on CD34+ Haematopoietic Progenitor Cells

Suzanne M Watt[1*], Angelo Cardoso[2,3], Jane Lee-Prudhoe[1], M. Ann Benton[1], Regis Doyonnas[1], Antoinette Hatzfeld[2], Helen Genevier[5], Colin Porter[5], Christoper Dennison[1] and Jacques Hatzfeld[2]

MRC Molecular Haematology Unit,
Institute of Molecular Medicine
John Radcliffe Hospital
Headington, Oxford. OX3 9DU, UK

Introduction

This European concerted action was established to promote collaborations and the exchange of information amongst European laboratories working on human haemopoietic stem cells. It is supported for three years by the European Commission, Directorate General XII under the direction of Dr Ch. Bardoux. There are, at present, over 56 contributing laboratories throughout Europe with 2 laboratories in Central and Eastern Europe joining through the PECO initiative in late 1994. The co-ordinator [Dr. S.M. Watt] is assisted in its management by Professor D. Weatherall and Dr. J. Hatzfeld and by an advisory board of 9 European Scientists whose names are listed in Table 1.

Collaborative projects cover 4 areas for which the participants possess considerable expertise. These are:

(i) Procedures for identifying the haemopoietic stem or very

* to whom correspondence should be addressed

2 Laboratoire de Biologie Cellulaire et Moleculaire des Facteurs de Croissance, CNRS, Villejuif, France

3 Division of Hematologic Malignancies, Dana-Farber Cancer Institute, Harvard Medical School, Boston, USA

4 Institute of Child Health, London, UK

5 Institute of Cancer Research, London, UK.

NATO ASI Series, Vol. H 94
Gene Technology
Edited by A. R. Zander, W. Ostertag, B. V. Afanasiev, F. Grosveld

primitive progenitor cells;

(ii) assays for human haemopoietic stem cells (such as *in vitro* culture assays and animal models) and *ex vivo* expansion of stem cells and their progeny;

(iii) the regulation of stem cells (e.g. by cytokines and extracellular matrix molecules) and the regulation of genes expressed by stem cells (e.g. cell surface adhesion and cytokine receptors; cell cycle components; signal transduction molecules and transcription factors);

(iv) clinical uses for stem cells (e.g. gene therapy for curing genetic inherited diseases and gene and cell or gene therapy for treating neoplastic diseases such as the leukaemias).

Table 1: E.U. CONCERTED ACTION

THE HUMAN HAEMATOPOIETIC STEM CELL

Co-ordinator : S Watt

Management Board : S. Watt, D. Weatherall and J. Hatzfeld

Advisory Board: TM Dexter, A Fauser, AM Gianni, F Grosveld, W Knapp, B Lowenberg, P Mannoni, Ottolenghi and J Visser

No. of Participating Laboratories :
56 throughout Europe (with 2 additional laboratories in Central and Eastern Europe joining through the PECO initiative in 1994)

The concerted action, because of its size, is now divided into 4 subgroups, to be co-ordinated as follows:

Subgroup 1 : Cytokine subgroup : Drs. J. Hatzfeld and G. Migliaccio

Subgroup 2 : Gene Transfer and Cell Therapy : Professors A.A. Fauser, A.M.Gianni, P. Mannoni, and W. Ostertag

Subgroup 3 : Cell Markers : Professor W. Knapp, D. Birnbaum, R. Ploemacher and J. Visser

Subgroup 4 : Cord Blood Banks : Professor E. Gluckmann, J. Hows and N. Testa

Subgroup 5: Animal Models: Dr B. Peault

Most of the papers being presented in the European Concerted Action Session of this Advanced Workshop relate to gene transfer and therapy. For these studies, we need to know:

(a) which tissue provides the best and most appropriate source of haemopoietic progenitor cells for the uses specified above;

(b) whether genes must be transferred into the haemopoietic stem cell or into their more mature progeny;

(c) how to identify the stem cell and its progeny and whether there is a need to purify these cells for gene transfer studies;

(d) how to target the appropriate gene into the appropriate cell;

(e) how the introduced genes are regulated;

(f) if the introduction of genes results in the disruption of normal genetic control mechanisms or induces an immune response.

There are several methods currently available for gene transfer and these are listed in Table 2, and these have recently been reviewed by Moritz & Williams, (1994). Both retroviral and receptor mediated gene transfer methods require interaction with molecules on the surface of the cells to be targeted. Within this concerted action, the molecules being studied include cell surface adhesion molecules, receptors involved in cell-cell, cell-matrix and cell-viral interactions, and cytokine receptors

Table 2 : GENE TRANSFER METHODS

- **Retroviruses**
- **Receptor Mediated**
- **Minichromosomes**
- **Liposomes**

(such as novel tyrosine kinase receptors). Molecules, within these families, that are of particular interest to our laboratories are CD34, Thy1, MDRI, ICAM-3, PECAM-1 or CD31, HLA-DR, CD33 and CD38 which allow identification and/or discrimination of primitive haemopoietic progenitor cell subsets.

This review will be divided into two parts. The first section will describe a panning protocol for obtaining highly enriched human haemopoietic precursors while maintaining the phenotypically most primitive haemopoietic subsets, while the second section will review three cell surface adhesion molecules, PECAM-1 (CD31), ICAM-3 (CD50) and CD33, that may be expressed on the most primitive haemopoietic stem cells.

(i) *The Isolation of $CD34^+$ Human Haemopoietic Progenitor Cells*

Inter-laboratory comparison of a two-step panning protocol for isolating $CD34^+$ cells

There is now a major demand for the isolation of a pure or at least a highly enriched population of human haemopoietic $CD34^+$ primitive progenitor cells by relatively simple techniques. Potential uses for these cells include (i) haemopoietic rescue without tumour cell contamination; (ii) their expansion *in vitro* with cytokines; (iii) as targets for gene expression studies; and (iv) as vehicles for gene therapy and genetic manipulation.

With the wide variety of uses for $CD34^+$ haemopoietic progenitor cells in both clinical and experimental research, there is an increasing interest in establishing reproducible and effective methods for the purification of this rare cell population, which comprises both the haemopoietic stem cell and immature progenitor cell compartments. Reproducibility is an indispensible requirement for developing a methodology for the purification of transplantable immature haemopoietic progenitors. In addition, it is necessary to isolate $CD34^+$ cells which allow both rapid and long term repopulation of haemopoiesis. In order to obtain highly enriched $CD34^+$ cells in a reproducible manner, the efficiency of a commercial two-step panning protocol was compared by evaluating the purity, recovery and clonogenic capacity of the purified cells in two independent European laboratories. Preliminary studies were conducted in order to standardise various inter-laboratory experimental parameters which were independent of the purification procedure. Such parameters included mononuclear cell separation by density centrifugation, labelling and flow cytometric analyses of the different fractions and the clonogenic assay and culture media. The CFU-GEMM assay was chosen to assess the clonogenic properties of the purified population. A defined, improved and standardised CFU-GEMM culture medium was produced in Villejuif for the analyses of clonogenic cells in semi-solid cultures and was used in all experiments. This StemGEM medium was extensively tested and compared in the previous European Community Concerted Action on Human Bone Marrow Cells, in which both groups participated, and was found to be reproducible (Hatzfeld, et al., 1994).

The protocol involved the separation of light density human bone marrow cells using isotonic Percoll (<1.068 g/ml) or Ficoll-hypaque (<1.077 g/ml) density centrifugation (Cardoso, et al., 1994; Watt, et al., 1993). This was followed by a two-step standardised procedure (Hatzfeld, et al., 1991; Lebkowski, et al., 1992), which included an enrichment by elimination of cells attaching to covalently immobilised soybean agglutinin (SBA; SBA CELLector flasks; Applied Immune Systems Inc. [AIS]; Santa Clara, CA, USA), followed by positive selection of $CD34^+$ cells using a polystyrene surface coated with the CD34 (Table 3) monoclonal antibody (MoAb), ICH-3 (CD34 CELLector flasks; AIS) essentially as described in the manufacturer's standard instructions.

Table 4 shows that the range of recoveries of the light density mononuclear cells (MNC) was similar in both laboratories, with mean cell recoveries of 22.4% and 23.3%. The enrichment for $CD34^+$ cells in this MNC fraction was also strikingly similar between the two laboratories, with the CD34 fluorescence profiles being equivalent to one another (data not shown).

Table 3: Monoclonal antibodies used for purification and phenotypic analyses

MoAb	Clone	Isotype	Form	Source
CD34	8G12	IgG1	FITC, PE	BD
	ICH3	IgG2a	Unconjugated	AIS
CD38	HB7	IgG1	PE	BD
anti-HLA-DR	L243	IgG2a	FITC, PE, Biotin	BD
anti-Thy1	F15-42-1	IgG1	FITC	J Fabre
anti-c-kit	95C3	IgG1	PE	IT
anti-MDR1	57	IgG2a	Unconjugated	M Cianfriglia
CD33	P67.6	IgG1	PE	BD

BD for Becton Dickinson; AIS for Applied Immune Sciences; IT for Immunotech;
PE : phycoerythrin; FITC : fluorescein isothiocyanate

In the first panning step of the purification procedure, the SBA^+ cells were eliminated. Both the cell recoveries and enrichments for $CD34^+$ cells were similar in both laboratories (Table 5). A reduction of the cell number by approximately 30% was accompanied by a 3 fold enrichment in the percentage of $CD34^+$ cells during the SBA depletion step, as measured by the Class III CD34 antibody, 8G12 or HPCA-2. No significant losses in the total numbers of $CD34^+$ cells in the MNC and SBA- fractions were evident when compared with the original bone marrow population(data not shown). An average of 0.8% of nucleated cells was recovered in the $CD34^+$ subset after CD34 panning for both groups (Table 5). A high but suboptimal enrichment for $CD34^+$ cells was obtained (Table 5), being on average 71.2% and 50.3% for the French and British group respectively. The CFU's were also highly enriched (Table 6).

The light scatter profiles for each fraction separated are shown in Figure 1. The purification results in the loss of cells with high side scatter characteristics and as expected, the concentration of cells with low side scatter and medium to high forward scatter characteristics.

Table 4: Percentage of cells recovered and expression of CD34 antigen in each cell fraction during the density and standard SBA panning procedure

	FRENCH LABORATORY			BRITISH LABORATORY		
Cell	Cell Recovery (%)			Cell Recovery (%)		
Fraction	n	mean±s.d.	Range	n	mean±s.d.	Range
Nucleated Cells	13	100	-	14	100	-
Mononuclear Cells	13	22.4±7.7	12.4-34.2	14	23.3±12.3	10.3-55.7
SBA^- Cells	13	6.3±3.4	2.6-12.6	11	7.4±3.4	0.9-15.3
	CD34 Expression (%)			CD34 Expression (%)		
	n	mean±s.d.	Range	n	mean±s.d.	Range
Mononuclear Cells	7	1.8±0.7	1.0-2.8	4	1.6±0.9	0.3-2.6
SBA^- Cells	6	5.3±1.4	3.9-7.8	4	5.1±1.9	2.5-7.2

CD34 expression detected with the FITC-8G12 (HPCA-2) antibody;
n = number of independent experiments

Modification of the two-step panning procedure for isolating $CD34^+$ cells

Using the manufacturer's protocol, the degree of purification was insufficient for our purposes. In order to increase its efficiency in the selection of immature progenitors, several modifications of the selection procedure were developed. These included shortening the separation times and more extensive washing at the CD34 panning stage. The protocol was as follows: the flask priming and incubation periods were shortened to 30

Table 5: Percentage of cells recovered and expression of CD34 antigen in the purified cell fraction using the manufacturer's protocol

	FRENCH LABORATORY			BRITISH LABORATORY		
$CD34^+$	Cell Recovery (%)			Cell Recovery (%)		
Fraction	n	mean±s.d.	Range	n	mean±s.d.	Range
From nucleated Cells	9	0.8±0.6	0.1-1.8	4	0.8±0.6	0.1-1.5
From MNC Fraction	9	1.9±1.1	0.8-4.1	4	2.9±2.0	0.7-5.6

	CD34 Expression (%)			CD34 Expression (%)		
	n	mean±s.d.	Range	n	mean±s.d.	Range
$CD34^+$ Fraction	3	71.2±12.7	56.7-80.1	4	50.3±17.0	35.0-74.1

CD34 expression detected by flow cytometry using FITC-8G12.
n = the number of individual experiments

minutes; the recovered SBA^- cells were washed once prior to being loaded into the CD34 CELLector flask; the HGG was exclusively used for the CD34 selection step; after the incubation, the CD34 CELLector flasks were washed several times with DPBSE to ensure elimination of the non-attached and weakly attached cells; the final SBA^-CD34^+ cells were then detached in Iscove's modified Dulbecco's medium with 10% fetal calf serum (IMDM/FCS) by a 2 hr incubation at 37°C. Cell viability was determined by trypan blue exclusion and propidium iodide incorporation and was shown to be systematically higher than 98%. In the different purification steps cells were recovered, concentrated and used for phenotypic analyses and clonogenic assays.

As indicated in Table 7, this technique generated a highly purified $CD34^+$ cell population (93.5% compared with 50.3-71.2 in the original protocol), ranging from 86.6 to 97% of cells isolated. The recovery of $CD34^+$ cells was estimated to be 65%.

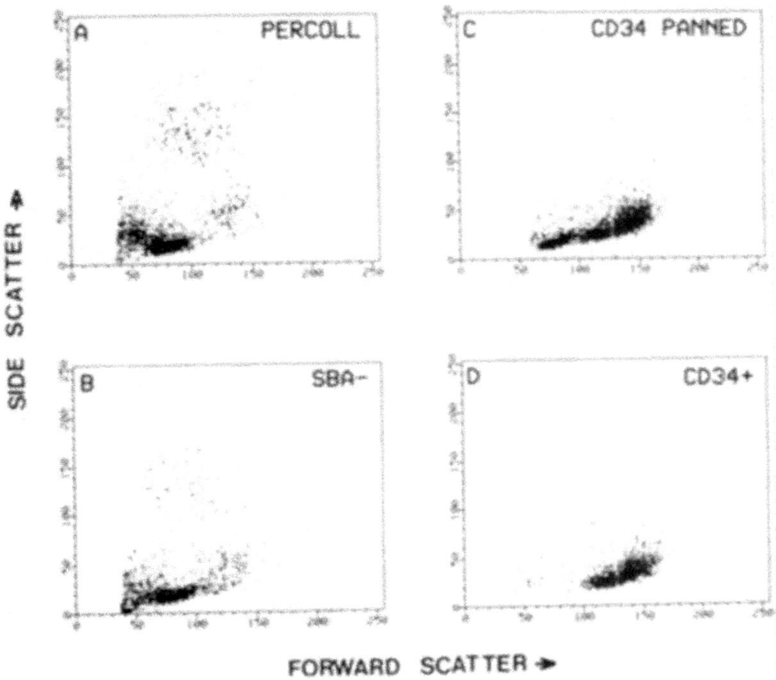

Figure 1: Light scatter patterns of adult human bone marrow cells on the FACScan after selection (A) of light density cells on Percoll (density 1.068 g/ml); (B) by SBA negative panning; and (C) by SBA^-CD34^+ panning (70% $CD34^+$). (D) shows the light scatter parameters of the $CD34^+$ cells from the SBA^-CD34^+ panned sample.

Figure 2 shows a representative histogram of the CD34 fluorescence in the isolated population with the modified procedure. There was a slightly higher number of cells expressing intermediate to high levels of the CD34 defined antigen on their surface and this was accompanied by removal of most of the $CD34^{dim}$ subset, and co-incident loss of some of the more mature BFU-E, CFU-G and CFU-M. A semi solid clonogenic assaywas used to evaluate the functional capacity of these cells to generate colonies derived from immature progenitors.

Table 6: Colony-forming cells detected in the different fractions of the standard purification procedure (colonies/10^5 cells)

A. FRENCH LABORATORY*

Cell Fraction	CFU-GEMM mean±s.d.	CFU-G/M** mean±s.d.	Total BFU-E mean±s.d.	Total CFU-C mean±s.d.	n
MNC	17±5	573±253	416±200	1006±458	4
CD34+	611±313	12363±4144	8999±4979	21032±4756	7

* Results are expressed as the number of (day 14) colonies (mean ± s.d.) of each type per 10^5 cells plated in methylcellulose colony assays

n = the number of independent experiments

B. BRITISH LABORATORY*

Cell Fraction	CFU-GEMM mean±s.d.	CFU-G/M mean ± s.d.	Total BFU-E mean±s.d.	Total CFU-C mean±s.d.	n
MNC	4±3	117 ± 10	81±102	202±114	3
SBA$^-$	36±19	816 ± 629	342±233	1193±796	5
CD34+	462±253	7667 ± 4324	3258±1322	11387±5661	5

* Results are expressed as the number of (day 14) colonies (mean ± s.d.) of each type detected in methylcellulose colony assays of 3 to 7 independent experiments, as indicated by n.

** CFU-G/M refers to the sum of day 14 colonies detected as CFU-G, CFU-M and CFU-GM

Higher enrichments for immature or large CFU-GEMM, BFU-E (i-BFU-E) and CFU-GM were observed with the modified protocol (Table 7). The enrichments averaged 45 fold for early CFU-GEMM, 37 fold for i-BFU-E and 35 fold for CFU-GM. An average of 27.5% (range 19.7-31.6%) of the purified cells generated day 14 colonies with the modified protocol compared with 21% (range 14.6-26) for the manufacturer's protocol. A high recovery of the more immature progenitors was also observed. In the four experiments presented, 65 to 82% of the CFU-GEMM, 55.4 to 69.8% of the primitive or i-BFU-E and 41.5 to 74.4% of the CFU-GM were recovered. A direct comparison of the manufacturer's method with the modified method was made with the same bone marrow

Table 7: Frequency and recovery of Day 14 colony-forming cells in the $CD34^+$ fraction* using the modified protocol

	CFU-GEMM	i-BFU-E	Total BFU-E	CFU-GM	CFU-G	CFU-M	Total CFU-C
CFU**	846 ± 176	5104 ± 1815	11200 ± 2804	1021 ± 105	7300 ± 1675	6898 ± 1835	27521 ± 5336
% CFU#	74 ± 7	62 ± 6	44 ± 7	62 ± 14	44 ± 20	44 ± 16	41 ± 13

* In 7 separate experiments 93.5 ± 3.4% of cells were $CD34^+$ when bone marrow cells were separated using the modified method of panning with 65.6% of $CD34^+$ cells being recovered when compared to the number of $CD34^+$ cells in the light density MNC fraction

** The number of day 14 colonies of each type per 10^5 cells plated in methylcellulose assays and represent the mean ± S.D. of 4 independent experiments. i-BFU-E or immature BFU-E were bursts consisting of more than 20 clusters

\# The number of each CFU type recovered in the $CD34^+$ fraction as a percentage of the number in the MNC fraction (mean ± S.D. of 4 independent experiments)

samples at the same time in the French Laboratory. Figures 3 and 4 show the frequency of $CD34^+$ cells and all day 14 CFU respectively.

Phenotypic characterisation of the purified $CD34^+$ cell population

The $CD34^+$ cells obtained using the modified protocol were analysed by two and three colour phenotypic analyses using antibodies listed in Table 3 and previously described methods (Cardoso et al., 1994). As indicated in Table 8, the enriched $CD34^+$ cell populations contain primitive haemopoietic progenitor

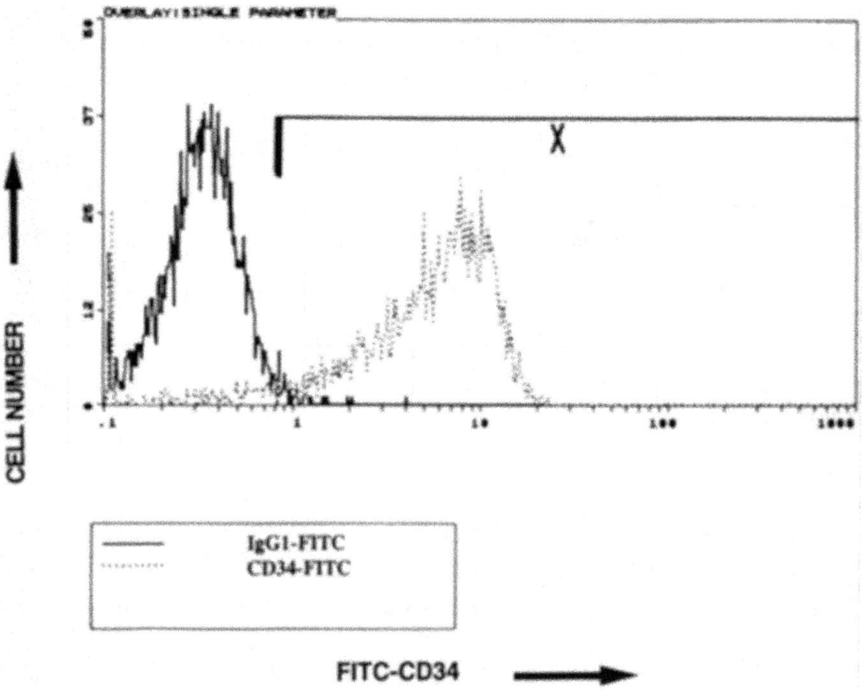

Figure 2: CD34 fluorescence histogram (using FITC-HPCA-2 or FITC-CD34) of the SBA^{-}CD34^{+} panned fraction from adult human bone marrow using the modified protocol. An isotype matched (IgG1-FITC) antibody served as the negative control. Cells were analysed on a Coulter Elite flow cytometer.

cell subsets defined as CD34^{+}CD38, CD34^{+}Thy1^{+}, CD34^{+}MDR1^{+} or CD34^{+}Rh123^{dull}. HLA-DR expression is controversial with some researchers identifying the most primitive haemopoietic cells as HLA-DR^{-} and others as HLA-DR^{+} or HLA-DR$^{lo/+}$. Our preliminary studies using 3-colour flow-cytometric analysis, indicate that the CD34^{+}CD38$^{lo/-}$ cells are mainly HLA-DR^{+} in adult human bone marrow (A. Cardoso and S. Watt, unpublished data).

Subsets of purified CD34^{+} cells could also be identified on the basis Rh123 positivity. Additional phenotypes are shown in Figure 5. Highlevels of purification were dependent on (i) the procedure used for collecting CD34^{+} cells from the panning flasks; (ii) analyses of CD34 expression with CD34 monoclonal antibodiesdirected to epitopes that were different to those recognised by the ICH-3 monoclonal antibody; and (iii) the use of defined, standardised culture media for optimal colony growth.

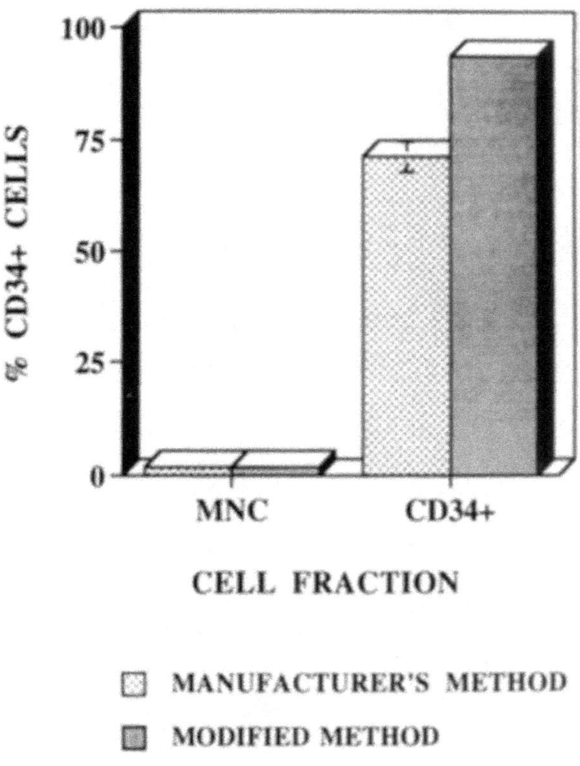

Figure 3: Percent $CD34^+$ cells obtained after SBA^-CD34^+ panning of adult bone marrow using the manufacturer's or a modified protocol compared to the MNC fraction and analysed as in Figure 2. Mean ± S.D.; n = 7 independent experiments for MNC and CD34 modified protocol and n = 3 for the CD34 manufacturer's protocol.

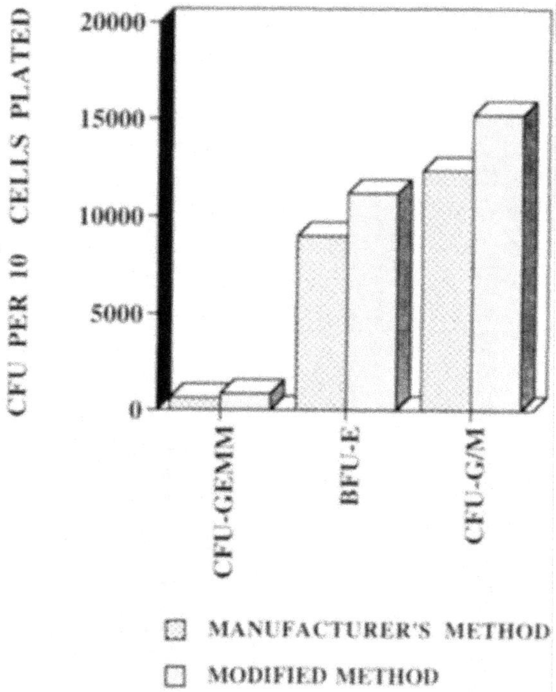

Figure 4: Frequency of CFU in the SBA⁻CD34⁺ adult bone marrow samples isolated using the manufacturer's or the modified panning protocol. Cultures were performed as described in Cardoso et al., (1994). Values are means of n = 4 independent experiments for the modified protocol and 7 for the manufacturer's protocol.

(ii) Adhesion molecules on primitive haemopoietic progenitor cells

Haemopoietic progenitor cells defined by their expression of the CD34 molecule differentiate and migrate within haemopoietic

Table 8: Phenotype of the CD34$^+$ purified fraction

	n	Percent Labelled Cells mean ± s.d.	Range
CD34$^+$CD38$^{lo/-}$	10	1.6 ± 0.5	0.7 - 3.7
CD34$^+$HLA-DR$^-$	6	11.75 ± 5.21	4.9 - 18.5
CD34$^+$Thy1$^+$	5	3.30 ± 1.93	1.1 - 5.4
CD34$^+$Rh123 dull	4	16.50 ± 9.43	6.1 - 29.3
CD34$^+$MDR1$^+$	4	6.20 ± 2.14	3.7 - 8.9
CD34$^+$c-kit$^+$	6	47.05 ± 7.94	37.2 - 58.8

n = number of individual experiments

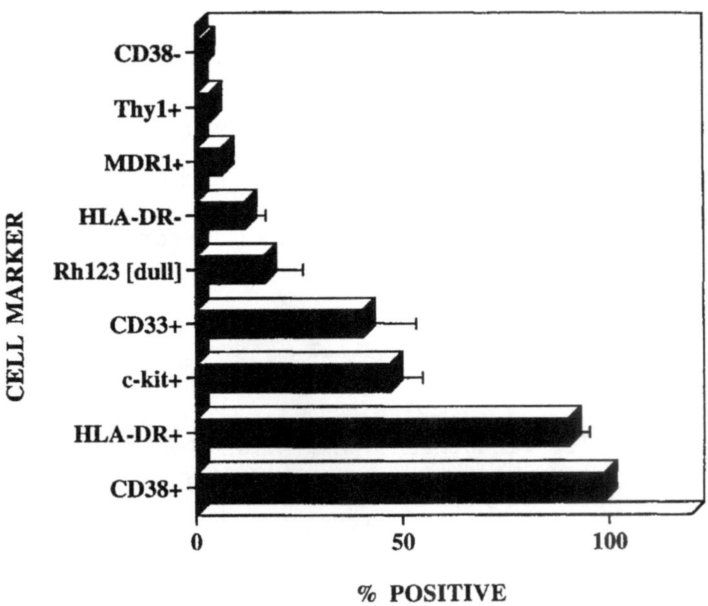

Figure 5: Phenotype of the CD34$^+$ cells using flow cytometry after separation by panning for SBA$^-$CD34$^+$ adult bone marrow cells. Methods are described in Cardoso et al., (1994). Mean ± S.D. where n is the number of independent experiments ($n \geq 4$)

tissues in a microenvironment containing a variety of stromal cell types and extracellular matrix molecules. The developmental potential of these progenitor cells is thought to rely on haemopoietic-stromal cell and cell-matrix interactions, both of which are mediated by a set of adhesion molecules. Recent studies by Olweus, et al., (1995) have examined the expression of a variety of cell surface molecules on $CD34^+$ haemopoietic progenitors from both human fetal and adult bone marrow, with the most primitive being classified as $CD34^+CD38^{-/lo}HLA\text{-}DR^+$. The phenotype obtained is indicated in Table 9. Some of these adhesion molecules are illustrated in Figure 6.

If the most primitive subset is represented by the $CD34^+CD38^{lo/-}$ $HLA\text{-}DR^+$ fraction, then a number of adhesion molecules appear to be expressed. The three that will be discussed are CD31, CD33 and CD50. Unlike CD34, these molecules are either haemopoietic cell specific (CD33) and/or continue to be expressed on more mature cells (CD31, CD33 and CD50) and may provide additional benefits for the introduction of genes into primitive haemopoietic progenitors or for regulated expression of introduced genes in more mature lineages via their attachment to promoter/enhancer regions of CD31, CD33 or CD50.

The expression and function of PECAM-1 (CD31) on $CD34^+$ cells

(a) PECAM-1 expression on mature cells

PECAM-1, the platelet-endothelial cell adhesion molecule, is a type I integral membrane protein initially identified on endothelial cells and platelets (reviewed by DeLisser, et al., 1994; Watt, et al., 1995). In normal peripheral blood, PECAM-1 is found on the majority of platelets, monocytes and neutrophilsand on approximately 50% of lymphocytes. The latter comprise the $CD16^+CD8^+$ large granular lymphocytes, the smIg positive B cells and 40% of the $CD3^+$ T cells. Of the $CD8^+$ T cells, approximately 70-90% are $CD31^+$ and these comprise the

Table 9: Phenotype of $CD34^+CD38^{lo/-}$HLA-DR$^+$ or HLA-DR$^-$ subsets of human fetal and adult bone marrow (Olweus et al., 1995)

Fetal Bone Marrow [HLA-DR$^+$ and HLA-DR$^-$]

Negative

CD2, 3, 5, 8, 14, 15, 16, 19, 20, 56, 64

Positive on DR$^+$ subset only

CD4, 10, 22, 33

Positive on DR$^-$ subset only

CD7 (h)*, 17, 30, 32(h), 36, 49e, 61, 106

Both DR$^+$ and DR$^-$ subsets positive

CD13, 15s(h), 18, 31, 32(h), 43**, 44**, 45, 48, 49b, 49d, 50**, 53, 54, 59, 62L(v), 71, 90, 99(h), 105

Adult Bone Marrow [HLA-DR$^+$]§

Negative

CD3, 5, 8, 14, 15, 16, 17, 19, 20, 32, 36, 49b, 49e, 56, 61, 64, 106

Positive

CD4, 7(h), 10, 13, 15s(h), 18, 22, 30, 31, 32(h), 33, 43, 44, 45, 48, 49d, 50, 53, 54, 59, 62L(v), 90, 99, 105

---

* CD7 : not determined on DR$^-$ cells

h ; heterogeneous staining; other staining homogeneous

** stains subsets of DR$^-$ cells

v: variable staining amongst different samples

Positive reactions varied from low to high levels of expression, but were above isotype matched controls

§ For adult bone marrow, there were insufficient DR$^-$ cells for phenotyping

naive CD45RA$^+$ cells and 50% of the memory CD45RA$^-$ cells. Of the CD4$^+$ T cells, 20% are CD31$^+$ with 50% of these being CD45RA$^+$. It has also been reported that cord blood CD3$^+$CD31$^+$ T cells are CD45RA$^+$ and that thymic CD31$^+$ T cells are CD45RA$^-$.

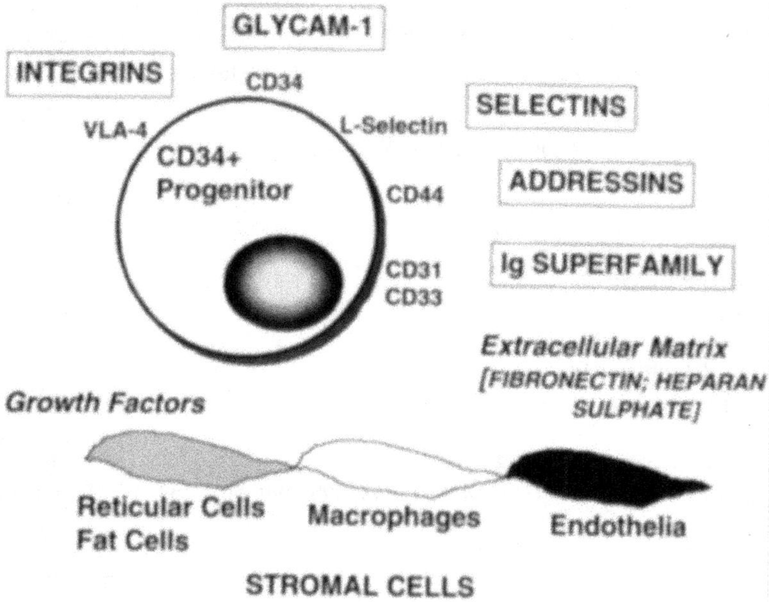

Figure 6: Adhesion molecule expression on CD34$^+$ cells.

(b) <u>Expression on CD34$^+$ Cells</u>

We have shown (Watt et al., 1993) that CD31/PECAM-1 is present on myeloid, lymphomyeloid and pre-pre-B cell lines. Within the B lymphoid lineage, PECAM-1 appears to be down-regulated during B lineage development but is re-expressed on plasma cells (Figure 7b; (Watt et al., 1993)). Using the CD34$^+$ human bone marrow preparations described above, essentially all (i.e. over 97%) of the CD34$^+$ cells were CD31$^+$ (Figure 8) and these encompass the day 14 clonogenic cells, the CFU-GEMM, CFU-G/M and BFU-E (Watt et al., 1993). More recently our own preliminary studies suggest that PECAM-1 may have a functional role on the more mature (week 1 to 2) compared to the more immature (week 5 to 8) cobblestone area forming cells (CAFCs; R. Ploemacher and S. Watt, unpublished data). Three colour flow cytometric analyses indicate that subsets of CD34$^+$CD31$^+$ cells can co-express CD10, CD19, CD33, CD71 or HLA-DR. It is present (Watt et al., 1993; Olweus et al., 1995) on CD34$^+$CD38$^{lo/-}$HLA-DR$^+$ and HLA-DR$^-$ cells in human bone marrow (Table 9).

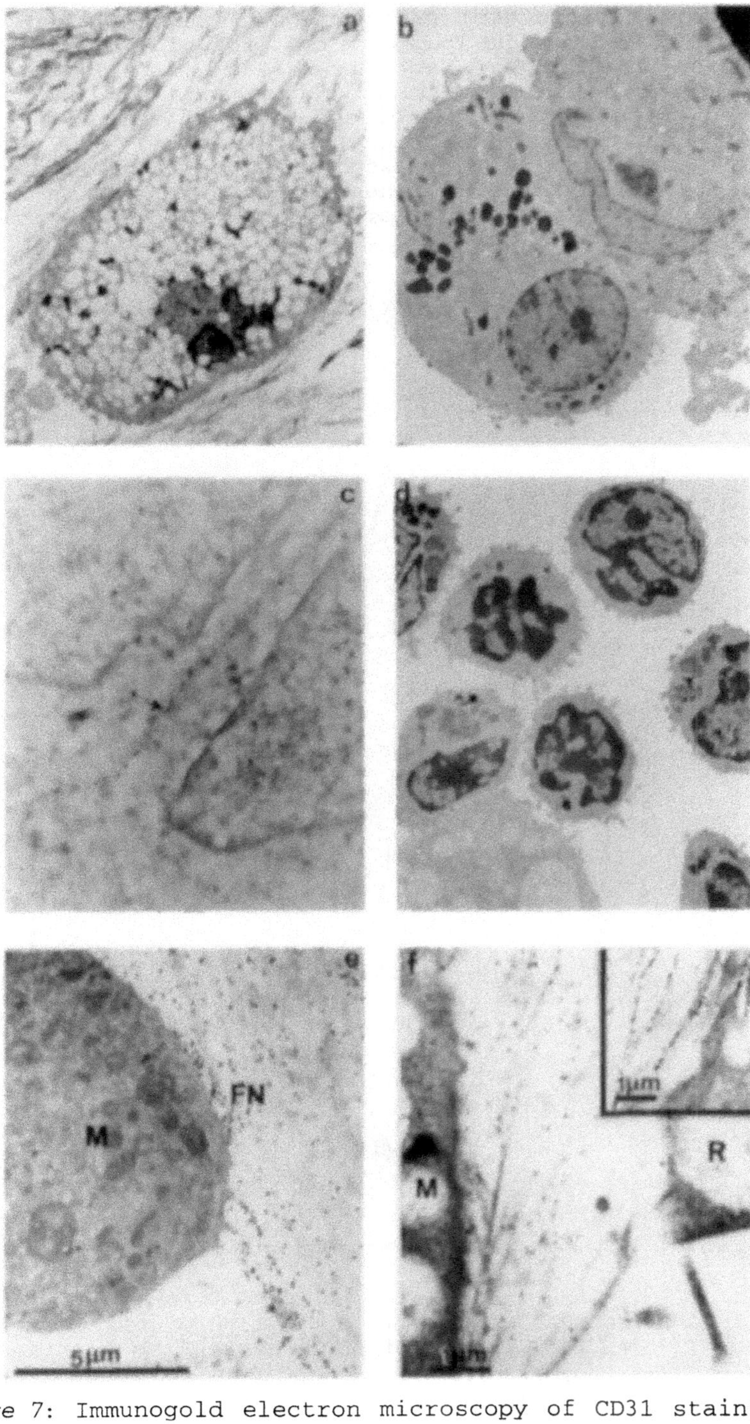

Figure 7: Immunogold electron microscopy of CD31 staining of stromal and haemopoietic cells and of extracellular matrix molecules in Dexter type long term cultures of human adult bone marrow as described by Watt et al., (1993, 1994b). a = macrophage; b = plasma cell and macrophage; c = endothelial cells; d = macrophage associated with 'lymphoid-like' or progenitor cells; e = fibronectin staining (FN) in the extracellular matrix (ECM) in association with a macrophage (M); f = perlecan staining in the ECM in association with a macrophage (M) and a reticular (R) cell. The inset shows perlecan forming fine fibrils.

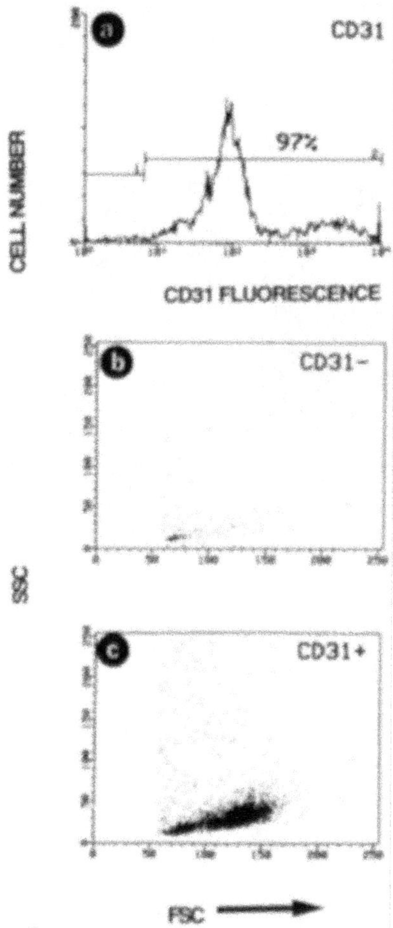

Figure 8: (a) The SBA$^-$CD34$^+$ panned adult bone marrow was stained with the CD31 MoAb (JC70/A) and FITC-rabbit F(ab)$_2$ anti-mouse Ig; light scatter of (b) the CD34 negative fraction [1 in (a)]; and (c) the CD34 positive fraction [2 in (a)] was determined on the FACScan.

(c) Stromal Cells

The haemopoietic region within human bone marrow is represented by two compartments: the haemopoietic and vascular compartments. Within the former, two major stromal cell types are present: the central macrophages which associate with maturing erythroid cells and the reticular cells which are associated with granulocyte development (reviewed by Watt et al., 1995). The blood supply to the haemopoietic spaces enters the long bones via the central artery which branches towards the bone and joins the vascular sinuses eventually draining into the central vein. The thin-walled sinuses are composed of a single layer of highly polarised endothelial cells, which surround the haemopoietic spaces. These endothelial cells, on their abluminal surfaces interact with megakaryocytes, peri-sinal macrophages, adventitial reticular cells and fat cells to form a blood-marrow barrier and so regulate the migration of mature cells from the haemopoetic spaces to the peripheral blood. This is illustrated diagrammatically in Figure 9.

Using the Dexter type long term culture system, we examined by immunogold electron microscopy, the distribution of PECAM-1 on stromal cells cultured for 2 to 5 weeks from adult human bone marrow. In these cultures, the two CD68$^+$ macrophage subsets were PECAM-1 positive. No staining was observed on reticular cells which contained actin microfilament bundles located at the periphery of the cells nor on fat cells which were reminiscent of lipid-laden reticular cells (Figure 7a-d and Watt et al., 1994b). Both the macrophages and reticular cells were associated with extracellular matrix molecules that stain positively for both fibronectin and the heparan sulphate proteoglycan, perlecan (Figure 7e-f). PECAM-1 could be observed to be concentrated on cytoplasmic processes extending from the macrophages towards these extracellular matrix molecules. PECAM-1 was also located between cells especially where endothelial cells adjoined endothelial cells and where macrophages adjoined macrophages, reticular cells, lymphoid-like or progenitor cells and plasma cells (Figure 7c-d) suggesting that it may be involved in both heterotypic and homotypic adhesion events.

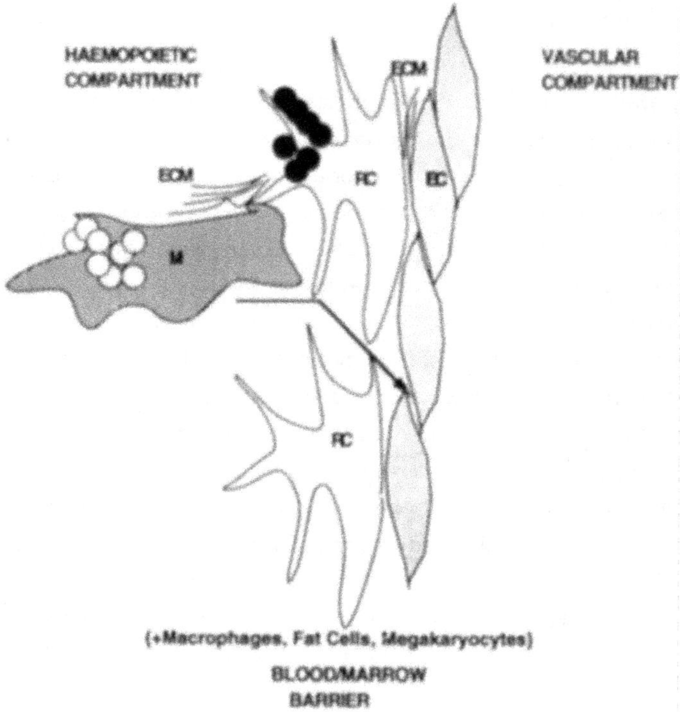

Figure 9: Diagrammatic representation of stromal cells in the haemopoietic compartment and those that form the blood-marrow barrier in adult bone marrow.

(d) The structure of human PECAM-1

The original isoform of PECAM-1 was identified as a single chain 115-140 kDa glycoprotein (reviewed by DeLisser et al., 1994; Watt et al., 1995), cloned both in Europe (Simmons, et al., 1992; Stockinger, et al., 1990) and the USA (Newman, et al., 1990) consisting of a 27 amino acid leader or signal peptide, that is cleaved to generate a 711 amino acid mature polypeptide. This comprises a 574 amino acid extracellular domain, a 19 amino acid hydrophobic transmembrane sequence and a cytoplasmic tail of 118 amino acids. The polypeptide and genomic structure are illustrated diagrammatically in Figure 10.

As a member of the Ig superfamily [IgSF], PECAM-1 contains six disulphide linked IgC2 SET domains in its extracellular region which share homology with adhesion molecules such as CEA (to Domains I, II, III and VI), MAG (to Domains IV and V) and N-CAM (to Domains I and IV).

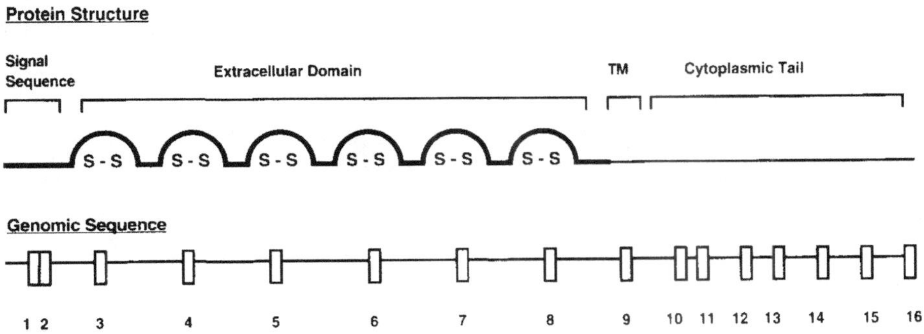

Figure 10: Schematic representation of the human PECAM-1 polypeptide and the exon (▭)/intron (–) structure of the gene. Shaded boxes indicate regions for splicing.

The genomic organisation of the PECAM-1 gene has recently been described (DeLisser et al., 1994) and comprises 16 exons which extend over more than 65 kb of DNA with the exons separated by introns of 86 bp to 12 kb in length. The signal sequence is encoded by exons 1 and 2, and each of the six IgC2 SET domains by a separate exon (exons 3 to 8). Exon 9 encodes the transmembrane (TM) region and exons 10 to 15 the cytoplasmic domains. The carboxy terminus containing the 3'-UTR is encoded by exon 16. Several other isoforms of human PECAM-1 have been identified with splicing of either exons 9, 13 or 14 (Baldwin, et al., 1994). Further heterogeneity is probably observed with differences in glycosylation variants (reviewed by Watt et al., 1995).

(e) Homotypic Adhesion

The adhesion functions of PECAM-1 on mature cells include involvement in transendothelial migration, angiogenesis, cardiovascular development, wound healing and inflammation and these have recently been reviewed (DeLisser et al., 1994; Watt et al., 1995). Our recent studies indicate that PECAM-1 on $CD34^+$ and $CD34^-$ cell lines representing lymphomyeloid progenitors (KG1B) pre-pre-B cells (MIK-ALL) and monocytic cells (U937) can bind to soluble chimaeric forms of PECAM-1 (Figures 11 and 12) containing all six IgC2 set domains linked to the Fc region of human IgG1 (Watt et al., 1993). Using domain variants of soluble PECAM-1 (Figure 11), maximal binding of both MIK-ALL cells and KG1B cells (data not shown) appeared to require the presence of all six IgC2 SET domains (Buckley et al. 1995). However, binding of the promonocytic cell line, U937, was more rapid, cation and particularly Mn^{2+} dependent and could partially proceed with the first IgC2-SET domain of PECAM-1 alone (Figure 12) suggesting an integrin adhesion mechanism rather than homotypic adhesion. Using antibodies (Fawcett, et al., 1994) that identify domains 1, 2 or 5 (Figure 13), the binding of PECAM-1 on $CD31^+$ Cos transfectants, on MIK-ALL cells [$CD31^+CD34^+$] and on U937 cells [$CD31^+CD34^-$] cells to the domain 1 to 6 PECAM-1-Fc (CD31-Fc) molecule could be inhibited, at least partially with antibodies [A and B] to Domain 2 but not with antibodies to Domains 1 or 5 (Figure 14). This suggests, as previously indicated by Simmons (1993), that PECAM-1 may bind in an anti-parallel and homotypic fashion for Cos-PECAM-1 transfectants and for $CD34^+$ cell lines such as MIK-ALL and KGIB (Figure 15). This contrasts with the N-N domain interactions previously observed with BGPc (Teixeira, et al., 1994; Watt, et al., 1994a; Watt, et al., 1995). Our more recent studies suggest that antibodies to Domain 2 may also partially inhibit the adhesion of week 1 to 2 CAFC's to a murine stromal cell line [R. Ploemacher and S. Watt, unpublished data].

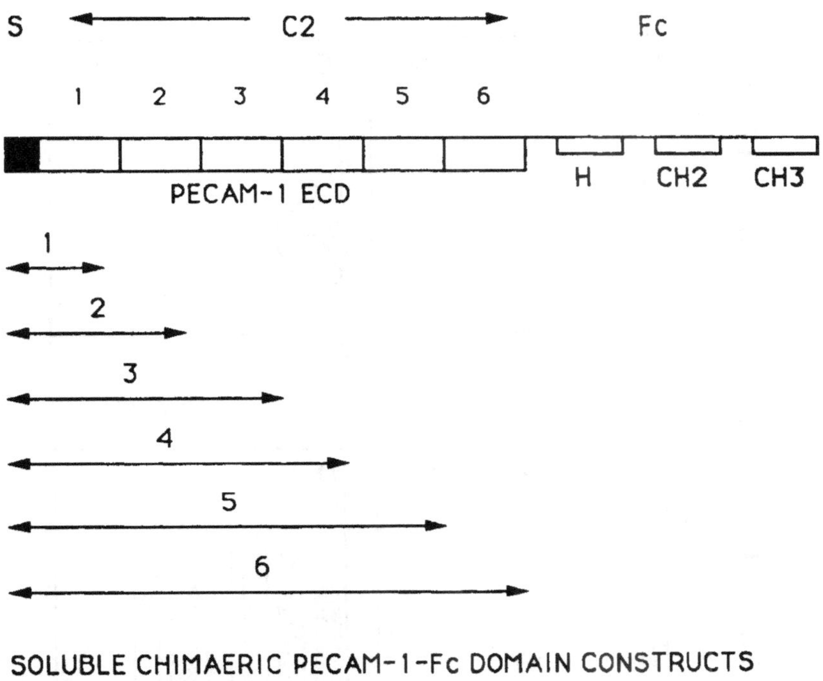

Figure 11: Schematic representation of the soluble chimaeric human PECAM-1-Fc domain variants used for the binding assays. The constructs used in Figure 12, containing domains (1 to 6) are indicated.

(f) Heterotypic adhesion via glycosaminoglycans

For U937, adhesion to PECAM-1-Fc (CD31-Fc) soluble constructs can be inhibited with anti-$\alpha V\beta 3$ inhibitory antibodies suggesting that domain I of U937 binds in part to this integin in a heterotypic Mn^{2+} dependent manner (Buckley et al 1995). Similar results have recently been obtained for murine PECAM-1 (Piali et al 1995). Sequence alignments of CD31 domains I and II with VCAM-1 domain I show aspartate or glutamate residues at exposed regions in their C-C' loops (Watt et al. 1995). These are thought to act as a common structural motif for integrin binding to IgSF members (Bergelson and Hemler, 1995; Loftus et al, 1994; Jones et al, 1995).

The second IgC2 SET domain of PECAM-1 also contains an X-B-B-X-B-X consensus sequence for glycosaminoglycans (GAGs), identified

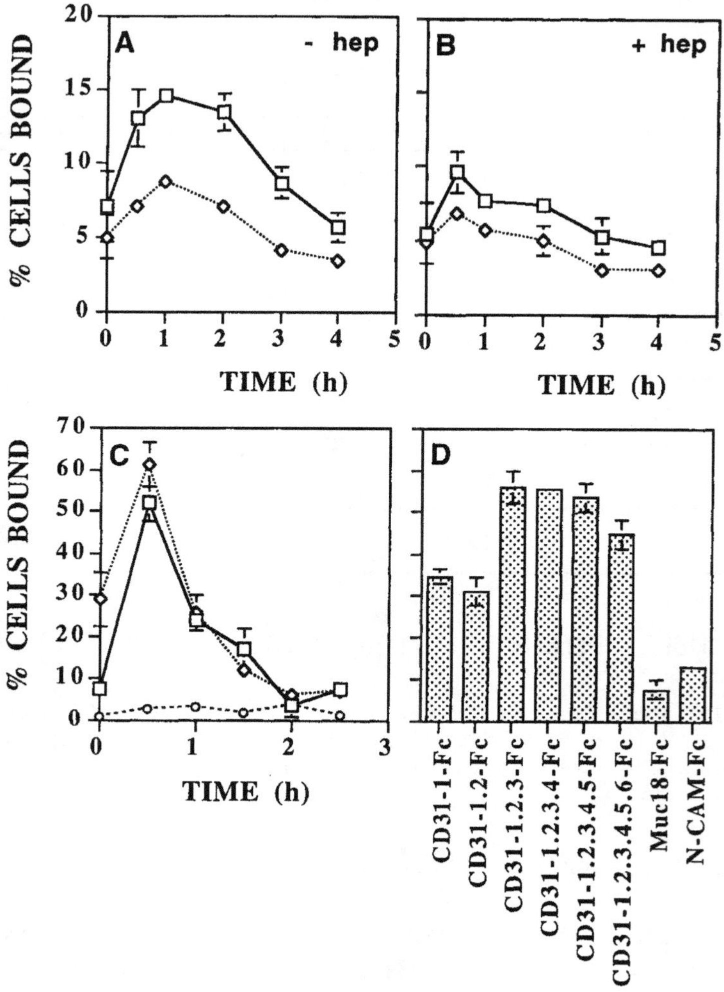

Figure 12: A & B show the time course of adhesion of MIK-ALL cells to CD31-1.2.3.4.5.6-Fc (-□-) or to Muc-18-Fc (◇) (A) in the absence of 100 μg/ml heparin (hep); (B) in the presence of 100 μg/ml heparin. C shows the time course of binding of U937 cells to CD31-1.2.3.4.5.6-Fc in the presence (◇) or absence (□) of heparin or to Muc-18-Fc, without heparin. D shows that U937 cells bind to all soluble CD31-Fc constructs that contain the first IgC2 SET domain, but not to Muc18-Fc or N-CAM-Fc. Values are means ± S.D. for six replicates. Methods are described in Teixeira et al. (1994).

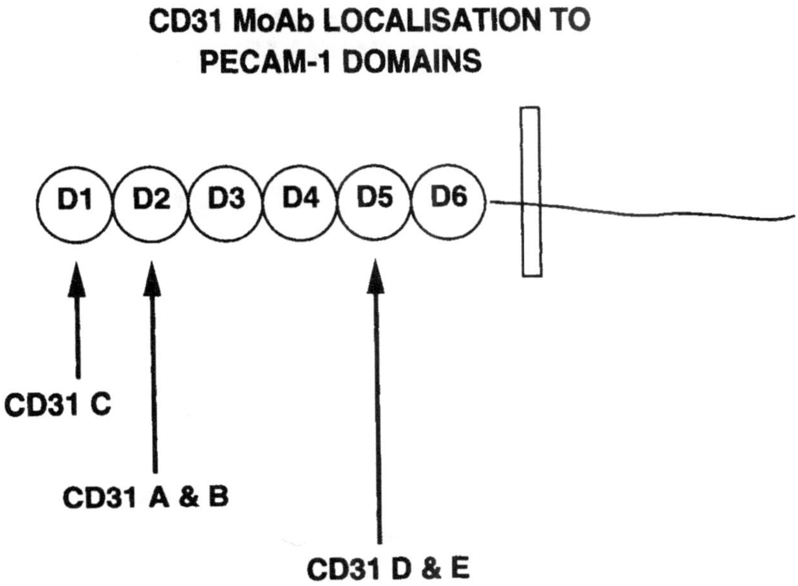

Figure 13: Diagrammatic representation of the location of monoclonal antibodies (MoAb) used to block adhesion assays described in Figure 12. MoAb A = L133.1; B = 9G11; C = JC70/A; D = HC1/6; E = 10B8. Some weak binding of MoAb D to constructs containing D4 has been detected (Fawcett et al., 1995).

at amino acids 150 to 155 as LKREKN (reviewed by DeLisser, et al., 1993a, 1993b, 1994; Watt et al., 1995) and occurring within the exposed loop of Domain II between the C' and Eβ strands.

Previous studies have shown that mouse L-cells transfected with human PECAM-1 can bind non transfected L-cells and this binding can be substantially inhibited by heparin, heparan sulphate or chondroitin-6-sulphate but not dermatan sulphate, dextran sulphate, keratan sulphate nor hyaluronic acid (DeLisser et al., 1993a, 1993b).

INHIBITION OF CD31+ CELL BINDING TO SOLUBLE CD31 DOMAIN VARIANTS

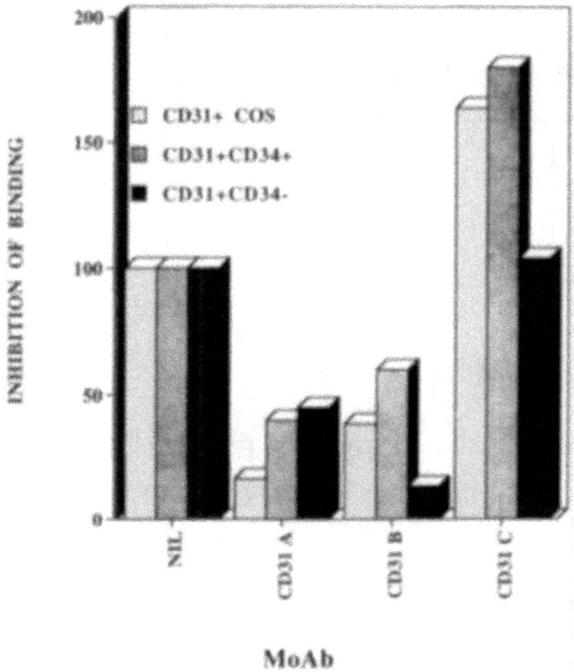

Figure 14: Inhibition of PECAM-1 transfected Cos ($CD31^+Cos$), MIK-ALL ($CD31^+CD34^+$) and U937 ($CD31^+CD34^-$) cell adhesion to soluble Domain 1 to 6 PECAM-1-Fc (CD31-Fc) proteins by antibodies shown in Figure 13. Values are mean $\pm$ S.D. ($n = 6$). Methods are those described in Figure 12.

The adhesion could also be reduced by pre-treatment with heparitinase, or chondroitinase ABC, but not hyaluronidase and requires both the 35 terminal amino acids of the cytoplasmic tail and domain VI for this heterotypic adhesion (DeLisser, et al., 1993a, 1993b, 1994). Our own studies (Watt et al., 1993) indicate that Domain 1-6 PECAM-1-Fc binds with low avidity to heparin sepharose, the binding being partially inhibited by heparan sulphate. In addition, initial studies suggest that the low level of adhesion of the $CD34^+$ MIK-ALL cell lines to CD31-1 to 6-FC is partially inhibited by heparin. With U937 there seems to be little effect of heparin addition (Figure 12). Further studies on CAFCs and $CD34^+$ cells from human bone marrow are in progress to analyse the binding of PECAM-1 to heparan sulphate proteoglycans from bone marrow stroma. A schematic representation of how some $CD31^+$ cells might bind to macrophages and endothelial cells expressing PECAM-1 or to heparan sulphate proteoglycans produced by reticular cells is shown in Figure 16.

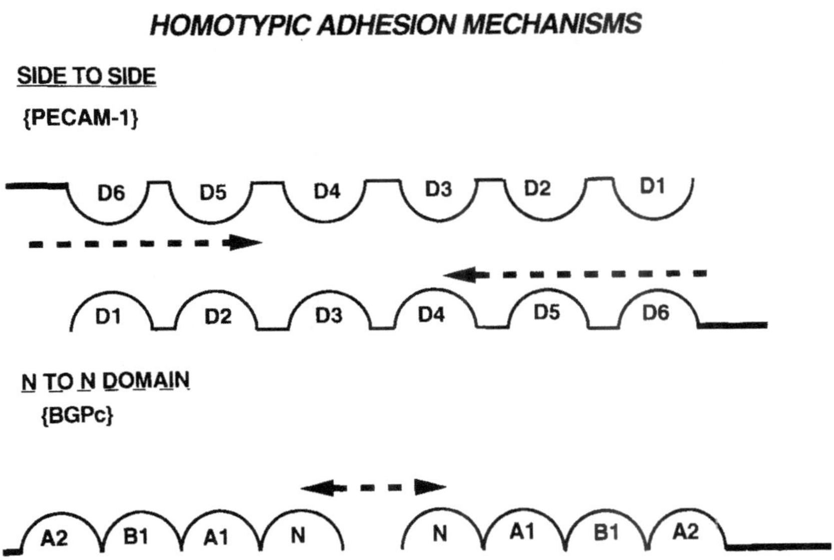

Figure 15: Schematic representation of the mechanisms of homotypic adhesion of PECAM-1 (side to side) and BGPc (N-domain).

Such adhesion would bring these progenitors into contact with heparan sulphate proteoglycans which concentrate growth factors or inhibitors and these interactions may activate VLA-4 and allow its adhesion to VCAM-1 as recently indicated by Leavesley, et al., (1994). Alternatively, PECAM-1 and the proteoglycans may reside on the same or apposing cells and may require modification or activation prior to adhesion functions.

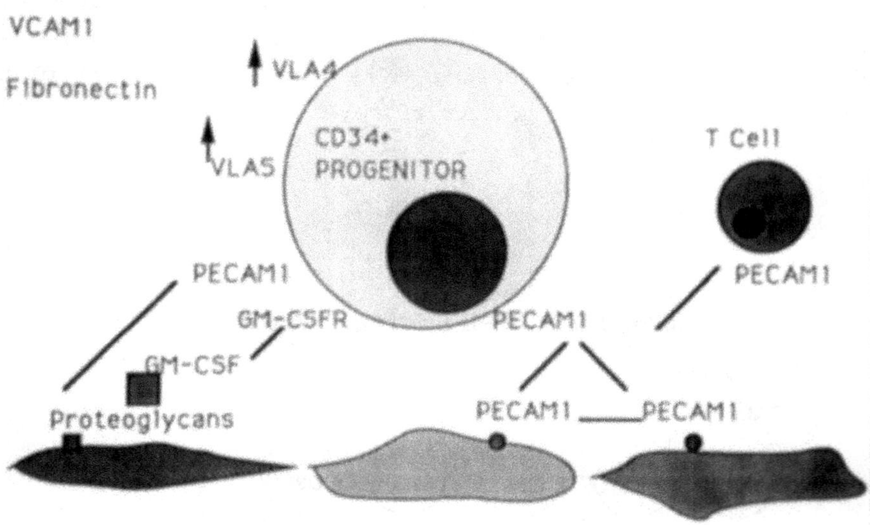

Figure 16: Possible mechanisms of homotypic and heterotypic adhesion of PECAM-1 on CD34$^+$ progenitors to PECAM-1 on stromal macrophages or vascular endothelial cells or to heparan sulphate containing proteoglycans produced by stromal reticular cells. Engagement of PECAM-1 may activate an adhesion cascade involving the β1 integrins.

CD33 : its expression and structure

(a) <u>CD33 Expression</u>

Under reducing conditions, CD33 monoclonal antibodies identify a 67 kDa type 1 integral membrane glycoprotein which is restricted in its expression to the haemopoietic lineage (Andrews, et al., 1983; Barclay, et al., 1993; Peiper, et al., 1989). On more mature cells, CD33 is expressed on the monocytic lineage. In the granulocytic lineage it occurs at higher amounts on promyelocytes, myelocytes and metamyelocytes and in low amounts on mature neutrophils. It is also found on multipotent progenitors, the CFU-GEMM, and on lineage restricted progenitors, the CFU-G/M and BFU-E. Recent studies by Olweus et al., (1994) indicate that it is expressed on both fetal and adult bone marrow primitive progenitor cells with the $CD34^{+}CD38^{lo/-}HLA\text{-}DR^{+}$ phenotype, but not on $CD34^{+}CD38^{lo/-}HLA\text{-}DR^{-}$ cells, and on human fetal liver progenitors with a high proliferative potential (Muench, et al., 1994). Myeloblasts from approximately 85% of patients with AML are $CD33^{+}$ (Scheinberg, et al., 1991).

(b) CD33 Structure

Both human and murine CD33 (hCD33 and mCD33 respectively) cDNAs have been cloned (Simmons & Seed, 1989; Tchilian, et al., 1994). In addition, preliminary studies have identified cosmids spanning the human genomic locus (Peiper, et al., 1988; M.A. Benton, J. Lee-Prudhoe, C. Dennison and S. Watt, unpublished data) and phage and cosmids (J. Lee-Prudhoe and S. Watt, unpublished data) spanning the murine genomic locus. These studies suggest that the murine Ig and transmembrane domains are encoded by separate exons and that at least two exons are present in the cytoplasmic domain. In the human, CD33 is localised to chromosome 19q13 (Peiper et al., 1988) and in the mouse to chromosome 7 in a region syntenic to human chromosome 19 (E. Carver, L. Stubbs and S. Watt, unpublished data).

Like CD31, CD33 is a member of the IgSF of adhesion molecules, but contains a signal peptide, one IgV SET N-terminal domain, one IgC2 SET domain, a transmembrane region and a cytoplasmic tail which, in the mouse, is differentially spliced. Two isoforms of murine CD33 have been identified (Tchilian et al., 1994) with the mature long CD33-L polypeptide comprising 387 amino acids and the mature short CD33-S form containing a cytoplasmic tail shortened by 83 nucleotides to yield a polypeptide of 318 amino acids. The alignment, domain structure and homology of murine with human CD33 are shown in Tables 10, 11 and 12, and Figure 17.

(c) Homology to MAG and CD22

Database searches revealed a high homology of mCD33 to hCD33 (Table 11 and Figure 17). At the amino acid level, a good alignment is observed with the extracellular domain IgV and IgC2 SET sequences (61% identity, 70% similarity) with the highest homology being evident in the IgC2 SET domain (75% identity, 80% similarity). Homology is observed between mCD33-S and hCD33 in the cytoplasmic region (32% identity, 42% similarity) but not with mCD33-L. Some similarity is also observed with two other adhesion molecules that are positioned on the same chromosome as CD33. The extracellular domains (signal, IgV and IgC2 SET domains) of CD33 show homology to both MAG (28% identity, 46%

Table 10: Domain Structure of Murine and Human CD33

Region of CD33	Amino Acid Sequence Murine CD33	Human CD33
ECD	1-240	3-237
Signal peptide	1-16	3-17
IgV-SET	17-130	18-130
IgC2-SET	152-232	152-232
TM	241-267	261-284
CD33-L	1-403	-
CD33-cyt-L	268-403	-
CD33-S	1-334	3-364
CD33-cyt-S	268-334	285-364

Alignment of individual domains and the full length amino acid sequences using the GCG BESTFIT programme revealed extensive similarity in length and sequence particularly to the extracellular domains (ECD). TM = transmembrane; CD33-cyt-L and CD33-cyt-S are cytoplasmic tails of the long and short forms of CD33.

Table 11: Sequence Similarity of Murine CD33 to Human CD33

Region	Amino Acid of sequence mCD33	Comparative Region sequence of hCD33	Comparative Amino-acid	% Identity	% Similarity	Quality Score	Ratio
mCD33-L	1-403	hCD33	3-363	49.6	63.6	273.3	0.757
mCD33-S	1-334	hCD33	3-361	53.5	64.9	278.7	0.834
mCD33IgV-SET	18-130	hCD33	18-130	53.6	65.2	93.0	0.823
mCD33IgC2-SET	152-232	hCD33	152-232	75.3	80.2	98.7	1.219
mCD33cyt-S	268-334	hCD33	286-361	31.8	42.4	32.7	0.495
mCD33ECD	18-240	hCD33	18-237	60.7	69.9	213.3	0.970

Sequence similarities are shown in Figure 17. Regions of sequence similarity, quality scores and ratio were obtained using the GCG programme BESTFIT from searches of the SWISSPROT and PIR databases using the local homology of Smith and Waterman alogarithms.

Table 12: Sequence Similarity of mCD33 Extracellular Domain to members of the IgSF.

Protein	Amino Acid Sequence	Comparative Region	Comparative Amino Acid Sequence	% Identity	% Similarity	Quality Score	Ratio
mCD33	1-221	mMAG	19-243	27.9	46.1	106.6	0.482
mCD33	3-223	mCD22	25-264	26.6	46.3	90.4	0.409
hCD33	2-241	hMAG	8-279	28.2	49.6	115.6	0.482
hCD33	1-238	hCD22	21-262	28.9	46.6	111.7	0.469

Sequence similarities are shown in Figure 17. Regions of sequence similarities were obtained from searches of the SWISSPROT and PIR databases using the local homology alogarithms of Smith and Waterman. A gap weight of 3.0 and a gap length weight of 0.1 were used with the GCG BESTFIT programme to obtain alignments, % identities, % similarities, quality scores and ratios. mMAG, mCD22, hMAG and hCD22 sequences were derived from results of Fujita, et al., 1989; Sato, et al., 1989; Stockinger et al., 1990; Torres, et al., 1992 respectively.

a

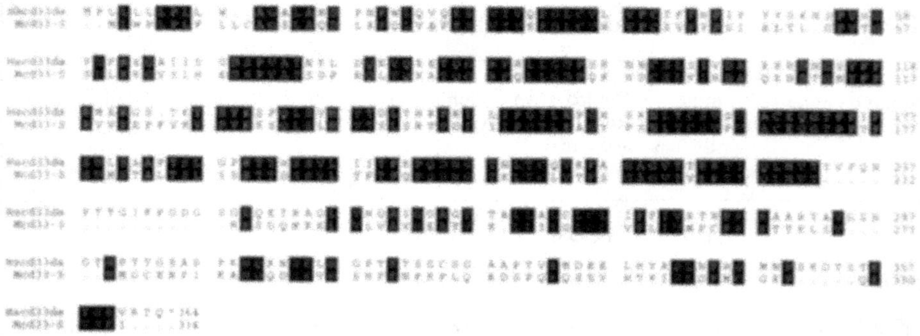

b

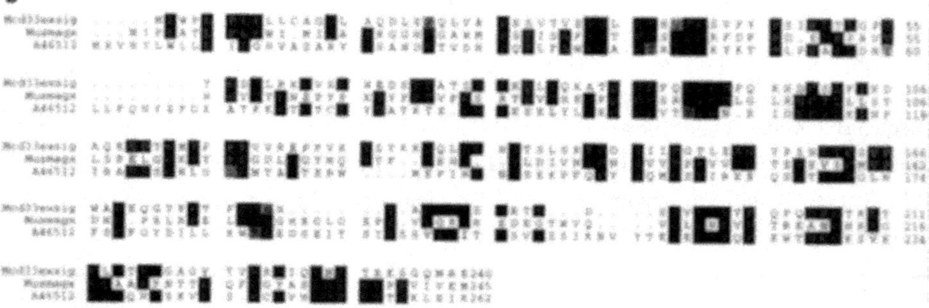

c

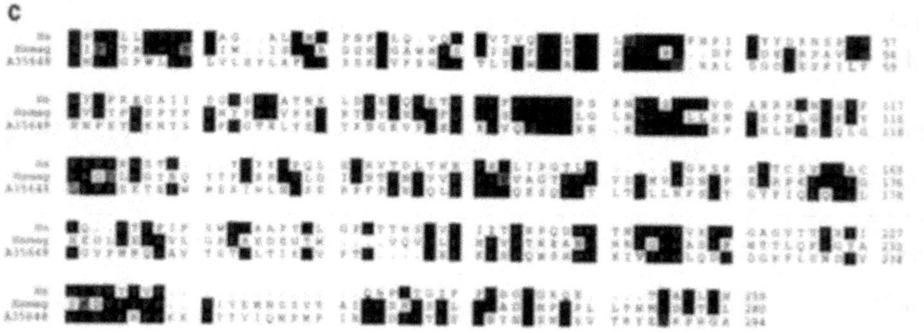

Figure 17: Alignment of amino acid sequences of human (Hscd33da) and murine (Mcd33-s) CD33 polypeptides (a) and of the extracellular domain sequences of mouse (Mcd33 exsig) and human (Hs) CD33 with the mouse MAG (Musmagx) and CD22 (A46512) or with the human MAG (Hsmag) and human CD22 (A35648) sequences. The sequences were aligned using PILEUP and displayed using the multi-alignment display programme, PRETTY BOX.

similarity for the mouse proteins; 28% identity, 50% similarity for the human proteins) and CD22 (27% identify, 46% similarity for the mouse proteins; 29% identity, 47% similarity for the human proteins) for the sequences shown in Table 12 and Figure 17. Homology of the mCD33 and hCD33 IgV and IgC2 set domains is observed with the IgV and first IgC2-SET domains of both CD22 and MAG.

For mCD33, the IgV and IgC2 SET domains contain 3 potential N-linked glycosylation sites (Asn-X-Ser/Thr) at positions 110, 160 and 230 and a possible GAG attachment site for heparan sulphate (Ser Gly Ala Gly or SGXG motif) at residue 217.

This is in contrast to the hCD33 sequence which has 5 potential N-linked glycosylation sites at residues 100, 113, 160, 209 and 230 and apotential chrondroitin sulphate or dermatan sulphate GAG attachment site at residues 246 to 253 carrying the sequence motif DXSG(X)E. Predicted GPI anchors are found with both mCD33 and hCD33 with the GPI signal sequences RKS at position 235 and DGS at position 248 respectively. The mCD33-S cytoplasmic region potentially contains two c-AMP/c-GMP phosphorylation sites at 269 and 270, two casein kinase 2 phosphorylation sites at 296 and 308 and two protein kinase C phosphorylation sites at 250 and 272.

The cytoplasmic tail of mCD33-L differs in having one casein kinase 2 phosphorylation site at residue 301 and four protein kinase C phosphorylation sites at residues 319, 344, 355 and 359. In contrast, hCD33 contains no c-AMP/c-GMP phosphorylation sites, one potential casein kinase 2 phosphorylation site at 354 and two putative protein kinase C phosphorylation sites at residues 284 and 307. In addition, hCD33 contains a consensus sequence for GAG binding identified as XBBXBX or THRPKI between residues 142 to 147.

(d) Isoforms and Function of CD33

Our studies (Tchilian et al., 1994) indicate that several isoforms of mCD33 exist. At least two forms of mRNA of 2 and 3.9 kb are evident by Northern blot analysis. The 3.9 kb form is found more abundantly in the bone marrow and thymus, to a lesser extent in the liver and brain (presumably in the Kupffer cells and microglia respectively) and in the P388 macrophage cell line, but not in the spleen. The 2 kb form appears to be the major transcript in bone marrow and at a lower level in spleen and in the multipotential or myelomonocytic progenitor cell lines, FDCP-mix, Wehi3B and M1. In addition, two cytoplasmic splice variants (mCD33-L and mCD33-S) are found by RT-PCR analyses in bone marrow, spleen, thymus, liver, brain and the multipotential and myelomonocytic progenitor cell lines. In the P388 cell line, only the smaller isoform is observed. The biological significance of these different isoforms is unclear. However, the generation of different phosphorylation sites in the two variants suggests that this may alter their signal transduction properties and consequently affect ligand affinity or adhesive functions. It is interesting that the mCD33-S cytoplasmic region contains a similar set of potential phosphorylation motifs as those observed in sialoadhesin (Crocker, et al., 1994; Kelm, et al., 1994), a macrophage specific lectin-like receptor that mediates sialic acid dependent adhesion. Indeed, recent studies (Freeman et al 1995) suggest that CD33 together with CD22 and sialoadhesin constitute a group of sialic acid dependent adhesion molecules although CD22 recognises sialic acids terminating in Neu5NAca2,6 Gal while sialoadhesin and CD33 recognise Neu5NAca2,3Gal moieties (Crocker et al., 1994; Engel, et al., 1995; Stamenkovic & Seed, 1990; Stamenkovic, et al., 1991; Wilson, et al., 1991, Powell & Varki 1995; Freeman et al 1995).

Alternative functions for CD33 could exist with the addition of heparan sulphate or chondroitin sulphate moieties to the predicted GAG recognition sequences allowing these molecules to bind appropriate adhesive ligands or to GAG binding growth factors or inhibitors (Gordon, et al., 1987; Roberts, et al., 1988).

ICAM-3 (CD50)

(a) Expression on Mature Cells

Intercellular adhesion molecule 3 (ICAM-3, CD50) is constitutively and highly expressed on most leukocytes, including antigen presenting cells, such as dendritic cells in lymphoid organs and Langerhans cells in the skin (Acevedo, et al., 1993; Cordell, et al., 1994; de Fougerolles, et al., 1993). Cordell et al. (1994) have reported that it is absent from many proliferating B cells in germinal centres of secondary lymphoid organs and from Reed-Sternberg cells in Hodgkin's disease but occurs on CLL, CML, AML, multiple myeloma and hairy cell leukaemia.

Initially, like CD33, ICAM-3 was thought to be haemopoietic specific (Acevedo et al., 1993; Cordell et al., 1994; de Fougerolles et al., 1993). However, recent studies (Doussis-Anagnostopoulou et al., 1993) indicate that it can be induced on vascular endothelium in disease states which include the myelomas and lymphomas.

(b) Expression on $CD34^+$ cells

We have examined the expression of ICAM-3 on both $CD34^+$ cell lines and human adult bone marrow cells using the CD50 MoAbs CH3.2 and CH3.3 which map to either domain 1 or the interface between domains 1 and 2 (Holness, et al., 1994). The results of binding to a set of cell lines are shown in Figure 18. Both antibodies react with the $CD34^+$ lymphomyeloid progenitor (KGIB), myeloid progenitor (KG1) and pre-pre-B lymphoid (TOM-1 and MIK-ALL) cell lines and with myelomonocytic (HL-60) and monocytic cell lines (U937, THN-1). The early T lymphoid cell line, CEM, is positive. The level of expression of ICAM-3, on the $CD34^+$ cell lines varied depending on the growth conditions with TOM-1 at times showing substantially higher expression than KG1 or KGIB. With the CH3.2 and 3.3 antibodies staining of more mature B cell lines (697, 207, REH, RAJI, DAUDI) and the erythroleukaemia line (K562) was not observed (Figure 18). The phenotype of the immature cell lines can be defined for KG1 and KG1B as $CD50^+$, $CD34^+$, $CD31^+$, $CD33^+$ and for TOM-1 and MIK-ALL as $CD50^+$, $CD34^+$, $CD31^+$, $CD33^-$.

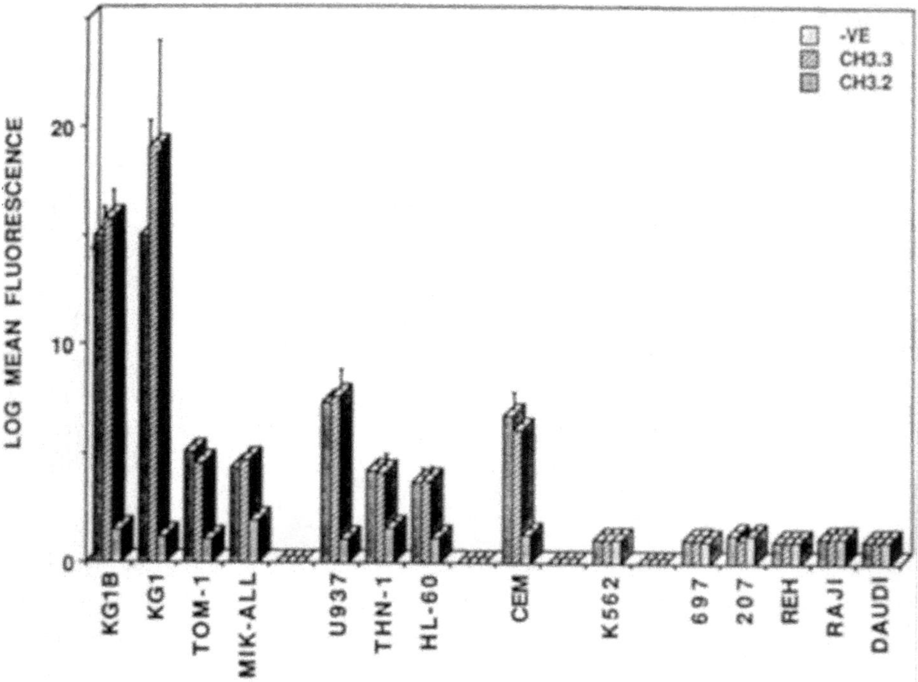

Figure 18: Mean fluorescence intensity values (± S.D. of triplicate samples) of CD50 (CH3.2 and CH3.3) binding to cell lines. Antibody binding was detected using FITC-rabbit $F(ab)_2$ anti-mouse Ig. Isotype matched negative antibodies were used in place of the CD31 MoAbs for the negative control (-ve). Cells were analysed on the FACScan.

The expression of ICAM-3 on human adult bone marrow cells was examined by two techniques: by panning and by flow sorting. The panning procedure involved separation of normal adult bone marrow cells on Ficoll-hypaque (denisty 1.077 g/ml) followed by depletion of cells on SBA Cellector plates (AIS; Santa Clara, CA, USA). The light density/non-adherent cells were labelled with the CH2.3 and CH2.2 MoAbs and allowed to adhere to goat anti-mouse Ig Cellector flasks (AIS, Santa Clara, CA, USA).

Table 13: Separation of Haemopoietic Progenitors using CD50 MoAb Selection

A: Panning SBA^- Adult Bone Marrow with CD50*

Fraction	% Cells	% Recovery of d14 CFU CFU-G/M	BFU-E	CFU-GEMM
SBA^-	100 ± 0	100	100	100
$SBA^-ICAM\text{-}3^{lo/-}$	31 ± 8	19 ± 7	37 ± 5	24 ± 6
$SBA^-ICAM\text{-}3^{++}$	25 ± 11	48 ± 9	50 ± 27	54 ± 36

* mean of 4 ± S.D. (independent experiments)

B: Panning SBA^- $CD34^+$ Adult Bone Marrow with CD50**

Fraction	% Cells	% Recovery of d14 CFU CFU-G/M	BFU-E	CFU-GEMM
$CD34^+$	100	100	100	100
$CD34^+ICAM\text{-}3^{lo/-}$	36.2	18	40	34
$CD34^+ICAM\text{-}3^{++}$	45.8	40	36	67

** mean of 2 independent experiments

C: Flow Sorting SBA^- Adult Bone Marrow with CD50‡

Fraction	% Cells	% Recovery of d14 CFU CFU-G/M	BFU-E	CFU-GEMM	Stromal Clones
SBA^-	100	100	100	100	100
$SBA^-ICAM\text{-}3^-$	10.8	0.3	0	0	17
$SBA^-ICAM\text{-}3^{lo}$	12.6	0.6	1.8	0	31
$SBA^-ICAM\text{-}3^+$	74.7	109.8	60.0	75.5	0

‡ mean of 2 independent experiments. Stromal clones were reminiscent of fibroblastic-like reticular cells. Recoveries were normalised to 100% and each ICAM-3 subfraction calculated as a % of the cell number or number of clonogenic cells in the SBA^- or $CD34^+$ fractions. An isotype matched irrelevant MoAb served as the negative control for these 2 experiments with 90.5% of cells occurring in the $ICAM\text{-}3^-$ fraction, 7.3% in the $ICAM\text{-}3^{lo}$ fraction and 0.5% in the $ICAM\text{-}3^+$ fraction.

The non-adherent cells ($SBA^{-}ICAM-3^{lo/-}$) were collected and the adherent cells then recovered by trypsinization prior to plating in StemGEM media. Recoveries are given in Table 13A which shows that, as for the $CD34^{+}$ flasks, the weakly adherent cellscontained a proportion of the day 14 progenitors. Similar results were obtained if $CD34^{+}$ cells were isolated prior to panning with ICAM-3 MoAb (Table 13B). However, the majority of both the SBA^{-} and $CD34^{+}$ cells isolated were $ICAM-3^{+}$ when analysed by flow cytometry (data not shown). For example, as in the experiments described below, analysis on the FACScan of one SBA^{-} bone marrow sample revealed that 91.8% of cells were $CD50^{+}$, although less than 50% of these bound in the ICAM-3/Goat anti-mouse Ig panning protocol. In order to determine if the day 14 progenitors expressed ICAM-3, the light density SBA^{-} cells from adult bone marrow were labelled with the CD50 MoAbs described above and sorted into 3 fractions such as those indicated in the histogram in Figure 19.

The majority of day 14 haemopoietic progenitors were recovered in the $ICAM-3^{+}$ fraction and these appeared to segregate from cells that formed stromal clones which resembled fibroblastic reticular cells and were present in the ICAM-3 negative or lo fractions (Table 13C).

As indicated in Table 9, these results are consistent with those of Olweus et al., (1995) and Waller et al., (1995) where CD50 also appears to be present on the most primitive haemopoietic progenitors but not on stromal progenitors, at least at high expression levels (Olweus et al., 1995).

c) <u>ICAM-3 structure and function</u>

ICAM-3 has been identified as a 120 kDa type 1 integral membrane protein belonging to the IgSF of ahesion molecules. c-DNA cloning studies (de Fougerolles et al., 1993; Fawcett, et al., 1992; Vazeux, et al., 1992) indicate that, like PECAM-1, all extracellular domains of ICAM-3 (in total five) belong to the IgC2-SET. ICAM-3 shares homology with the other two members of the ICAM family, ICAM-1 (52% amino acid identity) and ICAM-2 (37% amino acid identity) with Domain II of ICAM-3 sharing 77%

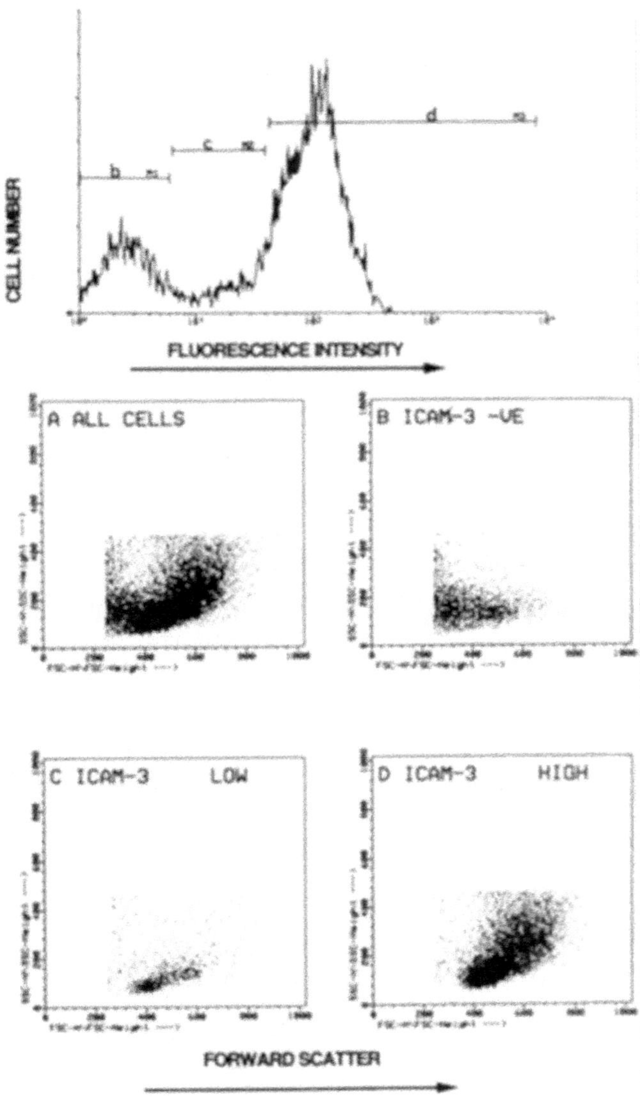

Figure 19: The light scatter distribution of SBA^- panned human adult bone marrow cells (separated using the manufacturer's method) within the progenitor cell window after staining cells with the CD50 MoAbs (CH3.2 and CH3.3) together with FITC rabbit $F(ab)_2$ anti-mouse Ig, analysing these on a FACStar plus and gating these on the relative fluorescence intensity (b, c or d gates) shown in the upper histogram. A: the whole SBA^-CD34^+ population; B: ICAM-3^- cells in fraction b; C: ICAM-3^{lo} cells in fraction c; D: ICAM-3^+ cells in fraction d. Cells were sorted into the 3 fluorescence fractions shown and analysed for CFU content as indicated in Table 14.

identity with ICAM-1 (de Fougerolles et al., 1993; Fawcett et al., 1992; Vazeux et al., 1992).

One major ligand for ICAM-3 on mature cells is LFA-1 (Acevedo et al., 1993; Bernard, et al., 1994; Campanero, et al., 1993; Cid, et al., 1994; Hernandez-Caselles, et al., 1993; Holness et al., 1995). Unlike PECAM-1, CD33 and BGP, adhesion of ICAM-3 to LFA-1 involves the first two IgC2-SET domains, Domains I and II of ICAM-3 (Holness et al., 1994). The functional significance of ICAM-3 on haemopoietic stem cells or progenitor cells is unknown. However, engagement of molecules such as CD45 on primitive haemopoietic precursors possibly with heparan sulphate moieties on stromal cells or in the extracellular matrix (Coombe, et al., 1994) may activate ICAM-3 on the haemopoietic progenitors inducing its adhesion to LFA-1 on stromal macrophages or endothelia. Such a mechanism would be reminiscent of that suggested by Bernard et al., (1994) for CD45 stimulated adhesion of ICAM-3 to LFA-1 on thymocytes. Additionally, like PECAM-1, engagement of ICAM-3 with its ligand may promote a cascade of adhesion events that are mediated via the β1 and β2 integrins. Such mechanisms have been described for mature T cells (Cid et al., 1994).

Conclusions

A method for isolating $CD34^+$ cells to a high degree of purity has been described. Three adhesion molecules, PECAM-1, CD33 and ICAM-3, are expressed in high amounts on all or on a proportion of these $CD34^+$ cells which appear to encompass the most primitive haemopoietic progenitors. Although several studies have shown that these molecules are functional on mature cells at least *in vitro*, very little information is available on the function of these molecules on haemopoietic progenitors, on where splice variants are expressed during haemopoietic commitment and if the molecules are altered by leukaemic transformation. These questions need to be addressed. In addition, regulatory elements of these genes may provide a useful means for regulating the expression of introduced genes into different subsets of cells within the haemopoietic lineage.

Acknowledgements

We are grateful to the European Union Biomed 1 programme, the Medical Research Council, the Leukaemia Research Fund, the Kay Kendall Leukaemia Fund, CNRS and Amgen-Europe for their support. The authors would like to thank Drs. J. Fawcett and D. Simmons for the soluble PECAM-1-Fc constructs, Drs. C. Buckley and C. Holness for advice on the binding assays, for antibodies and for helpful discussions, Mr. D. Davies for cell sorting and Mr. S. Gschmeissner for the electron microscopy. We also wish to particularly thank Mrs. E. Rose and L. Roberts for expert typing of the manuscript.

References

Acevedo A, del Pozo MA, Arroyo AG, Sanchez-Mateos P, Gonzalez-Amaro R, Sanchez-Madrid F (1993) Distribution of ICAM-3-bearing cells in normal human tissues. Expression of a novel counter-receptor for LFA-1 in epidermal Langerhans cells. Am. J. Pathol. 143: 774.

Andrews RG, Torok-Storb B, Bernstein ID (1983) Myeloid-associated differentiation antigen on stem cells and their progeny identified by monoclonal antibodies. Blood 62: 124.

Baldwin HS, Shen HM, Yan H-C, DeLisser HM, Chung A, Mickanin C, Trask T, Kirschbaum NE, Newman PJ, Albelda SM, Buck CA (1994) Platelet endothelial cell adhesion molecule-1 (PECAM-1/CD31): alternatively spliced, functionally distinct isoforms expressed during mammalian cardiovascular development. Development 120: 2539.

Barclay AN, Birkeland ML, Brown MH, Beyers AD, Davis SJ, Somoza C, Williams AF (1993). The Leukocyte Antigen Facts Book. London, UK: Academic Press.

Bergelson JM, Hemler ME (1995) Do integins use a "MIDAS touch" to grasp an Asp? Curr. Biol. 5: 615.

Bernard G, Zoccola D, Ticchioni M, Breittmayer JP, Aussel C, Bernard A (1994) Engagement of the CD45 molecule induces homotypic adhesion of human thymocytes through an LFA-1/ICAM-3 dependent pathway. J. Immunol. 152: 5161.

Buckley CD, Doyonnas, R, Newton JP, Simmons DL, Watt SM (1995) (submitted for publication). Identification of $\alpha v\beta 3$ as a heterotypic ligand for CD31/PECAM-1. J. Cell Science.

Campanero MR, del Pozo MA, Arroyo AG, Sanchez-Mateos P, Hernandez-Caselles T, Craig A, Pulido R, Sanchez-Madrid F (1993) ICAM-3 interacts with LFA-1 and regulates the LFA-

1/ICAM-1 cell adhesion pathway. J. Cell Biol. 123: 1007.

Cardoso A, Watt SM, Batard P, Li M-L, Hatzfeld A, Genevier H, Hatzfeld J (1995) An improved rapid panning technique for the selection of CD34 positive human hematopoietic progenitor cells with high recovery of early progenitors. Exp. Hematol. 23: 407.

Cid MC, Esparza J, Juan M, Miralles A, Ordi J, Vilella R, Urbano-Marquez A, Gaya A, Vives J, Yague J (1994) Signalling through CD50 (ICAM-3) stimulates T lymphocyte binding to human umbilical vein endothelial cells and extracellular matrix proteins via an increase in beta 1 and beta 2 integrin function. Eur. J. Immunol. 24: 1377.

Coombe DR, Watt SM, Parish CR (1994) The multipotential cell line $FDCP_{mix}$ A4 interacts with stromal fibroblastoid cells through Mac-1 and CD45. Blood 84: 739.

Cordell JL, Pulford K, Turley H, Jones M, Micklem K, Doussis IS, Tyler X, Mayne K, Gatter KC, Mason DY (1994) Cellular distribution of human leucocyte adhesion molecule ICAM3. J. Clin. Pathol. 47: 143.

Crocker PR, Mucklow S, Bouckson V, McWilliam A, Willis AC, Gordon S, Milon G, Kelm S, Bradfield B (1994) Sialoadhesin, a macrophage sialic acid binding receptor for haemopoietic cells with 17 immunoglobulin-like domains. EMBO J. 13: 4490.

de Fougerolles AR, Klickstein LB, Springer TA (1993) Cloning and expression of intercellular adhesion molecule 3 reveals strong homology to other immunoglobulin family counter-receptors for lymphocyte function-associated antigen 1. J. Exp. Med. 177: 1187.

DeLisser HM, Chilkstowsky J, Yan H-C, Daise ML, Buck CA, Albelda SM (1994) Deletions in the cytoplasmic domain of platelet-endothelial cell adhesion molecule-1 (PECAM-1, CD31) result in changes in ligand binding properties. J. Cell. Biol. 124: 195.

DeLisser HM, Muller W, Newman P, Buck CA, Albelda SM (1993a) PECAM-1 mediates heterophilic glycosaminoglycan-dependent cell-cell adhesion. J. Biol. Chem. 258: 16307.

DeLisser HM, Yan HC, Newman PJ, Muller WA, Buck CA, Albelda SM (1993b) Platelet/endothelial cell adhesion molecule-1 (CD31)-mediated cellular aggregation involves cell surface glycosaminoglycans. J. Biol. Chem. 268: 16037.

DeLisser HM, Newman PJ, Albelda SM (1994) Molecular and functional aspects of PECAM-1/CD31. Immunol. Today 15: 490.

Doussis-Anagnostopoulou I, Kaklamanis L, Cordell J, Jones M, Turley H, Pulford K, Simmons D, Mason D, Gatter K (1993) ICAM-3 expression on endothelium in lymphoid malignancy. Am. J. Pathol. 143: 1040.

Engel P, Wagner N, Miller AS, Tedder TF (1995) Identification of the ligand-binding domains of CD22, a member of the immunoglobulin superfamily that uniquely binds a sialic acid dependent ligand. J. Exp. Med. 181: 1581.

Fawcett J, Buckley C, Holness CL, Bird IN, Spragg JH, Saunders J, Harris A, Simmons DL (1995) Mapping of homotypic binding sites in CD31 and the role of CD31 adhesion in the formation of inter-endothelial cell contacts. J. Cell. Biol. 128: 1229.

Fawcett J, Holness CL, Needham LA, Turley H, Gatter KC, Mason DY, Simmons DL (1992) Molecular cloning of ICAM-3, a third ligand for LFA-1, constitutively expressed on resting leukocytes. Nature 360: 481.

Freeman SD, Kelm S, Barber EK, Crocker PR (1995) Characterization of CD33 as a new member of the sialoadhesin family of cellular interaction molecules. Blood 85: 2005

Fujita N, Sata SK T., Kuwano R, Sakimura K, Inuzuka T, Takahashi Y, Miyatake T (1989) cDNA cloning of mouse myelin-associated glycoprotein: a novel alternative splicing pattern. Biochem. Biophys. Res. Commun. 165: 1162.

Gordon MY, Dowding CR, Riley GR, Goldman JM, Greaves MF (1987) Altered adhesive interactions with marrow stroma of haemopoietic progenitor cells in chronic myeloid leukemia. Nature 328: 342.

Hatzfeld J, Li ML, Brown EL, Sookdeo H, Levesque JP, O'Toole T, Gurney C, Clark SC, Hatzfeld A (1991) Release of early human hematopoietic progenitors from quiescence by antisense transforming growth factor β1 or Rb oligonucleotides. J. Exp. Med. 174: 925.

Hatzfeld J, Batard P, Cardoso AA, Li M-L, Hatzfeld A (1994). Purification and *in vitro* assay of human hematopoietic progenitors. In Freshney RI, Pragnell IB, Freshney MG (Eds.), Culture of Hemopoietic Cells New York: John Wiley & Sons.

Hernandez-Caselles T, Rubio G, Campanero MR, de Pozzo MA, Muro M, Sanchez-Madrid F, Aparcio P (1993) ICAM-3, the third LFA-1 counter receptor, is a co-stimulatory molecule for both resting and activated T lymphocytes. Eur. J. Immunol. 23: 2799.

Holness CL, Bates PA, Littler AJ, Buckley CD, McDowall A, Bossy D, Hogg N, Simmons DL (1995) Analysis of the binding site of intercellular adhesion molecule 3 for the leukocyte function-associated antigen-1. EMBO J. 270: 877.

Jones EY, Harlos K, Bottomley MJ, Robinson RC, Driscoll PC, Edwards RM, Clements JH, Dudgeon TJ, Stuart DI (1995) Crystal structure of an integrin-binding fragment of vascular cell adhesion molecule-1 at 1.8A resolution. Nature 373: 539.

Kelm S, Pelz A, Schauer R, Filbin M, Tang S, Bellard MED, Schnaar R, Mahoney J, Hartnell A, Bradfield P, Crocker PR (1994) Sialoadhesin, MAG and CD22 define a new family of sialic acid-dependent adhesion molecules of the immunoglobulin superfamily. Curr. Biol. 4: 965.

Leavesley DI, Oliver JM, Swart BW, Berndt MC, Haylock DN, Simmons PJ (1994) Signals from platelet/endothelial cell adhesion molecule enhance the adhesive activity of the very late antigen-4 integrin of human $CD34^+$ hemopoietic progenitor cells. J. Immunol. 153: 4673.

Lebkowski JS, Schain LR, Ojrongly D, Levinsky R, Harvey MJ, Okarma TB (1992) Rapid isolation of human CD34 hematopoietic stem cells - purging of human tumor cells. Transplantation 53: 1011.

Loftus JC, Smith JW, Ginsberg MH (1994) Integrin-mediated cell adhesion: the extracellular face. J. Biol. Chem. 269: 25235.

Moritz T, Williams DA (1994) Gene transfer into the hematopoietic system. Current Opinion in Hematology 1: 423.

Muench MO, Cupp J, Polakoff J, Roncarolo MG (1994) Expression of CD33, CD38 and HLA-DR on $CD34^+$ human fetal liver progenitors with a high proliferative potential. Blood 83: 3170.

Newman PJ, Berndt MC, Gorski J, White GC II., Lyman S, Paddock C, Muller WA (1990) PECAM-1 (CD31) cloning and relation to adhesion molecules of the immunoglobulin superfamily. Science 247: 1219.

Olweus J, Lund-Johansen F, Terstappen LWMM (1995) Expression of cell surface markers during differentiation of $CD34^+$, $CD38^{-/lo}$ fetal and adult bone marrow cells. Immunomethods (in press).

Peiper SC, Ashmun RA, Look AT (1988) Molecular cloning, expression, and chromosomal localisation of a human gene encoding the CD33 myeloid differentiation antigen. Blood 72: 314.

Peiper SC, Leboeuf RD, Hughes CB, Prasthofer EF, Borowitz MJ, Dambuyant CD, Katz DR, Walker WS, Ashmun RA, Look AT (1989). Report on the CD33 cluster Workshop: Biochemical and genetic characterization of gp67. In Knapp W,Dorken B,Gilks WR,Rieber EP,Schmidt RE,Stein H, von dem Borne AEG Kr. (Eds.), Leucocyte Typing IV. White Cell Differentiation Antigens (pp. 814). Oxford: Oxford University Press.

Piali L, Hemme; P, Uherek C, Bachmann F, Gisler RH, Dunon D, Imhof BA (1995) CD31/PECAM-1 is a ligand for $\alpha v\beta 3$ integrin involved in adhesion of leukocytes to endothelium. J. Cell. Biol. 130: 451.

Powell LD, Varki A (1995) I-type lectins. Carbohydrate binding proteins of the immunoglobulin superfamily. J. Biol. Chem. (in press).

Roberts R, Gallagher J, Spooncer E, Allen TD, Bloomfield F, Dexter TM (1988) Heparan sulphate bound growth factors: a mechanism for stromal cell mediated haemopoiesis. Nature 332: 376.

Sato S, Fujita N, Kurihara T, Kuwano R, Sakimura K, Takahashi Y, Miyatake T (1989) cDNA cloning and amino acid sequence for human myelin-associated glycoprotein. Biochem. Biophys. Res. Commun. 163: 1473.

Scheinberg DA, Lovett D, Divgi CR, Graham MC, Berman E, Pentlow K, Feirt N, Finn RD, Clarkson BD, Gee TS, Larson SM, Oettgen HF, Old LJ (1991) A phase I trial of monoclonal antibody M195 in acute myelogenous leukemia: specific bone marrow targeting and internalization of radionuclide. Journal of Clinical Oncology 9: 478.

Simmons D, Seed B (1989) Isolation of a cDNA encoding CD33, a differentiation antigen of myeloid progenitor cells. J. Immunol. 141: 2797.

Simmons DL, Satterthwaite AB, Tenen DG, Seed B (1992) Molecular cloning of a cDNA encoding CD34, a sialomucin of human hematopoietic stem cells. J. Immunol. 148: 267.

Simmons DL (1993) Dissecting the modes of interactions amongst cell adhesion molecules. Development Suppl: 193.

Stamenkovic I, Seed B (1990) The B-cell antigen CD22 mediates monocyte and erythrocyte adhesion. Nature 345: 74.

Stamenkovic I, Sgroi D, Aruffo A, Sy MS, Anderson T (1991) The B lymphocyte adhesion molecule CD22 interacts with leukocyte common antigen CD45RO on T cells and α2-6 sialyltransferase, CD75, on B cells. Cell 66: 1133.

Stockinger H, Gadd SJ, Eher R, Majdic O, Schreiber W, Kasinrerk W, Strass N, Schnabl E, Knapp W (1990) Molecular characterisation and functional analysis of the leukocyte surface protein CD31. J. Immunol. 145: 3889.

Tchilian EZ, Beverley PCL, Young BD, Watt SM (1994) Molecular cloning of two isoforms of the murine homolog of the myeloid CD33 antigen. Blood 83: 188.

Teixeira AM, Fawcett J, Simmons DL, Watt SM (1994) The N-terminal domain of BGPc mediates homotypic adhesion: domain interactions and epitope analysis. Blood 84: 211.

Torres RM, Che-Leung L, Santos-Argumedo L, kirkham PA, Grabstein K, Parkhouse RME, Clark EA (1992) Identification and characterisation of the murine homologue of CD22, a B lymphocyte-restricted adhesion molecule. J. Immunol. 149: 2641.

Vazeux R, Hoffman PA, Tomita JK, Dickinson ES, Jasman RL, St John T, Gallatin WM (1992) Cloning and characterisation of a new intercellular adhesion molecule ICAM-R. Nature 360: 485.

Waller EK, Olweus J, Lund-Johansen F, Huang S, Ngugen M, Guo G-R, Terstappen L (1995) The "Common Stem Cell" hypothesis re-evaluated: human fetal bone marrow contains separate populations of hematopoietic and stromal progenitors. Blood 85: 2422.

Watt SM, Williamson J, Genevier H, Fawcett J, Simmons DL, Hatzfeld A, Nesbitt SA, Coombe DR (1993) The heparin binding PECAM-1 adhesion molecule is expressed by $CD34^+$ hematopoietic precursor cells with early myeloid and B-lymphoid phenotypes. Blood 82: 2649.

Watt SM, Fawcett J, Murdoch SJ, Teixeira AM, Gschmeissner SE, Hajibagheri NMAN, Simmons DL (1994a) CD66 identifies the biliary glycoprotein (BGP) adhesion molecule: cloning, expression and adhesion functions of the BGPc splice variant. Blood 84: 200.

Watt SM, Simmons DL, Teixeira AM, Tchilian EZ, Fawcett J, Grunert F, Nagel G, Waldmann H, Sala-Newby G (1994b). CD66 defined members of the carcinoembyronic antigen family: expression and function. In Epenetos A, Pignatelli M (Eds.), Cell Adhesion Molecules in Cancer and Inflammation (pp. in press). Switzerland: Harwood Academic Press.

Watt SM, Gschmeissner SE, Bates P (1995) PECAM-1: its structure and function on haemopoietic and endothelial cells. Leukaem. Lympho. 17: 229.

Wilson GL, Fox CH, Fauci AS, Kehrl JH (1991) cDNA cloning of the B cell membrane protein CD22: a mediator of B-B cell interactions. J. Exp. Med. 173: 137.

Hematopoietic stem cell: developmental potential and proliferative activity.

Chertkov JL, Drize NJ, Gan OI, Olovnikova NI, Belkina EV, Nikolaeva TL
Haematological Scientific Center
Russian Academy of Medical Science
Novosykovsky pr. 4a
125167 Moscow, Russia

The hematopoietic system is created by hierarchical progression of commitment events starting from the primitive hematopoietic stem cell (HSC). By strict definition, the primitive HSC is the founder cell for the entire hematopoietic system, i.e. it is capable of long-term support of hematopoiesis both *in vivo* and *in vitro*. It is generally assumed that the primitive HSC has the capacity both for differentiation and self-renewal (Chertkov, Gurevitch, 1984). Nevertheless, crucial question about capacity of HSC for inexhaustible proliferation remains unanswered. Many attempts to prove unlimited proliferative capacity of HSC by bone marrow cell passages always demonstrated the exhaustion of HSC potential after 3-5 transfers (Micklem et al, 1987). However, the technical reservation prevented from getting the conclusive evidence of inability of HSC to maintain itself. There is the other way to answer this intriguing question: if the founder cell has the capacity for self-renewal, hematopoiesis would be stochastic and many primitive HSCs would be functioning simultaneously. On the other hand, if the primitive HSC has no capacity to self-renew, hematopoiesis would be characterised by clonal succession. According to this model, primitive HSCs are used sequentially, one after the other, producing clones which replace each other (Rosendaal, Adam, 1984).

Reconstitution of irradiated mice with bone marrow cells carrying specific markers (HSC marked by retroviral integration) was used for characterisation of clonal succession (Williams et al, 1984; Lemischka et al, 1986; Snodgrass, Keller, 1987; Capel et al, 1989; Jordan, Lemischka, 1990). As a rule, after several months of " clonal fluctuations", mono-oligoclonal hematopoiesis occurs and longevity of clone functioning can be as long as 2.5 years. In contrast, the results obtained with HSC without transduced DNA sequences suggested that hematopoiesis is polyclonal both in normal and reconstituted animals (Micklem et al 1987; Harrisson et al, 1978).

NATO ASI Series, Vol. H 94
Gene Technology
Edited by A. R. Zander, W. Ostertag,
B. V. Afanasiev, F. Grosveld

Differences in experimental results concerning the clonality of HSC may be due to the prestimulation of HSC during gene transfer procedure with pharmacological concentrations of hematopoietic growth factors. It is possible that all primitive HSC may be induced to proliferate by cytokines. Furthermore, progenitors with selective proliferative advantage could gradually replace all others, and after a time, only this one or several cell clones will be preserved (mono-oligoclonal hematopoiesis). In order to verify this hypothesis and to study the possible clonal succession of HSC, the next conditions must be realized: clonal composition of the hematopoietic system must be studied sequentially in the same reconstituted mouse; the method used must be very sensitive to detect a great number of simultaneously functioning clones during clonal selection; and the method of gene transfer without prestimulation with pharmacological concentration of cytokines should also be used.

Here all these conditions were realized and hematopoietic cell clonal succession has been demonstrated conclusively. The proliferative potential and proliferative activity of more mature HSC, CFU-S-11, in animals reconstituted both with treated by exogenous cytokines and untreated hematopoietic cells were characterized.

Materials and Methods

Mice. Male and female (C57Bl/6 x DBA/2) F_1 mice, 3-5 months-old were used.

Long-term bone marrow culture. Femoral bone marrow was cultured in Fisher's medium supplemented with 20% of 2:1 mixture of horse and fetal bovine serum, 10^{-6} M hydrocortisone , L-glutamine and antibiotics as described (Chertkov et al, 1983).

CFU-S-11 assay, spleen repopulating ability and Hydroxyurea "suicide" of CFU-S-11. Recipients in the spleen colony-forming (CFU-S) assay were exposed to 1200 cGy ^{137}Cs irradiation (the dose rate 18 cGy/min, IPK irradiator). The dose was delivered into two equal fractions, given 3 h apart. Irradiated female mice were injected i.v. with 1 to 4 x 10^5 marrow from reconstituted mice. Individual macroscopic colonies were isolated under dissection microscope 11 days later and used both for DNA analysis and for determination of the self-renewal capacity (spleen repopulating ability /SRA/: number of daughter CFU-S-8 per 11-day colony). The colony cells (0.2 colony equivalent) were injected into irradiated secondary recipients. The number of daughter colonies were counted 8 days later and the number of colonies generated per one CFU-S-11 was calculated. Proliferative activity of CFU-S-11 was determined by hydroxyurea (HU) suicide. Bone marrow cells from reconstituted mice were incubated in RPMI 1640 media supplemented with 5% fetal bovine serum for 2 hours at 37°C with and without 1 mg/ml HU, washed trice with Hanks' balanced salt solution (HBSS) and injected (1-2 x 10^5) into irradiated recipients. Number of spleen colonies was determined following 8 days. Suicide (Su) was calculated by next equation:

$$Su\ \% = \frac{\text{num. of spleen col. without HU} - \text{num. of spleen col. with HU}}{\text{num. of spleen col. without HU}} \times 100$$

Retroviral vector and producing cell line. GP+E86 cell line producing a retrovirus containing human ADA cDNA was a generous gift of D.A. Williams. The details of the vector have been previously reported (Luskey et al, 1992). This cell line produces virus with a titer 1 x 10^6 CFU/ml. One day before use for infection of HSC, flasks with confluent producer cells were treated with 5 μg/ml of mitomycin C at 37^0C for 2.5 hours, repeatedly washed with HBSS, detached by trypsin-EDTA and split 1:2 in αMEM with 20% fetal bovine serum.

Bone marrow transfection. Two protocols of transfection were used. For prestimulation, a suspension of bone marrow cells from six 5-FU treated mice (150 mg/kg i.v., two days before cell collection), were cultured in 0.1% gelatine-treated T-25 flasks, at 37^0C in 10 ml αMEM with 20% fetal bovine serum and rhIL-6 (50-100 μg/ml) and rhSCF (50 μg/ml) (Protocol 1). In protocol 2 the similar cells were incubated over adherent cell layer of 15 Gy irradiated 3-4 weeks old long-term bone marrow cultures without exogenous cytokines. Following two days prestimulated bone marrow cells were transferred to a monolayer of producer cells in media with 4 μg/ml polybrene and with (protocol 1) or without (protocol 2) exogenous cytokines. After 2 days of infection, the hematopoietic cells were extensively washed from producer cells and used for reconstitution of irradiated mice. The doses used and the number of reconstituted mice are shown in table 1.

Table 1 Characteristics of mice reconstituted with retrovirally marked bone marrow cells

Group/ identifica- tion numbers	method of transduction	Cell dose	Number of mice	Survived 6 months	Survived 12 months
1/1-4	Protocol 1	10^5	4	3	2
2/5-8	(with cytokines)	10^6	4	4	1
3/9-12	Protocol 2	1,5ō10^5	4	3	2
4/13-27	(on ACL)	2,2ō10^6	15	15	11

Collection of hematopoietic cells for DNA analysis. Bone marrow samples were obtained at different time intervals from multiple groups of individual anaesthetized living mice in such a way to allow recovery between sample collection. Femoral bone marrow was aspirated by puncture through the knee joint by a syringe with a 22-gauge needle. Usually it was possible to obtain 5-15 x 10^6 bone marrow cells from a living mouse. Bone marrow cells from each transplanted mouse were injected into 6 irradiated female recipients for CFU-S and DNA analysis.

Southern blot analysis. DNA from bone marrow and individual spleen colonies was extracted, digested with EcoR1, electrophoresed through a 1% agarose gel, transferred to Hybond N filter and hybridized with ADA cDNA probe, prepared from 1.2 Kb PGK-ADA fragment. Digestion with EcoR1 permitted analysis of individual clones of hematopoietic cells, since only one EcoR1 restriction site is present within the LTRs of the vector used.

Determination of the origin of CFU-S in reconstituted mice. Only male mice were used as donors of bone marrow cells, recipients were always female. For identification of CFU-S origin a Y-chromosome probe (mouse Zfy-2) was used (a generous gift from Dr. L.G. Brown). Filters were rehybridized with this probe after stripping the ADA probe.

Results

The efficiency of gene transfer into CFU-S-11 was high, 52-83% of CFU-S derived colonies contained the human ADA cDNA (protocol 1). Without exogenous cytokines the efficiency of transduction was lower, only 18-32% of CFU-S derived colonies were marked (protocol 2). In many instances, several proviral copies per genome were integrated as demonstrated by Southern analysis (Fig. 1). Transduction efficiency into primitive HSC also appeared to be good since, as seen in table 2, marked clones were still present in CFU-S-11 from mice up to 1 year after reconstitution.

Analysis of clonal composition of mice reconstituted with cytokine stimulated HSC revealed that 2 and 4 months after reconstitution the integration patterns varied between different colonies of the same mouse, implying that the progeny of many different clones were functioning simultaneously. From 23 colonies studied in first 4 months after reconstitution 22 had different unique markers (Table 2).

Table 2. Clonal composition of the hematopoietic system at different times after reconstitution with bone marrow cells prestimulated with exogenous cytokines.

Months after reconstitution	No. of marked CFU-S/CFU-S studied	No. of different clones
2	10/15	10
4	13/16	12
6	6/8	3
8	25/50	9
12	4/5	2

These results demonstrate that multiple clones are existing early after transplantation. However, oligo-monoclonal hematopoiesis appeared to become established over a more prolonged period, since only 1-2 clones were observed 6-12 months after reconstitution in DNA obtained from CFU-S derived colonies.

In sharp contrast with these data were the results obtained with bone marrow cells prestimulated on ACLs without exogenous cytokines. No phase of polyclonal hematopoiesis was revealed in these mice (table 3). Efficiency of gene transfer into primitive HSC was lower and marked clones were observed only in 25-50% of reconstituted mice in each time interval studied. However, even in first months after reconstitution only one or two marked clones were revealed. No one of detected clones persisted more than 2-3 months. Disappearing clones were never observed again. As can be seen from table 3, the proportion of transduced CFU-S is about 10-15% of CFU-S studied. These observations suggest that about one-tenth of all bone marrow CFU-S are derived from marked HSC

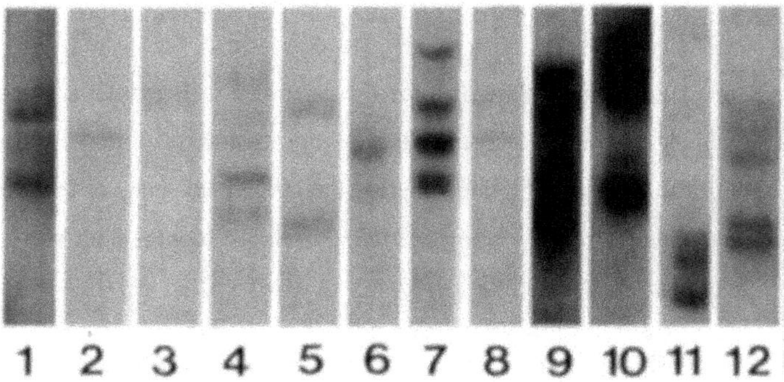

Fig. 1 Human ADA cDNA sequence integration pattern of primary day 11 CFU-S colonies derived from bone marrow cells prestimulated with exogenous cytokines or with adherent cell layer. Lanes 1-7, DNA from colonies, produced by cytokine prestimulated cells; lanes 8-11, DNA from colonies, produced by adherent cell layer prestimulated cells; lane 12, DNA from producer cell line GP+E86 (hADA).

Table 3. Clonal composition of hematopoietic system in different time after reconstitution with non-stimulated by exogenous growth factors bone marrow cells.

Months after reconstitution	No. of mice/No. of mice with marked CFU-S	Marked CFU-S/ CFU-S studied	No. of different clones
2	4/1	3/32	1
4	4/2	6/42	2
8	4/1	3/26	2

To confirm the donor origin of CFU-S we assessed the presence of Y-chromosome sequences in DNA obtained from spleen colonies in all experiments. All colonies derived from reconstitution (Fig. 2). Therefore, the procedure of gene transfer both with and without exogenous cytokines did not decrease the developmental potential of primitive HSC.

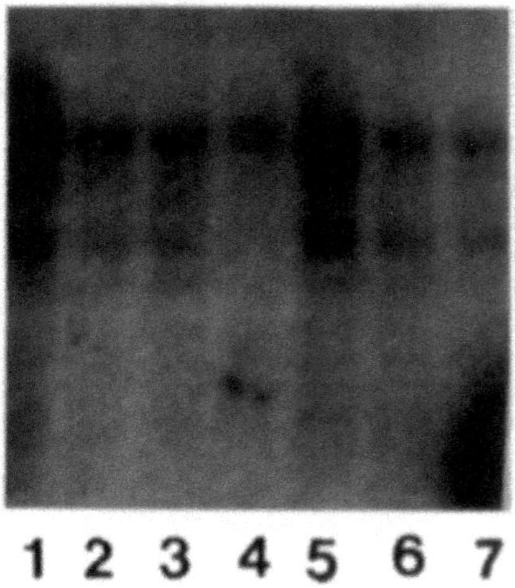

Fig. 2 Detection of Y-chromosome sequences in CFU-S-derived spleen colonies from female mice reconstituted with transduced male hematopoietic cells. Lanes 1-4, 6 months after reconstitution; lanes 5-7, 12 months after reconstitution.

Proliferative activity of CFU-S was studied in mice 6 months following reconstitution (Fig. 3). The results were essentially the same in groups reconstituted by bone marrow cells transduced both with and without exogenous cytokines prestimulation. In mice reconstituted with large cell inoculum suicide rate was about 10%, i.e. CFU-S were mainly out of cell cycle as during steady-state hematopoiesis in normal animals. Mice reconstituted with low cell dose demonstrated significant increase of suicide which was about 20%. CFU-S with integrated human ADA cDNA had 2-fold higher suicide rate than unmarked CFU-S. However, because of low amount of studied marked colonies (9 from 42) the difference is insignificant.

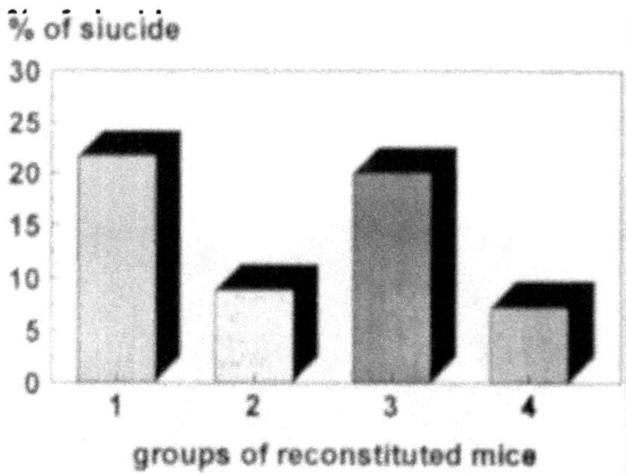

Fig. 3. CFU-S suicide in mice 6 months following reconstitution. 1,2, reconstitution according protocol 1, low and large dose of bone marrow cells, respectively; 3,4, reconstitution according protocol 2, low and large dose of bone marrow cells, respectively.

Spleen repopulating ability (SRA) of CFU-S-11 6 and 12 months after reconstitution was significantly decreased in all experimental groups of mice (Fig. 4). The average number of daughter CFU-S per 11 day colony was 20-60 as compared with 150 for normal bone marrow derived colony. In all mice reconstituted with cytokine prestimulated cells SRA gradually decreased during 12 months. On the other hand, two of five mice reconstituted with transduced cells without cytokine prestimulation had higher SRA 1 year after reconstitution than that after 6 months.

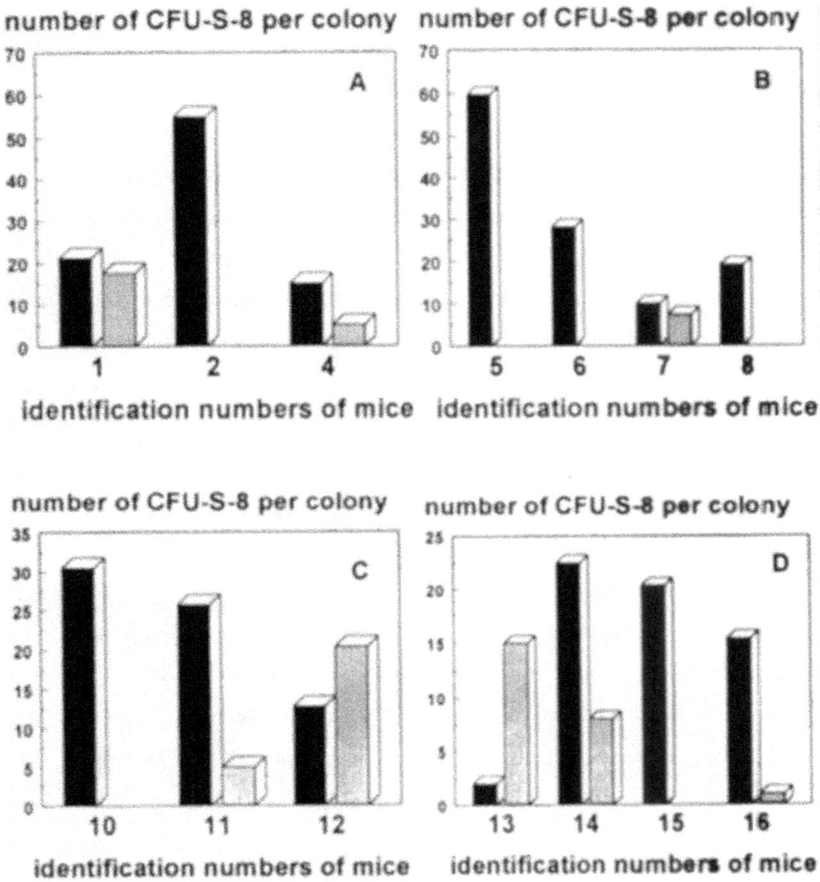

Fig. 4. Spleen repopulating ability of CFU-S-11 in individual reconstituted mice 6 (full bars) and 12 (open bars) months after reconstitution. A, B, reconstitution according protocol 1, low and large dose of bone marrow cells, respectively; C, D, reconstitution according protocol 2, low and large dose of bone marrow cells, respectively.

Discussion

The results presented here demonstrate that the difference in the developmental fate of primitive HSC, measured with traditional methods (competitive repopulation assay, serial bone marrow passages et al.), as compared with methods of retroviral gene transfer, could be explained by use of pharmacological doses of exogenous hematopoietic growth factors. A combination of cytokines (IL-6 + SCF) induced high efficiency gene transfer into HSC, what suggests that quiescent HSC had been stimulated for proliferation. Efficiency of gene transfer after more physiological prestimulation of HSC on irradiated ACL of long-term bone marrow culture was significantly lower, possibly because of low proportion of HSC that had been triggered in cell cycle. In mice reconstituted with bone marrow cells prestimulated with exogenous growth factors, many marked clones appeared simultaneously. Two to 4 months after reconstitution, essentially each marked CFU-S represents an unique clone (22 of 23 colonies were represented by unique marker). Statistically, the results indicate that in the first months after transplantation 200-400 clones are functioning simultaneously (confidence interval, 95%). CFU-S is a progenitor with relatively low proliferative potential and it can produce differentiated progeny not more than 4 - 5 weeks both *in vivo* (Bodine et al, 1991; Bodine et al, 1992) and *in vitro* (Chertkov et al, 1987). Therefore, during 2 - 4 months after reconstitution the revealed marked clones are progeny of primitive HSC, rather than initial CFU-S. Hence, many if not all primitive HSC are proliferative at least 4 months after transplantation. During this time, selection of clonogenic progenitor(s) with proliferative advantages is possible, and this clone(s) can replace all others.. As a result of such selection, all marked cells would have the same unique marker. Indeed, 6 - 12 months after reconstitution, hematopoiesis appeared to be oligo-monoclonal as shown here and by others (Rosendaal, Adam, 1984; Capel et al, 1989; Lim et al, 1989).

Essentially, different results were obtained after transplantation of cells not stimulated by exogenous cytokines. The efficiency of gene transfer was significantly lower, however the system was suitable for the study of the hematopoietic clonal composition after reconstitution. No phase of polyclonality was observed in this model with use of an ACL for prestimulation. From the first months after transplantation only 1 or 2 marked clones have been observed. The longevity of marked clones existence was shorter than 2 - 3 months because they were revealed only once during one year; analysis was performed each 2 - 3 months. It appears that no stem cell selection took place and hematopoiesis was maintained by clonal succession. If appear simultaneously. Hence, the cause of peculiarity of results obtained with gene transfer technique is the use of pharmacological concentration of exogenous cytokines. If it is possible to avoid such stimulation of HSC and to use a more physiological gene transfer, the same clonal succession and limited proliferative potential of primitive HSC would be observed. The other evidence, that transduced foreign sequences do not change the proliferative potential of HSC, is the same SRA in marked and non-marked colonies.

The clonal succession of hematopoietic cells was also supported by the data about increase of SRA in 12 months as compared with 6 months in the group of mice reconstituted with the physiologically prestimulated cells (protocol 2). It is difficult to understand how can SRA increase in the progeny of the same primitive HSC. In the case of clonal succession this effect could be explained by replacing the clones with low proliferating potential by clones which were less exhausted.

The incapacity of primitive HSC to maintain itself is supported by the data obtained here and earlier (Chertkov et al, 1993), that after reconstitution with large cell innoculum the proliferative potential of progenitors, as measured by SRA, is higher than after reconstitution with low cell dose. Of course, this could not happen if the primitive HSC has unlimited proliferative potential.

Proliferative activity of CFU-S, as measured by HU suicide, was increased only in groups reconstituted with low dose of hematopoietic cells. It is known, that clone size is bigger in animals transplanted with low doses of bone marrow cells (Micklem et al, 1987). The data suggest that this increase of clone size is due at least partially by expansion of CFU-S at the expense of increased proliferative rate.

On the whole, the results support the clonal succession model of hematopoiesis and mortal nature of primitive HSC. The data demonstrate the possibility of using this model for the sequential study of the clonal composition of the hematopoietic system and primitive HSC developmental potential.

Acknowledgements

The authors wish to thank Drs D.A.Williams and L.G.Brown for generous gift of the GP+E86 (hADA) cell line and the Y probe. This study was supported in part by ISF (Soros fund) and RFFR (Russian Fund for Fundamental Research).

References

Bodine DM, Crosier PS, Clark SC (1991) Effects of hematopoietic growth factors on the survival of primitive stem cells in liquid suspension culture. Blood 78:914-920

Bodine DM, Orlic D, Birkett NC, Seidel NE, Zsebo KM (1992) Stem cell factor increases colony-forming unit-spleen number in vitro in synergy with interleukin-6, and in vivo in Sl/Sl^d mice as a single factor. Blood 79:913-920

Capel B, Hawley R, Covarrusias L, Hawley T, Mintz B (1989) Clonal contribution of small numbers of retrovirally marked hematopoietic stem cells engrafted in unirradiated neonatal W/W mice. Proc Natl Avcad Sci USA 86:4564-4567

Chertkov JL, Drize NJ, Gurevitch OA, Udalov GA (1983) Hemopoietic stromal precursors in long-term culture of bone marrow. I. Precursors characteristics, kinetics in culture, and dependence on quality of donor hemopoietic cells in chimeras. Exp Hematol 11:231-242

Chertkov JL, Gurevitch OA (1984) Hematopoietic stem cell and its microenvironment. Moscow: Meditzina

Chertkov JL, Drize NJ, Deryugina EI, Udalov GA (1987) Individual clones of hemopoietic cells in murine long-term bone marrow cultures. Leukemia 1: 491-496

Chertkov Jl, Jiang S, Lutton JD, Harrison J, Levere RD, Tiefenthaler M, Abracham NG (1993) The hematopoietic stromal microenvironment promotes retrovirus-mediated gene transfer into hematopoietic stem cells. Stem Cells 11:218-227

Jourdan CT, Lemischka IR (1990) Clonal and systemic analysis of long-term hematopoiesis in the mouse. Genes Dev 4:220-

Harrison DE, Astle CM, Delaittre JA (1978) Loss of proliferative capacity in immunohemopoietic stem cells cause by serial transplantation rather then ageing. J Exp Med 147:1526-1531

Lemischka IR, Raulet DH, Mulligan RC (1986) Developmental potential and dynamic behaviour of hematopoietic stem cells. Cell 45:917-923

Lim B, Apperley JF, Orkin SH, Williams DA (1989) Long-term expression of adenosine deaminase in mice transplanted with retrovirus-infected hematopoietic stem cells. Proc Nat Acad Sci USA 86:8892-8895

Luskey BD, Rosenblatt M, Zsebo K, Williams DA (1992) Stem cell factor, interleukin-3, and interleukin-6 promote retroviral-mediated gene transfer into murine hematopoietic stem cells. Blood 80:396-405

Micklem HS, Lennon JE, Amsell JD, Gray RA (1987) Minbers and dispersion of repopulating cell clones in radiation chimeras as a function of injected cell dose. Exp Hematol 15:251-257

Rosendaal M, Adam J (1984) Haemopoiesis by clonal succession? Blood Cells 10:473-485

Snodgrass R, Keller G (1987) Clonal fluctuation within the hematopoietic system in mice reconstituted with retrovirus-infected cells. EMBO J 6:3955-3961

Williams DA, Lemischka IR, Nathan DG, Mulligan RC (1984) Introduction of new genetic material into pluripotent stem cells of the mouse. Nature 310:476-482

Abstract

Developmental fate of retrovirally marked hematopoietic stem cells (HSC) critically depends on the use of pharmacological concentration of growth factors during gene transfer procedure. Cytokines (interleukin-6 plus stem cell factor) stimulated HSCs proliferate at least 4 months after reconstitution. At this time hundred of hematopoietic clones are exist simultaneously. Later as a result of clonal selection only oligo-clonal hematopoiesis was detected. Vice verse, after more physiological prestimulation of HSC on irradiated adherent cell layer of long-term bone marrow culture no phase of polyclonal hematopoiesis has been observed. Hematopoiesis was according to clonal succession model and replacing each other hematopoietic cell clones were revealed. The longevity of clone existence was not longer than 2-3 months. The proliferative activity of CFU-S was higher in reconstituted with small dose of bone marrow cells mice. That could be connected with enlargement of individual clone size in these animals. The results support the clonal succession model of hematopoiesis and incapacity of primitive HSC to maintain itself. The data demonstrate the possibility of using this model for the sequential study of the clonal composition of the hematopoietic system and primitive HSC developmental potential

The Effect of Methylprednisolone on Bone Marrow and Spleen Clonogenic Fibroblasts in Humans.

Vladimirskaya E., Zamaraeva N., Osipova E., Volunetz M., Alexeev I., Kryzianovsky O., Rumiantsev A.

Vladimirskaya Elena B.
Deputy Director
Institute of Pediatric Hematology
Leninsky Prospect, 117
Moscow, 117513, Russia
Tel: (095)433-81-41
Fax: (095)434-86-78

Stromal microenvironment of hemopoietic organs creates a "homming effect" for hemopoietic cells. Stromal fibroblasts play an important role in this cell compartment (Fridenstain, Luria 1980; Fridenstain, 1982; Patt et al. 1982; Dexter et al. 1977). One of the ways of valuation of stromal function is the method of clonogenic cultivation in liquid medium. This method was established by Fridenstain and coworkers and adopted for human bone marrow and spleen fibroblasts (Fridenstain et al. 1973). The majority of stromal fibroblasts are out of cell circle but under optimal conditions in vitro become clonogenic and give rise to the colonies (Fridenstain et al. 1970; Fridenstain et al. 1973). The number of clonogenic fibroblasts (CFUf) and their proliferative potential (PP) can be the functional characteristics of stromal microenvironment of hemopoietic organs.

Regulation of proliferation and differentiation of hemopoietic cells by different cytokines, growth factors and hormons is studied well enough (Colin 1987; Dexter, 1990). The not numerous publications on regulation of proliferation of stromal fibroblasts are controversal. Most authors consider the growth factors of bone marrow clonogenic fibroblasts to be contained in thrombocytes and megakaryocytes. There is evidence (Dedhar et al. 1988; Valent et al. 1990) that IL-3 and GM-CSF stimulate the clonal growth and differentiation both of hemopoietic precursors and bone marrow fibroblasts, although other investigators didn't manage to confirm these data (Fridenstain et al. 1992; McIntyre and Bjornson, 1986).

Little is known about the influence of hormons on the microenvironment; among these hormons the corticosteroids are of special interest, because on the one hand, hydrocortizone is necessary for formation of stromal underlayer in long term culture, and on the other hand, it is widely used in clinic for stimulation of hemopoiesis without explanation of the mechanism of its action.

The influence of corticosteroids on the stromal fibroblasts was studied in different models, using various hormons and still remains vague (Juneja et al., 1984; Kamalia et al., 1992; McIntyre, Bjornson, 1986).

Our aim was to study the effect of a hormone widely used in clinical practice - methylprednisolone (MP) in physiological concentration (10^-7 M) on the CFUf and PP of bone marrow and spleen clonogenic fibroblasts in hematologically healthy children.

NATO ASI Series, Vol. H 94
Gene Technology
Edited by A. R. Zander, W. Ostertag,
B. V. Afanasiev, F. Grosveld

MATERIAL AND METHODS

Material: Bone marrow samples from 55 children (4-15 years) with apparently normal hemopoiesis (blood count and myelogramma were normal, cytostatics and hormons haven't been taken for 2 months before the examination) were obtained for differential diagnosis; pieces of extracted spleen were obtained from 7 children.

Methods: Bone marrow fibroblasts were cultured by method of Fridenstain and coworkers (Fridenstain et al. 1973) in our modification (Vladimirskaya et al. 1989). Bone marrow cells obtained by aspiration from sternum or iliacal crest were separated from erythrocytes by sedimentation, washed twice and explanted into culture flasks with bottom square of 25 cm^2 at the concentration of $1x10^5$ cells/mL in 199 mL culture medium. We used xenogenic feeder - rabbit's bone marrow cells previously irradiated with 30 Gy and stored at -20° C as the source of growth factors. These feeder cells were added in 24 hours of cultivation ($1x10^6$ cells per flask) after the replacement of culture medium with nonadherent, explantet cells in it and two-.times washing of adherent layer. Methylprednisolone (10^{-7} M) was added in different variants of culture. After 14 days of cultivation fibroblasts colonies were washed, fixated with 96% ethanol and stained.

We used 2 indices to analyze the functional condition of clonogenic fibroblasts: CFUf - colony forming units of fibroblasts - the number of colonies per 10^5 of explanted cells and PP - proliferative potential - the characteristic of the size of the colonies. PP was calculated as:

$$PP= \frac{(1xN1) + (2xN2) + (3xN3)}{100}$$

, where N1, N2 , N3 are percentages of small (50-250 cells), medium (250 - 1000 cells) and large (> 1000 cells) colonies respectively.

RESULTS AND DISCUSSION

We used two controls while studying the effect of MP on the bone marrow clonogenic fibroblasts growth: spontaneous colony formation (without change of the medium and without feeder) and standard colony formation (adding of the feeder and change of the medium after 24 hours of cultivation).

In the first series of experiments we used MP in the same conditions as the feeder - after change of the medium, removing of nonadherent cells and washing of adherent cells on the 2nd day of cultivation. The results are presented in Table 1.

Tabl 1.
CFUf and PP while cultivation of bone marrow fibroblasts with xenogenic feeder and MP in the same conditions.

Culture Indices	Spontaneous colonyformation	B.m. feeder	MP
N	8	8	8
CFU	17.7±3.9	53.1±4.7	59.0±7.1
PP	1.22	2.04	2.33

MP increases CFUf as well as feeder, and a tendency of enlarging the size of the colonies was observed. In other words, in these experiments MP acted as stimulator of the proliferation of clonogenic fibroblasts. The target of it's action was the adherent cells.

In other series of experiments we added MP for different periods of time to the whole mixture of bone marrow cells.

As it is clear from Table 2, MP shows a significant stimulating effect when added to the explanted cell mixture: CFUf was three or four times more than spontaneously colony formation and 1.5-2 times more than standard colony formation. The duration of the contact (1 day, 2 days, 14 days) was not important. This fact is the evidence that 1-day contact is enough for the realization of the stimulating effect of MP. In the third series of experiments we tried to estimate the stimulating effect of MP and feeder when using it consequently or simultaneously (adherent and nonadherent).

Tabl 2.
Adding of MP to the whole mixture of bone marrow cells for different periods.

Duration of contact Indices	1 day + replacement of media	2 days + replacement of media	14 days without replacement of media
N	35	12	15
CFUf	31.4±2.6	24.2±5.2	35.7±6.5
PP	1.69±0.08	1.63±0.33	1.73±0.15

Spontaneous colony formation: N=11; CFUf=8.7 ± 2.2; PP=1.23 ± 0.4
Standard colony formation: N=36; CFUf =17.0 ± 2.4; PP=1.49 ± 0.06

As it is clear from Table 3 CFUf was highest when we used 1-day contact of the whole mixture of explanted cells with MP.

Tabl 3.
Subsequent and simultaneous use of MP and feeder.

Cultivation Indices	MP for 1 day	MP for 1 day + feeder after change of medium	MP and feeder together after change of medium	Controls Standard.	 Spontan.
N	35	36	9	36	11
CFUf	31.4±2.6	24.7±2.5	10.4±2.0	17.0±2.4	8.7±2.2
PP	1.69±0.08	2.0±0.08	1.35±0.08	1.49±0.06	1.23±0.14

Tabl 4.
Adding of MP and feeder on different time of cultivation after removing of nonadherent cells on 2 day of culture.

Day of culture	Feeder		MP	
	CFUf	PP	CFUf	PP
2nd day	53.1±4.7	2.04±0.15	59.0±7.1	2.33±0.19
3rd day	48.3±9.1	1.97±0.16	57.7±11.3	2.65±0.19
5th day	37.5±7.3	1.99±0.11	56.0±9.7	2.70±0.21
7th day	43.2±11.7	2.01±0.17	49.3±3.3	2.67±0.22
9th day	27.0±5.9	1.68±0.09	23.7±4.8	2.01±0.18
11th day	19.3±7.8	1.07±0.06	11.3±0.9	1.09±0.14

Spontaneous colony formation: CFUf=17.7 ± 3.9; PP=1.22 ± 0.09

When we used bone marrow feeder after MP, CFUf did not change ($t=1.6$; $p>0.05$), but PP increased ($t=3.1$; $P<0.01$). When bone marrow feeder was used in standard conditions it's stimulating effect on CFUf and PP was lower. Using MP and bone marrow feeder simultaneously cancelled the stimulation of colony formation. The CFUf was as low as spontaneous CFUf.

We compared the efficiency of the action of MP and bone marrow feeder on clonogenic fibroblasts and studied the dynamic of their stimulating effect in the following experiments. MP and feeder were added to the same target cells in different time periods after removing of nonadherent cells. The results of these experiments are shown in Table 4.

We can make the following conclusions when analysing these data:

1) The character and dynamics of MP and feeder action are the same. CFUf was highest when the stimulators were added in the beginning of colony formation (1, 2 and 3 days). This fact may be the evidence of the promoting activity of the stimulators. The stimulating effect decreases after the 7th day of cultivation.

2) The stimulating effect of MP is higher than the feeder's till the 9th day of cultivation. After the 9th day bone marrow feeder possess more stimulating effect than MP.

3) The size of the colonies (PP) of bone marrow stromal fibroblasts is bigger while using MP during the whole period of cultivation.

Our experiments allow to draw the conclusion that MP (10^-7 M) acts as stimulator on the proliferation of human's bone marrow stromal fibroblasts as well as xenogenic feeder, but CFUf were higher when we used MP in comparison with feeder. Bone marrow adherent cells are the main target for the action of MP as well as bone marrow feeder. Both stimulators realize their promoting activity through the same fibroblast receptors and competing for them when used together.

Only bone marrow fibroblasts possess these characteristics. As it is clear from Table 5, neither bone marrow feeder nor MP (10^{-7} M) serve as stimulators for fibroblasts from amputated spleen.

So our proposal is to use MP as the stimulator of the proliferation of clonogenic human bone marrow fibroblasts instead of the xenogenic feeder we used before.

The optimal is adding MP (10^{-7} M) in 24 h of cultivation to washed adherent cells after the replacement of medium with nonadherent cells. We chose this concentration of MP after the experiments, presented in Tabl. 6, when we observed the maximal colony formation rate and the maximal size of colonies at this level of MP concentration. MP was added in standard culture condition instead of xenogenic feeder.

Tabl 5
Action of bone marrow feeder and MP on stromal bone marrow and spleen fibroblasts.

Indices Culture	Bone Marrow		Spleen	
	CFUf	PP	CFUf	PP
Spontan. colony formation	17.7±3.9	1.22±0.09	0.54±0.16	1.32±0.07
feeder 0.5x10 of irrad. cells	53.1±4.7	2.04±0.18	0.34±0.15	1.30±0.18
MP 10^{-7} M	59.0±7.1	2.33±0.17	0.38±0.13	1.37±0.11

Tabl. 6
CFUf and PP in different levels of MP concentration after 14d cultivation

MP concentration	10^{-5} M	10^{-6} M	10^{-7} M	10^{-8} M	10^{-9} M	0
CFUf	25.3±6.8	32.0±3.8	36.0±4.9	30.7±6.8	23.7±3.7	19.5±5.3
PP	1.9±0.14	2.1±0.3	2.3±1.4	1.9±0.03	1.9±0.07	1.8±0.03

Data confirming the stimulation effect of MP on functiona
l activity of bone marrow stromal fibroblasts may be the basis of including MP in treatment programs of bone marrow failures. Cultures of bone marrow fibroblasts with MP may be models

for the investigation of effectivities and mode of action of corticosteroids in several types of hemopoietic depressions.

REFERENCES

Assosian R.K., Grotendorst G.R., Miller D.M., Sporn M.B.(1984) Celllular transformation by coordinated action of three peptide growth factors from human platelets. Nature 309; 804.

Astahova B.S. (1982) Sravnitelnaya otsenka ksenofedera pri clonirovanii stromalnuch fibroblastov kostnogo mozga cheloveca. Bull. exper. biol. i med. 94; 111.

Bellows C.G., Heersche J.N., Aubin J.E. (1990) Determination of the capacity for proliferation and differentiation of osteoprogenitor cells in the presence and absence of dexamethasone. Develop. Biology 140; 132-138.

Castro-Malaspina H., Rabellino E., Gen A. (1981) Human megakaryocyte stimulation of proliferation of bone marrow fibroblasts. Blood 57; 781-787.

Colin A.S.(1987) Hematopoietic growth factors. J. Clin. Invest. 79; 1549-1557.

Dedhar S., Gaboury L., Eaves C. (1988) Human granulocyte-macrophage colony-stimulating factor is a growth factor active on a variety of cells of non hemopoietic origin. Proc. Natl. Acad. Sci USA 85; 9253-9257.

Dexter T.M., Allen T.D., Lajtha L.G. (1977) Conditions controlling the proliferation of hemopoietic stem cells in vitro. Cell. Physiol. 91; 335-349.

Dexter T.M. (eds) (1990) Haemopoietic growth factors Gardiner-Caldwell Communications Ltd; UK.

Fridenstain A.Ya., (1982) Stromalnye kletky kostnogo mozga i krovetvornoe mikrookrugenie. Archiv patol. 44; 3-11. (In Russian).

Fridenstain A.Ya., Chailichyan R.è., Lalykina è.S. (1970) Ä fibroblastnych kletkach v kulturach krovetvornoy tkani morskich svinok. Cytologia 12; 1147-1156. (In Russian).

Fridenstain A. Ya., DeriglasovaYu.F., Kulagina N.N. (1973) Klonirovanie kletok-predchestvennikov fibroblastov v monosloinich kulturach. Bull. exp. biol. i med. 76; 90-94. (In Russian)

Fridenstein A.Ya., Latzinik N.V., Gorskaya Yu.F., Luria E.A., Moskvina I.L. (1992) Bone marrow stromal colony formation requires stimulation by haemopoietic cells. Bone and Mineral 18; 199-213.

Fridenstain A.Ya., Luria E.A. (1980) Kletochnye osnovy hemopoeticheskogo mikrookrugenia. Terapevt. arch. 52; 67-71. (In Russian).

Hirata J., Kaneko S., Nishimura J. (1985) Effect of platelet-derived growth factor and bone marrow conditioned medium on the proliferation of human bone marrow-derived fibro-blastoid colony-forming cells. Acta Hematol. 74; 189-194.

Juneja H.S., Minquell J.J., Gardner F.H., Helmer R.E., Lee S. (1984) The effect of dexamethasone on the growth of bone marrow fibroblasts in aplastic anemia. Myelofibrosis and the Biology of Connective Tissue, NY, 265-274.

Kamalia N., McCulloch C.A., Tenebaum H.C., Limeback H. (1992) Dexamethasone recruitment of self-renewing osteoprogenitor cells in chick bone marrow stromal cell cultures. Blood 79; 320-326.

Mc Intyre A.P., Bjornson B.H. (1986) Human bone marrow stromal cell colonies: Response to hydrocortisone and dependence on platelet-derived growth factor. Exp. Hematol. 14; 833-839.

Patt H.M., Maloney M.A., Flannery L.M. (1982) Hematopoietic microenvironment transfer by stromal fibroblasts derived from bone marrow varying in cellularity. Exp. Hematol. 10, 728-732.

Rosenfeld M., Keating A., Bowen-Pope D. (1985) Responsiveness of the in vitro hemapoietic microenviroment to platelet-derived growth factor. Leuk. Res. V 9; 427-434

Valent R., Geissler K., Kahls P., Bettlheim P. (1990) Fibroblast colony stimulating activity of GM-CSF and Il-3 in human bone marrow cell cultures. Springer-Verlag, NY; 85-89.

Vladimirskaya E.B., Aytmanbetova Ç. î., Kolosova L. Yu., Tsurya é.î. (1989) Zavisimost proliferazii klonogennych fibroblastov kostnogo mozga cheloveka ot fidera. Exp. Onkologia 25; 7-9. (In Russian).

Umbilical cord blood: Morphological composition and functional characteristics.

Alexeev I., Volynetz M., Zamaraeva N., Osipova E., Loginov A., Vladimirskaya E., Rumjanzev A.

Zamarayeva Nina V.
Chief of Laboratory of Regulation of Hemopoiesis.
Institute of Pediatric Hematology
Leninsky Prospect, 117
Moscow, 117513, Russia
Tel: (095) 433-81-41
Fax: (095) 434-86-78

INTRODUCTION

During last decade umbilical cord blood attracts attention as an alternative source of hemopoieitic transplantable progenitors cells for treatment of various hematological diseases (Broxmeyer et al. 1992).

During the whole gestational period hemopoietic switch from one hemopoietic site to another through circulation occurs (Tavasolli M. 1991). Definitive localisation of hemopoiesis in bone marrow is observed from 23rd week (Kelemen et al. 1979). So, cell composition of umbilical cord blood is similar to peripheral blood composition of fetus during last stage of gestation.

There is considerable number of hemopoietic progenitor cells in umbilical cord blood. It has been shown that umbilical cord blood contained granulocyte-macrophage (CFU-GM), erythroid (CFU-E), and multipotential (CFU-GEMM) progenitors (Harris et al. 1994). However, data of different authors is incommensurable because it was obtained by different culture methods.

The aim of our study was to evaluate morphological composition of umbilical cord blood cells, composition and functional activity of cell pool of hemopoietic progenitors.

NATO ASI Series, Vol. H 94
Gene Technology
Edited by A. R. Zander, W. Ostertag,
B. V. Afanasiev, F. Grosveld

MATERIALS AND METHODS

Umbilical cord blood collection.

20 samples of human umbilical cord blood were obtained from normal full-term vaginal deliveries. Cord blood was collected after delivery of the infant and prior to expulsion of the placenta by 20 ml syringes into a 4 mL test-tube, containing 4 mg of 10 % solution of EDTA (for count by automatic analyzer) and into a test-tube, containing 1 mL of McCoy's medium with 300 units of heparin for each 10 mL of umbilical cord blood for clonogenic culture studies.

Morphological assay.

Morphological assay of umbilical cord blood was carried out by automatic counter "Cobas Argos 5 Diff" (Hoffmann La-Roche) and simultaneously by routine method with light microscope after Romanowsky-Giemsa staining of umbilical cord blood samples.

Cord blood cell separation.

The cord blood cells were centrifugated over density gradients (1.077 g/cm^3, Nycoprep, Nycomed, Oslo, Norway) during 30 minutes in 1000 G and 4 degrees centigrade. Low density cells were harvested, washed twice (400 G, 4 degrees centigrade) and resuspended in IMDM medium, containing 2% solution of fetal bovine serum (FBS). Suspension of obtained mononuclear cells was used for clonogenic culture assays.

Hemopoietic progenitor cell assays.

Hemopoietic progenitors were assayed by two cultural systems: "agar drop-liquid medium" system (Afanasiev B. et al., 1983) and using a modification of methylcellulose culture techniques (Broxmeyer H., et al., 1989).

In "agar drop-liquid medium" system mononuclear cells were placed into agar drop (0.5 mL 0.3% agar) in a concentration of 2 x 10^4 cells/mL (solid phase). A mixture of leukocytes from two healthy human donors as a source of colony-stimulating factors in concentration of 1 x 10^6 cells/ml was added to 2.5 mL complete McCoy's medium enriched with 15% fetal calf serum (liquid phase with standard feeder (Fst)). Spontaneous proliferative capacity was assayed by liquid phase without feeder cells (Fo).

The cells were plated in four 40 mm culture dishes and incubated at 37 degrees centigrade in 100% humidified atmosphere consisting of 7% CO2 in air during 7 days.

Cloning efficiency was calculated as a sum of colonies (cell aggregates of more than 20 cells) and clusters (aggregates of 5-20 cells) that have emerged from 1 x 10^5 plated cells.

In methylcellulose assay 3 x 10^4 mononuclear cells were suspended in 1 mL of culture medium consisting of 1.2 mL 2.3% methylcellulose, 300 µL FBS, 100 µL BSA, 10 µL 10 mM 2-ME, 10 µL 200 mM L-glutamine (Sigma, USA), 10 µL antibiotic antimicotic solution, 2.5 U of hr-Epo (Cilag, Switzerland), 1000 U hr-GM-CSF (Fenzime, USA) 50 ng hr-IL-3 (Fenzime, USA). The cells were plated in two 40 mm plastic culture dishes and incubated at 37 degrees centigrade in 100% humidified atmosphere in the presence of 5 % CO2 and 3% O2 in azote. Day 14 progenitors (CFU-GM, CFU-GEMM and BFU-E) were assayed by invert microscope. All assays were set up based on 1 x 10^5 mononuclear cells plated.

Statistical analysis was carried out by SuperCalc-4 computer program.

RESULTS

Umbilical cord blood differs by its morphological composition from peripheral blood of children and has the following characteristics (tabl. 1):

Table 1.
Common analyzis of umbilical cord blood.

N	Hb (g/L)	Er (10^12/L)	L (10^9/L)	tr (10^9/L)
1	138	4.60	13.7	283
2	139	4.29	15.5	34
3	160	4.84	18.2	306
4	179	5.52	22.0	130
5	152	4.54	8.9	298
6	153	4.51	22.0	518
7	-	-	-	-
8	144	4.10	18.8	191
9	175	5.39	18.9	322
10	154	4.90	8.1	332
11	143	4.57	24.0	318
12	142	4.67	21.9	321
13	165	4.79	14.7	283
14	176	5.46	21.9	300
15	162	5.41	15.4	364
16	170	5.12	11.1	157
17	-	-	-	-
18	151	4.43	16.5	247
19	167	5.10	17.8	220
20	145	4.93	15.5	167
Xm	156.4±3.1	4.84±0.10	16.9±1.1	266.2±4
n	18	18	18	18

- there is a high level of hemoglobin - 156.4±3.1 g/L, moderate leukocytosis - 16.91.1/L and thrombocytosis - 266.2±24.4/L.
- there is a shift in leukogram up to myelocytes; absolute monocytosis, prevalence of neutrophils upon lymphocytes and considerable normoblastosis.

Table 2.
Leukogram of umbilical cord blood by automatic counter and by light microscope.

N	Mye-lo-cytes	Meta-myel.	Band neutr	Segmn eutr	Eo-sino-cytes	Baso phils	Monoc ytes	Lymphoc ytes	Normobl. to 100 L	non-diff. cells
2*	0.5	-	5.0	60.0	2.0	0.5	6.0	26.0	0.5	-
	-	-	-	56.1	1.4	0.5	8.8	33.1	-	5.5
6*	0.5	1.5	4.5	38.0	8.0	2.0	13.0	33.5	7.5	-
	-	-	-	44.5	5.0	0.8	13.0	36.6	-	5.9
7*	3.5	2.0	7.0	43.0	10.0	0.5	9.0	25.0	1.5	-
	-	-	-	56.0	9.8	0.6	10.2	23.4	-	8.3
9*	0.5	0.5	3.0	35.0	10.0	0.5	14.9	36.0	3.0	-
	-	-	-	37.2	8.9	0.6	10.6	42.7	-	6.0
10*	2.0	2.0	5.0	42.5	2.5	0.5	12.5	33.0	4.5	-
	-	-	-	44.5	3.9	0.4	17.6	33.7	-	8.5
11*	2.5	3.5	6.5	27.5	2.5	-	11.5	46.0	10.0	-
	-	-	-	38.6	4.1	0.8	10.3	46.2	-	6.2
12*	0.5	2.0	8.5	47.0	2.0	-	11.5	29.0	36.0	-
	-	-	-	46.8	3.7	0.5	11.8	37.3	-	6.6
13*	1.0	2.0	5.0	43.0	2.0	-	9.5	37.5	1.5	-
	-	-	-	58.0	-	-	3.0	39.0	-	-
14*	0.5	3.0	10.0	33.0	2.5	1.5	12.5	37.0	19.5	-
	-	-	-	45.1	2.5	0.7	13.8	37.9	-	5.7

* - leukogram by light microscope.

Comparison of leukogram of umbilical cord blood performed by automatic counter and by light microscope (tabl. 2) showed that immature forms of neutrophils were recognized by counter as non-differentiated cells and normoblasts were recognized as lymphocytes. It demonstrates that automatic analysis alone can not be a reliable method in evaluation of morphological composition of umbilical cord blood.

Clonogenic culture of progenitor cells of umbilical cord blood in methylcellulose (tab. 3 and 4) demonstrated the presence of a great number of hemopoietic progenitors (total - 11826.1±2894.6 in 1 mL of umbilical cord blood) with prevalence of granulocyte and macrophage progenitors: 7880.2±2059.1 and 2706.8±642.1 respectively.

Table 3.
Content of progenitor cells in umbilical cord blood (culture in methylcellulose by $1x10^5$ explanted mononuclear cells)

N	CFU-G	CFU-E	CFU-M	CFU-GM	TOTAL
1	51.6	6.6	15.0	0	73.2
2	163.3	28.3	111.6	10.0	313.2
3	140.0	11.6	16.6	3.3	171.5
4	173.3	23.3	48.3	6.6	251.5
7	58.3	0	5.0	20.0	83.3
11	23.3	0	5.0	16.6	44.9
13	20.0	43.3	25.5	3.3	92.1
14	11.6	0	50.0	5.0	66.6
17	35.0	11.7	5.0	0	51.7
19	23.3	1.7	23.3	1.7	50.0
20	40.0	1.7	15.0	5.0	61.7
Xm	61.6±16.6	12.1±3.6	27.7±5.3	6.4±1.6	71 5±12.2

Table 4.
Content of progenitor cells in umbilical cord blood (culture in methylcellulose of 1 mL of blood)

N	Mononuclears in 1 mL of blood (x 10^6)	CFU-G	CFU-E	CFU-M	CFU-GM	TOTAL
1	5.2	2683.2	343.2	780.0	0	3806.4
2	13.5	21963.9	3806.4	15010.2	1345.0	42125.4
3	2.3	3255.0	269.7	385.9	76.7	3987.4
4	12.1	20882.7	2807.7	5820.2	795.3	30305.8
7	19.0	11077.0	0	950.0	3800.0	15827.0
11	19.9	4625.1	0	992.5	3295.1	8912.7
13	13.5	2700.0	5845.5	3442.5	445.5	12433.5
14	4.3	502.9	0	2167.5	216.8	2887.1
17	19.4	6790.0	2269.8	970.0	0	10029.8
19	14.8	3436.8	250.8	3436.8	250.8	7375.0
20	38.7	15472.0	657.7	5802.0	1934.0	23865.6
Xm	13.5±1.8	7880.2± 2059.1	1099.8±423 7	2706.8± 642.1	1183.2± 392.5	11826.1±28 94.6

Culture study of mononucleated fraction of cells of umbilical cord blood revealed great number of GM-progenitors - 6883.4±2484.5 in 1 mL of blood (tabl. 5 and 6). Comparison of results of clonogenic culture in methylcellulose and agar of the same samples of umbilical cord blood demonstrated that culture in methylcellulose (tabl. 6) discovered greater number of GM-progenitors.

Table 5.
Content of GM-precursors in umbilical cord blood (culture system "Agar drop-liquid medium")

N	Mononuclears in 1 mL of blood (x 10^6)	CFU-GM (x 10^5)	CFU-GM (in 1 mL.)	Clusters/ colonies	% of big colonies
3	2.3	89.0	2069.3	0.3	0
4	12.1	164.0	19762.0	0.5	0
7	19.0	63.0	11970.0	0.7	4.8
8	15.2	60.0	9120.0	0.8	13.3
11	19.9	458.0	90913.0	0.02	31.9
14	4.3	34.0	1473.9	0.5	1.0
15	11.1	29.0	3219.0	4.8	0
17	19.4	2.0	388.0	-	100.0
18	3.8	15.0	570.0	0.4	0
Xm	12.1±2.2	64.9±17.6	6883.4± 2484.5	1.1±0.5	12.7±56.0

Table 6.
Number of GM-precursors in different culture systems.

N	Mononuclears in 1 mL of Blood	Agar: CFU-GM (in 1 mL)	Methylcellulose: CFU-G+CFU-M+CFU-GM (in 1 mL)
3	2.3	2069.3	3717.7
4	12.1	19762.0	27498.1
7	19.0	11970.0	15827.0
14	4.3	1473.9	2887.1
17	19.4	388.0	7760.0
Xm	11.4±3.2	7132.6±3382.6	11538.0±4115.3

DISCUSSION

There are enough evidences to consider formation of fetal and embrionic hemopoiesis as migratory phenomenon. It was shown that hemopoietic progenitor cells migrate through circulation from yolk sac to liver (Migliaccio et al. 1986) and then some time before birth they appear in bone marrow which becomes the sole hemopoietic organ in postnatal life (Hann J.,et al., 1983). So, circulation which includes umbilical cord blood vessels is a connecting link between different hemopoietic sites. Umbilical cord blood cell composition is determined by migratory character of embrionic hemopoiesis and by start of intensive hemopoiesis in bone marrow some time before birth, which can explain lineal increasement of hemoglobin and hematocrite concentrations, as well as levels of thrombocytes and nucleated cells short time before birth. (Boulot P., et al. 1993). Our results confirmed previously reported data about levels of hemoglobine, erythrocytes, leukocytes and thrombocytes in umbilical cord blood after full-term deliveries (Boulot P., et al. 1993).

Light microscope leukogram showed the presence of all young myeloid cells starting by myelocytes and normoblasts in umbilical cord blood. Leukogram, made by automatic counter differs from it in great deal. Automatic counter does not recognize young myeloid cells up to band neutrophils which limits its usage for umbilical cord blood studies. We have found that total amount of early myeloid cells (without band neutrophils), eosinophils and basophils vary from 3 to 16% (an average - 8.3%). At the same time, other authors reported that total amount of these cells do not exceed 2% and do not vary during gestation (Davies N., et al. 1992).

Recent studies have demonstrated that human umbilical cord blood can be used as a source of transplantable stem cells (Broxmeyer H., et al. 1992). Although it is universally recognized that total amount of CD34+ cells in umbilical cord blood is approximately 1% from total amount of nucleated cells (Kinniburgh D., et al. 1993), the results obtained by clonogenic methods differ sometimes considerably. This can be explained by use of different cultural systems and sources of colonystimulating factors. By using two different cultural systems ("agar drop-liquid medium" and methylcellulose with hr-Epo, hr-GM-CSF and hr-IL-3) we have also obtained different results of GM-CFU concentration in the same samples of umbilical cord blood. We suggest that these cultural systems are favorable to colony growth from progenitors of different stages of maturation.

REFERENCES

Afanasiev B.V., Tiranova T.G., Kulibaba T.G., et al. (1983). Cloning of human hematopoietic cells in agar drop-liquid medium system. Ther. archiv. 8; 114.

Boulot P., Cattaneo A., Taib J., et al. (1993). Hematologic values of fetal blood obtained by means of cordocentesis. Fetal. Diagn. Ther. 8; 309.

Broxmeyer H., Cooper S., Yoder M., et al. (1992). Umbilical cord blood as a source of transplantable hematopoietic stem and progenitor cells. Curr. Top. Microbiol Immunol. 177; 195.

Broxmeyer H., Douglas G., Hangoc G., et al. (1989). Human umbilical cord blood as a source of transplantable hematopoietic stem/progenitor cells. Proc. Natl. Acad. Sci. USA 86; 3828.

Davies N., Buggins A., Snijders R., et al. (1992). Blood leucocyte count in the human fetus. Arch. Dis. Child. 67; 399.

Hann J., Bodger M., Hoffbrand A. (1983). Development of pluri-potent hematopoietic progenitor cells in the human fetus. Blood 62; 118.

Harris D.T., Schumacher M.J., Rychlik S., et al. (1994). Collection, separation and cryopreservation of umbilical cord blood for use in transplantation. Bone Marrow Transplantation 13; 135.

Kelemen E., Calvo W., Fliedner T.M. (1979). Atlas of human haemopoietic development. Berlin, Heidelberg, New York: Springer-Verlag; 266.

Kinniburgh D., Russell N. (1993). Comparative study of CD34-positive cells and subpopulations in human umbilical cord blood and bone marrow. Bone Marrow Transplant. 12; 489.

Migliaccio G., Migliaccio A., Petti S., et al. (1986). Human embrionic hemopoiesis kinetics of progenitors and precursors underlying the yolk sac - liver transition. J. Clin. Invest. 78; 51.

Tavasolli M. (1991). Embrionic and fetal hemopoiesis: an overview. Blood cells, 17; 269.

Stroma cell dependent leukemias

Growth of PGM-1 leukemia cells on stromal layers

Ulrich Dührsen
Medizinische Klinik
Abteilung für Onkologie und Hämatologie
Universitätskrankenhaus Eppendorf
Martinistraße 52
20246 Hamburg
Germany

Normal hematopoietic and leukemic cell populations resemble each other in structure and growth control (Griffin & Löwenberg, 1986; McCulloch et al., 1988; Löwenberg & Touw, 1993). Maintenance and expansion depend on an adequate number and function of stem cells that differentiate into committed progenitor cells, but also replenish the stem cell pool by means of self-renewing cell divisions. The factors regulating the growth of stem cells remain poorly understood, but there is very good evidence that the stromal bone marrow microenvironment is of crucial importance (Trentin, 1989; Mayani et al., 1992).

The radiation-induced murine PGM-1 leukemia (Dührsen et al., 1989) appeared to be a useful model for the study of the interactions between leukemic stem cells and bone marrow stroma cells. Similar to most human leukemias, PGM-1 cells are not viable in unstimulated in vitro culture and differentiate into non-leukemogenic, post-mitotic end cells in response to hematopoietic growth factors. In vivo, however, the majority of the cells maintain a blast-like morphology. Following subcutaneous inoculation in syngeneic C3H/HeJ recipients they form rapidly growing tumors at the injection site and invade the hematopoietic and lymphatic organs. Therefore we speculated that cell-cell interactions between leukemic cells and sessile regulatory cells provided the information required for self-renewal.

NATO ASI Series, Vol. H 94
Gene Technology
Edited by A. R. Zander, W. Ostertag,
B. V. Afanasiev, F. Grosveld

This hypothesis was tested in coculture experiments between PGM-1 cells and a selection of adherent cell lines. When it became clear that certain adherent lines did indeed promote the self-renewal of PGM-1 cells, the experiments were extended to primary bone marrow stroma layers from patients with hematological and non-hematological diseases. This approach allowed the detection of a regulatory abnormality in the hematopoietic microenvironment of patients with myeloid leukemia and other marrow-infiltrating disorders.

RESULTS

In vitro growth of PGM-1 cells. Figure 1 summarizes the fate of PGM-1 cells in vivo and in vitro. While subcutaneous transplantation was followed within 30 to 60 days by the development of local tumors and generalized enlargement of

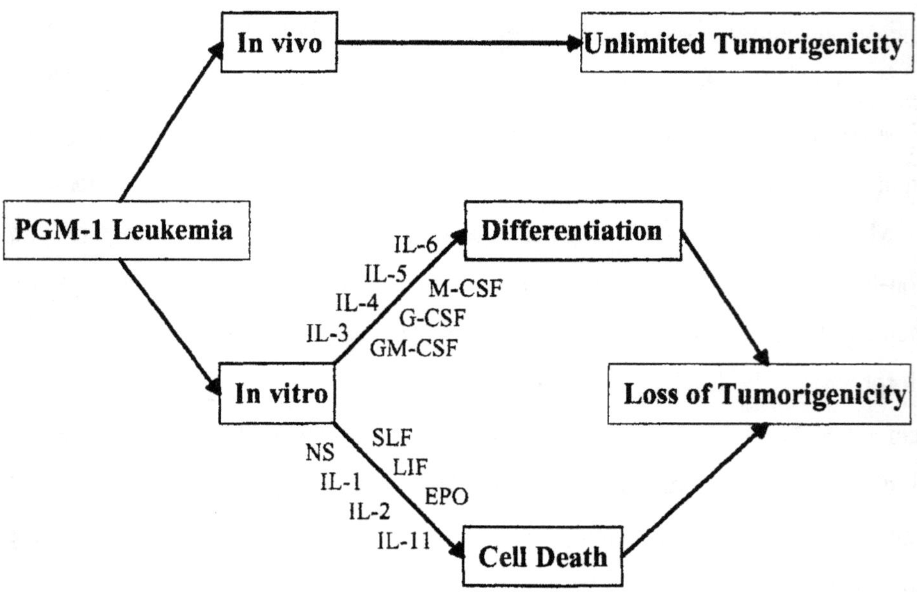

Figure 1: Fate of PGM-1 cells in vivo and in vitro.

the lymph nodes and the spleen, explanted PGM-1 cells lost viability within 48 hours in unstimulated suspension or agar cultures. The same behavior was observed in cultures stimulated with interleukin-1 (IL-1), IL-2, IL-11, Steel factor, leukemia inhibitory factor, or erythropoietin. By contrast, when the cultures were stimulated with IL-3, PGM-1 cells proliferated and gave rise to large numbers of mature granulocytes and macrophages ('PGM' denotes 'Progenitor cell for Granulocytes and Macrophages'; Dührsen et al., 1989). The percentage of cells able to form colonies in IL-3-stimulated agar cultures varied from donor animal to donor animal, but in many instances it approached 100 %. Table 1 shows that granulocyte-macrophage colony stimulating factor (GM-CSF), macrophage (M-)CSF, granulocyte (G-)CSF, IL-4, IL-5 and IL-6 also induced the development of granulocyte and macrophage colonies, although the percentage of cells responding to these cytokines was considerably lower

Table 1: Colony formation by PGM-1 cells in standard agar cultures in response to various growth factors - results of a typical experiment. M: macrophage colonies; G: granulocyte colonies; GM: mixed granulocyte macrophage colonies.

Stimulus	*Percent colony-forming cells*	*Differentiation*
None	0	0
Interleukin-3	81 %	M > GM
GM-CSF	63 %	M > GM
Interleukin-4	14 %	G > GM > M
M-CSF	9 %	M > GM
Interleukin-5	8 %	M > GM
Interleukin-6	2 %	M > GM
G-CSF	1 %	G > M > GM
Interleukin-1α	0	0
Interleukin-2	0	0
Interleukin-11	0	0
Steel Factor	0	0
Leukemia Inhibitory Factor	0	0
Erythropoietin	0	0

than that responding to IL-3. Irrespective of the growth factor used, PGM-1 colonies had no recloning potential, and cell suspensions stimulated for 14 days with GM-CSF or IL-3 were no longer leukemogenic (Dührsen et al., 1994). This suggested that the growth factors induced PGM-1 cells to undergo terminal differentiation.

Growth of PGM-1 cells in coculture with stromal cell lines. In order to ascertain whether PGM-1 self-renewal depended on interactions with other cell types, coculture experiments were performed using a variety of previously established bone marrow stroma cell lines (Novotny et al., 1990). Figure 2 shows the culture system used to detect effects of stroma cells on the survival and proliferation of immature PGM-1 cells. Adherent cell lines were grown to confluency in duplicate 6-well tissue culture trays containing 4 ml Iscove's Modified Dulbecco's Medium with 10 % fetal calf serum, and then overlaid with 10^5 PGM-1 cells. Three hours later the cells were removed from one of the wells, and the number of colony-forming cells able to be recovered was determined in

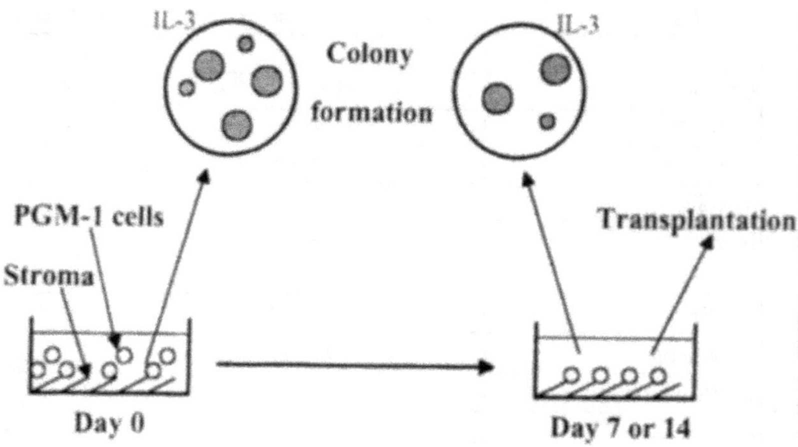

Figure 2: Assay system for the detection of stroma effects on survival and proliferation of colony-forming PGM-1 cells.

IL-3-stimulated agar cultures. The other well was maintained for a further 7 - 14 days and then analyzed for residual IL-3-responsive colony-forming cells (agar culture assay) and leukemogenicity (transplantation).

Figure 3 shows that cultures with no adherent layer failed to support the survival of colony-forming PGM-1 cells (top row). Murine 3T3 fibroblasts and AC 6.21 stroma cells induced the development of large numbers of maturing PGM-1 cells, but proved inefficient in maintaining the cells in a colony-forming

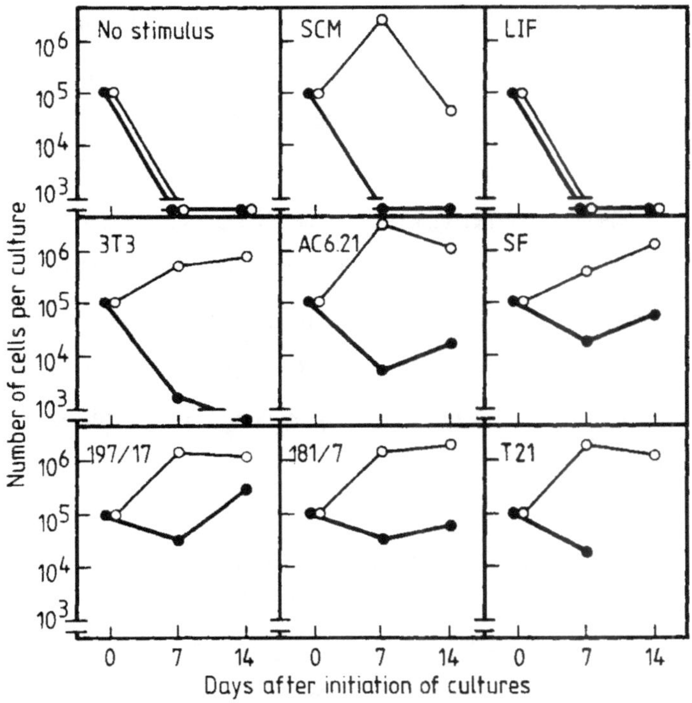

Figure 3: Number of total non-adherent cells (open circles) and colony-forming PGM-1 cells (filled circles) after 0, 7 and 14 days of 4-ml suspension culture (top row) or coculture with adherent cell lines (middle and bottom row). SCM: spleen cell conditioned medium (as a source of native IL-3); LIF: leukemia inhibitory factor; 3T3: mouse fibroblast line; AC 6.21: mouse bone marrow stroma cell line; SF: primary human foreskin fibroblasts; 197/17, 181/7 and T 21: human bone marrow stroma cell lines. (Reprinted by permission of *Leukemia 8 : 490-497, 1994*).

state (middle row). By contrast, human stroma cell lines supported the survival of PGM-1 colony-forming cells over prolonged periods of time (bottom row). The best performing cell line was line 197/17 which promoted not only survival, but also amplification of colony-forming cells. PGM-1 cells retained their leukemogenic potential even after 8 months of continuous passage on 197/17 cells which indicated a survival- or even proliferation-inducing activity on leukemic stem cells (Dührsen et al., 1994).

Molecular mechanism of the stroma-mediated support of immature PGM-1 cells. Human stroma cell lines provided much better survival conditions for colony-forming PGM-1 cells than did murine cell lines. One possible explanation was that human cells were unable to elaborate species-restricted growth factors with differentiation-inducing activity on murine PGM-1 cells. In order to test this hypothesis, murine growth factors were added to cocultures of PGM-1 cells and 197/17 cells. Table 2 shows that the species-restricted factors GM-CSF and IL-3 completely suppressed the survival of PGM-1 colony-forming cells, whereas M-CSF and G-CSF, i.e. factors with cross-species activity, had little if any effect on the colony-forming cells.

<u>*Table 2*:</u> PGM-1/197/17-stroma cell cocultures: inhibition by differentiation-inducing mouse active growth factors and effect of agar interlayers.

Coculture condition	*PGM-1-colony-forming cells after one week*	*Total non-adherent cells after one week*
No stimulus (control)	274	20.000
Interleukin-3	0	232.000
GM-CSF	0	260.000
M-CSF	18	2.010.000
G-CSF	286	168.000
1-ml agar interlayer	54	4.000
2-ml agar interlayer	11	1.000

Table 2 also shows that the interposition of an agar interlayer between 197/17 cells and PGM-1 cells was followed by a reduction, but not by a complete suppression of the stroma effect on colony-forming PGM-1 cells. This suggested that at least part of the effect was mediated by a soluble factor. Further experiments showed that conditioned medium from 197/17 cultures also provided support for colony-forming PGM-1 cells. Using a 5-step purification scheme McKinstry et al. (1993) have recently been able to identify a PGM-1 stimulating protein with an approximate molecular weight of 33 - 43 kD in 197/17 conditioned medium. Partial amino acid sequencing revealed 100 % homology to human complement factor Bb. Whether the PGM-1 supporting factor is identical to factor Bb, a proteolytic fragment thereof or a closely related, yet genetically different factor requires further investigation (N.A. Nicola & A.W. Boyd, personal communication).

Growth of PGM-1 cells in coculture with primary human stroma layers: evidence for abnormalities in the hematopoietic microenvironment in myeloid malignancies and other marrow-infiltrating disorders. Human leukemias arise through an irreversible genetic alteration in an early blood-forming cell. Evidence, however, is accruing that abnormalities in the bone marrow microenvironment contribute to the manifestation of the disease (Greenberger, 1992). Using colony-forming PGM-1 cells as a target for the effects of stroma cells, we studied the function of the hematopoietic microenvironment in 92 marrow samples from patients with different disorders (Dührsen et al., submitted).

To this end the samples were incubated using standard long-term marrow culture conditions (Gartner & Kaplan, 1980). Following establishment of the adherent layer the non-adherent cells were removed by repeated washing, and the cultures subsequently overlaid with freshly explanted PGM-1 cells. After 7 - 14 days of coincubation the PGM-1 cells were removed and cultured in IL-3-stimulated agar culture. The stroma was graded according to the proportion of residual colony-forming PGM-1 cells recovered, and ascribed 'low', 'intermediate' or 'high' PGM-1 supporting activity.

This experimental design allowed the detection of significant differences in the function of the adherent layer in different groups of patients. While 21 of 29 marrow cultures from healthy marrow donors or patients with non-malignant reactive marrow alterations exhibited 'low' activity, 13 of 24 cultures from patients with marrow infiltration by lymphoma cells displayed 'intermediate' activity, and 15 of 33 patients with myeloid malignancies showed 'high' activity. A sequential analysis of 4 patients with acute myeloid leukemia in different phases of the disease showed persistently 'high' activities irrespective of the remission status. No correlation was found between the PGM-1 supporting activities of the samples and the cellular composition of the initial marrow aspirate or the composition of the established stroma layer (Dührsen et al., submitted).

SUMMARY AND CONCLUSIONS

Murine PGM-1 cells are not viable in unstimulated culture and differentiate into non-leukemogenic granulocytes and macrophages in response to a variety of known growth factors. In coculture with bone marrow stroma cell lines they retain their colony-forming and leukemogenic potential. Human stroma cells provide better growth conditions for colony-forming PGM-1 cells than do murine stroma cells. This is partly explained by the inability of human cells to synthesize species-restricted growth factors such as GM-CSF or IL-3 which induce PGM-1 cells to undergo terminal differentiation. The beneficial effect of stroma cells on the survival and proliferation of colony-forming PGM-1 cells is at least partly mediated by a soluble factor which is closely related or identical to a fragment of complement factor Bb.

The fate of PGM-1 cells in primary human long-term cultures depends on the source of the marrow. Adherent layers in cultures from normal bone marrow provide no or very little support for colony-forming PGM-1 cells, whereas stroma layers from patients with myeloid malignancies tend to be very active in this respect. Whether these differences are related to differences in the production of complement factor Bb and whether the ability of stroma from a patient with leukemia to stimulate PGM-1 cells extends to his own leukemic cells, requires further investigation.

Acknowledgments. The data presented were generated in close collaboration with Drs. J. Novotny and A.W. Boyd from the Walter and Eliza Hall Institute of Medical Research in Melbourne, Australia, and Ms. G. Knieling from my own department in Hamburg, Germany. This work was supported by the Deutsche Forschungsgemeinschaft, grants Du 152/1-1, 2-1 and 2-2.

REFERENCES

Dührsen U, Metcalf D, Spangrude GJ, Nicola NA (1989) PGM-1: a transplantable murine leukemia of granulocyte and macrophage progenitor cells. Leukemia 3 : 796-803

Dührsen U, Novotny JR, Boyd AW (1994) Self renewal of a transplantable murine leukemia induced by coculture with human stromal cell lines. Leukemia 8 : 490-497

Dührsen U, Knieling G, Beecken W, Neumann S, Hossfeld DK (submitted) Chimeric cultures of human marrow stroma and murine leukaemia cells: Evidence for abnormalities in the haemopoietic microenvironment in myeloid malignancies and other infiltrating marrow disorders.

Gartner S, Kaplan HS (1980) Long-term culture of human marrow cells. Proc Natl Acad Sci USA 77 : 4756-4759

Greenberger JS (1992) Is the marrow stroma of AML patients a 'leukemic' stroma? Exp Hematol 20 : 1041-1042

Griffin JD, Löwenberg B (1986) Clonogenic cells in acute myeloblastic leukemia. Blood 68 : 1185-1195

Löwenberg B, Touw IP (1994) Hematopoietic growth factors and their receptors in acute leukemia. Blood 81 : 281-292

Mayani H, Guilbert LJ, Janowska-Wieczorek A (1992) Biology of the hemopoietic microenvironment. Eur J Haematol 49 : 225-233

McCulloch EA, Minden MD, Miyauchi J, Kelleher CA, Wang C (1988) Stem cell renewal and differentiation in acute myeloblastic leukemia. J Cell Sci Suppl 10 : 267-281

McKinstry WJ, Welch KL, Martin M, Moritz RL, Ward LD, Simpson RJ, Boyd AW, Nicola NA (1993) Purification of a fragment of human complement factor Bb having self-renewal activity on a murine hemopoietic progenitor cell line. Exp Hematol 21 : 1148 (abstract)

Novotny JR, Dührsen U, Welch K, Layton JE, Cebon JS, Boyd AW (1990) Cloned stromal cell lines derived from Whitlock/Witte-type long-term bone marrow cultures. Exp Hematol 18 : 775-784

Trentin J (1989) Hemopoietic microenvironments. Historical perspectives, status, and projections. In: Tavassoli M (ed.) Handbook of the hemopoietic microenvironment, pp. 1-86. Clifton, NJ: Humana Press

Part 2

Genes Crucial for Development and Disease

The Role of Non-receptor Protein Tyrosine Kinases in Receptor Mediated Cellular Activation and Immunodeficiency Disease.

Sandra J. Saouaf, Anne L. Burkhardt, and Joseph B. Bolen
Department of Molecular Biology
Bristol-Myers Squibb Pharmaceutical Research Institute
Princeton, New Jersey 08543
USA

Abstract

The non-receptor protein tyrosine kinases (PTKs) have been grouped into 10 different enzyme families based on predicted amino acid sequences. As the number of enzymes belonging to the non-receptor class of PTK is increasing, one challenge is to determine how these various classes of PTKs interact within the cell to promote signal transduction. Herein, the activation of four classes of non-receptor PTKs is discussed in relation to their interactions with each other as well as with other signaling molecules during the process of lymphocyte surface antigen receptor mediated activation. Recent findings of non-receptor PTK loss of function mutations in different immunodeficiency diseases has revealed the important contribution of this group of enzymes to lymphocyte development.

Non-receptor Protein Tyrosine Kinases

The non-receptor protein tyrosine kinases (PTKs) represent cellular enzymes grouped together based upon their lack of defined extracellular sequences. The currently identified non-receptor PTKs can be divided into ten different enzyme groups based upon predicted structural features. With the exception of the focal adhesion kinase (Fak) and the activated Cdc42Hs-associated kinase (Ack) which are the only known members of these two individual PTK groups, the remaining

NATO ASI Series, Vol. H 94
Gene Technology
Edited by A. R. Zander, W. Ostertag,
B. V. Afanasiev, F. Grosveld

PTKs appear to be members of distinct PTK families. They range in size from approximately 50 kDa for the c-src kinase (Csk) family to approximately 150 kDa for the Abl kinase family. All of these PTKs are likely to be involved in one or more signaling pathways that modulate growth, differentiation, and mature cell function.

Figure 1 presents a comparison of the recognized major common domains of the non-receptor PTK families. This comparison is significantly simplified and does not detail regions of the enzymes that might be unique for a given PTK type or PTK family. These unique sequences, which in some cases represent the bulk of the enzyme, are clearly important components of these kinases since they likely play a central role in mediating the specific interactions of the individual PTKs.

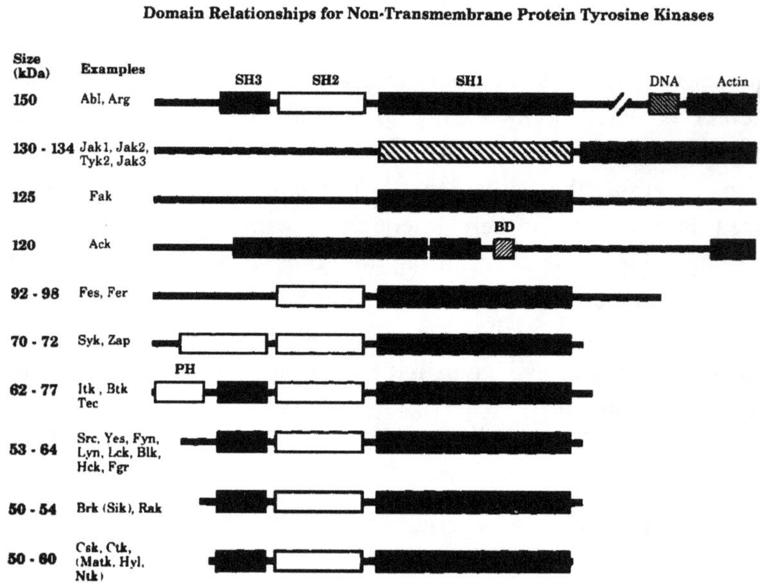

Figure 1. Common domain relationships for non-receptor protein tyrosine kinases.

The major area of sequence homology between the non-receptor PTKs and presumably the region of greatest shared structural identity is their catalytic domains (Hanks *et. al.*, 1988). The catalytic domains for the non-receptor PTKs are the SH1 (src homology 1) domains since this is the region between the kinases that shares the greatest sequence similarity with the c-src catalytic domain. The Janus kinases (Jak1, Jak2, and Tyk2) are interesting in that they have two catalytic-like domains (Wilks *et. al.*, 1991). However, close scrutiny of the sequences in each of these domains indicates that the catalytic activity of the Jaks is contained within the most carboxy-terminal catalytic-like domain (Wilks *et. al.*, 1991). The role of the second catalytic-like domain has not been determined. A site for enzyme tyrosine autophosphorylation is also common for most of the non-receptor PTKs corresponding to the tyrosine 416 residue of c-src. The Csk class of kinases does not possess a comparable autophosphorylation site (Nada *et. al.*, 1991) even though Csk and the other member of this family, Ctk, are capable of tyrosine autophosphorylation when expressed as recombinant proteins (Klages *et. al.*, 1994). The Syk/Zap class of kinases possess two tyrosines at the predicted autophosphorylation site (Chan *et. al.*, 1992).

With the exception of the Jak and Fak groups, all other non-receptor PTKs possess one or more SH2 domains and/or an SH3 domain (Mayer and Baltimore, 1993; Pawson and Gish, 1992). In most cases, the SH2 and SH3 domains are at the aminoterminal end of the enzyme with respect to the catalytic domain. The exception is the Ack PTK in which the SH3 domain of this enzyme lies carboxyterminal to the catalytic domain (Manser *et.al.*, 1993). In all cases where PTKs possess both SH3 and SH2 domains, the SH3 domain is positioned aminoterminal to the SH2 domain. The actual spacing between the SH2 and catalytic domains varies as does the spacing between SH3 and SH2 domains. The significance of these spacing differences is not known. SH2 as well as SH3 domains exist together and independently in a variety of cellular proteins, some of which possess enzymatic activity, while others are implicated in enhancing specific protein-protein interactions thought to be important in signaling regulation (Pawson and Schlessinger, 1993). It is well established that individual SH2 domains are capable of specific binding to selected phosphotyrosine containing proteins (Songyang *et. al.*, 1993) and are thought to play critical roles in the interactions between signaling components in PTK dependent pathways (Koch *et. al.*, 1991). The SH3 domains recognize proline-rich peptide ligands which in some cases are present in guanine nucleotide exchange factors and GTPase-activating proteins (Chardin *et. al.*, 1993; Ren *et. al.*, 1993). Both of these SH3 binding protein types are involved in the regulation of activity of small G proteins such as Ras/Rac/Rho. Recently the structures of several SH2 and SH3 domains with and without complexed phosphopeptides or peptides have been determined in solution by NMR and by X-ray crystallography (Booker *et. al.*, 1992; Eck *et. al.*, 1993; Musacchio *et. al.*, 1992; Overduin *et. al.*, 1992; Waksman *et. al.*, 1992; Waksman *et. al.*, 1993; Yu *et. al.*, 1992). These structures document the modular nature of the SH2 and SH3 domains and have defined important features of the specific interactions between these domains and their ligands.

Only the members of the src family, along with a single isoform of c-Abl/Arg (type IV), have an inherent mechanism for promoting their association with cell membranes. This feature of the src family and c-Abl type IV results from the posttranslational myristylation of a common glycine residue at position two (Buss *et. al.*, 1984; Daley *et. al.*, 1992; Jackson and Baltimore, 1989; Kamps *et. al.*, 1985; Resh, 1990). Several members of the src family also appear to be palmitylated in addition to being myristylated (Paige *et. al.*, 1993). Palmitylation of these enzymes occurs at one or more cysteines located at residue three and either residue five or six. Current evidence suggests that palmitylated src PTKs are located in regions of plasma membranes that enhances their capacity to associate with glycosyl-phosphatidylinositiol (GPI)-anchored proteins (Rodgers *et. al.*, 1994; Shenoy-Scaria *et. al.*, 1993; Shenoy-Scaria *et. al.*, 1992; Thomas and Samelson, 1992). Thus, these aminoterminal fatty acid modifications of the src PTKs distinguishes them (and c-Abl/Arg type IV) from the remaining non-receptor PTKs and promotes the likelihood that these kinases are located at cellular sites where a significant portion of PTK-mediated signaling reactions take place. It reasonable to assume that these are also the same sites where important physiologic substrates for the src enzymes exist.

Non-receptor Protein Tyrosine Kinase Activation Through Antigen Receptors

Antigen receptors on T, B, and mast cells are similar in that they consist of several associated protein subunits (Figure 2). The extracellular antigen binding domain mediates the interaction between the receptor and antigenic peptide complexed with MHC proteins for T cells, native antigen for B cells, or the Fc portion of IgE for mast cells or basophils. The antigen binding chain(s) of the multichain immune recognition receptors (MIRR) contain a very limited number of cytoplasmic amino acids, yet can transduce signals to the cell upon their engagement. The associated subunits however, have longer cytoplasmic domains with limited extracellular sequence. It is these receptor subunits which are responsible for signal propagation from the antigen binding domain to the cytoplasm of the cell. Through the use of chimeric receptors fusing the cytoplasmic domain of the receptor subunits with the extracellular and transmembrane domain of a heterologous receptor, it was shown that the cytoplasmic domains of the receptor are competent to induce the full array of signaling which takes place with the intact multichain receptor (Irving *et. al.*, 1993; Kim *et. al.*, 1993; Letourner and Klausner, 1992; Sanchez *et. al.*, 1993).

Phosphorylation of proteins on tyrosine, including the receptor cytoplasmic subunits, is one of the earliest detectable signaling events following MIRR surface ligation and is indicative of the activation of tyrosine kinases (Samelson *et. al.*, 1986; Qian *et. al.*, 1993; Gold *et. al.*, 1991; Paolini *et. al.*, 1991; Li *et. al.*, 1992). Mutation of the tyrosine residues in the B cell antigen receptor Igα subunit results in abrogation of cellular activation (Flaswinkel and Reth, 1994). Alignment of the cytoplasmic sequences of the receptor subunits of the MIRRs reveals a consensus sequence present in each of the proteins (Reth, 1989) (Figure 2). This consensus sequence is often referred to as a tyrosine activation motif (TAM) (Samelson and Klausner, 1992). The zeta chain of the T cell receptor contains three TAMs while the other receptor cytoplasmic subunits contain one copy of the TAM. The tyrosines within the TAM are phosphorylated during cellular activation. The spacing between the phosphotyrosine residues is conserved in the motif as is the position of leucine or isoleucine three amino acids from the tyrosine. The TAM has been shown to be required for the signal transduction function of T cell receptor chains (Letourner and Klausner, 1991; Romeo *et. al.*, 1992; Irving *et. al.*, 1993). In addition, the cytoplasmic subunits of the B cell antigen receptor transduce signals in transfected T lymphocytes,

MULTICHAIN IMMUNE RECOGNITION RECEPTORS

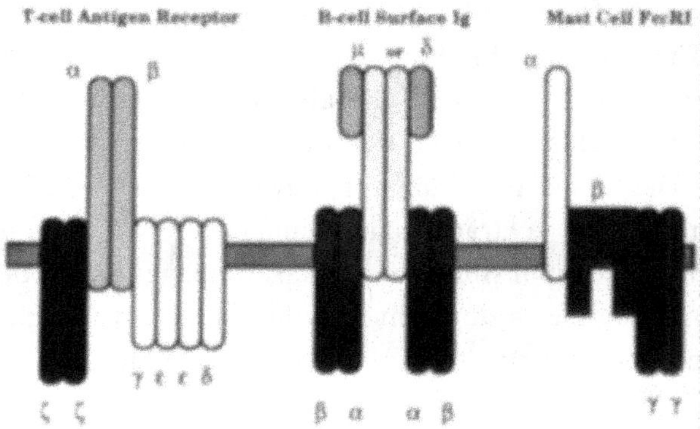

Alignment of Tyrosine Activation Motifs - TAMs

zeta	ADAYSDIGTKGERRRGKGHDGLYQGLSTATKDTYDALHMQTLAPR
ε gamma	RLKIRVDKADIASREKSDAVYTGLNTDNQETYETLKHEKPPQ
mb-1	RKRWQNEKFGVDMPDDYEDENDYEGLNLDDCSMYEDISRGLQGTYQDVGNLHIGD
b-29	DKDDGKAGMEEDHTYEGLNIDQTATYEDIVTLRTGEVKWSVGEHPGQE
ε **beta**	YRIGQELE-SKKVPDDRLYEEL-NVYSPIY-SELEDKGETSSPVDS
CD3 gamma	GQDGVRQSRASDKQTLLPNDQLYQPLKDREDDQY-SHLQGNQLRRN
CD3 delta	GHETGRLRGAADTQALLRNDQVYQPLRDRDDAQY-SHLGGNWARNR

TAM: DXXXXXXXDXXYXXLXXXXXXXXYDXL

12 amino acids

Figure 2. Structural features of multichain immune recognition receptors and alignment of the tyrosine activation motifs (TAMs) within the cytoplasmic domains of the TCR (zeta, CD3 gamma, CD3 delta), the BCR (Igα-mb-1, Igβ-b29) and the FcεRI (ε gamma, ε beta). The consensus TAM is shown.

indicating that although the MIRR chains differ in overall sequence, the common TAM is sufficient to couple signaling in diverse cell types (Burkhardt *et. al.*, 1994).

The MIRR chains do not contain endogenous tyrosine kinase activity. However, non-receptor PTKs have been found to co-immunoprecipitate with the MIRRs (Samelson *et. al.*, 1990; Burkhardt *et. al.*, 1991; Yamanashi *et. al.*, 1991; Eiseman and Bolen, 1992; Hutchcroft *et. al.*, 1992). Activation of the src family of non-receptor PTKs has been shown to occur rapidly after engagement of the MIRRs on T and B cells (Tsygankov *et. al.*, 1992; Burkhardt *et. al.*, 1994; Burkhardt *et. al.*, 1991; Saouaf *et. al.*, 1994). We have extended this observation to examine the temporal order of activation of several classes of non-receptor PTKs present in immune cells. The results are shown schematically in Figure 3. Upon engagement of the MIRR on B or T cells, the src family of non-receptor PTKs is maximally enzymatically active within seconds. This activation of src kinases correlated with the initial tyrosine phosphorylation of the Igα and Igβ subunits of the B cell antigen receptor and with the phosphorylation of the zeta chain of the T cell receptor (Saouaf *et. al.*, 1994; Burkhardt *et. al.*, 1994). The enzymatic activity of the src tyrosine kinases then decreased over time. Bruton's tyrosine kinase (Btk) was the next to reach its peak of activity at 5 minutes post antigen receptor engagement. The observed activation of Btk indicates that this newly identified protein tyrosine kinase plays a role in B cell receptor signaling. The Fer non-receptor PTK reached maximal activity 10 minutes post receptor crosslinking and then decreased rapidly in activity. Fer is expressed in a variety of cell types (Hao *et. al.*, 1989; Pawson *et. al.*, 1989) and its activation in lymphoid cells is the initial description of Fer involvement in a signal transduction pathway. The Syk/Zap family of PTKs were the last to reach maximal activity. Their activity remained high for as long as one hour following antigen receptor engagement. Syk and Zap formed complexes with the antigen receptor cytoplasmic subunits as a function of their activation (Saouaf *et. al.*, 1994; Burkhardt *et. al.*, 1994). The successful translocation of Syk and Zap to the plasma membrane is mediated by the interaction of their two SH2 domains with the tyrosine phosphorylated TAM of the receptor cytoplasmic subunits (Iwashima *et. al.*, 1994, Saouaf, *et.al.* unpublished results). The location of the src family of PTKs at the plasma membrane and the early kinetics of their

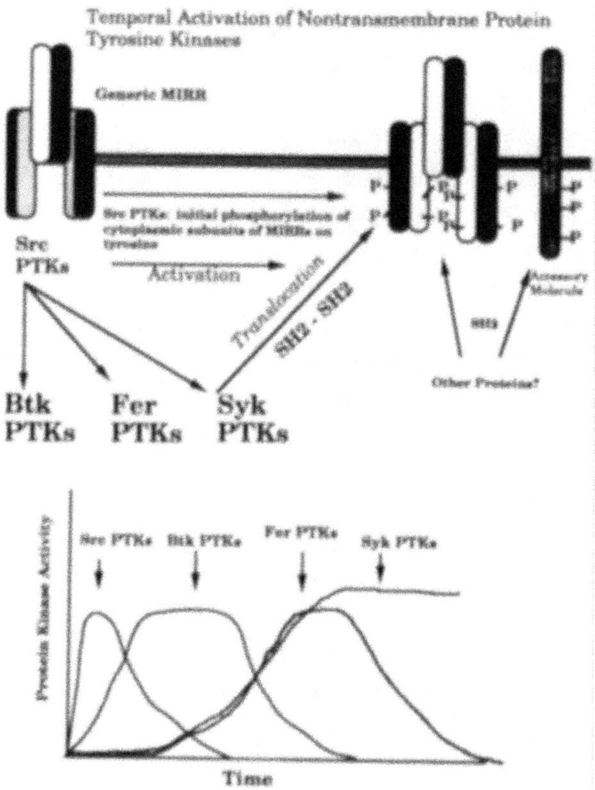

Figure 3. Model of signal transduction through multichain immune recognition receptors. The top portion illustrates activation of non-receptor protein tyrosine kinases and tyrosine phosphorylation of the receptor cytoplasmic sequences and the graph depicts the temporal order of activation of the protein tyrosine kinases.

activation make these enzymes likely candidates for the initial phosphorylation of the TAMs of the receptor. Activation of Btk and Syk is influenced by src kinases (unpublished results). Further activation of Syk may occur following its association with the receptor cytoplasmic subunits, enabling the kinase to act on other substrates localized to the plasma membrane. Other SH2 containing proteins may interact with the phosphorylated TAMs of the receptor subunits or with other tyrosine phosphorylated accessory molecules. Consequently, the net result is recruitment of many types of signaling molecules to the plasma membrane to propagate signals.

Protein Tyrosine Kinases in Immunodeficiency Disease

Btk was identified as the defective gene responsible for X-linked agammaglobulinemia (XLA) in humans (Vertie *et. al.*, 1993; Tsukada *et. al.*, 1993) and X-linked immunodeficiency (XID) in mice (Thomas *et. al.*, 1993; Rawlings *et. al.*, 1993). Identification of the XLA defect was the first report linking human genetic disease with mutation of a non-receptor PTK. XLA is characterized by a deficit of mature B cells, and patients therefore lack circulating immunoglobulin. Since the number of pre-B cells in the bone marrow of XLA patients is normal, the defect lies in the transition from pre-B to mature B cells (Pearl *et. al.*, 1978). Several mutations in the Btk gene have been described which result in the XLA phenotype. These mutations include point mutations in the pleckstrin homology, SH2, and kinase domains; a deletion in the carboxyl end of the SH3 domain, and mutations which destabilize the mRNA (Tsukada *et. al.*, 1994). The mutations which occur in the pleckstrin homology domain, SH2, and SH3 domains do not compromise the catalytic activity of Btk. These mutations likely disrupt Btk's interaction with other proteins, resulting in a similar block in B cell development as that which occurs upon inactivation of the enzyme's catalytic function. Thus, Btk and the proteins with which it interacts appear to be necessary for human B cell development. In mice, mutation of the pleckstrin homology domain of Btk at the same amino acid as the mutation resulting in XLA in humans produces a less severe phenotype, XID (Thomas *et. al.*, 1993; Rawlings *et. al.*, 1993). Mature B cells are present in XID, however, they do not produce antibody in response to polysaccharide antigens. The B cells of XID mice respond normally to T cell dependent antigens (Thomas *et. al.*, 1993). In mature murine B cells Btk is critical for proper signal transduction, however, these observations imply that Btk does not play as large a role in B cell development in mice as it does in humans.

The Zap-70 PTK has recently been shown to be defective in patients with selective T cell defect (STD) (Arpala *et. al.*, 1994; Elder *et. al.*, 1994; Chan *et. al.*, 1994). Mature T cells and thymocytes from these patients lack the $CD8^+$ subset. Peripheral $CD4^+$ T cells from these patients do not proliferate in response to mitogen stimulation or to antigen receptor crosslinking. It was found that the Zap-70 non-receptor PTK was not expressed in these patients due to mutations which destabilizes the protein (Arpala *et. al.*, 1994; Elder *et. al.*, 1994; Chan *et. al.*, 1994). This indicates that Zap-70 is required for development of the $CD8^+$ lineage of human T cells and is required for signal transduction in $CD4^+$ T lymphocytes.

The identification of Btk and Zap-70 as key enzymes in cellular responses resulting from surface engagement of lymphocyte MIRRs has furthered our understanding of signal tranduction mechanisms. The findings that loss of function mutations of these enzymes block not only signaling in the pathways noted above, but also alter lymphocyte differentiation, underscores the potentially multifunctional role of non-receptor PTKs in cellular physiology and development. It will be of interest to determine whether mutations in other non-receptor PTKs are associated with additional immunodeficiencies.

References

Arpala E, Sharar M, Dadi H, Cohen A and Roifman CM (1994) Defective T cell receptor signaling and CD8+ thymic selection in humans lacking Zap-70 kinase. Cell 76: 947-958

Booker GW, Breeze AL, Downing AK, Panayatou G, Gout I, Waterfield MD and Campbell ID (1992) Structure of an SH2 domain of the p85 alpha subunit of phosphatidylinositol-3-OH kinase Nature 358: 684-687

Burkhardt AL, Brunswick M, Bolen JB and Mond JJ (1991) Anti-immunoglobulin stimulation of B lymphocytes activates src-related protein-tyrosine kinases. Proc. Natl. Acad. Sci. USA 88: 7410-7414

Burkhardt AL, Costa T, Misulovin Z, Stealy B, Bolen JB and Nussenzweig MC (1994) Igα and Igβ are functionally homologous to the signalling proteins of the T-cell receptor. Mol. Cell. Biol 14: 1095-1103

Burkhardt AL, Stealey B, Rowley RB, Mahajan S, Pendergast M, Fargnoli J and Bolen JB (1994) Temporal regulation of non-transmembrane protein tyrosine kinase enzyme activity following T cell antigen receptor engagement. J. Biol. Chem. 269: 23642-23647

Buss JE, Kamps MP and Sefton BM (1984) Myristic acid is attached to the transforming protein of Rous sarcoma virus during or immediately after synthesis and is present in both soluble and membrane-bound forms of the protein. Mol. Cell. Biol. 4: 2697-2704

Chan AC, Iwashima M, Turck CW and Weiss A (1992) ZAP-70: a 70-kD protein tyrosine kinase that associates with the TCR ζ chain. Cell 71: 649-662

Chan AC, Kadlecek TA, Elder ME, Filipovich AH, Kuo W-L, Iwashima M, Parslow TG and Weiss A (1994) Zap-70 deficiency in an autosomal recessive form of severe combined immunodeficiency. Science 264: 1599-1601

Chardin P, Camonis J, Gale WL, VanAelst L, Schlessinger J, Wigler MH and Bar-Sagi D (1993) Human Sos1: a guanine nucleotide exchange factor for Ras that binds to GRB2. Science 260: 1338-1343

Daley GQ, Etten RAV, Jackson P, Bernards A and Baltimore D (1992) Nonmyristolated Abl proteins transform a factor-dependent hematopoietic cell line. Mol. Cell. Biol. 12: 1864-1871

Eck MJ, Shoelson SE and Harrison SC (1993) Recognition of a high-affinity phosphotyrosyl peptide by the Src homology-2 domain of $p56^{lck}$. Nature 362: 87-91

Eiseman E and Bolen JB (1992) Engagement of the high-affinity IgE receptor activates src protein-related tyrosine kinases. Nature 355: 78-80

Elder ME, Lin D, Clever J, Chan AC, Hope TJ, Weiss A and Parslow TG (1994) Human severe combined immunodeficiency due to a defect in Zap-70, a T cell tyrosine kinase. Science 264: 1596-1599

Flaswinkel H and Reth M (1994) Dual role of the tyrosine activation motif of the Igα protein during signal transduction via the B cell antigen receptor. EMBO J. 13: 83-89

Gold MR, Matsuuchi L, Kelly RB and DeFranco AL (1991) Tyrosine phosphorylation of components of the B-cell antigen receptors following receptor crosslinling. Proc. Natl. Acad. Sci. USA 88: 3436

Hanks SK, Quinn AM and Hunter T (1988) The protein kinase family: conserved features and deduced phylogeny of the catalytic domains. Science 241: 42-52

Hao QL, Heisterkamp N and Groffen J (1989) Isolation and sequence analysis of a novel human tyrosine kinase gene. Mol. Cell. Biol. 9: 1587-1593

Hutchcroft JE, Harrison ML and Geahlen RL (1992) Association of the 72-kDa protein-tyrosine kinase PTK72 with the B cell antigen receptor. J. Biol. Chem. 267: 8613-8619

Irving BA, Chan AC and Weiss A (1993) Functional characterization of a signal transducing motif present in the T cell antigen receptor zeta chain. J. Exp. Med. 177: 1093-1103

Iwashima M, Irving BA, Oers NSCv, Chan AC and Weiss A (1994) Sequential Interactions of the TCR with two distinct cytoplasmic tyrosine kinases. Science 263: 1136-1139

Jackson P and Baltimore D (1989) N-terminal mutations activate the leukemogenic potential of the myristolated form of c-Abl. EMBO J. 8: 449-456

Kamps MP, Buss JE and Sefton BM (1985) Mutation of NH2-terminal glycine of p60src prevents both myristylation and morphological transformation. Proc. Natl. Acad. Sci. USA 82: 625-4628

Kim KM, Alber G, Weiser P and Reth M (1993) Differential signaling through the Igα and Igβ components of the B cell antigen receptor. Eur. J. Immunol. 132: 125-146

Klages S, Adam D, Class K, Fargnoli J, Bolen JB and Penhallow RC (1994) Ctk: a protein-tyrosine kinase related to Csk that defines an enzyme family. Proc. Natl. Acad. Sci. USA 91: 2597-2601

Koch CA, Anderson D, Moran MF, Ellis C and Pawson T (1991) SH2 and SH3 domains: elements that control interactions of cytoplasmic signaling proteins. Science 252: 668-674

Letourner F and Klausner RD (1991) T-cell and basophil activation through the cytoplasmic domain of T cell receptor ζ family members. Proc. Natl. Acad. Sci. USA 88: 8905-8909

Letourner F and Klausner RD (1992) Activation of T cells by a tyrosine kinase activation domain in the cytoplasmic tail of CD3ε. Science 255: 79-82

Li W, Deanin GG, Margolis B, Schlessinger J and Oliver JM (1992) Fc epsilon R1-mediated tyrosine phosphorylation of multiple proteins, including phospholipase C gamma 1 and the receptor beta gamma 2 complex, in RBL-2H3 rat basophilic leukemia cells. Mol. Cell. Biol. 12: 3176-3182

Manser E, Leung T, Salihuddin, H, Tan, L., and Lim L (1993) A non-receptor tyrosine kinase that inhibits the GTPase activity of p21cdc42. Nature 363: 364-367

Mayer BJ and Baltimore D (1993) Signaling through SH2 and SH3 domains. Trends in Cell Biology 3: 8-13

Musacchio A, Noble M, Paulpitt R, Wierenga R and Saraste M (1992) Crystal structure of a Src-homology 3 (SH3) domain. Nature 359: 851-855

Nada S, Okada M, MacAuley A, Cooper JA and Nakagawa H (1991) Cloning of a complementary DNA for a protein-tyrosine kinase that specifically phosphorylates a negative regulatory site of $p60^{c\text{-}src}$. Nature 351: 69-72

Overduin M, Mayer B, Rios CB, Baltimore D and Cowburn D (1992) Secondary structure of Src homology 2 domain of c-Abl by heteronuclear NMR spectroscopy in solution. Proc. Natl. Acad. Sci. USA 89: 11673-11677

Paige LA, Nadley MJS, Harrison ML, Cassady JM and Geahlen RL (1993) Reversible palmitylation of the protein tyrosine kinase $p56^{lck}$. J. Biol. Chem. 268: 8669-8674

Paolini R, Jouvin MH and Kinet JP (1991) Phosphorylation and dephosphorylation of the high-affinity receptor for immunoglobulin E immediately after receptor engngement and disengagement. Nature 353: 855-858

Pawson T and Gish GD (1992) SH2 and SH3 domains: from structure to function. Cell 71: 359-362

Pawson T, Letwin K, Lee T, Hao AL, Heisterkamp N and Groffen J (1989) The FER gene is evolutionarily conserved and encodes a widely expressed member of the FPS/FES protein-tyrosine kinase family. Mol. Cell. Biol. 9: 5722-5725

Pawson T and Schlessinger J (1993) SH2 and SH3 domains. Current Biology 3: 434-442

Pearl ER, Vogler LB, Okos AJ, Crist WM, III ARL and Cooper MD (1978) B lymphocyte precursors in human bone marrow: an analysis of normal individuals and patients with antibody-deficiency states. J. Immunol. 120: 1169-1175

Qian D, Griswold-Prenner I, Rosner MR and Fitch FW (1993) Multiple components of the T cell antigen receptor complex become tyrosine phosphorylated upon activation. J. Biol. Chem. 268: 4488-4493

Rawlings DJ, Saffran DC, Tsukada S, Largaespada DA, Grimaldi JC, Cohen L, Mohr RN, Bazan JF, Howard M, Copeland NG, Jenkins NA and Witte ON (1993) Mutation of unique region of Bruton's tyrosine kinase in immunodeficient XID mice. Science 261: 358-361

Ren R, Mayer BJ, Cicchetti P, and Baltimore D (1993) Identification of a ten amino acid proline rich SH3 binding site. Science 259: 1157-1161

Resh MD (1990) Membrane interactions of $pp60^{v-src}$: a model for myristylated tyrosine protein kinases. Oncogene 5: 1437-1444

Reth M (1989) Antigen receptor tail clue. Nature 338: 383-384

Rodgers W, Crise B and Rose JK (1994) Signals determining protein tyrosine kinase and glycosyl-phosphatidylinositol-anchored protein targeting to a glycolipid-enriched membrane fraction. Mol. Cell. Biol. 14: 5384-5391

Romeo C, Amiot M and Seed B (1992) Sequence requirements for the induction of cytolysis by the T cell antigen/Fc receptor ζ chain. Cell 68: 889-897

Samelson LE and Klausner RD (1992) Tyrosine kinases and tyrosine-based activation motifs. Current research on activation via the T cell antigen receptor. J. Biol. Chem. 267: 24913-24916

Samelson LE, Patel MD, Weissman AM, Harford JB and Klausner RD (1986) Antigen activation of murine T cells induces tyrosine phosphorylation of a polypeptide associated with the T cell antigen receptor. Cell 46: 1083-1090

Samelson LE, Phillips AF, Leong ET and Klausner RD (1990) Association of the Fyn protein-tyrosine kinase with the T-cell antigen receptor. Proc. Natl. Acad. Sci. USA 87: 4358-4362

Sanchez M, Misulovin Z, Burkhardt AL, Mahajan S, Costa T, Franke R, Bolen JB and Nussenzweig M (1993) Signal transduction by immunoglobulin is mediated through Igα and Igβ. J. Exp. Med. 178: 1049-1055

Saouaf SJ, Mahajan S, Rowley RB, Kut S, Fargnoli J, Burkhardt AL, Tsukada S, Witte ON and Bolen JB (1994) Temporal differences in the activation of three classes of non-transmembrane protein tyrosine kinases following B-cell antigen receptor surface engagement. Proc. Natl. Acad. Sci. USA 91:

Shenoy-Scaria AM, Gauen LKT, Kwong J, Shaw AS and Lublin DM (1993) Palmitylation of an amino-terminal cysteine motif of protein tyrosine kinases p56^{lck} and p59^{fyn} mediates interaction with glycosyl-phosphatidylinositol-anchored proteins. Mol. Cell. Biol. 13: 6385-6392

Shenoy-Scaria AM, Kwong J, Fujita T, Olszowy MW, Shaw AS and Lublin DM (1992) Signal transduction through decay- accelerating factor. Interaction of glycosyl-phosphatidylinositol anchor and protein tyrosine kinases p56^{lck} and p59^{fyn}. J. Immunol. 149: 3535-3541

Songyang Z, Shoelson SE, Chaudhuri M, Gish G, Pawson T, Haser WG, King F, Roberts T, Ratnofsky S, Lechleider RJ, Neel BG, Birge RB, Fajardo JE, Chou MM, Hanafusa H, Schaffhausen B and Cantley LC (1993) SH2 domains recognize specific phosphopeptide sequences. Cell 72: 767-778

Thomas JD, Sideras P, Smith CIE, Vorechovsky I, Chapman V and Paul WE (1993) Colocalization of X-linked agammaglobulinemia and X-linked immunodeficiency genes. Science 261: 355-357

Thomas PM and Samelson LE (1992) The glycophosphatidylinositol-anchored Thy-1 molecule interacts with the p60^{fyn} protein tyrosine kinase. J. Biol. Chem. 267: 12317-12322

Tsukada S, Rawlings DJ and Witte ON (1994) Role of Bruton's tyrosine kinase in immunodeficiency. Cur. Opin. in Immunol. 6: 623-630

Tsukada S, Saffran DC, Rawlings DJ, Parolini O, Allen RC, Klisak I, Sparkes RS, Kubagawa H, Mohandas T, Quan S, Belmont JW, Cooper MD, Conley ME and Witte ON (1993) Deficient expression of a B cell cytoplasmic tyrosine kinase in human X-linked agammaglobulinemia. Cell 72: 279-290

Tsygankov AY, Broker BM, Fargnoli J, Ledbetter JA and Bolen JB (1992) Activation of tyrosine kinase p60fyn following T cell antigen receptor cross-linking. J. Biol. Chem. 267: 18259-18262

Vertie D, Vorechovsky I, Sideras P, Holland J, Davies A, Flinter F, Hammarstrom L, Kinnon C, Levinsky R, Bobrow M, Smith CIE and Bently DR (1993) The gene involved in X-linked agammaglobulinaemia is a member of the src family of protein-tyrosine kinases. Nature 361: 226-233

Waksman G, Kominos D, Robertson SC, Pant N, Baltimore D, Birge RB, Cowburn D, Hanafusa H, Mayer BJ, Overduin M, Resh MD, Rios CB, Silverman L and Kuriyan J (1992) Crystal structure of the phosphotyrosine recognition domain SH2 of v-src complexed with tyrosine-phosphorylated peptides. Nature 358: 646-653

Waksman G, Shoelson SE, Pant N, Cowburn D and Kuriyan J (1993) Binding of a high affinity phosphotyrosyl peptide to the Src SH2 domain: crystal structures of the complexed and peptide-free forms. Cell 72: 779-790

Wilks AF, Harpur AG, Kurban RR, Ralph SJ, Zurcher G and Ziemmeicki A (1991) Two novel protein-tyrosine kinases, each with a second phosphotransferase-related catalytic domian, define a new class of protein kinase. Molecular and Cellular Biology 11: 2057-2065

Yamanashi Y, Kakiuchi T, Mizuguchi J, Yamamoto T and Toyoshima K (1991) Association of B cell antigen receptor with protein tyrosine kinase Lyn. Science 251: 192

Yu H, Rosen MK, Shin TB, Seidel-Dugan C, Brugge JS and Schreiber SL (1992) Solution structure of the SH3 domain of Src and identification of its ligand-binding site. Science 258: 1665-1668

Directed Gene Trapping in Embryonic Stem Cells.

Forrester, L.M., Joyner, A.[1,2] Bernstein, A[1]. & Wurst, W.[1,3]

Centre for Genome Research,
University of Edinburgh,
Kings Buildings,
West Mains Rd. Edinburgh.

Introduction.

Gene Trapping in Embryonic Stem Cells.

Embryonic stem (ES) cell lines isolated from the inner cell mass of the mouse blastocyst (Evans & Kaufman, 1983), can be maintained as pluripotent stem cells and genetically manipulated *in vitro*. When re-introduced into a host embryo they can contribute to all tissues, including the germline of the resultant chimera (Beddington & Robertson, 1989). ES cells have provided the essential tool in the genetic manipulation of the mouse genome whereby an alteration to an allele of a loci can be made *in vitro*, introduced into the germline and then bred to homozygosity where the phenotypic effect can be examined.

---

1 Samuel Lunenfeld Research Institute, Mount Sinai Hospital, 600 University Avenue, Toronto, Ontario, Canada.

2 Present Address: New York University Medical Centre, Skirbal Institute of Biomolecular Medicine, Developmental Genetics Programme, 540 First avenue, New York, New York 10016, USA

3 Present Address: Institute fur Saugetiergenetik, Gesellschaft fur Strahlen und Umweltforschung mbH, Ingolstadter Landstr. 1, D-8042 Neuherberg, Germany.

NATO ASI Series, Vol. H 94
Gene Technology
Edited by A. R. Zander, W. Ostertag,
B. V. Afanasiev, F. Grosveld

Gene trapping in ES cells provides a method to identify and functionally characterise novel genes and has proven very powerful in the molecular analysis of embryonic development (Gossler *et al.*, 1989). The design and uses of the different types of gene trap constructs has been reviewed (Skarnes, 1993; Hill & Wurst, 1993). Briefly, the important features of a gene trap vector is an easily assayable reporter gene (eg. the gene encoding the bacterial enzyme, B-galactosidase (lacZ))) linked to an upstream splice acceptor site and a selectable marker (Figure 1). When the construct integrates into the intron of an active gene a fusion transcript is generated containing exons of the trapped gene that lie upstream of the integration site and the *lacZ* reporter sequence (Skarnes, Auerbach & Joyner, 1992). There are several consequences of this event. First, the reporter gene expression is under the control of the regulatory elements of the trapped endogenous gene and the simple analysis of LacZ expression can provide clues to the function of the trapped gene. Second, the endogenous gene can be cloned from the fusion transcript using RACE-PCR (rapid amplification of cDNA ends-polymerase chain reaction) and primers complementary to the known lacZ sequences. Sequence information can provide further clues as to the gene's cellular function particularly if the gene shows homology to known gene families. Finally, the insertional event has the potential to be mutagenic and so the introduction of the integration into the mouse germline and the subsequent breeding to homozygosity allows the functional role of the gene to be addressed *in vivo* .

Trapping vectors must also employ a selectable marker, most commonly the gene encoding neomycin phosphotransferase (neo^R), which can either be expressed from a constitutive promoter (eg. the promoter for the glycolytic enzyme phosphoglycerate kinase (PGK-1)), or coexpressed as a fusion product (βgeo) with LacZ (Freidrich & Soriano, 1991). In the latter case, the expression of the selectable marker is also under the control of the endogenous trapped gene's control elements. The advantage of using βgeo is that cell clones containing true gene trap events can be selected directly in the drug, G418. However, only integrations into genes that are expressed in undifferentiated ES cells will be selected. When an independently driven promoter is used for the expression of the selectable marker, all random integrations must be

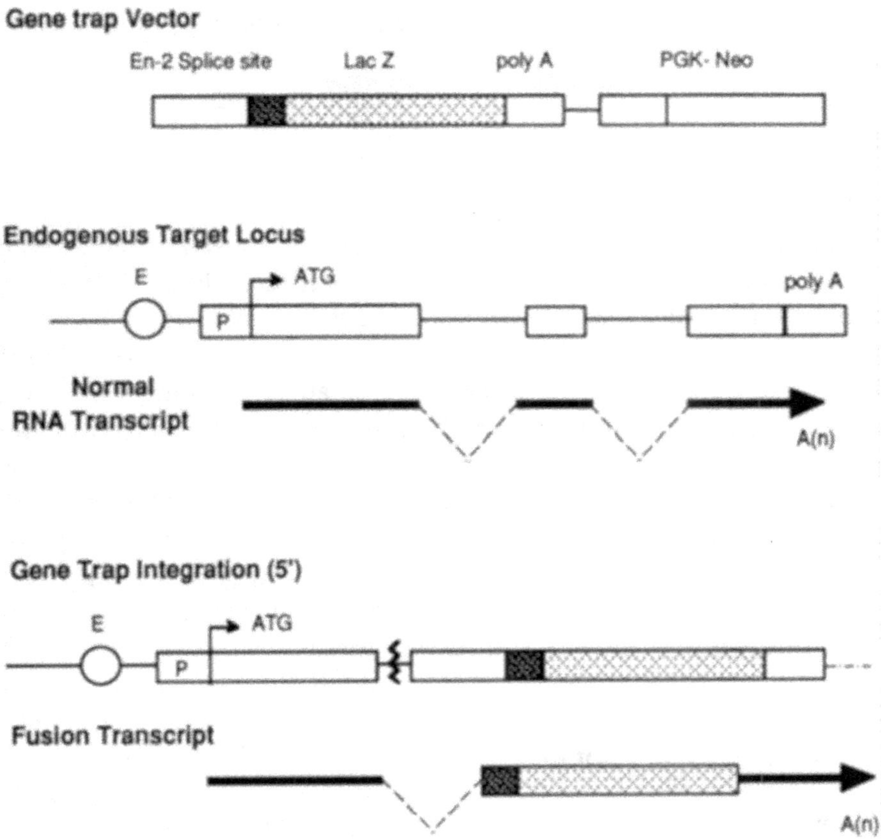

Figure 1.
Diagramatic representation of the activation of the reporter gene (lacZ) after integration into the intron of an active gene. The gene trap vector generates fusion transcripts through the use of the splice acceptor. The fusion transcript is regulated by enhancer (E) and promoter (P) elements of the disrupted target gene and terminated at a polyadenylation signal (pA) supplied by the vector.

screened to identify true gene trap events, the frequency of which can be as low as 2-5%. However, the advantage of this strategy is that it is possible to identify and fully characterise gene trap integrations into a class of genes that are expressed in differentiated, but not undifferentiated ES cells. This strategy was used in the study described here.

(a) Random Gene Trapping.

In a random gene trapping approach the ES cell clones were selected on the basis of reporter gene expression in undifferentiated ES cells and then screened for expression in the developing embryos. Integrations that result in the expression of the reporter gene in the tissue or developmental pathway of interest can then be selected for further study. In a large screen of over 250 gene trap integrations, 30% were constitutively expressed in the developing embryo, 30% were expressed in a developmentally-regulated manner and 30% were not expressed at the stages analysed (Wurst *et al*, 1994). This is a particularly time consuming and expensive *in vivo* screening strategy and clearly an *in vitro* pre-screen that could select for integrations into genes expressed in cell lineages of interest would be very useful.

(b) Directed Gene trapping.

ES cells can differentiate into a large number of different lineages *in vitro* (Doetschman *et al.* 1985). Therefore it should be possible to screen for gene trap integrations into an endogenous trapped gene that is expressed in a cell lineage of interest. We have performed a directed gene trap screen where gene trap integrations were selected *in vitro* on the basis of their response to the addition of the morphogen retinoic acid (RA). Addition of exogenous RA is known to have a profound effect on embryogenesis (reviewed by Eichele, 1989) and is known to activate, *in vitro* (Simeone *et al.* 1990, 1991) and *in vivo*, (Conlon & Rossant, 1992) well-characterised developmentally-regulated genes. Thus one would predict that genes that respond to RA would be expressed in a developmentally regulated manner.

Materials and Methods.

Gene Trap Vector.

The gene trap vector used was based on that described by Gossler *et al*, 1989. Briefly, PT1-ATG contained the En-2 splice acceptor site fused to the *lacZ* reporter gene which included an ATG start site. The bacterial neomycin resistance gene (neoR), driven by the phosphoglycerate kinase (PGK-1) promoter, was used as a selectable marker.

ES cell culture.

R1 ES cells (Nagy *et al.* 1993) were maintained on primary embryonic fibroblasts as described (Wurst & Joyner, 1993). After electroporation and selection in G418 for 8 days, resistant colonies were replica-plated (Hill & Wurst, 1993) and the filters were placed in ES cell medium (without LIF) containing 5% FCS and 10^{-6}M all-trans retinoic acid for 48 hours. Medium was replaced with fresh RA-containing medium 6 hours prior to staining. After staining for B-galactosidase activity, blue colonies were picked from the master plate, ES cell clones were expanded and retested for B-gal activity in the presence and absence of RA and clones were examined microscopically. Clones that were identified as either induced or repressed by RA were analysed quantitatively .

Quantitative B-galactosidase assay.

The time course and extent of induction of B-galactosidase activity was assessed for each cell line using a quantitative assay. ES cells ($2x10^5$) were plated onto 35mm gelatinised tissue culture dish in ES cell medium containing 15% FCS and LIF. After overnight culture, and then every 6 hours subsequently, the medium was changed on all plates such that cells were in the continuous presence of RA for varying time periods (6, 12, 24 or 48 hours). Control samples were in ES cell medium containing

5% FCS throughout the experiment. Cells were harvested by trypsinisation, washed in PBS and resuspended in 100µl 0.25M Tris pH 7.5. After 3 freeze/thaw cycles, samples were centrifuged at 13000rpm for 5 mins and the protein concentrations in the supernatant determined. The B-galactosidase assays were performed using equivalent amounts of protein (75-200µg) in 0.5ml Bgal buffer (60mM Na_2HPO_4, 40mM NaH_2PO_4, 10mM KCL, 1mM $MgCl_2$, 5mM B-mercaptoethanol) containing 0.2mg O-Nitrophenyl B-D-Galactosidase. Incubation was carried out overnight and the optical density at 420nM was determined.

Results.

Isolation of RA-responsive cell lines.

6.5 x 10^7 ES cells were electroporated with the PT-1 ATG vector and 3600 ES cell colonies were replica-plated, induced with RA and stained for B-galactosidase activity. 202 B-galactosidase-positive colonies were picked from the master plate into 24 well plates, expanded and lacZ expression retested in the presence and absence of RA. 9 colonies were identified in which the reporter gene was induced by RA and 11 that were repressed.

Quantitation of RA responsiveness.

Table 1 shows the quantitative induction of B-galactosidase activity in the RA-responsive cell lines at various times after exposure to RA. The results are expressed as a fold induction compared to the control sample (0) that was not treated with RA. The induced lines varied greatly in the level of induction and in the time course over which the induction was observed. For example, the level of reporter gene expression in ES cell lines I.23, I.75 and I.114 was induced two-fold after 48 hours, whereas in line I.210, over the same time period, β-gal activity was induced 16

Cell line	Time of RA exposure (hours) 0	6	12	24	48
I.9	1.0	0.7	0.6	0.6	11
I.23	1.0	0.9	1.2	1.3	2.1
I.75	1.0	1.1	1.1	1.2	1.9
I.114	1.0	1.1	0.8	0.7	2.8
I.134	1.0	1.7	2.1	2.1	20
I.163	1.0	0.8	0.8	1.1	5.0
I.193	1.0	0.8	0.8	0.9	8.6
I.210	1.0	0.6	0.6	1.3	16
I.214	1.0	0.9	0.7	3.2	16
R.24	1.0	0.9	0.9	0.7	0.6
R.68	1.0	0.9	0.9	0.7	0.4
R.77	1.0	0.9	0.9	0.9	0.8
R.108	1.0	0.9	1.0	0.9	0.8
R.121	1.0	0.9	0.9	0.7	0.5
R.140	1.0	1.0	1.0	1.0	0.9
R.170	1.0	0.9	1.0	0.7	0.4
R.179	1.0	1.0	1.1	0.8	0.5
R.194	1.0	1.4	0.9	0.6	0.4
R.213	1.0	0.9	0.8	0.4	0.2
R.124	1.0	1.3	1.2	0.6	0.4

Table 1.
Results of quantitative B-galactosidase assays on RA-responsive genetrap ES cell lines expressed as 'fold induction' compared to the control (0 hours) value.

fold. In lines I.163 and I.193, B-gal activity was induced 5 and 8.6 fold respectively after 48 hours of RA treatment. Two cell lines (I.134 and I.214) showed induction of the reporter gene at an early time point. In I.134, B-gal activity was induced 2 fold after 12 hours and subsequently induced 20 fold between 24 and 48 hours. Similarly, in line I.214, reporter gene expression was induced 3 fold after 24 hours and then 16 fold after 48 hours. In the ES cell lines in which the reporter gene was repressed by RA, the precise extent of repression was difficult to determine because of the unknown stability of the fusion proteins. However, repression was clearly observed after 48 hours in all lines and, in some, by 24 hours.

Embryonic Expression Patterns.

We have made chimeric embryos with some of the RA-responsive cell lines and preliminary results suggest that a high proportion (>70%) of the RA-responsive genes that have been trapped are expressed in a spatially and temporally restricted manner during embryogenesis. Further work is being done to confirm this result.

Discussion.

We describe the development of a gene trap screen in ES cells which pre-selects *in vitro* for integrations into genes that lie downstream of ligand/receptor-mediated signalling pathway. RA was used to test the general applicability of this method. We have successfully isolated 20 ES cell lines which carry gene trap vector integrations into RA responsive genes. The induction protocol was designed to identify genes that were either directly or indirectly responsive to RA. Gene trap lines were treated with RA for 48 hours to identify indirectly responsive genes (ie genes that are regulated as a consequence of RA-induced differentiation). We also added fresh RA-containing medium 6 hours prior to staining to identify genes that were possibly induced at an earlier time point and thus candidates as direct genes. The time courses of induction of the isolated lines suggest that the trapped, endogenous genes are not directly responsive to RA but rather lie further downstream in the RA-induced differentiation pathway. Gene that are directly induced by RA are normally induced within 6 hours and do not require protein synthesis for induction. However, the promoter region of the laminin B1 gene contains a retinoic acid binding element (RARE) but the induction of expression of this gene is not seen until 26 hours after RA treatment (Vasios *et al*, 1991). Obviously, the mode of RA regulation of the trapped gene can only be accurately assessed by a detailed analysis of their transcriptional regulatory regions.

We predict that a large proportion of these RA-responsive genes will be developmentally-regulated genes and our preliminary analysis supports this. We have introduced several of these ES cell lines into mouse embryos and have very specific temporal and spatial lacZ expression patterns.

This approach could be adapted to screen for genes that act downstream of any ligand-mediated pathway such as polypeptide growth factors. The success of this will be dependent on the expression of the appropriate receptors in ES cells but may be complicated by the autocrine expression of the growth factor in ES cell cultures. As ES cells can differentiate into haematopoietic lineages *in vitro* (Doetschman *et al.*, 1985, Wiles & Keller, 19) it may be possible to directly select gene trap integrations into genes that are expressed in haematopoietic lineages using this type of directed approach.

References.

Beddington, R.S.P. & Robertson, E.J. (1989) An assessment of the developmental potential of embryonic stem cells in the midgestation mouse embryo. Development 105: 733-737.

Conlon, R.A. & Rossant, J. (1992) Exogenous retinoic acid rapidly induces anterior ectopic expression of murine *Hox-2* genes *in vivo*. Development 116: 357-368.

Doetschman, T., Eistetter, H., Katz, M., Schmidt, W. & Kemler, R. (1985). The *in vitro* development of blastocyst-derived embryonic stem cell lines: formation of visceral yolk sac, blood islands and myocardium. J. Embryol. exp. Morph. 87:27-45.

Eichele, G. (1989) Retinoids: signalling molecules in vetrebrate pattern formation. Trends Genet. 5: 246-251.

Evans, M.J. & Kaufman, M.H. (1981) Establishment in culture of pluripotential cells from the mouse. Nature 292: 154-156.

Freidrich, G. & Soriano, P. (1991) Promoter traps in embryonic stem cells: a genetic screen to identify and mutate developmental genes in mice. Genes & Dev. 5: 1513.

Gossler, A., Joyner, A., Rossant, J. & Skarnes, W.C. (1989) Mouse embryonic stem cells and reporter constructs to detect developmentally regulated genes. Science 244: 463-465.

Hill & Wurst, W. (1993) Screening for Novel Pattern Formation Genes Using Gene Trap Approaches. Methods in Enzymology 225: 664-681.

Nagy, A., Rossant, J., Nagy, R., Abramow-Newerly, W. & Roder, J.C. (1993). Derivation of completely cell culture-derived mice from early-passage embryonic stem cells. Proc. Natl. Acad. Sci. USA 90: 8424-8428.

Simeone, A., Acampora, D., Arciono, L., Andrews, E., Boncinelli, E. & Mavilio, F. (1990) Sequential activation of HOX2 homeobox genes by retinoic acid in human embryonal carcinoma cells. Nature 346: 763-766.

Simeone, A. Acampora, D., Nigoro, V., Faiella, A., D'Esposito, M., Stornaiuolo, A., Mavilio, F. & Boncinelli, E. (1991) Differential regulation by retinoic acid of the homeobox genes of the four HOX loci in human carcinoma cells. Mech. Dev. 33: 215-228.

Skarnes, W.C. (1993). The identification of new genes: Gene trapping in Transgeneic Mice.. Current Opinion in Biotechnology 4:684-689.

Skarnes, W.C., Auerbach, A.B. & Joyner, A.L. (1992). A gene trap approach in mouse embryonic stem cells: the lacZ reporter is activated by splicing, reflects endogenous gene expression, and is mutagenic in mice. Genes Dev. 6: 903-918.

Vasios, G., Mader, S., Gold, J.D., Leid, M., Lutz, Y., Gaub, M-P., Chambon, P. & Gudas, L.J. (1991) The late retinoic acid induction of laminin B1 gene transcription involves RAR binding to the responsive element. EMBO J. 10: 1149-1158.

Wiles, M.V. & Keller, G. (1991) Multiple haematopoietic lineages develop from embryonic stem (ES) cells in culture. Development 111: 259-267.

Wurst, W. & Joyner, A.L. (1993). Production of targeted embryonic stem cell clones. In: Gene Targeting, A Practical Approach. ed. A.L.Joyner, IRL Press, Oxford.

Wurst, W., Rossant, J., Prideaux, V., Kownacka, M., Joyner, A.L., Hill, D.P., Guillemot, F., Gasca, S., Auerbach, A. & Ang, S-L. (1994). An embryonic expression screen for spatially restricted genes using a gene trap approach in ES cell chimeras. Genetics (in the press).

Molecular Basis of Genetic Disorders of Pigmentation in Humans and Mice

Richard A. Spritz
Departments of Medical Genetics and Pediatrics
University of Wisconsin
Madison, WI 53706
U.S.A.

Disorders of pigmentation were among the first genetic diseases recognized in humans. The distinctive phenotypes of oculocutaneous albinism (OCA) and piebaldism were known to the ancient Greeks and Romans, and the typical clinical features, modes of inheritance, and genetic heterogeneity of these disorders are apparent even in classical descriptions (Lucian, 1905; Pliny, 1942; Gellius, 1952). Similar phenotypes were recognized early in the mouse, and the availability of inbred lines carrying characterized mutations has made the mouse an invaluable tool for studies of genetic disorders of pigmentation. Absent catalytic activity of tyrosinase in the skin of albino animals was one of the first enzymatic deficiencies recognized (Durham, 1904), and as early as 1908 Garrod suggested that albinism might be an inborn error of metabolism (Garrod, 1908).

These genetic disorders of pigmentation have recently begun to yield to modern molecular analyses, and many of the associated genes have now been cloned. The molecular analysis of these genes, both of man and the mouse, have resulted in greatly improved understanding of the biochemical and molecular basis of these genetic disorders of pigmentation, as well as the development of a much more accurate diagnostic nosology than was possible based only on clinical and biochemical grounds. These advances provide opportunities for much more accurate diagnosis, carrier detection, and prenatal diagnosis of these disorders, especially in selected families and in high-risk ethnic groups. In addition, improved understanding of the biochemical and genetic basis of these disorders challenges us to develop novel and specific therapies -- biochemical, pharmacologic, and perhaps eventually DNA-based.

Genetic disorders of pigmentation can be classified into two general categories: disorders of melanocyte function and disorders of melanocyte development. Disorders of melanocyte

NATO ASI Series, Vol. H 94
Gene Technology
Edited by A. R. Zander, W. Ostertag,
B. V. Afanasiev, F. Grosveld

function are associated with generalized hypopigmentation resulting from reduced melanin per melanocyte, whereas disorders of melanocyte development are associated with heterogeneous pigmentation patterns resulting from abnormal distribution of normally pigmented melanocytes. In the following discussion, MIM numbers for each disorder are from Online Mendelian Inheritance in Man (1994).

THE MAMMALIAN PIGMENTARY SYSTEM

Melanocytes, which constitute only a minor population in the skin and eyes of mammals, are the only cells that produce melanin, which accounts for virtually all of the visible pigmentation of those tissues. Skin melanocytes are highly dendritic cells that originate during development as melanoblasts in the neural crest, whereas melanocytes in the retina derive from the optic cup. The melanoblasts subsequently migrate to the epidermal/dermal border of the skin, the hair bulbs in the dermis of the skin, and the iris, choroid, and retina of the eye (reviewed in Spritz and Hearing, 1995). For pigmentation to proceed normally, melanoblasts must thus differentiate appropriately as melanoblasts in the neural crest, the signal for migration must be given at the correct time, and movement of melanoblasts towards their eventual destination must begin. To achieve uniform pigmentation, melanoblasts must not only disperse correctly, but must subsequently receive the appropriate signal to stop migrating, establish themselves, proliferate and differentiate appropriately, and of course function correctly. The signals necessary to achieve all of these events necessarily depend on a number of genes, some expressed by the melanoblasts and/or melanocytes while others are expressed by other cell types. Since the ultimate patterns of pigmentation in the skin and hair depend on other cellular constituents, such as keratinocytes, genes which regulate those other cell types can also indirectly impact on pigmentation. More than 150 different mutations that affect pigmentation have been identified in the mouse, and these map to more than 60 distinct genetic loci (Silvers, 1979; Lyon and Searle, 1989). A number of these genes have now been cloned and will be discussed in the relevant sections that follow.

Within the melanocyte, the pigment biosynthetic machinery is confined to membrane-bound organelles termed melanosomes. Melanin is deposited on the internal fibers of the melanosome, and as melanization proceeds the melanosomes are moved steadily away from the perinuclear region down the dendritic processes of the melanocyte, and are eventually transported to keratinocytes, where they are subsequently degraded.

Melanin biosynthesis initiates from the amino acid tyrosine, and *in vitro* only a single enzyme is absolutely necessary to synthesize melanin: tyrosinase (monophenol monooxygenase; monophenol, L-dopa:oxygen oxidoreductase; EC 1.14.18.1). Tyrosinase is a copper-containing enzyme that catalyzes the first two steps of melanin biosynthesis, the hydroxylation of tyrosine to 3,4-dihydroxyphenylalanine (DOPA) and the subsequent oxidation of DOPA to DOPAquinone. The first reaction is the most critical since its spontaneous rate is negligible, whereas DOPA can readily autooxidize to DOPAquinone even in the absence of tyrosinase. DOPAquinone then continues through the melanin biosynthetic pathway via a series of reactions will take place spontaneously in a test tube, but *in vivo* involve several other enzymes, including DOPAchrome tautomerase and DHICA oxidase. These enzymes have been characterized biochemically, and the corresponding genes have been cloned and are discussed in detail below. Two different basic types of melanin can be produced in mammalian melanocytes, brown/black eumelanin and yellow/red pheomelanin. Relatively little is known about the controls that modulate biosynthesis of eumelanin versus pheomelanin. Melanin was initially thought to be an homogeneous high molecular weight biopolymer consisting exclusively of an indole-5,6-quinone backbone, but physiological melanins are now known to be much more heterogeneous, incorporating incorporate other intermediates in the melanin pathway as well.

DISORDERS OF MELANOCYTE FUNCTION: ALBINISM

Albinism is a heterogeneous group of disorders of melanocyte function, characterized principally by generalized hypopigmentation. In oculocutaneous albinism (OCA) pigment is reduced or absent in the skin, hair, and eyes, whereas in so-called "ocular albinism" (OA) visual involvement is accompanied by relatively normal pigmentation of the skin and hair. In fact, however, all forms of human albinism are oculocutaneous, involving both the eyes and the integument to at least some extent; thus, the distinction between OCA and OA is largely artificial. The role played by melanin in the developing visual system is not known. However, if retinal pigment production is reduced below some critical level, regardless of its genetic cause, there results a stereotypic set of defects of neuronal migration in the visual pathways, with consequent low vision, nystagmus, and strabismus in affected individuals. In general, the severity of the visual defects correlates with the severity of the pigmentation deficit in the eye. OCA has traditionally been classified as "tyrosinase-deficient" or "tyrosinase-positive" on the basis of biochemical assay of hairbulb tyrosinase, corresponding approximately to the current genetic designations, OCA1 and OCA2, although several additional tyrosinase-positive types of OCA have been distinguished on clinical grounds. Both X-linked and autosomal recessive forms of OA have been recognized. However, elucidation of the molecular basis of these disorders has led to improved understanding of their true relationships and to more accurate classification on molecular genetic grounds.

OCA1 (Tyrosinase-Deficient OCA; MIM #203100)

OCA1 is an autosomal recessive disorder resulting from deficient catalytic activity of tyrosinase, the enzyme that catalyzes the first two steps, and at least one subsequent step, in the melanin biosynthetic pathway (reviewed in Spritz and Hearing, 1995). As discussed above, tyrosinase is absolutely necessary for pigment production, although other control points in the pathway also regulate melanogenesis *in vivo*. Freshly epilated hairbulbs from patients with type OCA1 accumulate little or no melanin pigment on incubation in tyrosine or dopa *in vitro*, the traditional assay for this disorder (Kugelman and van Scott, 1961; reviewed in Witkop, 1989); however, this test is highly unreliable and is therefore not useful clinically. In humans at least two different subtypes of OCA1, OCA1A and OCA1B, have been distinguished on clinical grounds. In OCA1A ("tyrosinase-negative OCA"), the classic, most severe form of the disorder, tyrosinase activity is completely absent, and there is no detectable melanin pigment in the skin, hair, or eyes. Visual acuity is greatly decreased, usually to approximately 20/400, and nystagmus, strabismus, and photophobia are usually severe. In OCA1B ("yellow OCA"), initially recognized in the Amish (Nance et al., 1970), tyrosinase activity is greatly reduced, and there is little or no apparent melanin pigmentation at birth, although progressive melanization may occur during childhood and adult life. OCA1A and OCA1B have long been recognized to be allelic, based on the existence of apparent OCA1A/OCA1B compound heterozygotes who exhibit a phenotype approximately intermediate between those of the two homozygous forms (Hu et al., 1980).

Analyses of the human tyrosinase structural gene (*TYR*) in patients with OCA1 have demonstrated that an allelic series of tyrosinase gene mutations accounts for both OCA1A and OCA1B. The gene consists of five exons spanning more than 65 kbp of DNA on the long arm of human chromosome 11 (Giebel et al., 1991*a*). Tyrosinase is translated as a 529-amino acid polypeptide, from which an 18-amino acid hydrophobic leader peptide is cleaved to yield the

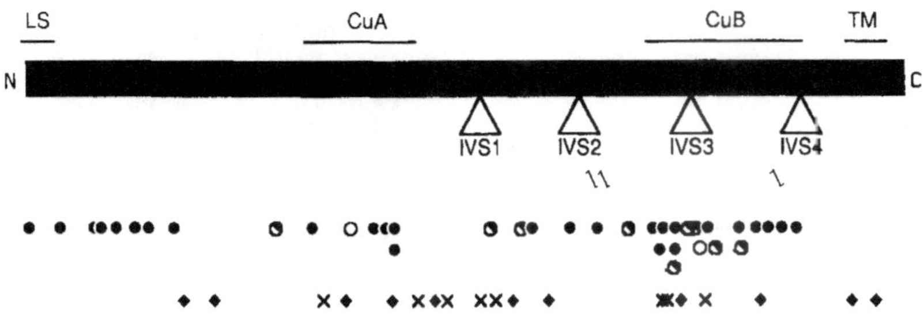

Figure 1. Locations of known *TYR* gene mutations associated with OCA1. The blackened box denotes the 529-amino acid tyrosinase polypeptide. The 18-amino acid amino-terminal leader sequence (LS), the two regions of sequence homology with hemocyanins (CuA and CuB), and the transmembrane domain (TM) are indicated. The positions of the four intervening sequences (IVSs) are indicated. Blackened circles indicate the sites of OCA1A missense substitutions; half blackened circles indicate the sites of OCA1B missense substitutions; unblackened circles indicate the sites of nonpathological amino acids 192 and 402. Blackened diamonds indicate the sites of frameshifts; x's indicate the sites of nonsense mutations; zig-zags indicate the sites of splice consensus mutations. Many of the indicated mutations represent our unpublished data.

mature tyrosinase (Kwon et al., 1987; Wittbjer et al., 1989). An apparent transmembrane domain near the carboxyl terminus of the polypeptide indicates that tyrosinase is membrane bound, most likely on the inner surface of the melanosome. Among Caucasians, two codons, 192 and 402, are commonly polymorphic (Giebel and Spritz, 1990; Tripathi et al., 1991), encoding either of two alternative amino acid residues at each site. As a result, there are four slightly different alternative normal tyrosinase isoforms, which differ significantly in their stabilities and thus also in their net catalytic activities (Tripathi et al., 1991). However, the occurrence of these four tyrosinase isoforms does not obviously correlate with the apparent pigmentation phenotypes among normal Caucasians.

OCA1 was genetically linked to the *TYR* gene by RFLP analysis of large families with OCA (Giebel et al., 1990; Spritz et al., 1990), and subsequently specific *TYR* gene mutations were demonstrated by DNA sequence analyses of *TYR* cDNAs cloned from cultured melanocytes (Tomita et al. 1989) and *TYR* genomic segments amplified from patient DNAs by the PCR (Giebel et al., 1990; Spritz et al., 1990). As shown in Figure 1, at least 61 different pathologic mutations of the *TYR* gene have been identified to date (reviewed in Spritz, 1994*a*;

Spritz and Hearing, 1995). These include missense and nonsense mutations, frameshifts, and mutations that interfere with normal pre-mRNA splicing. Surprisingly, deletions of the human *TYR* gene are quite rare; in analyses of more than 250 patients with OCA1 we have identified only a single gene deletion (unpublished data).

A number of these mutations have been subjected to functional analysis by expression in transfected cells, permitting correlation of the resultant tyrosinase enzymatic activity with the associated clinical phenotypes (reviewed in Spritz and Hearing, 1995). As expected, the nonsense, frameshift, and the missense mutations that completely abolish tyrosinase catalytic activity are associated with OCA1A. However, missense mutations that result in greatly reduced, but not abolished, enzymatic activity are specifically associated with OCA1B. In particular, the V275F, P406L, and R422Q missense substitutions, and also the R402Q polymorphism, result in unstable tyrosinase polypeptides, with temperature-sensitive catalytic activities that are greatly reduced at 37°C compared to 31°C, both *in vivo* and *in vitro*. As illustrated in Fig. 1, the nonsense and frameshift mutations are randomly distributed throughout the tyrosinase coding region. However, most of the missense substitutions associated with OCA1A cluster in two regions of the polypeptide (King et al., 1991; Tripathi et al., 1992): near the amino terminus of the mature tyrosinase polypeptide (codons 21-89) and near the center of the tyrosinase polypeptide (codons 371-448). This latter region corresponds closely to the so-called "Cu B" region of tyrosinase. This is one of two regions of the tyrosinase polypeptide that exhibit amino acid sequence homology to arachnid, crustacean, and molluscan hemocyanins (Lerch, 1988), proteins which, like tyrosinase, also complex with atomic copper. However, involvement of the Cu B region in copper binding by tyrosinase has yet to be demonstrated experimentally, and the only amino acid substitution we have found that involves one of the putative copper-binding histidines (Martinez et al., 1985) (H390D) is associated with a particularly mild OCA phenotype (unpublished data).

In Caucasians, no single mutant *TYR* allele accounts for a significant fraction of the total (reviewed in Spritz, 1994*a*; Spritz and Hearing, 1995); therefore, in the absence of parental consanguinity most patients are compound heterozygotes for different mutant alleles. Patients with OCA1A have two OCA1A alleles, whereas patients with OCA1B may have either two OCA1B alleles or one OCA1B and one OCA1A allele. However, the range of phenotypes associated with OCA1B mutations is very broad, even among patients with similar genotypes, probably because of epistatic phenomena that modulate pigmentation in the presence of even a small amount of tyrosinase activity. In fact, we have found that approximately 25 percent of Caucasian patients diagnosed with "tyrosinase-positive OCA" on the basis of clinical and biochemical criteria have pathologic *TYR* gene mutations that would be expected to only modestly reduce tyrosinase catalytic activity (Tripathi et al., 1992; unpublished data). Furthermore, we recently identified mutations of the *TYR* gene in two patients with apparent "autosomal recessive ocular albinism" (AROA; MIM #203310) (Fukai et al., 1995). Interestingly, both of these patients are compound heterozygotes for OCA1A mutant alleles and the R402Q polymorphic allele, which is associated with moderate thermoinstability of the corresponding tyrosinase polypeptide; thus these patients actually have very mild forms of OCA1B. Similarly, other cases of AROA appear to represent mild forms of OCA2 (see below). Thus, AROA does not appear to represent a distinct genetic entity. These observations underscore the need for molecular diagnosis to accurately distinguish these disorders.

We have studied several examples of OCA1 occurring in two successive generations of a kindred (Giebel et al., 1990, 1991*b*; unpublished data). All of these cases were found to result from pseudodominance, an affected individual having inadvertently married an unrelated carrier. It would be of great interest to determine whether this phenomenon might account for some or all of the reported cases of "autosomal dominant OCA", which has previously been considered to be a distinct entity (MIM #126070).

The lack of predominant mutant *TYR* alleles among Caucasian patients with OCA1 complicates efforts at molecular diagnosis or carrier detection, except in selected families. However, in certain ethnic subgroups one specific mutant allele strongly predominates (reviewed in Spritz, 1994*a*; Spritz and Hearing, 1995). In these populations, therefore, genetic screening procedures can be targeted for the appropriate predominant mutant *TYR* alleles. We have carried out DNA-based carrier detection of OCA1 in a number of instances, and recently carried out DNA-based prenatal diagnosis of OCA1A in an Israeli family (Falik-Borenstein et al., 1995).

OCA2 ("Tyrosinase-Positive OCA"; MIM #203200)

OCA2 is an autosomal recessive disorder with a phenotype similar to, but usually less severe than, OCA1, although the two disorders display considerable clinical overlap. Infants with OCA2 may have little or no apparent melanin pigment at birth. However, during early to mid childhood, most patients with OCA2 acquire modest amounts of pigment, often predominantly yellow-red pheomelanins. The visual deficits are also usually less severe in OCA2 than in OCA1; the visual acuity of patients with OCA2 is typically in the 20/60 to 20/200 range, with only moderate nystagmus. Furthermore, in OCA2 visual acuity and nystagmus may improve during childhood and even during adolescence, with ultimate visual acuity reaching approximately 20/40 to 20/70. Rarely, nystagmus may even disappear entirely and visual acuity may become normal (Summers et al., 1991).

As described above, molecular analysis has shown that a significant fraction of Caucasian patients thought to have "tyrosinase-positive" OCA actually have clinically mild forms of OCA1B. However, the existence of a true OCA2, genetically distinct from OCA1, has long been known on the basis of several instances of matings between parents, one with OCA1 and the other with OCA2, that produced normally pigmented children (Witkop et al., 1989). As will be discussed below, OCA2 is only one of perhaps several different types of "tyrosinase-positive OCA", although it is likely the most frequent type, especially in African and African-American populations.

Recently, we identified the gene responsible for OCA2 in humans and we have characterized mutations of this gene in a number of patients with OCA2 (Rinchik et al., 1993; Lee et al., 1994*a*,*b*) . The human type II OCA gene, *P* (Rinchik et al., 1993) proved to correspond to the pink-eyed dilution (*p*) locus of mouse (Gardner et al., 1992; Rinchik et al. 1993), which is associated with a mutant phenotype very similar to that of human OCA2 (reviewed in Silvers, 1979; Lyon and Searle, 1989). The human *P* gene is located in chromosome segment 15q11-q13, in agreement with mapping of the human OCA2 locus by genetic linkage (Ramsay et al., 1992). Furthermore, the *P* gene is located adjacent to the chromosomal region commonly deleted in patients with the Prader-Willi (PWS; MIM #176270) and Angelman (AS; MIM #105830) syndromes (reviewed in Nicholls, 1993), both of which are often associated with significant hypopigmentation (Butler, 1989). Approximately one percent of patients with PWS and AS also have OCA2, similar to the expected prevalence of OCA2 heterozygotes, suggesting that OCA2 in association with PWS and AS might result from hemizygosity (or, rarely, uniparental isodisomy) for inherited *P* gene mutations on the nondeleted chromosome, and that isolated OCA2 might result from point mutations of the *P* gene.

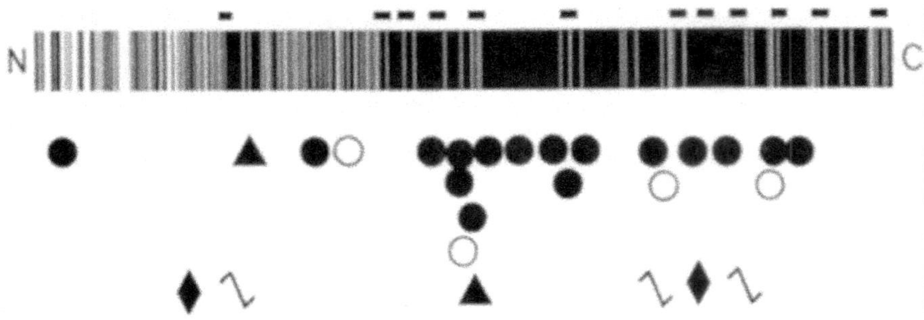

Figure 2. Locations of known *P* gene mutations associated OCA2. The box denotes the 838-amino acid P polypeptide. Black vertical bars indicate segments of amino acid identity between the human (Rinchik *et al.*, 1993) and mouse (Gardner *et al.*, 1992) P polypeptides. Horizontal bars overline indicate the twelve putative transmembrane domains. Blackened triangles indicate the sites of in-frame deletions; other symbols are as described in the legend to Figure 1. Many of the indicated mutations represent our unpublished data.

To test these hypotheses, we first studied one such patient with both PWS and OCA2, in whom we found a large *de novo* deletion of proximal 15q (including *P*) on the paternal chromosome, accounting for PWS, and a hemizygous maternally inherited mutant *P* allele containing a small partial gene deletion, accounting for OCA (Rinchik et al., 1993). Subsequently, to identify point mutations in patients with isolated OCA2, we characterized the *P* genomic locus, which consists of 25 exons spanning 250-600 kbp of DNA on proximal 15q (Lee et al., 1995). Using PCR primers based on these data, we have identified 25 additional different abnormalities of the *P* gene in other patients with OCA2, including two more patients with both OCA2 and PWS (Lee et al., 1994*a*,*b*). As shown in Fig. 2, these include large intragenic deletions, small in-frame deletions, frameshifts, splice junction mutations, and missense substitutions. In Caucasian and African-American patients none of these mutations are strongly predominant, although one substitution, V443I, is relatively frequent in both groups (Lee et al., 1994*a*,*b*; unpublished data). However, we found that one mutation, a 2.7-kb deletion encompassing only *P* exon 7 (Durham-Pierre et al., 1994; Lee et al., 1994b), is strongly predominant in Tanzanian patients with OCA2 (unpublished data). OCA2 is very frequent in this group, occurring with an incidence of approximately 1 per 1300 (Luande et al., 1985), and OCA2-associated skin cancers constitute a major public health problem in Tanzania. It thus seems likely that the repertoire of OCA2 mutant alleles in east Africa differs from that in west Africa, the population from which most African-Americans are derived. Interestingly, we have found *P* gene mutations in all twenty unrelated African-American (Lee et al., 1994*a*) and African (unpublished data) patients with "tyrosinase-positive OCA" we have studied, consistent with complete linkage of OCA2 to 15q markers in South African patients (Ramsay et al., 1992). However, we have found *P* gene mutations in about one-third of Caucasian patients (unpublished data). As discussed above, about 25% of these patients have mutations in the *TYR* gene, and, as will be discussed below, at least some of these patients may have mutations of a presumed third OCA locus. The reason for this unexpected difference between African-American and African versus Caucasian patients with "tyrosinase-positive OCA" is not known, but may be the result of historically different selection pressures on these two populations.

The range of clinical phenotypes associated with pathologic mutations of the *P* gene in humans is very broad. Individuals homozygous for mutations that would be expected to totally abolish P function exhibit a phenotype almost as severe as OCA1A (Lee et al. 1994*b*; unpublished data). At the other end of the range, mutations of *P* that would be expected to only partially reduce function appear to account for some cases of AROA (MIM #203310) (Lee et al., 1994*a,b*). AROA thus is heterogeneous in nature, in different patients from mild mutations of the *TYR*, *P*, and perhaps other genes involved in pigmentation.

Although the *P* gene product has not yet been characterized experimentally, it is expected to play an important role in eumelanotic pigmentation. In mouse, *P* mRNA is only expressed *in vivo* in melanocytes that synthesize brown-black eumelanin; melanocytes that specifically synthesize pheomelanin do not express *P* (Rinchik et al., 1993). The predicted human P polypeptide includes a total 12 transmembrane domains, in an arrangement characteristic of proteins that transport small molecules such as amino acids. This suggests that the P protein may be an integral component of the melanosomal membrane, possibly involved in the transport of tyrosine, the immediate precursor to melanin biosynthesis. This would be consistent with the observation that pigmentation of melanocytes from humans with OCA2 (reviewed in Witkop et al., 1989) and from homozygous *p* mutant mice (reviewed in Silvers, 1979) is greatly increased by incubation with excess exogenous tyrosine. More important, this suggests that treatment with exogenous tyrosine or other agents may offer an approach to pharmacologic therapy for patients with OCA2.

OCA3 (Brown OCA; MIM #203290)

Originally described in patients from Nigeria, "brown OCA" is associated with moderate hypopigmentation of the skin, hair, and eyes, and modest ocular dysfunction. Brown OCA has long been considered a possible human homologue of brown (*b*) mutant mice, which exhibit brown coat color (Silvers, 1979; Lyon and Searle, 1989). Patients with so-called "brown OCA" are tyrosinase-positive, and are not clinically distinguishable from patients with milder forms of OCA2. Thus, the issue of allelism between human "brown OCA" and OCA2 remains open. Recently, however, some evidence that abnormalities of the human *b* locus (*TYRP*; MIM #115501) might be associated with hypopigmentation in some patients has begun to emerge.

A cDNA for mouse *b* was the first member of the tyrosinase gene family to be (inadvertently) cloned (Shibahara et al., 1986), and was quickly shown represent the *b* locus, rather than tyrosinase itself (Jackson, 1988; Bennett et al., 1990). The amino acid sequence of the *b* gene product is very similar to that of tyrosinase, and hence was named "tyrosinase-related protein" (TRP-1); it corresponds to a melanosomal protein previously termed "gp75 (Vijayasaradhi et al., 1990). In the mouse, *b* locus mutations result in the production of brown rather than black eumelanin, and recent biochemical studies indicate that TRP-1 corresponds to 5,6-dihydroxyindole-2-carboxylic acid (DHICA) oxidase (Jiminez-Cervantes et al., 1994; Winder et al., 1994), an enzyme catalyzing a distal step in the eumelanin biosynthetic pathway.

The homologous human cDNA has been cloned (Cohen et al., 1990) and the gene (*TYRP*) mapped to chromosome segment 9p22-p23 (Chintamaneni et al., 1991; Murty et al., 1992), and Wagstaff (1993) demonstrated that *TYRP* is deleted in two patients with the 9p- syndrome associated with moderate hypopigmentation. Furthermore, Boissy and coworkers (1993) found virtually no immunologically detectable TRP-1 protein or mRNA in one patient with "brown OCA", and recently (Wildenberg et al., 1994) identified a frameshift mutation of the *TYRP* gene in this patient, who thus has "OCA3". However, the hypopigmentation phenotype of this patient is rather less severe than that of most patients with "tyrosinase-positive OCA", and it remains to be seen what fraction of such patients have mutations of *TYRP*.

X-Linked Recessive Ocular Albinism, Nettleship-Falls Type (OA1; MIM #300500)

Ocular albinism of the Nettleship-Falls type (OA1) is an X-linked recessive disorder clinically similar to AROA. In affected males mild cutaneous hypopigmentation is accompanied by hypopigmentation of the iris and retina and foveal hypoplasia, resulting in reduced visual acuity, photophobia, nystagmus, and strabismus. In some, but not all affected individuals, electron microscopy of skin demonstrates melanin macroglobules, often termed "macromelanosomes". Carrier females may exhibit a heterogeneous distribution of retinal pigmentation, sometimes referred to as a "tigroid" or "mud-spattered" pattern, which is thought to be indicative of X-inactivation.

Efforts at positional cloning of the *OA1* gene are far advanced. The gene was assigned to chromosome segment Xp22.3 by genetic linkage, with localization most likely between markers DXS143 and DXS85 (Bergen et al., 1990, 1991; Schnur et al., 1991; Charles et al., 1992). Sunohara *et al.* (1986) described a patient with three apparent X-linked recessive disorders; OA1, X-linked ichthyosis, and Kallmann syndrome, indicating that these three genes may be in close physical proximity and be deleted in this patient (Andria et al., 1987). Furthermore, the colocalization of both OA1 and "X-linked ocular albinism and deafness" (OASD; MIM #300650) to Xp22.3 suggests that OASD may also result from a microdeletion that includes the *OA1* gene (Winship et al., 1993). By detailed mapping of the X chromosomal deletions in the patient reported by Sunohara et al. (1986) and a second, similar patient, Wapennar *et al.* (1993) localized the *OA1* locus to a 200-kbp interval between markers 210B5-R and A187H6-41A. They assembled a 2.6-mbp YAC contig spanning this region, part of a larger contig spanning chromosome segment Xp22 (Schaefer et al., 1993). It seems likely that these YACs will be useful reagents for the eventual isolation of the *OA1* gene.

Interestingly, the *OA1* locus is located within 1-2 Mb of the gene for "microphthalmia with linear skin defects" (MLS; MIM #309801) (Wapenaar et al., 1993). Although there is not a clear homologue to human OA1 in the mouse, synteny between the human and mouse X-chromosomal maps suggests that the mouse "lined" mutation (*Li*) might be a chromosomal deletion encompassing MLS and possibly also OA1 (Lyon and Searle, 1989).

Unknown OCA Genes

Several additional human autosomal recessive genetic disorders, including Chediak-Higashi syndrome, Hermansky-Pudlak syndrome, and Cross syndrome, to name but three, are characterized by complex phenotypes that include moderate to severe manifestations of OCA. The genes responsible for these disorders have not yet been identified. Even more intriguing, we have recently obtained evidence for the existence of at least one additional human OCA locus. As discussed above, many patients with "tyrosinase-positive OCA" genes do not have identifiable mutations of either the *TYR* or *P* genes. Recently, we studied one of these large families by genetic linkage analysis, and excluded abnormalities of *TYR*, *P*, and also *TYRP* (unpublished data). This indicates that tyrosinase-positive OCA in at least this kindred results from abnormalities of a different, as-yet unknown, locus. Identification and analysis of this and perhaps other human OCA genes will thus constitute major challenges as our understanding of these most ancient human genetic disorders continues to progress.

DISORDERS OF MELANOCYTE DEVELOPMENT: PIEBALDISM AND WAARDENBURG SYNDROME

Piebaldism and Waardenburg syndromes are distinct inherited disorders characterized in part by striking areas of congenital depigmentation. These disorders contrast with albinism, in which relatively homogeneous hypopigmentation results from defects of melanocyte function, and with vitiligo, in which acquired areas of depigmentation result from reduced melanocyte survival. In piebaldism the abnormal distribution of pigment results from abnormal distribution of melanocytes during development, most likely due to defective proliferation of melanoblasts prior to migration to the dermis during embryogenesis. In piebaldism the pigmentary anomalies constitute the principal clinical manifestation, whereas in Waardenburg syndrome deafness and dystopia canthorum (lateral displacement of the eyes) are also of major clinical importance. The elucidation of the molecular defects underlying piebaldism and Waardenburg syndrome provide additional examples of the use of human and mouse genetics in combination to dissect the molecular basis of genetic diseases.

Piebaldism (MIM #172800)

Piebaldism is an autosomal dominant disorder of melanocyte development characterized clinically by congenital patches of white skin (leukoderma) and white hair (poliosis), principally located on the scalp, forehead, ventral chest and abdomen, and extremities (Keeler, 1934; Froggat, 1951; Cooke, 1952). In contrast to vitiligo, with which it is frequently confused, in piebaldism the depigmented patches are present from birth and are generally static in shape and distribution, although limited filling in sometimes occurs, especially in milder cases. Hirschprung disease is occasionally also present, but most affected individuals display only the pigmentary anomaly. Piebaldism is relatively rare, although its actual frequency is not known. Because of its distinctive phenotype, piebaldism, sometimes incorrectly called "partial albinism", was one of the first autosomal dominant genetic disorders recognized (Morgan, 1786; Lucian, 1905), and was one of the first genetic disorders for which a pedigree was presented (reviewed in Froggat, 1951). Histologically, the depigmented patches lack most or all melanocytes (Breathnach *et al.*, 1965; Jimbow *et al*, 1975), although areas of increased pigmentation may occur at the boundaries of, or even within, the regions of hypopigmentation. Melanocytes in the skin and hairbulbs derive embryologically from the neural crest, and piebaldism has thus been considered to be a lineage-specific disorder of neural crest development, most likely involving defective melanoblast proliferation, migration, or survival (Murphy *et al.*, 1992; Steel *et al.*, 1992).

The critical clues to the molecular basis of human piebaldism came from studies of a similar disorder of mice, "dominant white spotting" (*W*) (reviewed in Spritz, 1993, 1994*b*). Mouse dominant white spotting, which is associated with defects of pigmentation, hematopoiesis, and germ-cell development (reviewed in Silvers, 1979; Lyon and Searle, 1989), was found to result from deletions or point mutations of the c-*kit* protooncogene (Chabot *et al.*, 1988; Geissler *et al.*, 1988; Reith *et al.*, 1990; Tan *et al.*, 1990; reviewed in Morrison-Graham and Takahashi, 1993). C-*kit*, originally identified in the genome of the HZ4-feline sarcoma virus (Besmer *et al.*, 1986), encodes the cell-surface receptor for an embryonic growth factor, variously called “steel factor” (SLF), “mast cell growth factor", "stem cell factor", "KL" (kit ligand) in the literature (Copeland *et al.*, 1990; Flanagan and Leder, 1990; Huang *et al.*, 1990; Zsebo *et al.*, 1990; Murphy *et al.*, 1992; reviewed in Morrison-Graham and Takahashi, 1993). The KIT receptor is expressed by melanocytes, whereas SLF is expressed by other, as yet undefined, cell types. SLF is the product of the steel (*Sl*) locus of mice, and both *Sl* and *W* mutant mice display similar piebald-like phenotypes. However, *Sl* mutant melanocytes develop normally when transplanted to normal skin (since normal ligand is available) whereas *W* mutant melanocytes (with an inherently

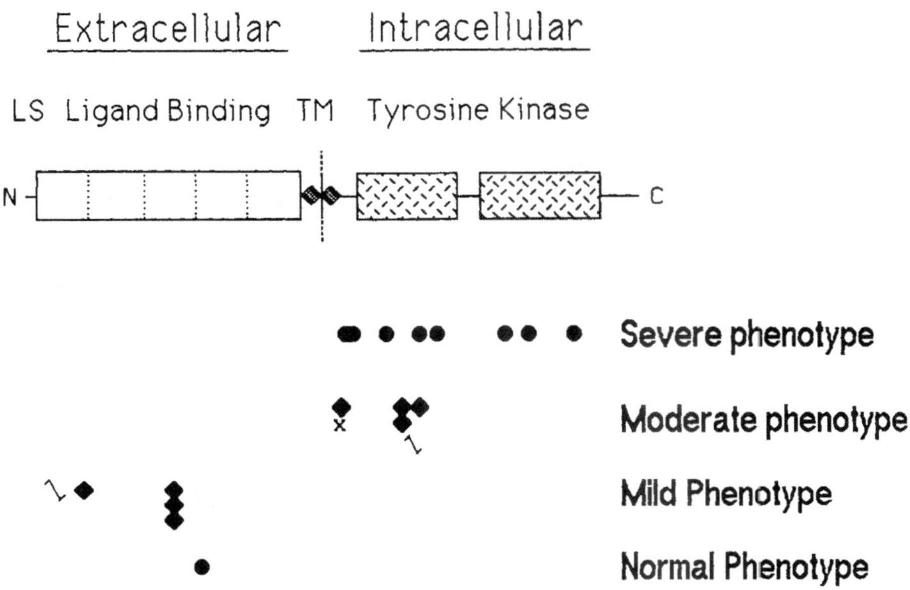

Figure 3. Locations of known *KIT* gene mutations associated with piebaldism. The amino-terminal extracellular leader sequence (LS) and pentarepetitive ligand-binding domain, the transmembrane domain (TM), and the intracellular bipartite tyrosine kinase domain are indicated. Symbols are as described in the legend to Figure 1.

defective receptor) do not. The human piebaldism locus was mapped to chromosome segment 4q12 on the basis of several patients with piebaldism and *de novo* chromosomal translocations or deletions involving this region (Funderburk and Crandall, 1974; Lacassie *et al.*, 1977; Hoo *et al.*, 1986; Yamamoto *et al.*, 1989). Subsequently, the human *KIT* gene was mapped to approximately the same chromosomal location (Yarden *et al.*, 1987; d'Auriol *et al.*, 1988), suggesting that human piebaldism might, like mouse "dominant white spotting", also result from abnormalities of the *KIT* gene.

The KIT protein is a member of the tyrosine kinase family of transmembrane receptors. As illustrated in Figure 3, the KIT polypeptide consists of an amino-terminal extracellular ligand-binding receptor domain composed of five immunoglobulin-type repeats, a short transmembrane domain, and an intracellular domain consisting of a bipartite tyrosine kinase domain followed by a carboxyl-terminal tail. The stoichiometry of interaction between SLF and the extracellular domain of the KIT receptor is not certain, but is thought to be monovalent (Lev *et al*, 1992). On binding SLF, the KIT receptor dimerizes within the cell membrane (Lev *et al.*, 1992), activating its intracellular tyrosine kinase. This, in turn, results in autophosphorylation of specific tyrosine residues within the KIT kinase domain, enhancing the binding of various proteins, including phosphatidylinositol 3' kinase and phospholipase Cγ1, which act as downstream mediators of the mitogenic signal in the KIT-dependent pathway of signal transduction (reviewed in Morrison-Graham and Takahashi 1993). In the mouse, KIT function is required both immediately prior to melanoblast migration and also postnatally (Nishikawa *et al.*, 1991), and we have shown that KIT function is required for proliferation, although not survival, of human melanocytes in culture (Spritz et al., 1994*a*). It seems likely, therefore, white spotting in both human piebaldism and mouse *W* results deficient proliferation of melanoblasts prior to migration during development.

The human *KIT* gene consists of 21 exons spanning more than 70 kb at chromosome segment 4q12 (Giebel *et al.*, 1992; Vandenbark *et al.*, 1992), part of a cluster of genes encoding type III receptor tyrosine kinases, organized *PDGFRA-KIT-KDR* (Spritz *et al.*, 1994*b*). The first direct evidence for the involvement of *KIT* mutations in human piebaldism came from Giebel and Spritz (1991), who studied an extended kindred with piebaldism and identified a missense mutation, G664R, within the highly conserved intracellular tyrosine kinase domain, and demonstrated linkage with a lod score in excess of 6 with no recombinants. Further support came from analyses of two additional patients, one with a cytogenetic deletion of 4q12-q21.1 (Yamamoto *et al.*, 1989), in both of whom *KIT* and *PDGFRA* were found to be deleted (Fleischman *et al.*, 1991; Spritz *et al.*, 1992). As shown in Figure 3, a total of 18 different point mutations of the *KIT* gene have now been identified in different families with piebaldism. Interestingly, as also shown in Figure 3, these mutations can be classified into four general groups, each group tending to be associated with piebald phenotypes of differing severity. Thus, a hierarchical paradigm of pathologic human *KIT* mutations appears to account for a graded series of dominant phenotypes in human piebaldism.

The first group of *KIT* gene mutations consists of missense substitutions, all but one of which have been located within the highly conserved tyrosine kinase domain and almost all of which involve amino acid residues that have been particularly conserved. Several of these human *KIT* substitutions correspond closely to the positions of similar mutations in various strains of *W* mutant mice (reviewed in Morrison-Graham and Takahashi, 1993), underscoring the importance of these sites to function of the KIT receptor. In fact, the human E583K mutation corresponds precisely to the mouse *W*[37] mutation (Fleischman, 1992). All of the *KIT* missense substitutions in the kinase domain are associated with relatively severe piebald phenotypes. This is a consequence of the fact that the KIT kinase is only activated on dimerization of the receptor, and KIT receptor heterodimers consisting of one normal KIT polypeptide and one abnormal KIT polypeptide are inactive. Thus, patients heterozygous for these "dominant-negative" *KIT* missense mutations have only one-fourth of the normal amount of KIT receptor dimer, and accordingly they exhibit relatively severe piebald phenotypes.

In contrast, the second group of *KIT* mutations consists of proximally located nonsense, frameshift, and splice junction mutations that completely eliminate the production of KIT protein by the mutant gene. Patients heterozygous for these "loss of function" mutant alleles thus express half of the normal amount of KIT receptor, resulting in haploinsufficiency for KIT-dependent signal transduction. These *KIT* mutations are thus associated with relatively mild piebald phenotypes, in which the depigmented patches are usually small and poliosis is often absent, although affected individuals frequently experience early graying of the hair. In fact, some members of families with *KIT* mutations of this type have been so mildly affected that the clinical diagnosis was only made subsequent to DNA diagnosis. Interestingly, we detected one of these proximal frameshifts, codons 250-251 ΔCAGT, in three different, unrelated families with mild piebaldism (Spritz *et al.*, 1993), suggesting that this recurrent *KIT* gene mutation may account for a significant fraction of human piebaldism, especially among clinically milder cases.

The third group of mutations consist of frameshifts, nonsense, and splice junction mutations located distally, within the intracellular tyrosinase kinase domain. These mutations tend to be associated with a rather variable phenotype, ranging from extremely mild to quite severe, even among affected members of an individual family. All of these mutations would result in premature termination of translation, truncating the nascent KIT polypeptide distally, within the intracellular tyrosine kinase domain. Clearly, these mutations would abolish expression of normal KIT polypeptide from this allele. However, the truncated KIT receptors apparently can still bind SLF and even form dimers, dominant negatively inhibiting function of the normal KIT polypeptide (Lev *et al.*, 1992). It is likely, however, that both the truncated KIT polypeptides and the incompletely translated *KIT* mRNA are relatively unstable.

Therefore, these mutations probably reduce KIT function to an amount between one-fourth and one-half of normal, accounting for the intermediate and highly variable piebald phenotype.

The fourth group of *KIT* mutations consists of only a single example, a missense substitution located within the extracellular ligand-binding domain of the KIT receptor. This substitution is not associated with any abnormality, and thus appears to constitute a normal polymorphic variant with no pathologic consequences (Ezoe et al., 1995). As noted above, this region of the KIT polypeptide consists principally of five immunoglobulin-like repeat domains, and it is therefore possible that this region may be at least in part functionally redundant. Thus, many amino acid substitutions in this portion of the KIT polypeptide may have little or no effect on ligand binding, and thus may have little or no phenotypic effect. However, the extracellular portion of the KIT polypeptide is also thought to contain the segments that mediate receptor dimerization, and it seems likely that pathologic amino acid substitutions of residues involved in this process may be encountered in the future.

Overall, we have identified pathologic point mutations or large deletions of the *KIT* in approximately 75% of the patients with piebaldism studied to date (Ezoe et al., 1995). Of the remaining patients, none have abnormalities of the *MGF* gene (the human *Steel* gene homologue), suggesting that, in contrast with mice, mutations of this gene may not result in the piebald phenotype in humans (Ezoe et al., 1995). Genetic linkage analyses in two of these families showed complete linkage of the piebald trait to *KIT* intragenic markers (Ezoe et al., 1995), and some of these patients likely have occult *KIT* point mutations or gene deletions not detected by PCR-based SSCP/heteroduplex screening. However, it is also possible that some might have abnormalities in the adjacent *PDGFRA* and *KDR* genes. The mouse "patch" (*Ph*) mutation, which results in a dominant white spotting phenotype similar to that of *W*, comprises a deletion that includes the *pdgfra* gene but not c-*kit* (Smith *et al.*, 1991; Stephenson *et al.*, 1991). Although white spotting in *Ph* mice might result either from deletion of the *pdgfra* gene itself or from inhibitory position effects of the large chromosomal deletion on expression of the nearby c-*kit* gene, it thus seems possible that some cases of human piebaldism might result from mutations in *PDGFRA*.

Waardenburg Syndrome (MIM #193500)

Waardenburg syndrome, first described in 1951 by a Dutch ophthalmologist (Waardenburg, 1951), is an autosomal dominant disorder characterized clinically by piebald-like pigmentary anomalies of the skin and hair, pigmentary abnormalities of the iris (heterochromia irides), lateral displacement of the inner canthi of the eyes (dystopia canthorum), and sensorineural deafness (Waardenburg, 1951; DiGeorge *et al.*, 1960). Waardenburg syndrome occurs with an overall frequency of 1 to 2 per hundred thousand, and accounts for at least 0.5 percent of cases of congenital deafness. All of the abnormalities in Waardenburg syndrome involve the neural crest, and both Hirschsprung disease (Currie *et al.*, 1986; Ariturk *et al.*, 1992) and neural tube defects (Bergleiter and Harris, 1992; Carezani-Gavin *et al.*, 1992; Chatkupt *et al.*, 1993; Kromberg and Krause, 1993) occur at increased frequencies among affected individuals. Waardenburg syndrome has thus been considered a more general disorder of neural crest development than piebaldism. Three subtypes of Waardenburg syndrome have been distinguished on clinical grounds; Waardenburg syndrome type I (WS1) is the classic form, Waardenburg syndrome type II (WS2; MIM #193510) lacks dystopia canthorum, and Waardenburg syndrome type III (WS3; Klein-Waardenburg syndrome; MIM #148820) is associated with limb abnormalities as well as dystopia canthorum.

Elucidation of the molecular basis of human WS1 was facilitated by studies of a similar disorder of mouse, "Splotch" (*Sp*) (reviewed in Pierpont and Erickson, 1993; Spritz, 1993). Splotch mutations typically result in piebald-like patches of unpigmented skin and hair (Epstein *et al.*, 1991, 1993; Goulding *et al.*, 1991; Moase and Trasler, 1992; Goulding *et al.*, 1993; Vogan *et al.*, 1993), probably due to delayed migration of melanoblasts or to a reduction in their number so that melanoblasts fail to reach the affected regions before hair follicles develop. The murine *Splotch* locus encodes a developmental gene known as *Pax-3*, one of a family of so-called "paired box" genes involved in embryological development (Burri *et al.*, 1989). Although the exact function of the *Pax-3* gene product is not yet known, it is thought to be a transcription factor critical for activating melanoblasts to begin migration from the neural crest. The human PAX3 polypeptide contains four principal structural motifs: the "paired-box" domain, a homeobox domain, a conserved octapeptide, and a serine-threonine-proline-rich carboxyl segment (Figure 4) The 128-amino acid paired-box domain, from which the name of the gene is derived, and the 60-amino acid homeobox domain are both highly conserved amino acid sequence motifs present in a number of mammalian and *Drosophila* genes involved in controlling segmentation. These motifs form a helix-turn-helix structure and are both thought to mediate DNA binding.

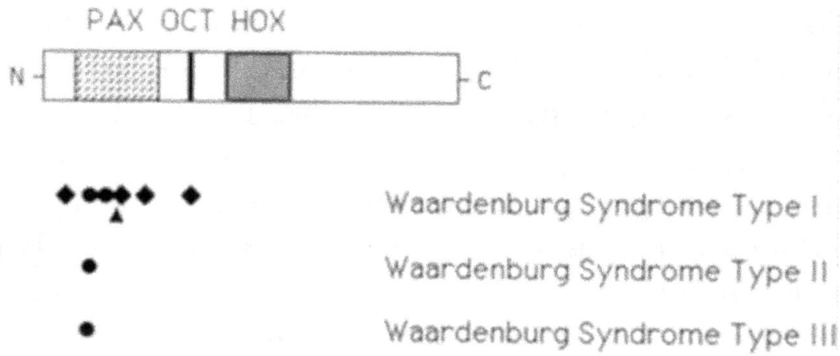

Figure 4. Locations of known *PAX3* gene mutations associated with Waardenburg syndrome. The paired-box domain (PAX), conserved octapeptide (OCT), and the homeobox domain (HOX) are indicated. Symbols are as described in the legend to Figure 1.

The murine *Splotch* locus was mapped to chromosome 1, in a region that is homologous to part of human chromosome 2q, and the mouse Sp^{2H} mutation was then found to consist of a deletion in the paired box domain of *Pax-3* (Epstein *et al.*, 1991). The human WS1 gene was first localized to the distal long arm of chromosome 2 on the basis of a patient with an chromosomal inversion of 2q35-q37.3 (Ishikiriyama *et al.*, 1989), and subsequent genetic linkage analyses demonstrated linkage between WS1 and markers in this region (Foy *et al.*, 1990; Asher *et al.*, 1991), suggesting that human WS1 and mouse Splotch might be homologous. Although only a portion of the human *PAX3* gene had been characterized, Tassabehji and coworkers (1992) and Baldwin and coworkers (1992) quickly identified point mutations in several patients with WS1, and demonstrated that these mutations were genetically linked to the WS1 phenotype in these families.

At least seven different mutations of the *PAX3* gene have now been reported in patients with WS1, including an in-frame deletion within the paired-box domain (Tassabehji *et al.*, 1992), three frameshifts (Morell *et al.*, 1992, 1993; Tassabehji *et al.*, 1993), and two missense substitutions, both located within the paired box domain (Baldwin *et al.*, 1992; Hoth *et al.*, 1993). Furthermore, Hoth *et al.* (1993) identified an additional missense mutation, N47H, in a family with so-called WS3, demonstrating that WS1 and WS3 are allelic, resulting from different abnormalities of *PAX3*. By genetic linkage analysis of 41 families, Farrer and coworkers (1992) determined that abnormalities on distal chromosome 2, presumably in the

PAX3 gene, account for about half of cases of WS1. This implies the existence of at least one additional gene for Waardenburg syndrome located elsewhere in the genome. In addition, they found no evidence of linkage between chromosome 2 markers and WS2, indicating that this disorder does not result from mutations of *PAX3*. However, Tassabehji *et al.* (1993) reported a *PAX3* missense substitution in a family that they considered to have WS2, although clinical photographs of this family are suggestive of mild dystopia canthorum and thus WS1.

Two groups reported refined map data for the human *PAX3* gene. Tsukamoto and coworkers (1992) showed that the distal chromosome 2q inversion associated with Waardenburg syndrome type I, described above, interrupted exons 2 and 3 of the *PAX3* gene, and they suggested that the *PAX3* gene is located within chromosome segment 2q35. Similarly, Lu-Kuo *et al.* (1993) showed that the *PAX3* gene is deleted in a patient with Waardenburg syndrome and a distal chromosome 2 deletion (Kirkpatrick *et al.*, 1992), and they mapped the PAX3 gene to 2q36.1-q36.2. These two map assignments are actually quite close, probably within the limits of cytogenetic resolution.

Farrer and coworkers (1992) found no evidence of genetic linkage between chromosome 2 markers and WS2, indicating that WS2 does not usually result from mutations of *PAX3*. Recently, Hughes and coworkers (1994) reported localization of the gene for WS2 to chromosome segment 3p12-p14.1, in close proximity to *MITF*, the human homologue to the mouse *microphthalmia* gene (Tachibana et al., 1994). The phenotype of *mi* mutant mice includes both pigmentary abnormalities and hearing loss, as well as microphthalmia, various abnormalities of neural crest development, deficient mast cells, and osteopetrosis (Silvers, 1979; Lyon and Searle, 1989), suggesting that *MITF* is a very good candidate for the human WS2 gene. The *MITF* gene product is a novel member of the basic-helix-loop-helix-leucine zipper of transcription factors (Hodgkinson et al., 1993; Tachibana et al., 1994)--proteins that have a variety of roles in regulating gene expression, cellular proliferation and differentiation--consistent with the pleiotropic phenotypic effects seen in *mi* mutant mice and humans with WS2. Thus, it seems likely that at least some cases of human WS2 result from mutations of *MITF*.

In a surprising turn of events, Barr and coworkers (1993) showed that the *PAX3* gene is rearranged in the chromosome 2;13 translocation consistently observed in cases of alveolar rhabdomyosarcoma (MIM #268220), an aggressive solid tumor of childhood. In three such tumors the chromosome 2 breakpoints are located within the fourth intervening sequence of the *PAX3* gene, juxtaposing the upstream PAX3 DNA binding motifs, including the paired box domain, conserved octapeptide, and homeobox domain, to a specific gene located on chromosome 13q14. As the result, an abundant novel mRNA is transcribed, most likely encoding a novel fusion protein that interferes with the normal controls of striated muscle cell proliferation. Thus, the *PAX3* gene appears to lead a double life: inherited loss-of function mutations result in Waardenburg syndrome and acquired gain of function mutations result in alveolar rhabdomyosarcoma.

ACKNOWLEDGMENTS

This work was supported by a grant from the National Institutes of Health (AR-39892) and By a Clinical Research Grant from the March of Dimes Birth Defects Foundation (6-0281). This is paper number 3426 from the Laboratory of Genetics, University of Wisconsin.

REFERENCES

Andria G, Ballabio A, Parenti G (1987) X-linked ichthyosis due to steroid sulfatase deficiency associated with hypogonadism and anosmia. Ann Neuro. 22:98.

Ariturk E, Tosyali N, Ariturk N (1992) A case of Waardenburg syndrome and aganglionosis. Turk J Pediatr 34:111-114.

Asher JH Jr, Morell R, Friedman TB (1991) Waardenburg syndrome (WS): The analysis of a single family with a WS1 mutation showing linkage to RFLP markers on human chromosome 2q. Am J Hum Genet 48:43-52.

Baldwin CT, Hoth CF, Amos JA, Da Silva EO, Milunsky A (1992) An exonic mutation in the *HuP2* paired domain gene causes Waardenburg's syndrome. Nature 355:637-638.

Barr FG, Galili N, Holick J, Biegelo JA, Rovera G, Emanuel BS (1993) Rearrangement of the PAX3 paired box gene in the paediatric solid tumor alveolar rhabdomyosarcoma. Nature Genetics 3:113-117.

Bennett DC, Huszar D, Laipis PJ, Jaenisch R, Jackson IJ (1990) Phenotypic rescue of mutant brown melanocytes by a retrovirus carrying a wild-type tyrosinase-related protein gene. Development 110:471-475.

Bergen AAB, Samanns C, Van Dorp DB, Ferguson-Smith MA, Gal A, Bleeker-Wagemakers EM (1990). Localization of the X-linked ocular albinism gene (OA1) between DXS278/DXS237 and DXS143/DXS16 by linkage analysis. Ophthalmic Paediat Genet 11:165-170.

Bergen AAB, Samanns C, Schuurman EJM, Van Osch L, Van Dorp DB, Pinckers AJLG, Bakker E, Gal A, Van Ommen GJB, Bleeker-Wagemakers EM (1991) Multipoint linkage analysis in X-linked ocular albinism of the Nettleship-Falls type. Hum Genet 88:162-166.

Bergleiter ML, Harris DJ (1992) Waardenburg syndrome and meningocele. Am J Med Genet 44:541.

Besmer P, Murphy JE, George PC, Qiu F, Bergold PJ, Lederman L, Snyder HW Jr, Brodeur D, Zuckerman EE, Hardy WS (1986) A new acute transforming feline retrovirus and the relationship of its oncogene v-kit with the protein kinase gene family. Nature 320:415-421.

Boissy RE, Zhao H, Austin LM, Nordlund JJ, King RA (1993) Melanocytes from an individual with brown oculocutaneous albinism lack expression of TRP-1, the product of the human homologue of the murine brown locus. Am J Hum Genet 53: Suppl., 160.

Breathnach AS, Fitzpatrick TB, Wyllie LMA (1965) Electron microscopy of melanocytes in a case of human piebaldism. J Invest Dermatol 45:28-37.

Burri M, Tromvoukis Y, Bopp D, Frigerio G, Noll M (1989) Conservation of the paired domain in metazoans and its structure in three isolated human genes. EMBO J 8:1183-1190

Butler MG (1989) Hypopigmentation: a common feature of Prader-Labhart-Willi syndrome. Am J Hum Genet 45:140-146.

Carezani-Gavin M, Slarren SK, Steege T (1992) Waardenburg syndrome associated with meningomyelocele. Am J Med Genet 42:135

Chabot B, Stephenson DA, Chapman VM, Besmer P, Bernstein A (1988) The proto-oncogene c-*kit* encoding a transmembrane tyrosine kinase receptor maps to the mouse *W* locus. Nature 335:88-90.

Chakraborty AK, Orlow SJ, Pawelek JM (1992) Evidence that DOPAchrome tautomerase is a ferrous iron-binding glycoprotein. FEBS Lett 302:126-128.

Charles SJ, Moore AT, Yates JRW (1992) Genetic mapping of X linked ocular albinism: linkage analysis in British families. J Med Genet 29:552-554.

Chatkupt S, Chatkupt S, Johnson WG (1993) Waardenburg syndrome and myelomeningocele in a family. J Med Genet 30:83-84.

Chintamaneni CD, Ramsay M, Colman MA, Fox MF, Pickard RT, Kwon BS (1991) Mapping the human CAS2 gene, the homologue of the mouse brown (b) locus, to human chromosome 9p22-pter. Biochem Biophys Res Commun 178:227-235.

Cohen T, Muller RM, Tomita Y, Shibahara S: Nucleotide sequence of the cDNA encoding human tyrosinase-related protein. Nucl Acids Res 18:2807-2808 (1990).

Cooke JV (1952) Familial white skin spotting (piebaldness) ('partial albinism') with white forelock. J Pediatr 41:-12.

Copeland NG, Gilbert DJ, Cho BC, Donovan PJ, Jenkins NA, Cosman D, Anderson D, Lyman SD, Williams DE (1990) Mast cell growth factor maps near the steel locus on mouse chromosome 10 and is deleted in a number of steel alleles. Cell 63:175-183.

Currie ABM, Haddad M, Honeyman M, Boddy S-A M (1986) Associated developmental abnormalities of the anterior end of the neural crest: Hirschsprung's disease-Waardenburg's syndrome J Pediatr Surg 21:248-250.

d'Auriol L, Mattei MG, Andre C, Galibert F (1988) Localization of the human c-kit protooncogene on the q11-q12 region of chromosome 4. Hum Genet 78:374-376.

DiGeorge AM, Olmsted RW, Harley RD (1960) Waardenburg syndrome. J Pediatrics 57:649-669.

Durham FM (1904) On the presence of tyrosinases in the skins of some pigmented vertebrates. Proc R Soc Lond 74:310-313.

Epstein DJ, Vekemans M, Gros P (1991) Splotch (Sp2H), a mutation affecting development of the mouse neural tube, shows a deletion within the paired homeodomain of Pax-3. Cell 67:767-774.

Epstein DJ, Vogan KJ, Trasler DG, Gros P (1993) A mutation within intron 3 of the Pax-3 gene produces aberrantly spliced mRNA transcripts in the splotch (Sp) mouse mutant. Proc Natl Acad Sci USA 90:532-536.

Ezoe K, Holmes SA, Ho L, Bennett CP, Bolognia JL, Brueton L, Burn J, Falabella R, Gatto EM, Ishii N, Moss C, Pittelkow MR, Thompson E, Ward KA, Spritz RA (1995) Novel mutations and deletions of the *KIT* (steel factor receptor) gene in human piebaldism. Am J Hum Genet, in press.

Falik-Borenstein TC, Holmes SA, Borochowitz Z, Spritz RA (1995) DNA-based carrier detection and prenatal diagnosis of tyrosinase-negative oculocutaneous albinism (OCA1A). Prenatal Diagnosis, in press.

Farrer LA, Grundfast KM, Amos J, Arnos KS, Asher JH Jr, Beighton P, Diehl SR, Fex J, Foy C, Friedman TB, Greenberg J, Hoth C, Marazita M, Milunsky A, Morell R, Nance W, Newton V, Ramesar R, San Agustin TB, Skare J, Steens CA, Wagner RG Jr, Wilcox ER, Winship I, Read AP (1992) Waardenburg syndrome (WS) type I is caused by defects at multiple loci, one of which is near ALPP on chromosome 2: First report of the WS consortium. Am J Hum Genet 50:902-913.

Flanagan JG, Leder P (1990) The kit ligand: a cell surface molecule altered in steel mutant fibroblasts. Cell 63:185-194.

Fleischman RA, Saltman DL, Stastny V, Zneimer S (1991) Deletion of the c-kit protooncogene in the human developmental defect piebald trait. Proc Natl Acad Sci USA 88:10885-10889.

Fleischman RA (1992) Human piebald trait resulting from a dominant negative mutant allele of the c-*kit* membrane receptor gene. J Clin Invest 89:1713-1717.

Foy C, Newton VE, Wellesley D, Harris R, Read AP (1990) Assignment of WS1 locus to human 2q37 and possible homology between Waardenburg syndrome and the Splotch mouse. Am J Hum Genet 46:1017-1023.

Froggat P (1951) An outline with bibliography of human piebaldism. Ir J Med Sci 398:86-94.

Fukai K, Holmes SA, Lucchese NJ, Siu VM, Weleber RG, Schnur RE, Spritz RA (1995) Autosomal recessive ocular albinism associated with a functional significant tyrosinase gene polymorphism. Nature Genet 9:92-95.

Funderburk SJ, Crandall BF (1974) Dominant piebald trait in a retarded child with a reciprocal translocation and small intercalary deletion. Am J Hum Genet 26:715-722.

Gardner JM, Nakatsu Y, Gondo Y, Lee S, Lyon MF, King RA, Brilliant MH (1992) The mouse pink-eyed dilution gene: association with human Prader-Willi and Angelman syndromes. Science 257:1121-1124.

Garrod AE (1908) The Croonian lectures on inborn errors of metabolism: lecture I. Lancet 2:1-7.

Geissler EN, Ryan MA, Housman DE (1988) The dominant-white spotting (W) locus of the mouse encodes the c-kit proto-oncogene. Cell 55:185-192.

Gellius (1952) The Attic Nights of Aulus Gellius. Book 9. Rolfe JC, trans. Cambridge, Harvard University Press.

Giebel LB, Spritz RA (1990) RFLP for MboI in the human tyrosinase (TYR) gene detected by PCR. Nucl Acids Res 18:3103.

Giebel LB, Spritz RA (1991) Mutation of the c-kit (mast/stem cell growth factor receptor) proto-oncogene in human piebaldism. Proc Natl Acad Sci USA 88:8696-8699.

Giebel LB, Strunk KM, King RA, Hanifin JM, Spritz RA (1990) A frequent tyrosinase gene mutation in classic, tyrosinase-negative (type IA) oculocutaneous albinism. Proc Natl Acad Sci USA 87:3255-3258.

Giebel LB, Strunk KM, Spritz RA (1991*a*) Organization and nucleotide sequences of the human tyrosinase gene and a truncated tyrosinase-related segment. Genomics 9:435-445.

Giebel LB, Tripathi RK, King RA, Spritz RA (1991*b*) A tyrosinase gene missense mutation in temperature-sensitive type I oculocutaneous albinism. A human homologue to the Siamese cat and the Himalayan mouse. J Clin Invest 87:1119-1122.

Giebel LB, Strunk KM, Holmes SA, Spritz RA (1992) Organization and nucleotide sequence of the human *KIT* (mast/stem cell growth factor receptor) proto-oncogene. Oncogene 7:2207-2217.

Goulding MD, Chalepakis G, Deutsch U, Erselius JR, Gruss P (1991) Pax-3, a novel murine DNA binding protein expressed during early neurogenesis. EMBO J 10:1135-1147.

Goulding MD, Sterrer S, Fleming J, Balling R (1993) Analysis of the *Pax-3* gene in the mouse mutant splotch. Genomics 17:355-363.

Hodgkinson CA, Moore KJ, Nakayama A, Steingrimsson E, Copeland NG, NA, Arnheiter H (1993) Mutations at the mouse microphthalmia locus are associated with defects in a gene encoding a novel basic-helix-loop-helix-zipper protein. Cell 74:395-404.

Hoo JJ, Haslam RHA, van Orman C (1986) Tentative assignment of piebald trait gene to chromosome band 4q12. Hum Genet 73:230-231.

Hoth CF, Milunsky A, Lipsky N, Sheffer R, Clarren SK, Baldwin CT (1993) Mutations in the paired domain of the human PAX3 gene cause Klein-Waardenburg syndrome (WS-III) as well as Waardenburg syndrome type I (WS-I). Am J Hum Genet 52:455-462

Hu F Hanifin JM, Prescott GH, Tongue AC (1980) Yellow mutant albinism: cytochemical, ultrastructural, and genetic characterization suggesting multiple allelism. Am J Hum Genet 32:387-395.

Huang E, Nocka K, Beler DR, Chu TY, Buck J, Lahm HW, Wellner D, Leder P, Besmer P (1990) The hematopoietic growth factor KL is encoded by the *Sl* locus and is the ligand of the c-*kit* receptor, the gene product of the *W* locus. Cell 63:225-233.

Hughes AE, Newton VE, Liu XZ, Read AP (1994) A gene for Waardenburg syndrome type 2 maps close to the human homologue of the *microphthalmia* gene at chromosome 3p12-p14.1. Nature Genetics 7:509-512.

Ishikiriyama S, Tonoki H, Shibuya Y, Chin S, Harada N, Abe K, Niikawa N (1989) Waardenburg syndrome type I in a child with de novo inversion (2)(q35q37.3). Am J Med Genet 33:505-507.

Jackson IJ (1988) A cDNA encoding tyrosinase-related protein maps to the brown locus in mice. Proc Natl Acad Sci USA 85:4392-4396.

Jimbow K, Fitzpatrick TB, Szabo G, Hori Y (1975) Congenital circumscribed hypomelanosis: a characterization based on electron microscopic study of tuberous sclerosis, nevus depigmentosus, and piebaldism. J Invest Dermatol 64:50-62.

Jiminez-Cervantes C, Solano F, Kobayashi T, Hearing VJ, Lozano JA, Garcia-Borron JC (1994) The DHICA oxidase activity of mouse melanoma tyrosinases. J Biol Chem, in press.

Keeler CE (1934) The heredity of a congenital white spotting in Negroes. J Am Med Assoc 103:179-180.

King RA, Mentink MM, Oetting WS (1991) Nonrandom distribution of missense mutations within the human tyrosinase gene in type I (tyrosinase-related) oculocutaneous albinism. Mol Biol Med 8:19-29.

Kirkpatrick SJ, Kent CM, Laxova R, Sekhon GS (1992) Waardenburg syndrome type I in a child with deletion (2)(q35q36.2). Am J Med Genet 44:699-700.

Kromberg JGR, Krause A (1993) Waardenburg syndrome and spina bifida. Am J Med Genet 45:536-537.

Kugelman TP, van Scott EJ (1961) Tyrosinase activity in melanocytes of human albinos. J Invest Dermatol 37:73-76.

Kwon BS, Haq AK, Pomerantz SH, Halaban R (1987) Isolation and sequence of a putative cDNA clone for human tyrosinase that maps at the mouse c-albino locus. Proc Natl Acad Sci USA 84:7473-7477.

Lacassie Y, Thurmon TF, Tracy MD, Pelias MZ (1977) Piebald trait in a retarded child with interstitial deletion of chromosome 4. Am J Hum Genet 29:641-642.

Lee S-T, Nicholls RD, Schnur RE, Guida LC, Lu-Kuo J, Spinner NB, Zackai EH, Spritz RA (1994) Diverse mutations of the *P* gene among African-Americans with type II (tyrosinase-positive) oculocutaneous albinism (OCA2). Hum Molec Genet 3:2047-2051

Lee S-T, Strunk KM, Bundey S, Laxova R, Musarella M, Spritz RA (1994) Mutations of the *P* gene in oculocutaneous albinism, ocular albinism, and Prader-Willi syndrome plus albinism. New Engl J Med 330:529-534.

Lerch K (1988) Protein and active-site structure of tyrosinase. Adv Pigment Cell Res, Prog Clin Biol Res 256:85-98.

Lee, S-T, Nicholls RD, Jong MTC, Spritz RA (1995) Organization and sequence of the human P gene and identification of a new family of transport proteins. Genomics, in press.

Lerch K (1988) Protein and active-site structure of tyrosinase. Adv Pigment Cell Res, Prog Clin Biol Res 256:85-98.

Lev S, Yarden Y, Givol D (1992) Dimerization and activation of the kit receptor by monovalent and bivalent binding of the stem cell factor. J Biol Chem 267:15970-15977.

Lucian (1905) The Works of Lucian of Samosata, vol. 1 Fowler HW, trans], Oxford, Clarendon Press.

Lu-Kuo J, Ward DC, Spritz RA (1993) Fluorescence *in situ* hybridization mapping of 25 markers on distal human chromosome 2q surrounding the human Waardenburg syndrome, type I (WS1) locus (*PAX3* gene). Genomics 16:173-179.

Lyon M, Searle AG (1989) Genetic Variants and Strains of the Laboratory Mouse. Oxford University Press, Oxford.

Martinez JH, Solano F, Garcia-Borron JC, Iborra JL, Lozano JA (1985) The involvement of histidine at the active site of Harding-Passey mouse melanoma tyrosinase. Biochem Int 11:729-738.

Moase CE, Trasler DG (1992) Splotch locus mouse mutants: models for neural tube defects and Waardenburg syndrome type I in humans. J Med Genet 29:145-151.

Morell R, Friedman TB, Moelijopawiro S, Hartono, Soewito, Asher JH Jr (1992) A frameshift mutation in the HUP2 paired domain of the probable human homolog of murine *PAX-3* is responsible for Waardenburg syndrome type I in an Indonesian family. Hum Mol Genet 1:243-247.

Morell R, Friedman TB, Asher JH Jr (1993) A plus-one frameshift mutation in PAX3 alters the entire deduced amino acid sequence of the paired box in a Waardenburg syndrome type 1 (WS1) family. Hum Molec Genet 2:1487-1488.

Morgan J (1786) Some accounts of motley colored or pye Negro girl and mulatto boy. Trans Am Philos Soc (Phila) 2:392-395.

Morrison-Graham K, Takahashi Y (1993) Steel factor and c-kit receptor: from mutants to a growth factor system. BioEssays 15:77-83.

Murphy M, Reid K, Williams DE, Lyman SD, Bartlett PF (1992) Steel factor is required for maintenance, but not differentiation, of melanocyte precursors in the neural crest. Devel Biol 153:396-401.

Murty VVVS, Bouchard B, Mathew S, Vijayasaradhi S, Houghton AN (1992) Assignment of the human *TYRP (brown)* locus to chromosome region 9p23 by nonradioactive *in situ* hybridization. Genomics 13:227-229.

Nance WE, Jackson CE, Witkop CJ Jr (1970) Amish albinism: a distinctive autosomal recessive phenotype. Am J Hum Genet 22:579-586.

Nicholls RD (1993) Genomic imprinting and uniparental disomy in Angelman and Prader-Willi syndromes: A review. Am J Med Genet 46:16-25.

Nishikawa S, Kusakabe M, Yoshinaga K, Ogawa M, Hayashi S-I, Kunisada T, Era T, Sakakura T, Nishikawa S-I (1991) *In utero* manipulation of coat color formation by a monoclonal anti-*c-kit* antibody: two distinct waves of c-kit-dependency during melanocyte development. EMBO J 10:2111-2118.

Online Mendelian Inheritance in Man, OMIM (TM) [database online] (1994) The Johns Hopkins University, Baltimore, Maryland.

Pierpont JW, Erickson RP (1993) Facts on PAX. Am J Hum Genet 52:451-454.

Pliny (1942) The Natural History of Pliny: Plinius Secundus the Elder. Book 7. Rackman H, trans. Cambridge, Harvard University Press.

Ramsay M, Colman M-A, Stevens G, Zwane E, Kromberg J, Farrall M, Jenkins T (1992) The tyrosinase-positive oculocutaneous albinism locus maps to chromosome 15q11.2-q12. Am J Hum Genet 51:879-884.

Reith AD, Rottapel R, Giddens E, Brady C, Forrester L, Bernstein A (1990) W mutant mice with mild or severe developmental defects contain distinct point mutations in the kinase domain of the c-kit receptor. Genes & Devel 4:390-400.

Rinchik EM, Bultman SJ, Horsthemke B, Lee S-T, Strunk KM, Spritz RA, Avidano KM, Jong MTC, Nicholls RD (1993) A gene for the mouse pink-eyed dilution locus and for human type II oculocutaneous albinism. Nature 361:72-76.

Schaefer L, Ferrero GB, Grillo A, Bassi MT, Roth EJ, Wapenaar MC, Van Ommen GJB, Mohandas TK, Rocchi M, Zoghbi HY, Ballabio A (1993) A high resolution deletion map of human chromosome Xp22. Nature Genetics 4:272-279.

Schnur RE, Nussbaum RL, Anson-Cartwright L, McCowell C, Worton RG, Musarella MA (1991) Linkage analysis in X-linked ocular albinism. Genomics 9:605-613.

Shibahara S, Tomita Y, Sakakura T, Nager C, Chaudhuri B, Muller R (1986) Cloning and expression of cDNA encoding mouse tyrosinase. Nucl Acids Res 14:2413-2427.

Silvers WK (1979) The Coat Colors of Mice. Springer-Verlag, New York.

Smith EA, Seldin MF, Martinez L, Watson ML, Choudhury GG, Lalley PA, Pierce J, Aaronson S, Barker J, Naylor SL, Sakaguchi AY (1991) Mouse platelet-derived growth factor receptor α gene is deleted in W^{19H} and patch mutations on chromosome 5. Proc Natl Acad Sci USA 88:4811-4815.

Spritz, RA (1993) Molecular basis of piebaldism and Waardenburg's syndrome. Current Opinion in Dermatology 1:78-93.

Spritz RA (1994*a*) Molecular genetics of oculocutaneous albinism. Hum Molec Genet 3:1469-1475.

Spritz RA (1994*b*) The molecular basis of human piebaldism. J Invest Dermatol, in press.

Spritz RA and Hearing VJ Jr (1995) Genetic disorders of pigmentation. Ann Rev Genet 22:1-45

Spritz RA, Strunk KM, Giebel LB, King RA (1990) Detection of mutations in the tyrosinase gene in a patient with type IA oculocutaneous albinism. New Engl J Med 322:1724-1728.

Spritz RA, Droetto S, Fukushima Y (1992) Deletion of the *KIT* and *PDGFRA* genes in a patient with piebaldism. Amer J Med Genet 44:492-495.

Spritz RA, Holmes SA, Berg SZ, Nordlund JJ, Fukai K (1993) A recurrent deletion in the *KIT* (mast/stem cell growth factor receptor) gene is a frequent cause of human piebaldism. Hum Molec Genet 2:1499-1500.

Spritz RA, Ho L, and Strunk KM (1994*a*) Inhibition of proliferation of human melanocytes by a *KIT* antisense oligodeoxynucleotide: Implications for human piebaldism and mouse dominant white spotting (*W*). J Investig Derm 103:148-150.

Spritz, RA, Lu-Kuo, JM, Le Paslier, DL, Altherr, MR, Dorman, T, Moir, DT, Strunk, KM, and Ward, DC (1994*b*) A 2-Mb YAC contig spanning a cluster of genes encoding type III receptor tyrosine kinases (*PDGFRA-KIT-KDR*) is located at 4q12. Genomics 22:431-436.

Steel KP, Davidson DR, Jackson IJ (1992) TRP2/DT, a new early melanoblast marker, shows that steel growth factor (c-kit ligand) is a survival factor. Development 115:1111-1119.

Stephenson DA, Mercola M, Anderson E, Wang C, Stiles CD, Bowen-Pope DF, Chapman VM (1991) Platelet-derived growth factor receptor α-subunit gene (*Pdgfra*) is deleted in the mouse patch (*Ph*) mutation. Proc Natl Acad Sci USA 88:6-10.

Summers CG, Creel D, Townsend D, King RA (1991) Variable expression of vision in sibs with albinism. Am J Med Genet 40:327-331.

Sunohara N, Sakuragawa N, Satoyoshi E, Tanae A, Shapiro LJ (1986) A new syndrome of anosmia, ichthyosis, hypogonadism, and various neurological manifestations with deficiency of steroid sulfatase and arylsulfatase C. Ann Neurol 19:174-181.

Tachibana M, Perez-Jurado LA, Nakayama A, Hodgkinson CA, Li X, Schneider M, Miki T, Fex J, Francke U, Arnheiter H (1994) Cloning of *MITF*, the human homolog of the mouse *microphthalmia* gene, and assignment to human chromosome 3, region p14.1-p12.3. Hum Molec Genet, in press.

Tan JC, Nocka K, Ray P, Traktman P, Besmer P (1990) The dominant W^{42} *spotting* phenotype results from a missense mutation in the c-*kit* receptor kinase. Science 247:209-212.

Tassabehji M, Read AP, Newton VE, Harris R, Balling R, Gruss P, Strachan T (1992) Waardenburg's syndrome patients have mutations in the human homologue of the *Pax-3* paired box gene. Nature 355:635-636.

Tassabehji M, Read AP, Newton VE, Patton M, Gruss P, Harris R, Strachan T (1993) Mutations in the *PAX*3 gene causing Waardenburg syndrome type 1 and type 2. Nature Genetics 3:26-30.

Tomita Y, Takeda A, Okinaga S, Tagami H, Shibahara S (1989) Human oculocutaneous albinism caused by single base insertion in the tyrosinase gene. Biochem Biophys Res Commun 164:990-996.

Tripathi RK, Giebel LB, Strunk KM, Spritz RA (1991) A polymorphism of the human tyrosinase gene that is associated with temperature-sensitive enzymatic activity. Gene Expression 1:103-110.

Tripathi RK, Strunk KM, Giebel LB, Weleber RG, Spritz RA (1992) Tyrosinase gene mutations in type I (tyrosinase-deficient) oculocutaneous albinism define two clusters of missense substitutions. Amer J Med Gen 43:865-871.

Tsukamoto K, Tohma T, Ohta T, Yamakawa K, Fukushima Y, Nakamura Y, Niikawa N (1992) Cloning and characterization of the inversion breakpoint at chromosome 2q35 in a patient with Waardenburg syndrome type I. Hum Molec Genet 1:315-317.

Vandenbark GR, DeCastro CM, Taylor H, Dew-Knight S, Kaufman RE (1992) Cloning and structural analysis of the human c-*kit* gene. Oncogene 7:1259-1266.

Vijayasaradhi S, Bouchard B, Houghton AN (1990) The melanoma antigen gp75 is the human homologue of the mouse b (brown) locus gene product. J Exp Med 171:1375-1380.

Vogan KJ, Epstein DJ, Trasler DG, Gros P (1993) The splotch-delayed (Spd) mouse mutant carries a point mutation within the paired box of the Pax-3 gene. Genomics 17:364-369.

Waardenburg PJ (1951) A new syndrome combining developmental anomalies of the eyelids, eyebrows and nose root with pigmentary defects of the iris and head hair with congenital deafness. Am J Hum Genet 3:195-253.

Wagstaff J (1993) A translocation-associated deletion defines a critical region for the 9p-syndrome. Am J Hum Genet 53, Suppl., 619.

Wapenaar MC, Bassi MT, Schaefer L, Grillo A, Ferrero GB, Chinault AC, Ballabio A, Zoghbi HY (1993) The genes for X-linked ocular albinism (OA1) and microphthalmia with linear skin defects (MLS): cloning and characterization of the critical regions. Hum Molec Genet 2:947-957.

Wildenberg SC, Boissy RE, Oetting WS, Fryer JP, Zhao H, King RA (1994) Identification of a mutation in the tyrosinase related protein 1 (TRP1) gene associated with brown oculocutaneous albinism (OCA3). Am J Hum Genet 55:Suppl. A5.

Winder AJ, Kobayashi T, Tsukamoto K, Urabe K, Aroca P, Kameyama K, Hearing VJ (1994) The tyrosinase gene family: Interactions of melanogenic proteins to regulate melanogenesis. Biochem J, in press.

Winship IM, Babaya M, Ramesar RS (1993) X-linked ocular albinism and sensorineural deafness: linkage to Xp22.3. Genomics 18:444-445.

Witkop CJ Jr., Quevedo WC Jr., Fitzpatrick TB, King RA (1989) Albinism. In, Scriver CR, Beaudet AL, Sly WS, Valle D, eds. The Metabolic Basis of Inherited Disease. New York, McGraw-Hill, pp. 2905-2947.

Wittbjer A, Dahlback B, Odg G, Rosengren A-M, Rosengren E, Rorsman H (1989) Isolation of human tyrosinase from cultured melanoma cells. Acta Derm Venereol (Stokh) 69:125-131.

Yamamoto Y, Nishimoto H, Ikemoto S (1989) Interstitial deletion of the proximal long arm of chromosome 4 associated with father-child incompatibility within the Gc-system: probable reduced gene dosage effect and partial piebald trait. Am J Med Genet 32:520-523.

Yarden Y, Kuang W-J, Yang-Feng T, Coussens L, Munemitsu S, Dull TJ, Chen E, Schlessinger J, Francke U, Ullrich A (1987) Human proto-oncogene c-*kit*: a new cell surface receptor tyrosine kinase for an unidentified ligand. EMBO J 6:3341-3351.

Zsebo KM, Williams DA, Geissler EN, Broudy VC, Martin FH, Atkins HL, Hsu RY, Birkett NC, Okino KH, Murdock DC, Jacobsen FW, Langley KE, Smith KA, Takeishi T, Cattanach BM, Galli SJ, Suggs SV (1990) Stem cell factor is encoded at the *Sl* locus of the mouse and is the ligand for the c-*kit* tyrosine kinase receptor. Cell 63:213-224.

MDR-1 GENE EXPRESSION IN CHRONIC MYELOGENOUS LEUKEMIA: PROGNOSTIC MARKER OF THE DISEASE

I.Stuif, A.Sominskaya, T.Bykova, O.Frolova, A.Zaritskey

BMT CENTRE, Centre for Advanced Medical Technologies
Dynamo 3
197042, St.Petersburg
Pavlov Medical University,
L.Tolstoy, 6/8
197089, St.Petersburg, Russia

SUMMARY

The prognosis of the disease was studied in 31 patients with Ph-positive chronic myelogenous leukemia (CML) in relation to the expression of the MDR-1 gene of peripheral blood cells. 24 patients were in chronic phase (CP) and 7 patients were in accelerated phase (AP).

In situ hybridization was used as a method of detecting MDR-1 gene expression. It appeared that much more patients in AP than in CP expressed MDR-1/mRNA in their peripheral blood cells. In CP MDR-1 gene expression appeared to be of prognostic significance in relation to further progression of the disease. Prognostic significance depended on the period of the onset of the disease to the date of testing.

INTRODUCTION.

The MDR-1 gene is responsible for resistance to some cytostatic agents (Gottesman M. et al. 1988, Moscow I. et al. 1988). The majority of data related to primary or drug induced MDR-1 overexpression was obtained on cell lines of acute leukemia (AL), chronic lymphocytic leukemia (CLL), multiple myeloma (MM), (Dalton W. 1989, Epstein G. et al. 1989, Shustik C. et al. 1992). CML, especially chronic phase (CP), seems to be poorly studied in relation to MDR-1 gene expression. MDR-1-positivity (MDR+) of CML cells, if any, should be used as an additional marker of the leukemic clone, which can be of practical value for the treatment of advanced stages of the disease.

It is well known that the progression of CML is poorly drug regulated (Devita V. et al. 1993). Earlier studies of oncogene expression in CML neglected the analysis of the duration of the period from the diagnosis to the date of testing and from the date of testing to the beginning of the blastic phase of the disease (BP) (Kuwazuru Y. et al. 1990). This questions the real prognostic significance of the earlier obtained results.

Our study analyzes MDR-1 gene expression as a prognostic factor according to the history of pre- and post-study duration of the disease. This study seems to be novel because previous publications dealt only with blastic phase of CML (Goldstein L. et al. 1989, Maslak P. et al. 1994). Moreover, the duration of pre- and post-study period has not been previously analyzed.

NATO ASI Series, Vol. H 94
Gene Technology
Edited by A. R. Zander, W. Ostertag,
B. V. Afanasiev, F. Grosveld

MATERIALS AND METHODS.

1.Patients. 31 CML Ph-positive patients were included in the study, 24 in chronic phase (CP), 7 in accelerated phase (AP). The treatment consisted of hydroxyurea, myleran, interferon.

2.The procedure is based on the method of in situ hybridization (Kanz L. et al. 1988) with modifications. Briefly, light density fractions of peripheral blood cells (d < 1.077 g/ml) were fixed with ethanol/acetic acid (3:1) on the slide circular areas followed by dehydration with increasing concentrations of ethanol (70.95%). Then cells were treated with hydrochloric acid and sodium citric salt. Acetylation of the slides was performed before hybridization with triethanolamine and acetic anhydride.

3.Probe. The 1000-base pair MDR-1 specific DNA probe was labeled by standard nick translation reaction, according to the protocol supplied with nick translation kit. We used P33 - dCTP.

4.Prehybridization and hybridization. These steps were performed according to L. Kanz (Kanz L. et al. 1988). After posthybridization the slides were covered by emulsion 3M. Exposure of the slides was carried out 24 h at 4° C in the dark. Thereafter the slides were Giemsa stained. The evaluation of the level of expression of studied gene was carried out according to the following criteria: 0-5 silver grains per cell were equal to background value or to the number of grains per peripheral blood cell of healthy donor; 6-15 grains per cell were considered as low value, it was slightly more or equal to the number of grains per peripheral blood cell of healthy donor. These cells were considered as MDR-1-negative (MDR-). The expression exceeding 16-30 grains per cell was set as intermediate or high (> 30) and was referred to as MDR-1-positive (MDR+).

RESULTS AND DISCUSSION

The background of this study arises from the differences in the frequency of MDR-1 gene expressing cells of patients in CP and AP (Table 1). AP group consisted of patients who have never been treated with drugs known to induce MDR-1 gene expression.

Table 1. Expression of MDR-1 GENE in CML Patients (CP and AP)

Phase of CML	Expression of MDR-1 gene		
	MDR+	MDR-	
CP	9	17	P < 0,05
AP	6	1	

It appeared that more patients in AP were MDR-1-positive in comparison to patients in CP.

The main principle for further analysis of CP patients was the duration of the disease from the onset to the date of study and further duration of CP. At first we have analyzed the prognostic significance of gene expression in patients with 1,5 year or less duration of the disease (Table 2, 3 and 4)

Table 2. Prognostic significance of MDR-1 gene expression in CP CML patients in relation to the development of BP (less than half a year)

	Progression to BP in 3 months or less after the date of investigation	Progression to BP after 1 year from the date of investigation
MDR+	2	2
MDR-	0	7
	$P < 0.05$	

Table 3. Prognostic significance of MDR-1 gene expression in CP CML patients in relation to the development of BP. (less than a year).

	Progression to BP in 6 months or less after the date of investigation	Progression to BP after 1.5 year from the date of investigation
MDR+	4	2
MDR-	0	9
	$P < 0.01$	

Table 4. Prognostic significance of MDR-1 gene expression in CP CML patients in relation to the development of BP. (The history to date of study less than 1.5 years).

	Progression to BP in 6 months or less after the date of investigation	Progression to BP after 1.5 year from the date of investigation
MDR+	5	2
MDR-	0	11
	$P < 0.01$	

It appeared that patients with high MDR-1 gene expression are more prone to have BC.(Table 2, 3 and 4).

We have compared the percent age of MDR-1 positive patients with different duration of the disease to the date of testing in intervals 0-1.5 years, 1.5-3 and more than three years. There was no stepwise increase in frequency of MDR-1-positive individuals with increased duration of CP to the date of testing (2/3, 1/3 and 2/3). MDR-1-positivity seems to be coincident with transformation to the next stage of the disease (AP), (Table 1).

The significance of MDR-1-positivity is rather low (Table 5) or absent (Table 6) in the patients with duration of CP from the diagnosis to the date of study of 1.5-3 years or longer.

Table 5. Prognostic significance of MDR-1 gene expression in CP CML patients in relation to the development of BP. (more than 1,5 years and less than 3 years).

	Progression to BP in 18 months or less after the date of investigation	Progression to BP after 2 years from the date of investigation
MDR+	1	0
MDR-	0	3
	$P < 0.05$	

Table 6. Prognostic significance of MDR-1 gene expression in CP CML patients in relation to the development of BP (The history to date of study more than 3 years).

	Progression to BP in 12 months or less after the date of investigation	Progression to BP after 2 years from the date of investigation
MDR+	2	0
MDR-	2	1
	$P > 0.05$	

Conclusions: High MDR-1 gene expression when detected early in CP appeared to be a negative prognostic factor related to early BP. Absence of any correlations in patients with long lasting CP proposes different mechanisms of the disease progression.

REFERENCES.

Dalton W. (1989) Drug resistance in multiple myeloma and Non-Hodgkin's lymphoma. J. Clin. Oncol. 17: 415-424.

Devita V., Hellman J., Rosenberg S. (1993) In Cancer:.Principles and Practice of Oncology. Philadelpia.

Epstein G., Xiao H., Oba B. (1989) P-glycoprotein expression in plasma-cell myeloma is associated with resistance to VAD. Blood 47: 913-917.

Goldstein L., Galski H., Fojo A., (eds) (1989) Expression of Multidrug Resistance Gene in Human Cancer. J. Natl. Cancer Inst. 81: 116-124.

Gottesman M., Pastan I. (1988) The multidrug transporter, a double-edged sword. J.Biol.Chem. 263: 12163-12166.

Kuwazuru Y., Yoshimura A., Hamada S. (eds.) (1990) Expression of the multidrug transporter, P-glycoprotein, in CML-cells in blastic crises. Brit. J. Haem. 74: 24-29.

Kanz L., Mielke R., Lohr G., Fauser A. (1988) Detection of messenger RNAs within single hemopoietic cells by in situ hybridization on small slide areas. Exp. Hematol. 16: 394-399.

Maslak P., Hegewisch-Becker S., Godfrey L., Andreeff M. (1994) Flow cytometric determination of the multidrug- resistant phenotype in acute leukemia. Cytometry 17: 84-93.

Moscow I., Covan K. (1988) Multidrug resistance, J. Natl. Cancer Inst. 80: 257-281.

Shustik C., Groulx N., Gros P. (1992) Analyses of multidrug resistance (MDR-1) gene expression in chronic lymphocytic leukemia (CLL). Brit. J. of Hematol. 79: 50-56.

Savignano C., Geromin A., Michielli M. (eds.) (1993) The expression of Multidrug resistance related glycoprotein in adult acute lymphoblastic leukemia. Haematologica 78: 261-263.

Part 3

Methods of Gene Transfer, Applications for Gene Therapy

Towards Gene Therapy for Haemophilia B

Ann Gerrard
Sir William Dunn School of Pathology
University of Oxford
South Parks Road
Oxford OX1 3RE

Haemophilia B is an X-linked bleeding disorder affecting about 1 in 30,000 males. It is caused by mutations in the gene for blood clotting factor IX, which results in reduced amounts of functional factor IX in plasma. This leads to failure of the blood clotting cascade, which results in life-threatening bleeding episodes. Current treatment of haemophiliacs involves the intravenous injection of factor IX which has been purified from pooled human donor plasma. Although the haemostatic effectiveness of this treatment is well established and there is now a minimal risk of HIV infection, there is still the possibility that hepatitis B and C may be transmitted. Furthermore, treatment is given in response to bleeding episodes and not prophylactically. Gene therapy for haemophilia B would provide the patient with a permanent supply of factor IX, so that bleeds would be prevented rather than merely treated. Also frequent injections would no longer be necessary, thus improving the life-style of patients.

Gene therapy for haemophilia B involves introducing the factor IX cDNA into the cells of patients, either directly *in vivo* or by an *ex vivo* approach in which cells from the patient are removed, the cDNA is introduced into cells in culture and the modifed cells are then transplanted back into the patient. Whether an *in vivo* or an *ex vivo* approach is used depends largely on the method used to introduce the cDNA into cells and on the particular cell type.

So far in gene therapy studies for haemophilia B, both viral and non-viral methods have been investigated. Viral methods have included the use of retroviruses, adenoviruses and Herpes Simplex viruses, and non-viral methods have involved complexing DNA to polypeptide ligands which can recognise specific cell surface receptors. A major advantage of retroviruses is that they can stably integrate into the host genome, ensuring a long-lasting genetic

NATO ASI Series, Vol. H 94
Gene Technology
Edited by A. R. Zander, W. Ostertag,
B. V. Afanasiev, F. Grosveld

modification of cells. However, retroviruses require dividing cells for integration, and since many cells in the body are not actively proliferating they must be removed from the patient and grown in culture, at which time the cDNA is introduced. It is therefore often difficult to use retroviruses *in vivo*. Adenoviruses, Herpes Simplex viruses and the non-viral methods do not require replicating cells for the entry of DNA, so can be used *in vivo*. However, because the transferred DNA does not integrate into the chromosomal DNA, the inserted DNA is usually lost within a few months or even weeks. There might also be an immune response against cells expressing the viral proteins.

A wide variety of cells types have been studied for their use in gene therapy for haemophilia B. Although factor IX is normally synthesized in hepatocytes, it is secreted into the bloodstream where it functions in the clotting cascade. This means that many cell types in the body are potential targets for gene therapy as long as they have the ability to correctly post-translationally modify factor IX so that it is functional, and secrete it to the bloodstream. So far hepatocytes, skin fibroblasts, endothelial cells and myoblasts have been investigated in gene therapy studies (see below). We have studied the use of keratinocytes, the major cell type of the epidermis (Gerrard *et al.,* 1993). Keratinocytes may be advantageous as they are amongst the most accessible cells in the body, they can be grown easily and quickly in tissue culture, and grafting procedures are well-practiced and successful for regenerating epidermis in burn patients. The graft could be easily monitored and removed if necessary. The ease of grafting and monitoring of keratinocytes is not shared by other cells types. Keratinocytes also retain the ability for continuous self-renewal throughout adult life through the proliferative potential of a stem cell population.

We have used an *ex vivo* approach in our studies, in which a retroviral vector (based on the Moloney murine leukaemia virus) was used for introducing the human factor IX cDNA into keratinocytes in tissue culture. This vector, in which the factor IX cDNA was driven by the promoter/enhancer in the LTR and a neomycin resistance marker gene was driven by the SV40 early promoter, was introduced into the packaging cell lines ψ2 and PA317 by standard techniques. Defective factor IX virus from the best PA317 producer clone was used to transduce human primary keratinocytes.

Transduced keratinocytes, which had been selected for on G418, were found to secrete human factor IX into the culture medium at rates of about 600 ng/10^6 cells/24 hours (tested for by ELISA). The human factor IX was then purified from the culture medium and tested for biological activity in a clotting assay. It was found to be somewhere between 40% and 70% active.

Transduced keratinocytes were then transplanted into nude mice (which avoids an immune response to human factor IX). The grafting procedure involved trypsinising keratinocytes and layering them onto a collagen/fibroblast matrix which had been set up on the previous day. After a 2 day incubation, the keratinocyte/collagen layer was transplanted under a skin flap on the mid-back of a mouse, with the collagen matrix resting on the body wall. (The graft was transplanted under the skin solely for experimental purposes and would be transplanted in its correct epidermal position in humans). Four mice received implants and for each mouse blood samples were taken from the tail veins at daily intervals and tested for human factor IX antigen. Plasma was tested for antigen by ELISA. In each mouse human factor IX could be detected in the bloodstream for about one week. Maximum concentrations were reached after one day and they then gradually reduced over the course of the week.

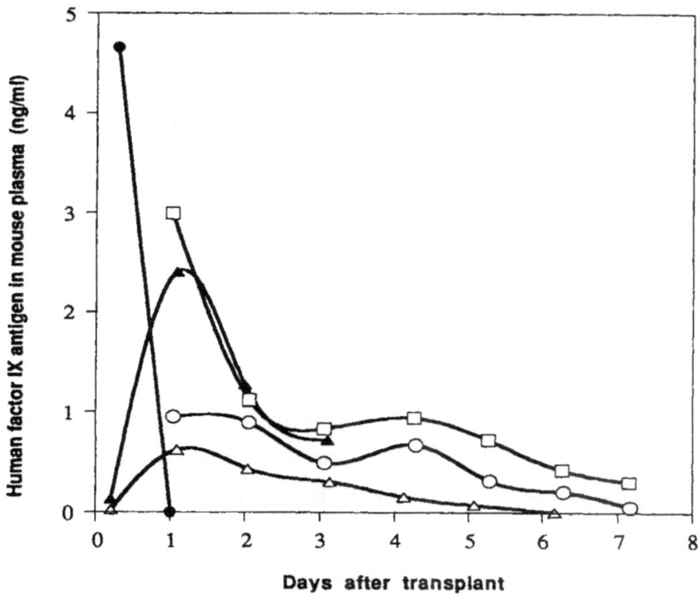

❑ ▲ O Δ represent 4 different mice in which transduced keratinocytes were transplanted . ● is a control mouse which received a collagen matrix 'spiked' with human plasma-derived factor IX.

Possible reasons for the gradual reduction in amounts of factor IX might have been due to suppression of the MoMLV 5'LTR promoter regulating the factor IX cDNA. Alternatively, even though keratinocytes were clearly visible by histological examination 7 days after grafting, there might have been gradual cell death. The amounts of factor IX in plasma were low, only 1-3 ng/ml was detected, which is about 0.1% of normal concentrations. This is too low to be of therapeutic benefit.

It was necessary to understand why the amounts of factor IX were low, so that appropriate improvements could be made in future experiments. It was calculated that 24 hours after transplantation only 1-3% of the factor IX expected to be secreted by the graft (based on amounts secreted in culture) circulated in the bloodstream. Two possible explantations for such low recoveries in blood included inefficient transport of factor IX from the skin to the blood and a reduced half-life of human factor IX in nude mice. (The half-life in humans is about 18 - 30 hours). In order to investigate this further, the transport of human factor IX (purified from pooled human plasma) was studied in nude mice (Gerrard *et al.*, 1992). To test the efficiency of transport from the skin to the bloodstream, 3 mice were injected subcutaneously with factor IX (in a position equivalent to that of the graft) and 3 mice were injected intravenously into the tail veins. All 6 mice received the same dose. In all mice blood samples were taken from the tail veins at intervals over 5 days and tested for human factor IX antigen. Results showed that at all time points, the average concentration of factor IX in plasma from mice which had received the subcutaneous injections was always about 30 - 40% of the concentrations reached in the mice which had received the intravenous injections. This result showed that transport of factor IX from the skin to the blood was reasonably efficient. The half-life of human factor IX in nude mice was about 12 hours. Both the degradation of factor IX and some loss of factor IX during its transport to the bloodstream could partly, but not totally, account for the low recoveries in blood in the transplantation experiments.

Further experiments were carried out in order to determine whether subcutaneously injected factor IX passed directly into the blood vessels at the site of injection, or whether it entered the bloodstream indirectly via the lymphatics, in which case it would enter the lymphatic capillaries at the site of injection and enter the bloodstream at the thoracic duct. The thoracic duct of a rat was cannulated and human factor IX was injected subcutaneously into the rat. Lymph samples from the thoracic duct and blood samples from the tail veins were collected and tested for human factor IX by ELISA. Factor IX was easily detectable in lymph after 10 mins, whereas it was undetectable in blood for at least 1 hour. These results showed that subcutaneously injected factor IX was transported to the bloodstream at least partly via the lymphatics.

The presence of human factor IX in the lymph (which was also detected after intravenous injection, results not shown), demonstrated that factor IX was not only distributed to the bloodstream. Taken together, these results show that the low recoveries in blood were caused by a number of factors including some degradation of factor IX, distribution to lymph and interstitial fluid and some loss of factor IX during transport to the blood. The amount lost during transport is not so great to suggest that other cell types which have more direct access to the bloodstream such as endothelial cells, would have a significant advantage over keratinocytes.

In other gene therapy studies retroviral vectors have been used to introduce human or canine factor IX cDNA into fibroblasts (Palmer *et al.*, 1989), endothelial cells (Axelrod *et al.*, 1990; Yao *et al.*, 1991), hepatocytes (Armentano *et al.*, 1990) and myoblasts (Yao and Kurachi, 1992). Results have shown that the rates of secretion of factor IX from these other cell types were similar to those observed with keratinocytes. Transduced fibroblasts and myoblasts have been transplanted into mice, and in one study retroviral vectors were directly introduced into the liver of dogs *in vivo* (Kay *et al.*, 1993). However, as found with keratinocytes, the amounts of factor IX in plasma were too low to be of therapeutic use and expression was also transient. In a recent study, in which tissue-specific enhancers were used in myoblasts, factor IX was detected at low concentrations in mice for about 6 months (Dai *et al.*, 1992).

Haemophilia B gene therapy studies *in vivo* using adenoviruses (Smith *et al.*, 1993; Kay *et al.*, 1994) and Herpes Simplex viruses (Miyanohara *et al.*, 1992) and also non viral methods (Perales *et al.*, 1994) have resulted in high initial concentrations of factor IX in plasma, which would be therapeutic. However, expression has decreased to baseline levels after a few weeks or months, usually caused by the loss of the DNA.

To summarise, studies so far have shown that many cell types have the capacity to synthesize and secrete biologically active factor IX, which is secreted to the blooodstream. With retroviruses, although there has been some progress towards achieving long-term expression of factor IX, the concentrations in plasma have always been too low to be of therapeutic benefit. Studies with adenoviruses, Herpes Simplex viruses and also the non-viral methods have resulted in very high initial concentrations of factor IX, but these are not long-lasting. Before gene therapy becomes a reality both long-term expression and increased factor IX production in cells must be achieved.

Armentano D, Thompson AR, Darlington G and Woo SLC (1990) Expression of human factor IX in rabbit hepatocytes by retrovirus-mediated gene transfer: potential for gene therapy of haemophilia B. Proc Natl Acad Sci USA 87: 6141-6145

Axelrod JH, Read MS, Brinkhous KM and Verma IM (1990) Phenotypic correction of factor IX deficiency in skin fibroblasts of haemophiliac dogs. Proc Natl Acad Sci USA 87: 5173-5177

Dai Y, Roman M, Naviaux RK and Verma IM (1992) Gene therapy via primary myoblasts: long term expression of factor IX protein following transplantation *in vivo.* Proc Natl Acad Sci USA 89:10892-10895

Gerrard AJ, Austen DEG and Brownlee GG (1992) Subcutaneous injection of factor IX for the treatment of haemophilia B. Br J Haematol 81: 610-613

Gerrard AJ, Hudson DL, Brownlee GG and Watt FM (1993) Towards gene therapy for haemophilia B using primary keratinocytes. Nature Genetics 3: 180-183.

Kay MA, Rothenberg S, Landon CN *et al* (1993) *In vivo* therapy of haemophilia B: sustained partial correction in factor IX-deficient dogs. Science 262: 117-119

Kay MA, Landon CN, Rothenberg S *et al* (1994) *In vivo* hepatic gene therapy: complete albeit transient correction of factor IX deficiency in haemophilia B dogs. Proc Natl Acad Sci USA 91: 2353-2357

Miyanohara A, Johnson PA, Elam RL, Dai Y, Witztum JL, Verma I and Friedmann T (1992) Direct gene transfer to the liver with herpes simplex virus type 1 vectors: transient production of physiologically relevant levels of circulating factor IX. New Biol 4: 238-246

Palmer TD, Thompson AR and Miller AD (1989) Production of human factor IX in animals by genetically modified skin fibroblasts: potential therapy for haemophilia B. Blood 73: 438-445

Perales JC, Ferkol T, Beegen H, Ratnoff OD, Hanson RW (1994) Gene transfer *in vivo:* sustained expression and regulation of genes introduced into the liver by receptor-targeted uptake. Proc Natl Acad Sci USA 91:4086-4090

Smith TAG, Mehaffrey MG, Kayda DB *et al* (1993) Adenovirus mediated expression of therapeutic plasma levels of human factor IX in mice. Nature Genetics 5: 397-402

Yao SN, Wilson JM, Nabel EG, Kurachi S, Hachiya HL and Kurachi K (1991) Expression of human factor IX in rat capillary endothelial cells: towards somatic gene therapy for haemophilia B. Proc Natl Acad Sci USA 88: 8101-8105

Yao SN and Kurachi K (1992) Expression of human factor IX in mice after injection of genetically modified myoblasts. Proc Natl Acad Sci USA 89:3357-3361

Inhibition of autocrine TGF-ß in CD34+ human progenitor cells reveals their potentiality to engraft adults and improves gene transfer efficiency.

Jacques Hatzfeld, Béatrice Panterne, Pascal Batard, Patricia Sansilvestri, Ma-Lin Li, Angelo A. Cardoso and Antoinette Hatzfeld.
Cytokines Molecular and Cell Biology Laboratory,
UPR 9044, CNRS-IRSC,
7, rue Guy-Mocquet, 94801
FRANCE

ABSTRACT

Using antisense TGF-ß1 oligonucleotides (ODN) or anti-TGF-ß serum in single cell and clonal culture of enriched progenitors from normal human bone marrow cells and umbilical cord blood, we have demonstrated that early hematopoietic progenitors produce autocrine TGF-ß1. These cells could not be triggered into cycle by optimal concentrations of growth factors available, which included IL-3, IL-6, G-CSF, SL and Epo. Using medium with this cytokine cocktail supplemented with either antisense TGF-ß1 ODN or anti-TGF-ß serum, we first estimated the proliferative capacity and ability to generate early progenitors in long-term cultures of bone marrow and umbilical cord blood cells. The results showed that the capacity for long-term engraftment of a 100-200 ml sample of umbilical cord blood is greater than that of the volume of bone marrow typically used to engraft an adult. The release from quiescence of the early progenitor cells by anti-TGF-ß serum was also used to improve gene transfer efficiency, since retroviral gene transfer requires a proliferative status of the target cell. High efficiency (95%) of gene transfer was obtained in early colony-forming unit-granulocyte-erythroid-megakaryocyte-macrophage (CFU-GEMM) with a stable gene expression. Development and size of the colonies were not affected by the retrovirally-mediated gene transfer.

NATO ASI Series, Vol. H 94
Gene Technology
Edited by A. R. Zander, W. Ostertag,
B. V. Afanasiev, F. Grosveld

Among the normal hematopoietic progenitor populations, the earliest pluripotent cells are largely quiescent while a larger proportion of later, lineage-restricted cells are in active phases of the cell cycle (Lajtha and Schofield, 1974). Using antisense TGF-ß1 oligonucleotides (ODN) or anti TGF-ß serum in single cell and clonal culture of enriched progenitors from normal human bone marrow cells, we have demonstrated that early colony forming cells produce autocrine TGF-ß1. Indeed, antisense TGF-ß ODN or anti TGF-ß serum enhance colony formation from early progenitor cells such as CFU-GEMM and early BFU-E: we obtained twice more mixed colonies and erythroid bursts than in the control cultures.These colonies were larger (Hatzfeld *et al.*, 1991). These cells could not be triggered into cycle by optimal concentrations of growth factors available, which included IL-3, IL-6, G-CSF and Epo. We have further shown that, although inclusion of SF increases the frequency of colony formation by early progenitor cells, it still does not reverse the effects of TGF-ß1 in culture and many cells remain quiescent (Li *et al.*, 1994). We have therefore used cultures with our standard cytokine cocktail supplemented with SF and either antisense TGF-ß1 ODN or anti-TGF-ß serum:

1) to compare the proliferative capacity and ability to generate early progenitors in long-term cultures of bone marrow and umbilical cord blood cells.

2) to improve gene transfer efficiency since retroviral gene transfer requires a proliferative status of the target cell.

To evaluate the proliferative capacity of bone marrow and umbilical cord blood progenitors, cells obtained from iliac crest of healthy individuals or from umbilical cord blood with the appropriate consent of the donors were purified as previously described, using the AIS procedure (SBA and CD34 CELLector flasks; Applied Immune Sciences Inc., Santa Clara, CA) (Cardoso *et al.*, 1995). The isolated CD34+ population obtained by this procedure was 95 ± 3% pure. Cells were then cultured with the cocktail of cytokines in the presence of 21mers phosphorothioate TGF-ß1 ODN or turkey anti-TGF-ß blocking serum, either in liquid cultures for expansion or in clonogenic assays to evaluate the types of progenitors. The CD34+ CD38- population was purified by FACS.

When cultured in the presence of IL-3, IL6, GM-CSF, SF, and Epo, colony formation by early progenitors (CFU-GEMM, HPP-CFC, and early BFU-E) in the cord blood $CD34^{+}CD38^{-}$ fraction, but not in the

$CD34^+CD38^+$ fraction, was significantly (1.5 to 2 fold) enhanced by antisense TGF-ß1 ODN. Late progenitors (CFU-E, CFU-M, and CFU-G) in the same preparation were unaffected by the antisense ODN. These results suggest that some subpopulations of CD34+ CD38- cells are under the control of autocrine TGF-ß. Cumulative evaluation of the respective number of colonies obtained after more than 35 days of culture leads to the conclusion that, although a typical bone marrow sample prepared for adult bone marrow transplant contains significantly more $CD34^+$ cells than a typical cord blood sample (100-200 ml), the ability to generate early progenitors in primary culture are quite comparable for both samples. This results from the higher frequency of $CD34^+$ $CD38^-$ cells in the cord blood preparations and from the higher proliferative potential of these cells in culture. Indeed, the $CD34^+$ $CD38^-$ cord blood fraction yielded similar number of CFU-GEMM but almost 3 fold more early BFU-E and 2 fold more CFU-GM than did the corresponding bone marrow cell preparation (Cardoso *et al.*, 1993).

We took advantage of releasing the early progenitors from quiescence to improve transfer efficiency since the proliferative status of the target cell is a prerequisite for an efficient transfer. Purified cord blood CD34+ cells were prestimulated in the presence of cytokines and anti TGF-ß serum. They were then transfected by cocultured for 40 h on 40% confluent NB16, a packaging cell line which express the E. Coli ß-galactosidase, kindly provided by Drs J.M. Heard and O. Danos (Ferry *et al.*, 1991). Transduced CD34+ cells were then tested in clonogenic assays and colonies were stained with X-Gal as substrate (Strair *et al.*, 1990). Results showed that addition of anti-TGF-ß serum increased the transfer efficiency from 24% without anti-TGF-ß serum up to 47.3% in the presence of anti-TGF-ß serum. Interestingly, the effect of anti-TGF-ß is most effective on the most immature progenitors, which develop into large mixed colonies (CFU-GEMM) with $1\text{-}2 \times 10^5$ cells. Anti-TGF-ß serum pretreatment increases gene transfer in CFU-GEMM from 54 to 95%. It augments significantly the stability of gene expression in subpopulations of mixed colonies and the expression of the stably integrated recombinant provirus does not reduce their size.

In conclusion, the elucidation of the mechanisms controlling cell proliferation, especially the role of inhibitors like TGF-ß1, will be of interest in clinical applications for transplantation with cell expansion as well as for gene therapy.

ACKNOWLEDGEMENTS: this research is supported by a European Concerted Action on Human Haematopoietic Stem Cells, by Direction des Recherches Etudes et Techniques (DRET), the Institut National de la Santé et de la Recherche Médicale-CNAM, the Agence Nationale de Recherches sur le SIDA (ANRS) and the Ministère de l'Enseignement et de la Recherche (MESR).

REFERENCES

Cardoso, A.A., Li, M.L., Batard, P., Hatzfeld, A., Brown, E.L., Levesque, J.-P., Sookdeo, H., Panterne, B., Sansilvestri, P., Clark, S.C., and Hatzfeld, J. (1993). Release from quiescence of CD34+ CD38- human umbilical cord blood cells reveals their potentiality to engraft adults. Proc. Natl. Acad. Sci. USA **90**, 8707-8711.

Cardoso, A.A., Watt, S., Batard, P., Li, M.-L., Hatzfeld, A., Genevier, H., and Hatzfeld, J. (1995). An improved panning technique for the selection of CD34-positive human bone marrow hematopoietic cells with high recovery of early progenitors. Exp. Hematol. **In press**,

Ferry, N., Duplessis, O., Houssin, D., Danos, O., and Heard, J.-M. (1991). Retroviral-mediated gene transfer into hepatocytes *in vivo*. Proc. Natl. Acad. Sci. USA **88**, 8377-8381.

Hatzfeld, J., Li, M.-L., Brown, E.L., Sookdeo, H., Levesque, J.-P., O'Toole, T., Gurney, C., Clark, S.C., and Hatzfeld, A. (1991). Release of early human hematopoietic progenitors from quiescence by antisense transforming growth factor ß1 or Rb oligonucleotides. J. Exp. Med. **174**, 925-929.

Lajtha, L.G., and Schofield, R. (1974). On the problem of differenciation in haemopoiesis. Differenciation **2**, 313-320.

Li, M.-L., Cardoso, A.A., Sansilvestri, P., Hatzfeld, A., Brown, E.L., Sookdeo, E., Levesque, J.-P., Clark, S.C., Hatzfeld, J., and . (1994). Additive effects of Steel Factor and antisense TGF-ß1 oligonucleotide on CD34+ hematopoietic progenitor cells. Leukemia **8**, 441-445.

Strair, R.K., Towle, M., and Smith, B.R. (1990). Retroviral mediated gene transfer into bone marrow progenitor cells : use of beta-galactosidase as a selectable marker. Nucleic Acids Res. **18**, 4759-4762.

GENE TRANSFER INTO HEMOPOIETIC PROGENITORS: IMPLICATIONS FOR BONE MARROW TRANSPLANTATION

Helen E Heslop, Cliona M. Rooney, Donna R Rill, and Malcolm K Brenner

Division of Bone Marrow Transplantation (HEH, DRR, MKB), Department of Hematology-Oncology, and Department of Virology and Molecular Biology (CMR,), St. Jude Children's Research Hospital, Memphis, TN 38105; the Department of Pediatrics (HEH, MKB), and Department of Pathology (CMR) University of Tennessee, Memphis, Memphis, TN 38163

INTRODUCTION

Hemopoietic progenitor cells have been considered an ideal target for gene therapy of many single gene disorders currently treated by bone marrow transplantation as they are easily obtained and manipulated ex vivo. Furthermore genetic modification of stem cells could be sufficient to repopulate the hemopoietic and lymphoid system of an individual for their entire life. However, the level of gene transfer obtained with current gene transfer techniques as wells as concerns about the expression and regulation of new genes in such cells has delayed the use of gene therapy in most patients with these conditions (Miller, 1992; Anderson, 1992). In contrast gene marking studies allow clinically biological questions to be addressed even with the limited efficiency of gene transfer available with current vectors. In our initial studies in autologous bone marrow transplantation (AMBT), we used gene marking techniques to determine the source of relapse after ABMT and to learn more about the biology of normal marrow reconstitution. In follow up studies we are evaluating the efficacy of purging and the effect of growth factors on reconstitution. We are also using gene marking to monitor the efficacy of an adoptive transfer approach in patients at high risk of developing EBV lymphoproliferation following allogeneic bone marrow transplantation.

GENE MARKING STUDIES AFTER AUTOLOGOUS BMT

1. Source of Relapse After Autologous Bone Marrow Transplantation

While autologous bone marrow transplantation appears to produce therapeutic benefit in a number of malignant diseases, including acute myeloid leukemia (Santos et al, 1989; Spitzer et

NATO ASI Series, Vol. H 94
Gene Technology
Edited by A. R. Zander, W. Ostertag,
B. V. Afanasiev, F. Grosveld

al, 1992; Jones, 1993), relapse remains the major cause of treatment failure. The possibility that reinfused leukemic cells may contribute to relapse has led to extensive evaluation of techniques for purging marrow to eliminate putative residual malignant cells (Gorin et al, 1990; Yeager et al, 1986). However it has been unclear whether purging is necessary. One approach to resolving this issue is to mark the marrow at the time of harvest and then find out if the marker gene is present in malignant cells at the time of a subsequent relapse.

The first generation studies using bone marrow of patients with AML and neuroblastoma began in September 1991 and January 1992, respectively and closed in March 1993. Nucleated bone marrow ($>1.5 \times 10^8$ cells/kg of body weight) was taken from the posterior iliac crest and two thirds were cryopreserved immediately. The remaining third was separated on a Ficoll gradient to produce a mononuclear cell fraction which was transduced with either the LNL6 or the closely related G1N retroviral vector for 6 hours as previously described (Rill et al, 1992). Both these vectors encode the neomycin resistance gene which can subsequently be detected in transduced cells by PCR or because it confers resistance to the neomycin analogue G418. At the time of transplantation, both transduced and unmanipulated marrow cells were thawed and reinfused through each patient's central venous line.

To assess the efficiency of gene transfer and expression in marrow progenitor cells both pre- and post-transplantation, we obtained mononuclear cells from peripheral blood or bone marrow, separated them on Ficoll-Hypaque gradients and cultured them in methylcellulose as previously described (Rill et al, 1992). Transduced cells were cultured with or without G418. PCR was used to detect the transferred Neo^R gene in individual colonies or in bulk populations as previously described (Rill et al, 1992). The PCR-amplified products were analyzed by gel electrophoresis (1.5% agarose) and Southern blotting with a ^{32}P-labeled Neo-specific probe. Semiquantitative PCR was also used, with 25 cycles of amplification corresponding to linear signal strength.

In the AML study, four of twelve patients have relapsed. In two, the malignant cells contained the marker gene (Brenner et al, 1993b). The third patient was uninformative as marrow mononuclear cells had low levels of Neo^R by PCR but no blast colonies grew in G418 and his blasts did not have a leukemia specific marker so the source of the PCR signal could be determined. The fourth patient is currently being evaluated. To definitively prove that the marker gene is in leukemia cells it is necessary to have a collateral leukemia specific marker so it is possible to demonstrate both this marker and the marker gene in the same cell. For example patient 2 was found to be in relapse on a routine bone marrow at 6 months. This patient had two leukemia specific markers. First her blasts co-expressed CD34 and CD56 a combination not found on normal hemopoietic cells. Second she had a complex t(1:8:21) translocation resulting in generation of a AML1/ETO fusion transcript that could be identified by PCR. We were therefore able to sort blasts expressing CD34 and CD56 and show co-expression in a single clonogenic cell of both a leukemia-specific marker (the AML:ETO fusion protein) and the transferred neomycin gene (Brenner et al, 1993b).

Similarly, five neuroblastoma patients have relapsed, and in 4 cases, gene marked neuroblastoma cells were detected (Rill et al, 1994). In these patients, identification of marked neuroblastoma cells was confirmed by detection of co-expression of neuroblastoma-specific antigens (for example GD2) together with the transferred marker gene. Four of the neuroblastoma patients relapsed in their marrow, but the fifth had disease recurrence in an extramedullary site in his liver. Biopsy of this extra-medullary site showed the presence of gene marked neuroblasts.

We used a limiting cycle PCR technique to estimate the proportion of resurgent neuroblasts that were marker gene positive in each of the first three patients (Rill et al, 1994). The neomycin-resistance gene was present in 0.05% to 1% of the $GD2^+$ $CD45^-$ cell population at sites of relapse. Since only 1% of clonogenic neuroblasts present in the autologous marrow transplant were genetically marked the proportion of marked malignant cells in vivo at the time of relapse was similar to the proportion marked ex vivo and infused with the autologous marrow transplant. These data show that residual tumor cells in marrow have considerable clonogenic potential in vivo and suggest that they make a substantial contribution to resurgent disease.

These data show definitively that marrow harvested in apparent clinical remission in two pediatric malignancies may contain residual tumorigenic cells and that these cells can contribute to disease recurrence at both medullary and extra-medullary sites.

Gene Transfer to Normal Cells

Two of the twenty one patients died (of disease progression and sepsis) within 21 days of transplantation and could not be further assessed. In the remaining 19 patients, there were no indications that laboratory manipulation of marrow specimens had adversely affected engraftment. None of the patients has shown evidence of malignant transformation or other complications that could be attributed to the gene transfer process.

The presence of the gene in hemopoietic progenitor cells in vivo was confirmed by clonogenic assays that showed gene-marked progenitor cells in 15/19 patients at one month after ABMT. We obtained evidence for the contribution of infused marrow to long-term hemopoietic recovery from the 17 patients who have been followed for 6 months or more after transplantation. By clonogenic assay, G418-resistant progenitor cells persisted for six months in 16/17 patients, for one year in 11/11, for 18 months in 6/6 and at 24 months in the two patients evaluable thus far. Multilineage GEMM colonies became apparent 6 months after transplantation in 3/9 patients (Brenner et al, 1993a). The marker gene was also present and expressed long term in the mature progeny of marrow precursor cells, including peripheral blood T and B cells and neutrophils. The level of transfer varied and was highest in marrow clonogenic hemopoietic progenitors where the percentage of G418 resistant myeloid colonies ranged from 0-29%. In peripheral blood cells expression was variable between the different lineages and was higher by an order of magnitude in myeloid cells than in T lymphocytes. The lowest level of expression was seen in B lymphocytes. These levels of transfer are higher than predicted from animal models and may be attributed to the fact that marrow was harvested during regeneration after intensive chemotherapy, when a higher than normal proportion of stem cells are in cycle.

These data also support a substantial contribution to marrow reconstitution from autologous transplants and suggest that this contribution includes long-lived multipotent stem cells. It should be possible to use the genetic marker technique we describe to compare the regenerative potential of marrow versus peripheral blood-derived progenitor cells and hence discover their relative value for hemopoietic rescue.

Purging Studies

We have begun second generation studies of marrow purging using two gene markers to compare marrow purging versus no purging or two different purging techniques. We are using two closely related vectors, G1N and LNL6, which can be discriminated by virtue of differing fragment sizes they produce after PCR amplification. In the AML study 1/3 of the marrow is frozen unpurged as a safety backup. The remaining marrow is split into two aliquots which are marked with G1Na or LNL6 and then randomly assigned to purging with 4HC or IL2 (Brenner et al, 1994). At the time of transplant both aliquots are reinfused. If the patient should subsequently relapse detection of either marker will allow us to learn if either of these purging techniques is effective. Eight patients have been treated on this protocol as of November 1994. None have relapsed, but transfer has been seen in normal progenitors albeit at a lower level than in the studies using unpurged marrow. A similar protocol for neuroblastoma will begin accrual imminently (Brenner et al, 1993).

Ex vivo expansion studies

Genetic marking could also be used to determine directly which ex vivo or in vivo combination of cytokines will increase the entry of long-term marrow repopulating cells into cell cycle and thereby reduce the period of marrow hypoplasia and immunodeficiency that follows autologous stem cell transplantation. A number of animal models have suggested that ex vivo treatment with growth factor combinations, with or without the addition of marrow stromal cells or extracellular matrix, can induce stem cells to enter cell cycle and thereby increase both their number and the efficiency with which they can be transduced (Bodine et al, 1989; Moritz et al, 1994). In humans, it certainly appears possible to use growth factors such as IL-1, IL-3 and stem cell factor to increase the numbers of hemopoietic progenitor cells by 10 to 50 fold and to increase the efficiency of gene transfer to levels which may exceed 50%. Unfortunately, it is not certain that such ***ex vivo*** data will be reflected by results ***in vivo***. In primate studies, transplantation of marrow treated ex vivo with the growth factor combinations shown to greatly augment both progenitor numbers and gene transfer rates, have been followed by disappointingly low levels of long term gene expression in vivo. Similarly, in human studies in which SCF, IL-3 and IL-6 were used to stimulate marrow progenitor cells before reinfusion, expansion of CFUs and a high gene-transfer rate ex vivo were associated with in vivo expression that was both low and attenuated (Dunbar et al, 1994). The likeliest explanation for this apparent paradox is that many of the growth factors intended only to induce cycling in marrow stem cells also induce their differentiation and the loss of their self renewal capacity.

Without any proven ***ex vivo*** surrogate method for studying the effects of growth factors on stem cell expansion and transducibility, it is possible to use the marker-gene technique to evaluate whether any increase in progenitor cell numbers and transducibility produced by growth factor combinations and cell culture devices ex vivo has an effect in vivo. By using two distinguishable vectors to mark each patient's marrow, we can compare treatment regimens **within** a patient. Because each patient acts as their own control, it should be possible to discern the effects of any given growth factor regimen on stem cells, even in a cohort as small as 10 patients.

GENE MARKING STUDIES AFTER ALLOGENEIC BMT

Gene Marking of Adoptively Transferred EBV Specific CTLs

We have also begun a marking study in recipients of allogeneic bone marrow obtained from matched unrelated or mismatched family member donors. In these individuals there is greater genetic disparity between donor and recipient and a higher risk of GVHD. In vitro T cell depletion of donor marrow is an effective means of reducing the risk of GVHD but recipients of depleted marrow have delayed immune recovery and an increased incidence of viral infections (Koehler et al, 1994). One such infection is Epstein-Barr virus lymphoproliferative disease (EBV-LPD). EBV usually persists lifelong after primary infection, by a combination of chronic replication in the mucosa and latency in peripheral blood B cells. EBV infected B cells are highly immunogenic and outgrowth only occurs in severely immunocomprimised patients. This complication occurs in 5-30% of patients receiving marrow from mismatched family members or unrelated donors and until recently was almost invariably lethal.

Recently several groups have evaluated the effects of infusing donor T cells to treat this complication (Papadopoulos et al, 1994; Servida et al, 1993; Heslop et al, 1994a). Such therapy can induce disease regression, although with a high risk of inducing significant graft versus host disease (Servida et al, 1993; Heslop et al, 1994a). To avoid this latter complication, we are currently administering infusions of EBV specific donor T cells to high risk patients in a Phase I prophylaxis study (Heslop et al, 1994b). These specific T cells are generated by culturing donor T cells with donor derived EBV infected lymphoblastoid cell lines. To determine whether these cells persist and whether they cause adverse effects they are marked with the neomycin resistance gene before administration. As of November 1994 ten patients have been enrolled on this protocol. The level of marking of infused CTL as assessed by semi-quantitative PCR ranges from 0.5-5%, levels which previous experience suggested would be adequate to enable the survival and behavior of these cells to be followed. At the lowest dose level of 10^7 cells/m^2, CTL will comprise 1 in 10,000 mononuclear cells immediately post infusion which is within the detection limits of PCR. The Neo^R gene was detected in peripheral blood lymphocytes for up to 18 weeks or more after injection of gene marked cells and in EBV specific CTL lines regenerated form these patients for up to 9 months.

We also have evidence that these transferred cells were functional in vivo (Rooney et al, 1994). In two patients, it has been possible to show anti-viral efficacy in vivo with a dramatic fall in EBV DNA following infusion. In an additional patient, both anti-viral and anti-lymphoproliferative activity were demonstrable. In this individual, cervical, hilar and retroperitoneal lymphadenopathy developed before the CTL were given. Immunohistology was typical of post-transplant lymphoproliferative lymphoma. Administration of CTL was associated with resolution of the lymphadenopathy over a 6 week period; to date there has been no resurgence of his disease (Rooney et al, 1994).

It should be possible to use a similar approach to transfer T cells specific for other malignancies with tumor restricted antigens (Brenner et al, 1991).

CONCLUSION

Genetic marking can be used to evaluate the contribution of residual tumorigenic cells in marrow to relapse after autologous bone marrow transplantation. Marker studies should also prove valuable for assessing the efficacy of purging and for learning about the impact of hemopoietic growth factor treatment of harvested marrow or peripheral blood stem cells on subsequent progenitor cell growth and development in vivo. Finally gene marking should help evaluate the toxicity and efficacy of T cell adoptive transfer in recipients of allogeneic BMT.

ACKNOWLEDGEMENTS

This work was supported in part by NIH Cancer Center Support CORE Grant P30 CA21765, CA 61384, CA 20180 and the American Lebanese Syrian Associated Charities (ALSAC). We would like to thank all our clinical and laboratory associates at St Jude Children's Research Hospital who helped make this work possible and Nancy Parnell for word processing.

REFERENCES

Anderson WF (1992) Human gene therapy. Science 256:808-813.

Bodine DM, Karlsson S, Nienhuis AW (1989) Combination of interleukins 3 and 6 preserves stem cell function in culture and enhances retrovirus-mediated gene transfer into hematopoietic stem cells. Proc Natl Acad Sci,USA 86:8897-8901.

Brenner M, Santana V, Bowman L, Furman W, Heslop H, Krance R, Mahmoud H, Boyett J, Moss T, Moen RC, et al (1993) Use of marker genes to investigate the mechanism of relapse and the effect of bone marrow purging in autologous transplantation for stage D neuroblastoma. Hum Gene Ther 4:809-820.

Brenner MK, Heslop HE (1991) Graft-versus-host reactions and bone marrow transplantation. Curr Opin Immunol 3:752-757.

Brenner MK, Rill DR, Holladay MS, Heslop HE, Moen RC, Buschle M, Krance RA, Santana VM, Anderson WF, Ihle JN (1993a) Gene marking to determine whether autologous marrow infusion restores long-term haemopoiesis in cancer patients. Lancet 342:1134-1137.

Brenner MK, Rill DR, Moen RC, Krance RA, Mirro J, Jr., Anderson WF, Ihle JN (1993b) Gene-marking to trace origin of relapse after autologous bone marrow transplantation. Lancet 341:85-86.

Brenner MK, Krance R, Heslop HE, Santana V, Ihle J, Ribeiro R, Roberts WM, Mahmoud H, Boyett JM, Moen RC, et al (1994) Assessment of the efficacy of purging by using gene marked autologous marrow transplantation for children with AML in first complete remission. Hum Gene Ther 5:481-499.

Dunbar CE, Bodine DM, Sorrentino B, Donahue R, Cottler-Fox M, O'Shaughnessy J, Cowan K, Carter C, Doren S, Cassell A, et al (1994) Gene transfer into hematopietic cells: Implications for cancer therapy. Ann N Y Acad Sci 716:216-224.

Gorin NC, Aegerter P, Auvert B, Meloni G, Goldstone AH, Burnett A, Carella A, Korbling M, Herve P, Maraninchi D, et al (1990) Autologous bone marrow transplantation for acute myelocytic leukemia in first remission: A European survey of the role of marrow purging. Blood 75:1606-1614.

Heslop HE, Brenner MK, Rooney CM (1994a) Donor T cells as therapy for EBV lymphoproliferation post bone marrow transplant. N Eng J Med (In press).

Heslop HE, Brenner MK, Rooney C, Krance RA, Roberts WM, Rochester R, Smith CA, Turner, Sixbey J, Moen R, Boyett JM (1994b) Administration of neomycin resistance gene marked EBV specific cytotoxic T lymphocytes to recipients of mismatched-related or phenotypically similar unrelated donor marrow grafts. Hum Gene Ther 5:381-397.

Jones RJ (1993) Autologous bone marrow transplantation. Curr Opin Oncol 5:270-275.

Koehler M, Hurwitz CA, Krance RA, Coustan-Smith E, Williams LL, Santana V, Ribeiro RC, Brenner MK, Heslop HE (1994) XomaZyme-CD5 immunotoxin in conjunction with partial T cell depletion for prevention of graft rejection and graft-versus-host disease after bone marrow transplantation from matched unrelated donors. Bone Marrow Transplantation 13:571-575.

Miller AD (1992) Human gene therapy comes of age. Nature 357:455-460.

Moritz T, Patel VP, Williams DA (1994) Bone marrow extracellular matrix molecules improve gene transfer into human hemopoietic cells via retroviral vectors. J Clin Invest 93:1451-1457.

Papadopoulos EB, Ladanyi M, Emanuel D, MacKinnon S, Boulad F, Carabasi MH, Castro-Malaspina H, Childs BH, Gillio AP, Small TN, et al (1994) Infusions of donor leukocytes to treat Epstein-Barr virus-associated lymphoproliferative disorders after allogeneic bone marrow transplantation. N Engl J Med 330:1185-1191.

Rill DR, Moen RC, Buschle M, Bartholomew C, Foreman NK, Mirro JJ, Krance RA, Ihle JN, Brenner MK (1992) An approach for the analysis of relapse and marrow reconstitution after autologous marrow transplantation using retrovirus-mediated gene transfer. Blood 79:2694-2700.

Rill DR, Santana VM, Roberts WM, Nilson T, Bowman LC, Krance RA, Heslop HE, Moen RC, Ihle JN, Brenner MK (1994) Direct demonstration that autologous bone marrow transplantation for solid tumors can return a multiplicity of tumorigenic cells. Blood 84:380-383.

Rooney CM, Smith CA, Ng C, Loftin S, Li C, Krance RA, Brenner MK, and Heslop HE (1994) Use of virus-specific cytotoxic T lymphocytes to control Epstein-Barr virus-related lymphoproliferation. Lancet (In Press).

Santos GW, Yeager AM, Jones RJ (1989) Autologous bone marrow transplantation. Annu Rev Med 40:99-112.

Servida P, Rossini S, Traversari C, Ferrari G, Bonini C, Nobili N, Vago L, Faravelli A, Vanzulli A, Mavilio F, et al (1993) Gene Transfer into peripheral blood lymphocytes for in vivo immunomodulation of donor anti-tumor immunity in a patient affected by EBV-induced lymphoma. Blood 10(Suppl 1):214a

Spitzer G, Velasquez W, Dunphy FR, Spencer V (1992) Autologous bone marrow transplantation in solid tumors. Curr Opin Oncol 4:272-278.

Yeager AM, Kaizer H, Santos GW, Saral R, Colvin OM, Stuart RK, Braine HG, Burke PJ, Ambinder RF, Burns WH, et al (1986) Autologous bone marrow transplantation in patients with acute nonlymphocytic leukemia, using ex vivo marrow treatment with 4-hydroperoxycyclophosphamide. N Engl J Med 315:141-147.

ERYTHROPOIETIN AND THE ENV gp55 OF THE SPLEEN FOCUS FORMING VIRUS (SFFV) INTERACT DIFFERENTLY WITH ERYTHROID CELLS IN THE MOUSE AND IN TISSUE CULTURE

Ursula Just, Norbert Ahlers, Gökhan Arman, Nicholas Hunt, Wolfram Ostertag, and Joachim Nowock

Heinrich-Pette-Institut für Experimentelle Virologie und Immunologie an der Universität Hamburg, Martinistr. 52, 20251 Hamburg, Germany

Introduction

Erythropoietin (Epo) is a glycoprotein hormone, which is the major regulator of mammalian erythropoiesis. Epo promotes the survival, proliferation and differentiation of erythroid progenitor cells by binding to and activation of the specific cell surface receptor (Epo-R) (D'Andrea et al., 1989; reviewed by Youssoufian et al., 1993). The recently cloned cDNA of the murine Epo-R encodes a protein of 507 amino acids that contains a single hydrophobic membrane spanning domain. The Epo-R is a member of the cytokine receptor family, which include the receptors for granulocyte colony-stimulating factor (G-CSF), interleukin-3 (IL-3), IL-4, granulocyte-macrophage colony-stimulating factor (GM-CSF) and several other growth factors (D'Andrea et al., 1990; Bazan, 1990). Receptor polypeptides in this family share a set of four conserved Cys residues that stabilize the structure of the membrane-distal sub-domain and a five-residue motif located close to the transmembrane domain, Trp-Ser-X-Trp-Ser (WSXWS), that was predicted to be an essential component of the ligand-binding site of cytokine receptors (D'Andrea et al., 1990; Bazan, 1990). For several members of the cytokine receptor family, generation of high-affinity receptors requires the formation of hetero-oligomers (reviewed by Sato and Miyajima, 1994). While the oligomeric structure of the Epo-R on the cell surface is not yet unequivocally established, various lines of evidence indicate that Epo-R exists as a multimeric complex (Mayeux et al., 1991; Yoshimura and Lodish, 1992; Watowich et al., 1992; Kishi et al., 1993; Hermine et al., 1994; Watowich et al., 1994). Epo is thought to generate signals for both differentiation and proliferation, however little is known regarding the differentiation signals (Noguchi et al., 1988; Liboi et al., 1993; Chiba et al., 1993; Maruyama et al., 1994). Two cytoplasmic regions of opposing function were defined: a membrane-proximal serin-rich region that is sufficient for transduction of proliferative signals, and a distal region that downmodulates intracellular growth signals (D'Andrea et al., 1991; Quelle and Wojchowski, 1991). In contrast to many other hematopoietic growth factors, Epo acts primarily on relatively late, committed, erythroid precursor cells in adult bone marrow or fetal liver (Sawada et al., 1990). Because the Epo-R gene is expressed

NATO ASI Series, Vol. H 94
Gene Technology
Edited by A. R. Zander, W. Ostertag,
B. V. Afanasiev, F. Grosveld

in embryonic stem cells as well as multipotential cell lines, it may also affect proliferation and/or differentiation of multiple types of cells (Heberlein et al., 1992). However, ectopic expression and activation of the Epo-R in pluripotent progenitors has no major influence on their commitment to the erythroid lineage (Dubart et al., 1994; Pharr et al., 1994).

Constitutive (dysregulated) overexpression of Epo causes a proliferative syndrome resulting in erythrocytosis and splenomegaly (Semenza et al., 1989; Hoatlin et al., 1990; Villeval et al., 1992). In mice a similar pathological form of erythropoiesis is induced by the Friend virus (reviewed by Ostertag et al., 1987; Kabat, 1989; D'Andrea et al., 1992). Friend virus is an acutely transforming murine erythroleukemia retrovirus that is comprised by a replication-defective spleen focus-forming virus (SFFV) and a replication-competent Friend murine leukemia virus (F-MuLV). The pathogenic activity for induction of erythroleukemia resides in the viral *env* gene of the SFFV which encodes the recombination/deletion membrane glycoprotein gp55 (Bilello et al., 1980). The gp55 binds to and activates the Epo-R, resulting in growth of host cells irrespective of the presence of Epo (Aizawa et al., 1990; Li et al., 1990). While the bulk of gp55 binds to the Epo-R in the endoplasmic reticulum, some of the gp55-Epo-R complex is expressed on the cell surface; the latter fraction is thought to be essential for the transforming ability of SFFV (Li et al., 1987). Productive interaction between the gp55 and the Epo-R requires both the transmembrane and the exoplasmatic region of the two polypeptides (Chung et al., 1989; Zon et al., 1992). The binding of gp55 does not interfere with the binding of Epo to the Epo-R, implying that the Epo-R has discrete binding sites for gp55 and Epo (Casadevall et al., 1991).

Although the Epo-R can be activated by binding of either Epo or gp55, it is thus far not clear whether this elicits identical responses. To date, mutational and biochemical analyses were performed in systems that only showed a proliferative response (D'Andrea et al., 1992; Kozak et al., 1993). We therefore were interested in determining to what extent gp55 mimics the biological effects of the normal receptor ligand Epo. To do this, we constructed selectable retroviral vectors on the basis of SFFV that contained either the gp55-encoding *env* gene (SFenv) or the cDNA coding for monkey Epo (SFepo) and used two distinct approaches to compare these vectors. In the first, we monitored the survival, proliferation, and differentiation of erythroid cells in *in vivo* assays. The results of this study show that both vectors induce a polycythemia and splenomegaly in susceptible mice however with vector-specific differences in disease phenotype and progression. The SFepo caused a greater increase of hematocrit and erythrocytes, whereas the SFenv increased the pool of late erythroid progenitors (CFU-E), resulting in a more severe splenomegaly. The distribution of cytologically recognizable erythroid precursors was shifted more towards immature cell types by the SFenv than it was by SFepo.

In the second approach, we infected the multipotential progenitor cell line FDCP-mix with either SFepo or SFenv. FDCP-mix cells have a normal karyotype, are non-leukemic and can undergo multilineage differentiation in response to hematopoietic growth factors or when cultured in association with marrow stromal cells (Spooncer et al., 1986; Heyworth et al., 1990; Just et al., 1991). For their survival and proliferation they require the continuous presence of interleukin-3 (IL-3); in absence of IL-3 they rapidly die by a mechanism involving apoptosis (Spooncer et al., 1986; Williams et al., 1990). The results of this study show that under conditions that favour differentiation rather than self-renewal Epo-expressing FDCP-mix cells underwent erythroid differentiation. This was not observed with SFenv infected cells, unless Epo was exogenously supplied. Neither of the transferred genes could substitute for IL-3 in sustaining self renewal of the FDCP-mix cells. The results of our studies suggest that Epo acts on a wide spectrum of committed precursor cells, whereas only later stages of erythroid differentiation respond to gp55. Furthermore, gp55 appears to promote proliferation over differentiation, whereas Epo preferentially drives differentiation.

Results

Selectable SFFV-based retroviral vectors

To directly compare the biological effects of gp55 with Epo, we engineered selectable retroviral vectors (Fig. 1) on the basis of a MPSV/SFFV hybrid construct expressing either gp55 (SFenv) or Epo (SFepo). The SFenv construct consisted of 5'LTR, gag, 3'pol and the neomycin resistance gene sequences from the neo-MPSV (Stocking et al., 1985; Ostertag et al., 1986) and the env gene and 3'LTR from the F-SFFVp (Lilly-Steeves strain; Linemeyer et al., 1980). Replacing the env gene of SFenv by cDNA sequences of the monkey (Macaca fascicularis) Epo gene (Lin et al., 1986) generated the SFepo vector (Fig. 1). Reverse orientation of the Epo gene yielded the control vector (Fig. 1). Due to the replication strategy of retroviruses the U3 region of the 5'LTR is derived from the F-SFFVp after one round of replication. Both viruses were of substantial titre and biologically functional (data not shown). To allow for viral spread of the defective SFenv and SFepo vectors in infected mice, clones of amphotropic GA12 (Markowitz et al., 1988) SFenv or SFepo producing cells were superinfected with replication competent ecotropic helper virus (F-MuLV clone 643/22N; Ostertag et al., 1980). Since instability was observed with a similar vector by others (Joyner and Bernstein, 1983) we wanted to know if SFenv was stable on transfer. Serial passage of the virus in cell culture did not reveal apparent alterations of genome size and structure. Further, expression of neo and gp55 did not segregate (Fig. 2).

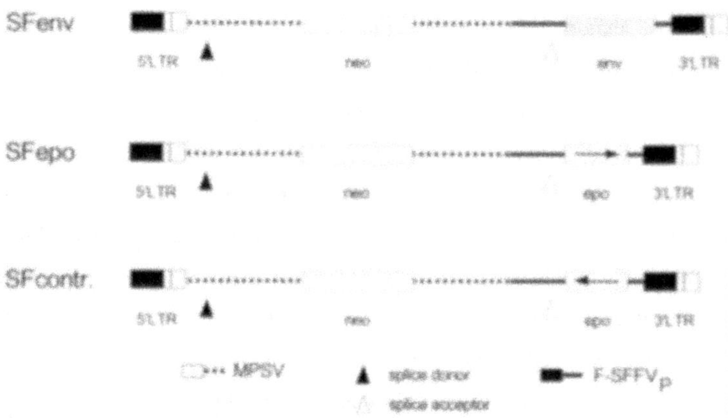

FIG. 1. Selectable SFFV-based retroviral vectors. The vectors are hybrid constructs that contain MPSV sequences and a neo^R gene in the 5' part and sequences of F-SFFV$_P$ (Lilly-Steeves strain) at the 3' end. The latter encompass the splice acceptor, *env* gene, and 3' LTR in the case of SFenv. In SFepo, the *env* gene is replaced by cDNA of the monkey Epo gene. Inverse orientation of these sequences generated the control vector SFcontr.

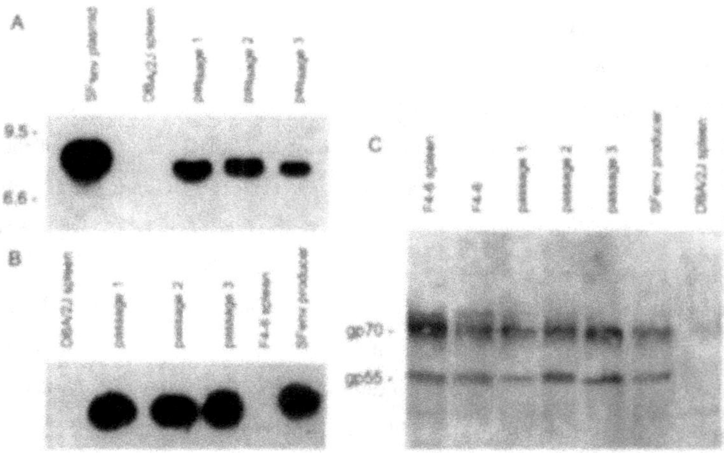

FIG. 2. Serial passage of SFenv in vivo. (A) Southern blot analysis of spleen DNA after serial passage of SFenv in DBA/2J mice. DNA (10 μg) was digested with SacI and then analyzed by Southern blotting and hybridization to a neo gene specific probe. The position of the molecular size markers (in kilodaltons) are indicated on the left. (B) Neomycin phosphotransferase assay. F4-6, Friend cell line (Pragnell et al., 1977). (C) Western blot analysis of serially passaged spleen cell and other cell lines. Protein extracts were separated by SDS-PAGE before being transferred to GeneScreen Plus and probed with a goat anti-Friend gp70 antiserum.

SFepo and SFenv cause acute erythroid polycythemia with different disease phenotypes

To test whether gp55 mimics the effect of overexpression of Epo *in vivo*, either SFenv or SFepo viruses were injected into DBA/2J mice. Similar titers, ranging from 5 x 10^1 to 6.5 x 10^4 G418-transfer units (GTU) were injected per animal. Survival, spleen weight, hematologic parameters, Epo titer and erythroid progenitor cells were monitored up to 3 months after virus injection and compared with those of uninfected and control virus injected animals. Both, SFepo and SFenv induced an acute erythroid polycythemia, however with differences in disease progression and phenotypic expression.

Hematological abnormalities

The striking feature of SFepo was a highly elevated hematocrit that was sustained for more than two months after virus injection (Fig. 3 and Fig. 4). Already at day 10, SFepo infected mice showed a significant increase of hematocrit and erythrocytes. This was paralleled by an increase of Epo up to 60 mU/ml in the plasma of infected mice. At later time points (days 27 and 36) Epo levels decreased to about 30 mU/ml, but it sustained high erythrocyte and reticulocyte counts and a high hemoglobin content during this time period (Fig. 3). After 3 months however, surviving SFepo treated mice showed a decrease in hematocrit levels (Fig. 4).

After SFenv infection, hematocrit and erythrocytes increased to lower levels as compared to SFepo. Since the hemoglobin content did not vary significantly from control infected and uninfected animals, red blood cells were hypochromic. Spleen weight progressively increased up to 2.5 g. In contrast, SFepo caused a less severe splenomegaly and at later time points the spleen size regressed (Fig. 3). Animals injected with the control virus did not show pathological symptoms.

Viral spread and disease progression

Viral spread was monitored during the course of infection using the spleen as indicator. The selection marker neo^R served as a tag in a quantitative PCR analysis. Estimates of infection rates showed that already 10 days after injection of SFenv nearly all spleen cells harbored a provirus. The spread of SFepo was only moderate with 15% at day 10, and this value decreased to low levels (3 %) at day 36. The infection rate of SFcontr reached maximally 0.5%.

The differences in viral spread could be due to differences in vertical (proliferation potential of infected cells) or horizontal transmission (infection). To test whether the different phenotypes of polycythemia induced by SFepo or SFenv were not a mere consequence of varying viral spread, animals were injected with increasing doses. Similar high hematocrit values with moderate increases in spleen weight were obtained with SFepo, independent of the amount of virus injected, i.e. between 2.5 x 10^4 and 1 x 10^2 GTU, although the onset of disease was delayed in mice injected with a low dose of virus (Fig. 4). In addition, at a comparable hematocrit, SFenv infected mice had substantially larger spleens than SFepo treated mice; and inverse, at comparable spleen weight, SFepo induced a higher hematocrit than SFenv. Thus, over a wide range of input titres of virus the disease phenotypes were distinct. Because of this, differences in viral spread may rather be a consequence of the nonidentical effects of Epo and gp55.

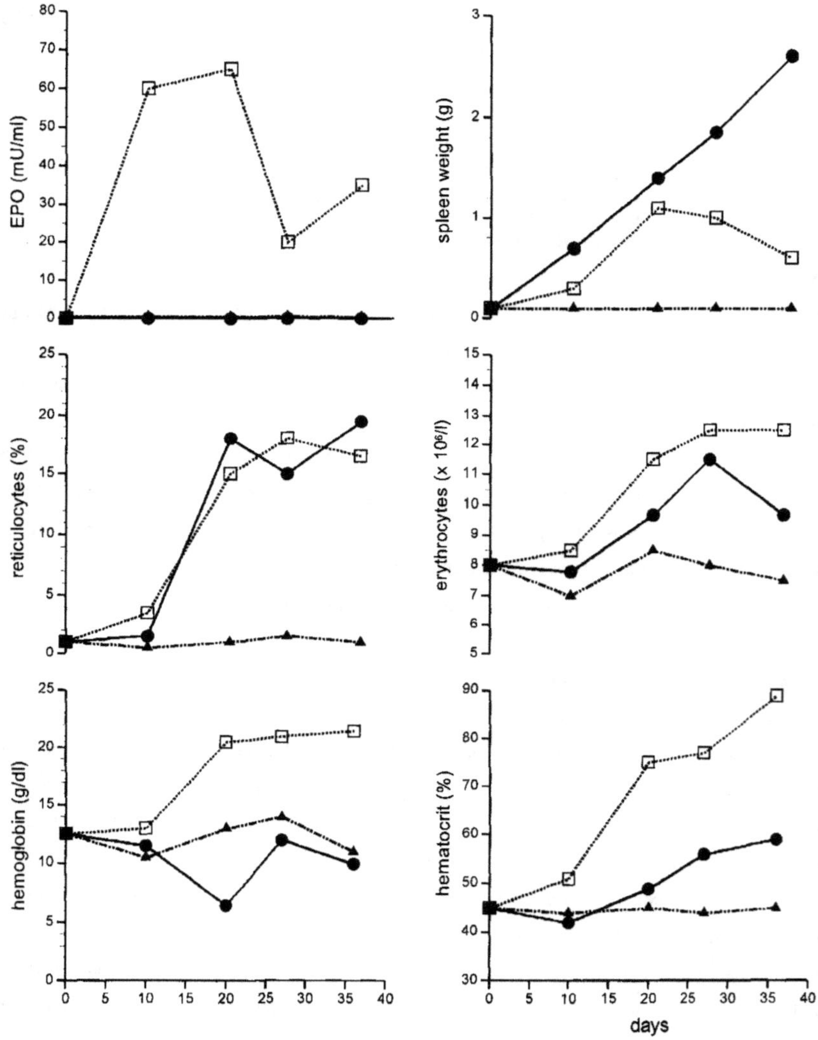

FIG. 3. Analysis of spleen weight and erythroid indices in peripheral blood of mice injected with the virus SFenv, SFepo, or SFcontr. Female DBA/2J mice were injected intravenously with matched titers of SFenv, SFepo, and SFcontr. At increasing intervals after injection, the animals were analyzed for changes in spleen weight and hematologic parameters. Data represent means (n = 3). Open squares, SFepo; closed circles, SFenv; closed triangles, SFcontr.

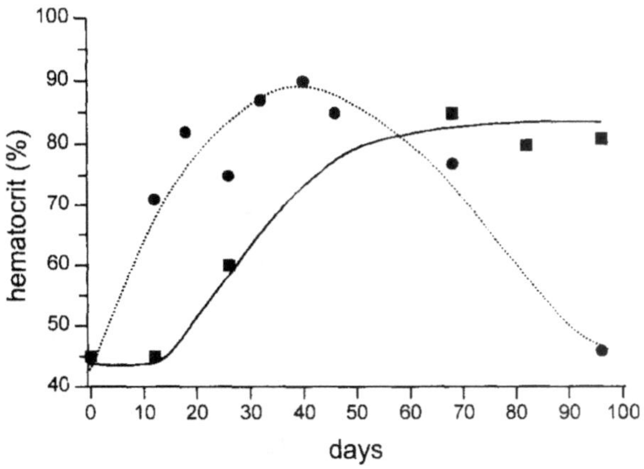

FIG. 4. Increase of hematocrit in peripheral blood of mice injected with various titers of SFepo. Female DBA/2J mice were injected intravenously with 2.5×10^4 GTU (circles, dashed line), or 1×10^2 GTU of SFepo (squares, solid line). At increasing intervals after injection, the animals were analyzed for hematocrit levels. Data represent means (n = 3).

The disease induced by SFenv was progressive and resulted in death of all infected animals within 5-7 weeks (Fig. 5). In contrast however, diseased mice infected with SFepo survived considerably longer but the disease surprisingly was also progressive (Fig. 5). The difference in survival between SFenv and SFepo infected animals was observed over a wide range of input titers of virus (2.5×10^4 to 1×10^2 GTU).

A target cell for a virus can be a cell in which the virus multiplies or in which viral gene expression alters the cell's behaviour. This change in cell behaviour can be a direct or indirect effect of viral gene expression. It has been shown that the MPSV U3 region is crucial for generating the MPSV-specific myeloproliferative disease (Stocking et al., 1985). The persistence of the SFepo virus leading ultimately to disease and death of infected animals (Fig. 4 and Fig. 5) is a surprising result when compared with MPSV-based vectors expressing GM-CSF or Multi-CSF that did not result in disease and were undetectable within 2 days after beeing injected into mice (C. Stocking, unpublished). It is possible that Epo has a higher transformation potential than GM-CSF or Multi-CSF, however the most likely explanation is that the SFFV-LTR provides a more efficient enhancer/promotor. Transient transcription assays performed in multipotential cells (C. Baum, unpublished) support the latter hypothesis.

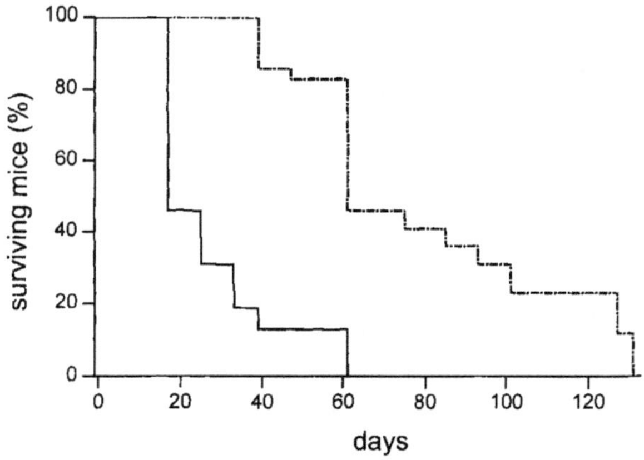

FIG. 5. Survival of DBA/2J mice after infection with SFepo or SFenv. Female DBA/2J mice were injected intravenously with 2.5 x 10^4 GTU of SFepo (dashed line) or SFenv (solid line).

Expansion of erythroid progenitors in spleen and bone marrow

Since the major targets of Epo and gp55 are late erythroid progenitor cells, we determined the relative number of CFU-E in spleen and bone marrow of infected mice. SFenv caused a high increase in CFU-E (Fig. 6). Nearly all colonies were G418 resistant, i.e. contained the provirus. This was consistent with the PCR estimates of infection rate. Further, most if not all of the infected cells grew independently of Epo. Although SFepo also increased the number of CFU-E, the increase was to a much lower extent (Fig. 6). Only a small fraction of the colonies were derived from virus-infected progenitors and these cells grew in an autocrine fashion. This indicated that the majority of CFU-E was recruited into the progenitor pool by the exocrine action of SFepo containing cells. CFU-E generated by exogenously administered or by endogenously produced Epo were of similar appearance. The SFenv-induced CFU-E however differed from those as they were larger but less hemoglobinized. The differences in CFU-E content and appearance of SFepo and SFenv infected mice were observed over a wide range of input titers of virus.

Up to 30% of bone marrow BFU-E of SFenv infected mice contained provirus as determined by G418-resistance. However, colony growth and differentiation was dependent on exogenous Epo (Table 1). No G418-resistant bone marrow BFU-E could be detected in SFepo infected animals. This is most likely a consequence of decreased spread of this vector. Although these mice contained high Epo titers, the BFU-E population was unchanged compared to controls. This suggests that early erythroid progenitors (BFU-E) are not affected by Epo or gp55.

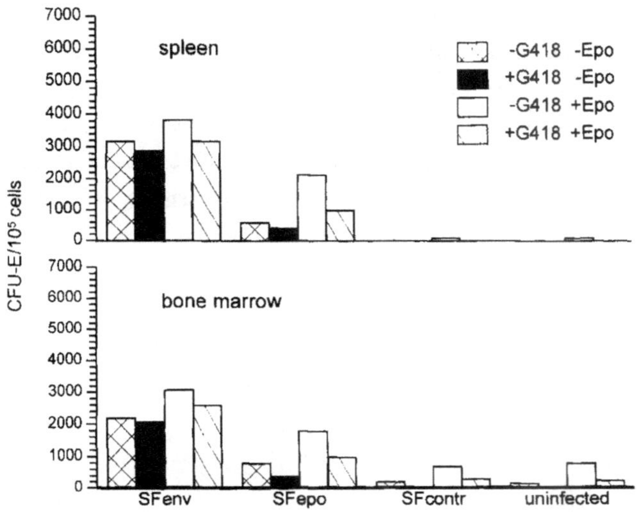

Fig. 6. Increase of CFU-E in spleen and bone marrow after infection of mice with SFepo, SFenv, and SFcontr. Animals were analyzed 10 days after injection. Average spleen weights were 1.31 and 0.29 g for SFenv- and SFepo-infected mice, respectively. The hematocrit was 54% for both groups. SFcontr.-infected animals were without symptoms. Spleen and femural bone marrow cells were suspended and cloned at a density of 10^5 cells per ml in 0.9% methylcellulose under different conditions (with or without 1.6 mg of G418 per ml and with or without 1U of Epo per ml). After 48 h, benzidine-positive colonies were recorded. Data represent means (n = 3).

TABLE I

BFU-E in bone marrow of mice injected with SFepo, SFenv, and SFcontr

Virus injected	BFU-E / 10^6 bone marrow cells (mean ± SD) under indicated conditions		
	+ G418, - Epo	- G418, + Epo	+ G418, + Epo
SFepo	0	138 ± 51	40 ± 22
SFenv	0	111 ± 11	0
SFcontr	0	130 ± 18	0
None	0	120	0

Femural bone marrow from animals described in Fig. 6 was suspended and cloned at a density of 10^6 cells per 3 ml in 0.9 % methylcellulose under different conditions (with or without 1.6 mg of G418 per ml; with or without 1 U of Epo per ml). Benzidine-positive bursts were scored after 9 days of incubation.

Distribution of morphologically recognizable erythroid cells

In conjunction with the analysis of BFU-E and CFU-E, which were identified only indirectly by their progeny in colony assays, we monitored the effect of SFepo and SFenv on morphologically recognizable erythroid cells in the same set of animals. Both viruses increased the relative number of nucleated erythroid cells in spleen and bone marrow, however the effect was more pronounced with SFenv (Fig. 7). Erythroid expansion occurred mainly at the expense of lymphoid cells. When corrected for the increase in spleen weight however, the total number of lymphocytes appeared unaltered. The effect on differential cell counts was milder in the bone marrow, reflecting the major extension of hemopoiesis in the spleen (Fig. 7).

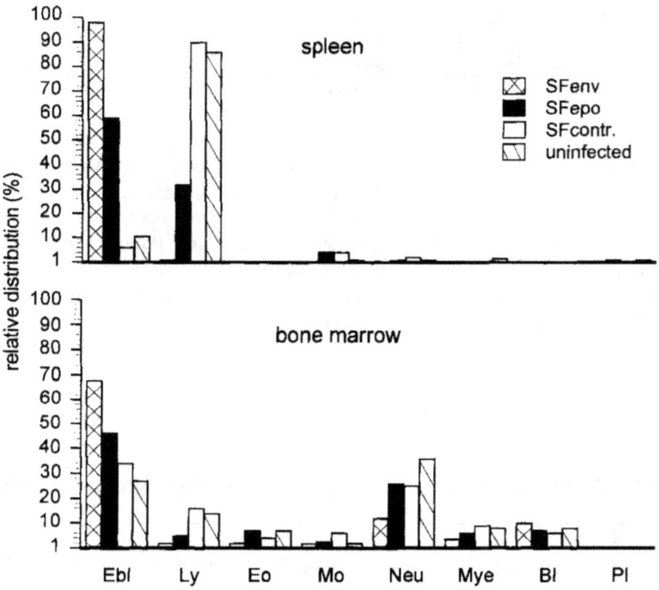

FIG. 7. Differential cell counts in spleen and bone marrow of mice injected with SFenv, SFepo, and SFcontr. Mice were analyzed 10 days after injection. Spleen and bone marrow cells were concentrated on slides by cytocentrifugation and stained with benzidine and May-Grünwald Giemsa. Data are means (n = 3). Abbreviations: Ebl, erythroblasts; Ly, lymphocytes; Eo, eosinophils; Mo, monocytes; Neu, neutrophils; Mye, myelocytes; Bl, blast cells; Pl, plasma cells.

Nucleated erythroid cells were further discriminated by their differentiation state. In uninfected animals, the ratio between more mature benzidine positive (orthochromatic and polychromatic normoblasts) and benzidine negative cells (proerythroblasts and basophilic erythroblasts) were on the average 4.66 in spleen and 3.57 in bone marrow. Infection with SFepo led to a shift to more immature cell types, the ratios being 2.75 and 2.30 for spleen and bone marrow respectively. This shift was even more pronounced with SFenv, where the ratios reached values close to one (1.14 in bone marrow and 1.29 in spleen). Thus, although SFenv induced a more enhanced amplification of erythroid progenitors, the maturation of these cells was more retarded as compared to mice infected with SFepo.

SFepo and SFenv have different effects on the multipotential progenitor cell line FDCP-mix

Our *in vivo* results suggests that erythropoiesis supported by the physiological ligand Epo has different proliferation and differentiation dynamics than that driven by the pathological effector gp55. A useful *in vitro* model in which effects on factor-dependent proliferation and erythroid differentiation can be analysed are the FDCP-mix cell lines. FDCP-mix cells have a normal karyotype, are non-leukemic and can undergo multilineage differentiation in response to hematopoietic growth factors or when cultured in association with marrow stromal cells (Spooncer et al., 1986; Heyworth et al., 1990; Just et al., 1991). For their survival and proliferation they require the continuous presence of interleukin-3 (IL-3); in absence of IL-3 they rapidly die. Infection of FDCP-mix cells by either SFepo or SFenv thus presented us with an opportunity to further clarify the differential cellular responses to Epo and gp55.

Erythroid differentiation of FDCP-mix cells after infection with SFepo

To obtain clones of FDCP-mix cells that exhibit ectopic expression of Epo, the FDCP-mix cell line A4 was infected with SFepo by cocultivation with lethally irradiated amphotropic helper cells, producing pseudotyped retroviral vectors, and then selected in the presence of G418 and IL-3. Cultures of SFepo-infected clones in which the intact viral genome was confirmed by Southern analysis of genomic DNA, were then tested for Epo release. Conditioned medium of all clones contained detectable amounts of biologically active Epo, ranging from 20 mU/ml to 300 mU/ml. In the presence of a high concentration of IL-3 most clones showed blast cell morphology with some differentiation into erythroid cells. The parental FDCP-mix cells can be induced to differentiate along the erythroid lineage by supplementing their growth medium with fetal calf serum (FCS) instead of horse serum (HS), reducing the level of IL-3 and addition of Epo. The SFepo infected cells were thus cultured in the presence of FCS and a reduced amount of IL-3. A time course of the SFepo infected FDCP-mix cells revealed a progressive and, for some clones, almost synchronous increase of erythroid differentiating cells during 4 to 8 days of culture in absence of exogenous Epo, that was almost identical to that found in presence of added Epo (Fig. 8). Withdrawal of IL-3 from the SFepo infected cultures resulted in survival of a few erythroid cells and a complete loss of viability within 4 days.

SFenv expression does not substitute for Epo in FDCP-mix cells

To obtain clones of FDCP-mix cells that express gp55, the FDCP-mix cell line A4 was infected with SFenv and then selected in the presence of G418 and IL-3. Cultures of SFenv-infected clones in which the intact viral genome was confirmed by Southern analysis of genomic DNA, were then tested for expression of the gp55 protein by western analysis using a polyclonal goat antibody against F-MuLV gp70 as probe (Bilello et al., 1980). Clones expressing gp55 and showing release of functional SFenv virus (data not shown) were then used for further analysis. In contrast to the SFepo infected cells, the SFenv infected cells did not differentiate into erythroid cells after induction of differentiation by FCS and low IL-3 in absence of Epo, although these cells were still responsive to exogenous Epo (Fig. 8). Withdrawal of IL-3 from SFenv infected cultures resulted in complete loss of viability within 2-3 days.

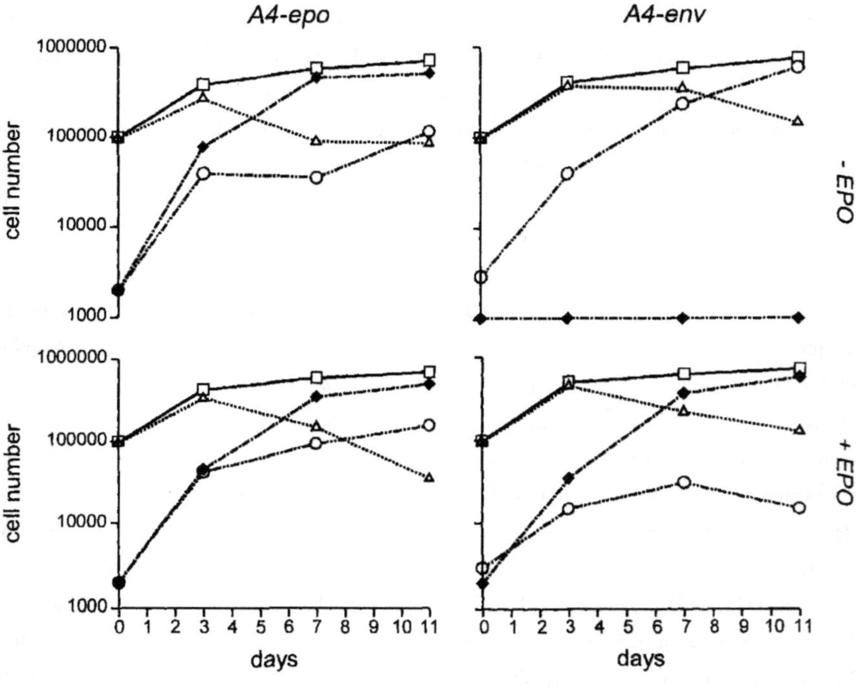

FIG. 8. Proliferation and differentiation of FDCP-mix cells infected with either SFepo or SFenv. SFepo-infected FDCP-mix clone #4 cells (left panels) or SFenv-infected FDCP-mix clone #5 cells (right panels) were washed and exposed to new medium containing 20 % FCS and 1% Wehi-3B-CM as a source of IL-3 with (lower panels) or without (upper panels) 2 U per ml recombinant murine Epo. Open squares, total number of cells; open triangles, blast cells; open circles, granulocytes and macrophages; closed rhomb, erythroid cells.

Summary and conclusions

In the first part of our study we have directly compared the natural Epo-R ligand Epo and its pathological agonist gp55 in *in vivo* assays. The genes coding for Epo and gp55 were transferred by selectable retroviral vectors into hemopoietic cells. The use of the neo^R gene as a selectable marker thus allowed the determination of viral spread and the analysis of infected cells and their progeny. Our results showed that both vectors induced a polycythemia and splenomegaly in susceptible mice, but with differences in disease phenotype and progression.

The disease induced by SFenv was progressive and resulted in death of all infected animals, whereas diseased mice infected with SFepo survived considerably longer. SFepo caused a greater increase in mature cells, whereas SFenv increased immmature erythroid cells (CFU-E) and produced a more severe splenomegaly. Most CFU-E and spleen cells were infected by SFenv whereas spread of SFepo was only moderate. Neither vectors influenced the number of BFU-E.

Evidence for different cellular responses to Epo and gp55 was also obtained from the second part of our study, the infection of multipotential FDCP-mix cells with either SFepo or SFenv. Under conditions that favour differentiation rather than self-renewal, Epo-expressing FDCP-mix cells underwent erythroid differentiation in the absence of added Epo. This was not observed with SFenv infected FDCP-mix cells, unless Epo was exogenously supplied. Neither Epo nor gp55 could substitute for IL-3 in sustaining self renewal of FDCP-mix cells. It thus appears that gp55 promotes proliferation over differentiation, whereas Epo preferentially drives differentiation. Although the SFenv was expressed in the multipotential FDCP-mix cells, we were unable to detect any effect of gp55 expression on differentiation or proliferation of these cells. In contrast however, ectopic expression of Epo in FDCP-mix cells led to differentiation into erythroid cells. Thus, the target cells for SFepo appears somewhat broader and distinct from those of SFenv, which appears to be the late CFU-E.

How can the differential effects of Epo and gp55 be explained on a mechanistic level? Epo and gp55 have distinct binding sites on the Epo-R (Casadevall et al., 1991; Kozak et al., 1993). In addition, various lines of evidence indicate that the Epo-R exists as a multimeric complex (Casadevall et al., 1991; Mayeux et al., 1991; Yoshimura and Lodish, 1992; Kishi et al., 1993; Hermine et al., 1994; Watowich et al., 1994). Thus, one possibility could be that gp55 competes with a putative second Epo-R subunit or with associated proteins. The amount of the second subunit present in a specific cell type would then determine gp55 binding and activation of the EpoR. Recent evidence has shown that the Epo-R transmits signals for both differentiation and proliferation (Noguchi et al., 1988; Liboi et al., 1993; Chiba et al., 1993; Maruyama et al., 1994). Signalling generated by activation of a putative subunit or associated proteins could possibly be involved in differentiation. Competition of gp55 with this subunit would then influence the balance between proliferation and differentiation. Further analysis of the molecular mechanisms underlying the differential effects of

Epo and gp55 in FDCP-mix cells should help to understand not only leukemic processes but also normal hematopoiesis.

Acknowledgements

This work was supported by the Deutsche Forschungsgemeinschaft (Ju 197/3-1, Ju 197/3-2, No 89/5-4 and Os 31/14). The Heinrich-Pette-Institut is financially supported by Freie und Hansestadt Hamburg and the Bundesministerium für Gesundheit.

References

Aizawa S, Suda Y, Furuta Y, Yagi T, Takeda N, Watanabe N, Nagayoshi M, Ikawa Y (1990) *Env*-derived gp55 gene of Friend spleen focus-forming virus specifically induces neoplastic proliferation of erythroid progenitor cells. EMBO J 9:2107-2116

Bazan JF (1990) Structural design and molecular evolution of a cytokine receptor superfamily. Proc Natl Acad Sci USA 87:6934-6938

Bilello JA, Colleta G, Warnecke G, Koch G, Frisby D, Pragnell IB, Ostertag W (1980) Analysis of the expression of spleen focus-forming virus (SFFV)-related RNA and gp55, a Friend and Rauscher virus-specific protein. Virology 107:331-344

Casadevall N, Lacombe C, Müller O, Gisselbrecht S, Mayeux P (1991) Multimeric structure of the membrane erythropoietin receptor of murine erythroleukemia cells (Friend cells) - Cross-linking of erythropoietin with the spleen focus-forming virus envelope protein. J Biol Chem 266:16015-16020

Chiba T, Nagata Y, Machide M, Kishi A, Amanuma H, Sugiyama M, Todokoro K (1993) Tyrosine kinase activation through the extracellular domains of cytokine receptors. Nature 362:646-648

Chung SW, Wolff L, Ruscetti SK (1989) Transmembrane domaine of the envelope gene of a polycythemia-inducing retrovirus determines erythropoietin-independent growth. Proc Natl Acad Sci USA 86:7957-7960

D'Andrea AD, Lodish HF, Wong GG (1989) Expression cloning of the murine erythropoietin receptor. Cell 57:277-285

D'Andrea AD, Fasman GD, Lodish HF (1990) Erythropoietin receptor and interleukin 2 receptor beta chain: A new receptor family. Cell 58:1023-1024

D'Andrea AD, Yoshimura A, Youssoufian H, Zon LI, Koo J-W, Lodish HF (1991) The cytoplasmic region of the erythropoietin receptor contains nonoverlapping positive and negative growth-regulatory domains. Mol Cell Biol 11:1980

D'Andrea AD, Moreau J-F, Showers MO (1992) Molecular mimicry of erythropoietin by the spleen focus-forming virus gp55 glycoprotein: the first stage of Friend virus-induced erythroleukemia. Biochim Biophys Acta 1114, 31-41

Dubart A, Feger F, Lacout C, Goncalves F, Vainchenker W, Dumenil D (1994) Murine pluripotent hematopoietic progenitors constitutively expressing a normal erythropoietin receptor proliferate in response to erythropoietin without preferential erythroid cell differentiation. Mol Cell Biol 14, 4834-4842

Heberlein C, Fischer K-D, Stoffel M, Nowock J, Ford A, Tessmer U, Stocking C (1992) The gene for erythropoietin receptor is expressed in multipotential hematopoietic and embryonal stem cells: evidence for differentiation stage-specific regulation. Mol Cell Biol 12:1815-1826

Hermine O, Dong YJ, Goldwasser E (1994) Effects of butyrate on the erythropoietin receptor of cell line IW201. Blood 84:811-814

Heyworth CM, Dexter TM, Kan O, Whetton AD (1990) The role of hemopoietic growth factors in self-renewal and differentiation of IL-3-dependent multipotential stem cells. Growth Factors 2:197-211

Hoatlin ME, Kozak SL, Lilly F, Chakraborti A, Kozak CA, Kabat D (1990) Activation of erythropoietin receptors by Friend viral gp55 and by erythropoietin and downmodulation by the murine FV-2^{r} resistance gene. PNAS 87:9985-9989

Joyner DL, Bernstein A (1983) Retrovirus transduction: Segregation of the viral transforming function and the herpes simplex virus tk gene in infectious Friend spleen focus forming virus thymidine kinase vectors. Mol Cell Biol 3:2191-2202

Just U, Stocking C, Spooncer E, Dexter TM, Ostertag W (1991) Expression of the GM-CSF gene after retroviral transfer in hematopoietic stem cell lines induces synchronous granulocyte-macrophage differentiation. Cell 64:1163-1173

Kabat D (1989) Molecular biology of Friend viral erythroleukemia. Curr Top Microbiol Immunol 148:1-42

Kishi A, Chiba T, Sugiyama M, Machide M, Nagata Y, Amanama H, Taira H, Katsumata T, Todokoro K (1993) Erythropoietin receptor binds to Friend virus gp55 through other membrane components. Biochem Biophys Res Commun 192:1131-1138

Kozak SL, Hoatlin ME, Ferro FE, Majumdar MK, Geib RW, Fox MT, Kabat D (1993) A Friend virus mutant that overcomes Fv-2^{rr} host resistance encodes a small glycoprotein that dimerizes, is processed to cell surfaces, and specifically activates erythropoietin receptors. J Virol 67:2611-2620

Liboi E, Carroll M, D'Andrea AD, Mathey-Prevot (1993) Erythropoietin receptor signals both proliferation and erythroid-specific differentiation. Proc Natl Acad Sci USA 90:11351-11355

Li JP, Bestwick RK, Spiro C, Kabat D (1987) The membrane glycoprotein of Friend spleen focus-forming virus: Evidence that the cell surface component is required for pathogenesis and that it binds to a receptor. J Virol 61:2782-2792

Li JP, D'Andrea AD, Lodish HF, Baltimore D (1990) Activation of cell growth by binding of Friend spleen focus-forming virus gp55 glycoprotein to the erythropoietin receptor. Nature 343:762-764

Lin FK, Lin C-H, Lal P, Browne JK, Ergie JC, Smalling R, Fox GM, Chen KK, Castro M, Suggs S (1986) Monkey erythropoietin gene: Cloning, expression and comparison with the human erythropoietin gene. Gene 44:201-209

Linemeyer DL, Ruscetti SK, Menke JG, Scolnik M (1980) Recovery of biologically active spleen focus-forming virus from molecularly cloned spleen focus-forming virus pBR 322 circular DNA by cotransfection with type C retroviral DNA. J Virol 35:710-721

Markowitz D, Goff S, Bank A (1988) Construction and use of a safe and efficient amphotropic packaging cell line. Virology 167:400-406

Maruyama K, Miyata K, Yoshimura A (1994) Proliferation and erythroid differentiation through the cytoplasmic domain of the erythropoietin receptor. J Biol Chem 269:5976-5980

Mayeux P, Lacombe C, Casadevall N, Chretien S, Dusanter I, Gisselbrecht S (1991) Structure of the murine erythropoietin receptor complex. J Biol Chem 266:23380-23385

Noguchi T, Fukumoto H, Mishina Y, Obinata M (1988) Differentiation of erythroid progenitor (CFU-E) cells from mouse fetal liver cells and murine erythroleukemia (TSA8) cells without proliferation. Mol Cell Biol 8:2604-2609

Ostertag W, Vehmeyer B, Fagg IB, Pragnell IB, Paetz MC, leBousse F, Smadja-Joffe B, Klein C, Jasmin C, Eisen H (1980) Myeloproliferative virus, a cloned murine sarcoma virus with spleen focus-forming properties in adult mice. J Virol 33:573-582

Ostertag W, Seliger B, Kollek R, Stocking C, Bergholz U, Smadja-Joffe F (1986) The myeloproliferative sarcoma virus retains transforming functions after introduction of a dominant selectable marker gene. J Gen Virol 67:1361-1371

Ostertag W, Stocking C, Johnson GR, Kluge N, Kollek R, Franz T, Hess N (1987) Transforming genes and target cells of murine spleen focus-forming viruses. Adv Cancer Res 48:193-355

Pharr PN, Ogawa M, Hofbauer A, Longmore GD (1994) Expression of an activated erythropoietin or a colony-stimulating factor 1 receptor by pluripotent progenitors enhances colony formation but does not induce differentiation. Proc Natl Acad SciUSA 91:7482-7486

Pragnell IB, Ostertag W, Paul J (1977) The expression of viral and globin genes during differentiation of the Friend cell. Exp Cell Res 108:269-278.

Quelle FW, Wojchowski DM (1991) Localized cytosolic domains of the erythropoietin receptor regulate growth signaling and down-modulate responsiveness to granulocyte-macrophage colony-stimulating factor. Proc Natl Acad Sci USA 88:4801-4805

Sato N, Miyajima A (1994) Multimeric cytokine receptors: common versus specific functions. Curr Biol 6:174-179

Sawada K, Krantz SB, Dai CH, Koury ST, Horn ST, Glick AD, Civin CI (1990) Purification of human blood burst forming units-erythroid and demonstration of the evolution of the erythropoietin receptor. J Cell Physiol 142:219-230

Semenza GL, Traystman MD, Gearhart JD, Antonarakis SE (1989) Polycythemia in transgenic mice expressing the human erythropoietin gene. Proc Natl Acad Sci USA 86:2301-2305

Spooncer E, Heyworth CM, Dunn A, Dexter TM (1986) Self-renewal and differentiation of interleukin-3-dependent multipotent stem cells are modulated by stromal cells and serum factors. Differentiation 31:111-118

Stocking C, Kollek R, Bergholz U, Ostertag W (1985) Long terminal repeat sequences impart hematopoietic transformation property to the myeloproliferative sarcoma virus. Proc Natl Acad Sci USA 82:5746-5750

Villeval JL, Metcalf D, Johnson GR (1992) Fatal polycythemia induced in mice by dysregulated erythropoietin production by hematopoietic cells. Leukemia 6:107-115

Watowich S, Yoshimura A, Longmore G, Hilton D, Yoshimura Y, Lodish HF (1992) Homodimerization and constitutive activation of the erythropoietin receptor. Proc Natl Acad Sci USA 89:2140-2144

Watowich SS, Hilton DJ, Lodish HF (1994) Activation and inhibition of erythropoietin receptor function: role of receptor dimerization. Mol Cell Biol 14:3535-3549

Williams GT, Smith CA, Spooncer E, Dexter TM, Taylor DR (1990) Hemopoietic colony stimulating factors promote cell survival by suppressing apoptosis. Nature 343:76-79

Yoshimura A, Lodish HF (1992) *In vitro* phosphorylation of the erythropoietin receptor and an associated protein, pp130. Mol Cell Biol 12:706-715

Youssoufian H, Longmore G, Neumann D, Yoshimura A, Lodish HF (1993) Structure, function, and activation of the erythropoietin receptor. Blood 81:2223-2236

Zon Li, Moreau J-F, Koo J-W, Mathey-Prevot B, D'Andrea AD (1992) The erythropoietin receptor transmembrane region is necessary for activation by the Friend spleen focus-forming virus gp55 glycoprotein. Mol Cell Biol 12:2949-2957

CONTRIBUTION OF THE RETROVIRAL LEADER TO GENE TRANSFER AND EXPRESSION IN PACKAGING CELLS AND MYELOID STEM AND PROGENITOR CELLS

Christopher Baum, Hans-Georg Eckert and Wolfram Ostertag
Heinrich-Pette-Institut für Experimentelle Virologie und Immunologie
an der Universität Hamburg
Martinistr.52
D-20251 Hamburg
Federal Republic of Germany

Introduction

Retroviral vectors designed to transduce hemopoietic stem cells have to be optimised with respect to basic requirements for gene transfer and expression. (i) The cis-regulatory elements of the genomic vector RNA interacting with retroviral proteins supplied from packaging cells must allow for optimal packaging, reverse transcription and integration into the genome of the target cell. (ii) The cis-regulatory elements of the integrated 'proviral' vector DNA interacting with the cellular transcriptional machinery must guarantee efficient and long-lasting gene expression in the target cell and its progeny.

Conventional vectors rely on cis-regulatory elements of the murine type C retrovirus Moloney murine leukemia virus (MoMuLV), and its relative, Moloney murine sarcoma virus (MoMuSV) [review in Miller, 1992a]. The cis-regualtory elements incorporated into retroviral vectors generally include the U3 regions of the long terminal repeat (LTR) and leader sequences. The U3-region displays the vector's enhancer/promotor setting, greatly contributing to tissue tropism of vector mediated gene expression [review in Stocking et al., 1993]. The leader can be referred to as the complete 5' untranslated part of the vector RNA that is necessary to fulfill the retroviral life cycle. Hence, it contains the R and U5 region of the LTR, the primer binding site (PBS) for the tRNA-primer initiating reverse transcription and the packaging region mediating efficient pseudotyping of the vector RNA in packaging cells. State of the art MoMuLV/MoMuSV-based vectors allow for efficient and safe production of high-titre stocks in safety-modified packaging cells of fibroblast origin [review in Miller, 1992a]. This has been achieved by inclusion of an extended leader containing gag-coding sequences that were altered to eliminate residual expression of viral protein [Bender et al., 1987; Armentano et al., 1987]. However, when targeting hemopoietic cells, MoMuLV-based vectors suffer from low gene

NATO ASI Series, Vol. H 94
Gene Technology
Edited by A. R. Zander, W. Ostertag,
B. V. Afanasiev, F. Grosveld

expression levels in primitive stem cells and derived myeloid lineages [review in Stocking et al., 1993; Moritz and Williams, 1994]. This can be attributed to a lack of transactivation from lymphotropic MoMuLV-type U3 regions as well as to transcriptional inhibition mediated by a yet unidentified repressor binding at the primer binding site (PBS) of the retroviral leader [Baum et al., submitted].

The seemingly stem cell specific block involving the PBS for $tRNA^{Pro}$ of MoMuSV/MoMuLV can be relieved by usage of a PBS for $tRNA^{Gln}$, derived from dl587rev, a recombined virus arosen from a defective MoMuLV and an endogenous murine retrovirus [Colicelli and Goff, 1987, Grez et al., 1990]. A leader fragment containing this permissive PBS and downstream located sequences extending into the packaging region of the retroviral leader has been used to generate the murine embryonic stem cell virus (MESV), the first retroviral vector cloned mediating successfull gene transfer and expression in primitive embryonic stem cells (ES) [Grez et al., 1990]. Usage of the PBS from dl587rev/MESV is also associated with ability of long-term gene expression from retroviral vectors in hemopoietic cells *in vivo* [Hawley et al., 1992; 1993]. Hence, MESV-based vectors have been suggested to be promising candidates for use in somatic gene therapy protocols involving retroviral gene transfer to hemopoietic cells [review in Stocking et al., 1993; Moritz and Williams, 1994].

Here, we provide evidence that sequences of MoMuSV/MoMuLV located in the retroviral packaging region also contain an inhibitory element blocking retroviral gene expression in a stage and lineage-specific manner within the hemopoietic system. Again, this block can be abolished by utilization of corresponding sequences of dl587rev/MESV. Thus, leader-sequences formerly believed to be solely involved in interaction with *viral* proteins in vector packaging might also interact with *cellular* proteins contributing to vector transcription. However, sequence degeneration in the packaging region might contribute to slightly suboptimal packaging of MESV-based vectors.

Methods and Results

Homolgy comparison of the leader sequences of MESV and MoMuLV.

Our MESV-based vectors contain leader sequences from dl587rev between Kpn1 (bp 37 of the leader) and Pst1 (bp 574) [sequence numbering according to Colicelli and Goff, 1987]. The remaining vector backbone is derived from PCMV (PCC4 embryonal carcinoma cell passaged myeloproliferative sarcoma virus [Franz et al., 1986; Hilberg et al., 1987]). Leader sequence comparison between the MESV-based vectors and MoMuLV-based vectors reveals a number of point mutations, insertions and deletions contributing to less than 80% homology at the DNA level in the exchanged leader fragment (Fig 1). 5 point mutations convert the PBS for $tRNA^{Pro}$

of MoMuLV/MoMuSV to the PBS for $tRNA^{Gln}$ of dl587rev/MESV. The PBS for $tRNA^{Pro}$ overlaps in 17 of its 18 bp with the binding site for the yet unidentified stem-cell specific repressor also active in hemopoietic stem cells, therefore it was also referred to as repressor binding site [Kempler et al., 1993]. Downstream of the PBS follows the packaging region (Ψ). Sequences absolutely necessary for packaging are located between Msc1 (bp 216) and Xma3 (bp355). Removal of that fragment is sufficient to exclude packaging of retroviral transcripts, a feature used for construction of safety-modified packaging cells like GP&env86 or GP&envAM12 [Markowitz et al., 1988a; 1988b]. Within this fragment, a stretch of remarkably low sequence conservation (less than 50%) is observed, named nonhomology region 1 [Collicelli and Goff, 1987]. Evidence obtained from analysis of oncogenic avian retroviruses suggests that a 254-bp sequence immediately downstream of the PBS, hence included in the exchanged fragment and spanning the nonhomology region 1, resembles an RNA enhancer [Boerkoel and Kung, 1992]. Sequences downstream of Pst1 (bp 574) are identical between MESV-based vetors and MoMuLV/MoMuSV-based vectors like LNL6 [Miller and Rosman, 1989]. Thus, MESV-based vectors contain the extended leader including 418 bp 5'gag coding region whose AUG (bp627) has been mutated to TAG to preclude translation initiation. This region is required for optimal packaging [Bender et al., 1987; Armentano et al., 1987].

We further studied the implications of the sequence degeneration in the packaging region of MESV for vector transfer and expression.

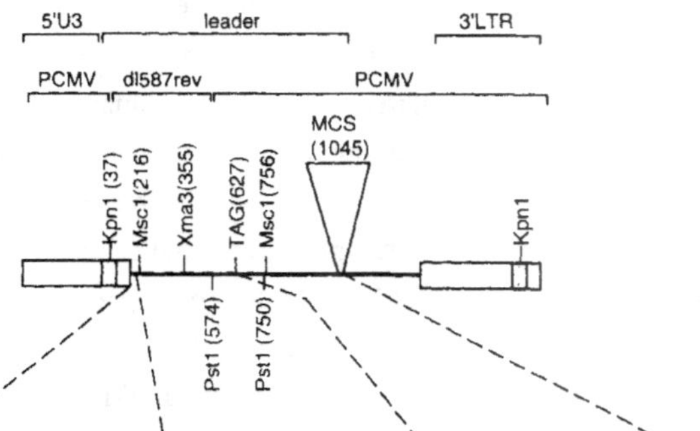

	PBS	Ψ	gag
size (bp)	18 (146-163)	463 (164-626)	418 (627-1044)
source	dl587rev	dl587rev (164-573) PCMV=MoMuLV (574-626)	PCMV=MoMuLV
function	primer binding site repressor binding site	packaging region RNA enhancer (?)	gag coding region packaging extension
homology to MoMuLV[1]	13/18 (72%)	365/463 (79%)	418/418 (100%)

Fig.1 : Homology comparison of leader regions from MESV-based vectors and MoMuLV-based vectors. [1]Homology comparison according to Colicelli and Goff, 1987.

A second repressor function is located in the MoMuSV/MoMuLV-leader downstream of the primer binding site; in contrast, the corresponding region of MESV is permissive.

PCME-CAT (PCMV-LTR followed by MESV-PBS and truncated at Msc1 (bp 216 of the leader), originally named PCMV(BX)/Bal587-CAT) and PCMO-CAT (PCMV-LTR followed by MoMuSV-PBS and truncated at Msc1 (bp 216), originally named PCMV(BX)/Balp-CAT) have been described previously [Grez et al., 1990]. In order to elucidate the influence of the leader region located downstream of the PBS for vector expression, CAT-reporter gene vectors were cloned containing the U3 region of PCMV followed by the complete leader of MESV (PCMEME-CAT) or MoMuSV/MoMuLV (PCMOMO-CAT), respectively. To seperate the influence of downstream sequences from that of the PBS, CAT-vectors were cloned containing the MESV-PBS followed by the complete downstream leader sequences of MoMuSV/MoMuLV (PCMEMO-CAT) and, vice versa, the PBS of MoMuSV followed by downstream leader sequences of MESV (PCMOME-CAT). This was achieved by exchanging a 560 bp Msc1 fragment of the MESV leader with the corresponding 528 bp fragment of V-MDR (a MoMuLV/MoMuSV based vector kindly provided by A. Deisseroth, MD Anderson Cancer Center, Houston, Texas). These vectors were transfected into the pluripotent murine myeloid stem cell line FDC-Pmix 15S and granulocyte/macrophage precursor line FDC-P1, the human T-lymphoblast CCRF-CEM as well as in the murine fibroblast NIH3T3 cells as previously described [Baum et al., 1994]. The results are shown in Fig.2.

As reported elsewhere, PCME-CAT, containing the MESV-PBS, has the same activity as a construct bearing the PCMV-U3 only, consequently the PBS of MESV is completely permissive in all cell lines studied, irrespective of developmental stage or lineage [Baum et al., submitted]. The same is true for the complete MESV-based leader contained in PCMEME-CAT. However, the MoMuSV-PBS restricts gene expression in a stage and lineage specific manner, being most repressive in multipotent myeloid stem cells. Interestingly, the fusion of the Msc1-leader fragment of MoMuSV/MoMuLV to the permissive MESV-PBS restricts CAT-accumulation in a similar tissue-specificity as the MoMuSV-PBS per se: silencing activity can be detected in myeloid cells, but not in fibroblasts or T-lymphoblasts. However, the Msc1-leader fragment of MoMuSV does not add significantly to the repressor activity of the MoMuSV-PBS, what can be concluded from the comparison of PCMO-CAT and PCMOMO-CAT; nor does the Msc1-leader fragment of MESV relieve the repression triggered by the MoMuSV-PBS, if PCMO-CAT and PCMOME-CAT are compared.

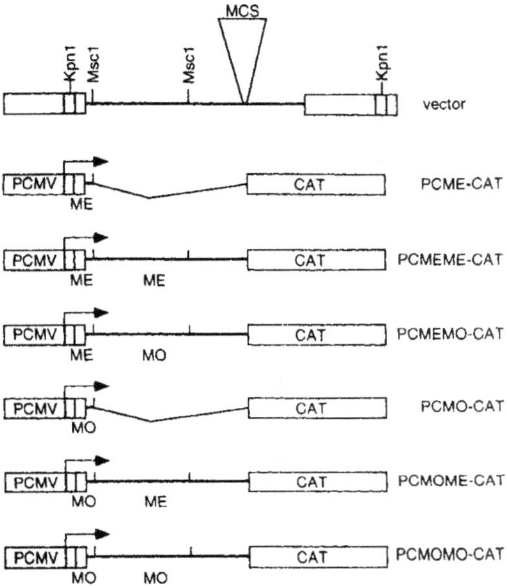

	relative CAT-activity[1] in			
	FDC-Pmix myeloid stem cell	FDC-P1 G/M progenitor cell	CCRF-CEM T-lympho-blastoma	NIH3T3 embryonal fibroblast
PCME-CAT	1	1	1	1
PCMEME-CAT	1.06	0.95	0.92	1.2
PCMEMO-CAT	0.51	0.33	0.83	1.1
PCMO-CAT	0.3	0.43	0.93	0.97
PCMOME-CAT	0.37	0.3	nd	nd
PCMOMO-CAT	0.27	0.23	nd	nd

Fig.2 : Influence of retroviral leader on gene expression in different developmental lineages. [1]CAT-activity was determined with the CAT-Elisa Kit (Boehringer Mannheim), and corrected for transfection efficiency measured by cotransfected galactosidase-activity. Values shown represent mean values of at least two independent transfections and are given as ratio to PCME-CAT, whose activity was set as 1. ME, sequence derived from MESV. MO, sequence derived from MoMuSV/MoMuLV.

Influence of the MESV-leader on packaging.

Taking into consideration the relatively low degree of conservation in the leader sequences necessary for packaging of the vector's genomic RNA, the feasibility of the MESV-leader to obtain high-titre producer clones has to be investigated. We therefore established large numbers of amphotropic producer clones for MESV-based retroviral vectors via retroviral infection of amphotropic packaging cells GP&envAM12 with supernatants from ecotropic producers.

The vectors transduce the dominant selectable marker human multiple drug resistance 1 (h*mdr*1) [review in Gottesman and Pastan, 1993]. The basic vector architecture is comparable, ie 5'U3-leader-h*mdr*1-3'LTR. In the MESV based vectors, the LTRs are of PCMV, myeloproliferative sarcoma virus (MPSV), Harvey murine sarcoma virus (HaMuSV) or spleen focus forming virus (SFFVp) origin [Baum et al., submitted]. In comparison to the MoMuLV/MoMuSV-based V-MDR1, the MESV-based vectors consistently yield slightly lower titres or lower incidence of high-titre clones (Table 1). Titres were determined by infection of ecotropic fibroblast packaging cells GP&env86 with supernatants from amphotropic producer clones and subsequent selection in the presence of 20 μg/ml colchicine and are given as *mdr*1-transduced units (MTU) per ml supernatant.

vector	V-MDR	HA-MDR	MP-MDR	SF-MDR	PC-MDR
U3	MoMuLV	HaMuSV	MPSV	SFFVp	PCMV
leader	MoMuSV/MoMuLV	MESV	MESV	MESV	MESV
RTA[1]	1	1	1.1	0.8	0.6
mean MTU/ml[2]	7×10^5	1.7×10^5	2.4×10^5	1×10^5	3.2×10^4
clones analysed[3]	36	37	49	21	12
10^2-10^3/ml	0	3	4	5	4
10^3-10^4/ml	0	9	7	5	3
10^4-10^5/ml	0	6	4	5	3
10^5-10^6/ml	24	17	30	6	2
$>10^6$/ml	12	2	4	0	0

Table 1: Incidence of high-titre clones as well as mean titer is only slightly suboptimal in usage of MESV-based vectors. [1]RTA, relative transcriptional activity in comparison to MoMuLV/MoMuSV-based vectors in a fibroblast background, as determined by CAT-Elisa and Northern Blot (not shown). [2]mean MTU/ml, mean titer of all clones analysed per ml supernatant. [3]clones analysed, numer of clones and distribution according to titre/ml supernatant.

For PC-MDR and SF-MDR, the reduced titers can partially be explained by a weaker activity of the U3 region of PCMV or SFFVp in fibroblasts [Baum et al., submitted]. However, even MESV-vectors like MP-MDR or HA-MDR containing U3-regions with equal activity as MoMuLV in a fibroblast background showed a reduced incidence of high-titre clones in comparison to V-MDR, the vector based on the MoMuSV/MoMuLV leader. The mean titer was about five-fold reduced. Still individual clones can be selected that yield titres high enough for gene transfer into hemopoietic cells. Since the MESV-leader does not disturb gene expression in a fibroblast background (see above), transcription is most propably not the limiting event. Likewise, vector-rearrangements as a possible reason for lower titres have not been detected in the amphotropic clones by Southern blot analysis (not shown). Therefore, sequence degeneration in the packaging region of MESV might result in slightly suboptimal packaging

Retroviral vectors cloned to elucidate the influence of the retroviral leader on vector expression and packaging.

Since the *mdr*1-vectors based on the leaders of MESV or MoMuSV/MoMuLV also differ in respect to the multiple cloning site and the length of the residual env-sequences, we cloned a series of completely identical vectors differing only with respect to the retroviral leader region and enharboring the neomycin-resitance gene (neoR) as a dominant selectable marker (Fig.3). These contain either the U3-region of SFFVp (myelotropic) or the U3-region of MoMuLV (lymphotropic). The leader is either of MESV-type or of MoMuSV/MoMuLV-type. One SFFVp-driven vector contains a hybrid leader composed of the MESV-sequences including the PBS up to Msc1 (bp 216) followed by the 532 bp Msc1-fragment of MoMuSV/MoMuLV identical to the assembly in PCMEMO-CAT (see above).

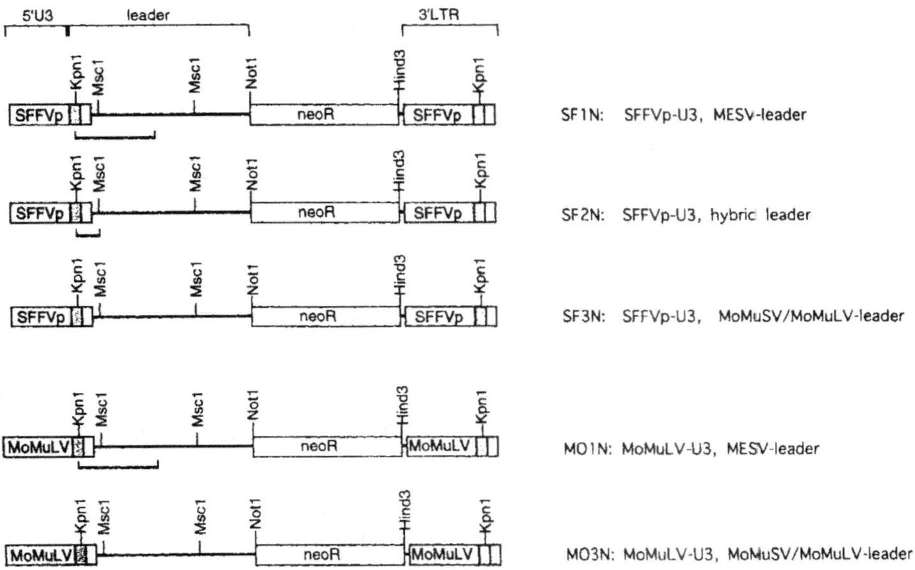

Fig.3: A series of retroviral vectors cloned to further elucidate the impact of the retroviral leader on gene transfer and expression. Brackets below vector figures indicate sequences derived from dl587rev.

Mass cultures were established in ecotropic GP&env86 packaging cells stably transfected with these proviral constructs by electroporation [Baum et al., 1994]. We are in the process of establishing amphotropic producer clones by retroviral transduction of GP&envAM12 cells with supernatants of the ecotropic producers. These will then be scored for release of functional viruses in conventional GTU-assays. Subsequently, retroviral transduction of hemopoietic cells belonging to different developmental compartments should enable us to determine the influence of the leader on retroviral gene expression in more detail.

Discussion

Murine retroviruses and derived retroviral vectors have been used for a number of years to study developmental properties of hemopoietic cells. More recently, the boom in somatic gene therapy studies contributed to an enhanced demand for retroviral vectors that are both safe and highly efficient for long-term gene expression in lymphoid or myeloid cell. However, the efficiency of retroviral gene expression obtained by conventional MoMuLV-based vectors is quite low. Insufficient transactivation as well as progredient extinction of the activity of retroviral cis-regulatory elements seems to compromise their use in a variety of approaches relying on high levels and long-term maintenance of gene expression [reviews in Miller, 1992b; Stocking et al., 1993; Moritz and Williams, 1994]. Since the level as well as the duration of vector-mediated gene expression is determined by interactions of the vector's cis-elements with trans-acting proteins of the host cell, a systematic analysis of the molecular prerequisites for efficient vector function is needed.

In this report, we focus on the impact of the retroviral leader on vector function. A negative-acting element triggered by the PBS of MoMuLV/MoMuSV has already been reported that inhibits high levels as well as long-term maintenance of vector expression in myeloid cells [Hawley et al., 1992; 1993; Baum et al., submitted]. A similar seemingly stem cell specific repressor function of the MoMuLV/MoMuSV PBS has been noticed for murine embryonal carcinoma cells and ES cells [Weiher et al., 1987; Grez et al., 1990]. In this study, we obtained evidence hat sequences located downstream of the PBS in the putative packaging region of the leader also contain cis-acting elements with inhibitory function in myeloid cells. In contrast, both the PBS and the packaging region of MESV are completely permissive for U3-driven gene expression in myeloid cells. Again, similar observations have been made in ES cells, resulting in usage of a leader fragment containing both the PBS and the packaging region of the endogenous retrovirus dl587rev to assemble MESV [Grez et al., 1990].

The influence of the leader could be at the level of transcription or post-transcriptionally (RNA secondary structure, RNA half life, translation). The remarkable tissue specificity of the repressory effect does not rule out one of these possibilities. Interestingly, sequences located in a 254 bp fragment of the retroviral leader located immediately downstream of the PBS might be involved in a positive-feedback activation of the 5'LTR and consequent downregulation of the 3'LTR, as was suggested from studies in avian retroviruses. The mechanism might resemble that of an RNA enhancer, in that it is dependend on the position as well as the orientation in respect to the 5'LTR [Boerkoel and Kung, 1992]. Interestingly, in the corresponding region of MESV, the greatest degree of sequence degeneration is observed (nonhomology region 1). A more precise analysis of the repressor element(s) is desirable to elucidate the nature of the repressor as well as to design optimised leaders for retroviral vectors.

The latter task has to take into consideration the role of the leader region in packaging of genomic vector RNA. The sequences necessary for packaging have initially been mapped in a small fragment of the MoMuLV-leader located immediately downstream of the retroviral splice donor [Mann et al., 1983]. Hence they were excluded from gag/pol and env -expression constructs incorporated into safety-modified packaging cells [Danos and Mulligan, 1986; Markowitz et al., 1989]. Later studies emphasized the necessity of a far extended leader extending into the 5' portion of the gag-coding region to obtain high-titre packaging [Bender et al., 1987; Armentano et al., 1987]. Vectors containing these sequences to optimize packaging are addressed as gag+vectors [Bender et al., 1987]. MESV-vectors are identical to MoMuLV/MoMuSV-vectors in regard to gag+, however, both the fragment absolutely necessary for packaging and the subsequent packaging region are less than 80% conserved in comparison to MoMuLV/MoMuSV-vectors. This could account for a slightly suboptimal packaging, a possible explanation for our observations that the first generation of MESV-vectors does not as routinely yield high-titre producer clones as conventional MoMuLV/MoMuSV vectors do. On the other hand, degeneration of vector-leader sequences in comparison to corresponding sequences contained in transgenes of packaging cell lines makes sense for safety reasons, because it reduces the possibility of homologous recombination necessary to create replication competent viruses out of replication-incompetent packaging cell transgenes. Therefore, dissection of the leader's cis-elements involved in packaging and transcription is an important prerequisite to develop fully permissive retroviral vectors for gene therapy applications.

Acknowledgements

This work was supported by the Deutsche Krebshilfe. The Heinrich-Pette-Institut is financially supported by Freie und Hansestadt Hamburg and the Bundesministeriumfür Gesundheit.

References

Armentano D, Yu SF, Kantoff PW, von Ruden T, Anderson WF and Gilboa E (1987) Effect of internal viral sequences on the utility of retroviral vectors. J. Virol., 61:1647-1650

Baum C, Forster P, Hegewisch-Becker S, and Harbers K (1994) An optimised electroporation protocol applicable to a wide range of cell lines. BioTechniques, 17: in press

Baum C, Hegewisch-Becker S, Eckert H-G, Stocking C and Ostertag W (submitted) Retroviral vectors for efficient gene expression in myeloid cells.

Bender MA, Palmer TD, Gelinas RE and Miller AD (1987) Evidence that the packaging signal of Moloney murine leukemia virus extends into the gag region. J. Virol. 61: 1639-1646

Boerkoel CF and Kung H-J (1992) Transcriptional interaction between retroviral long terminal repeats (LTRs): Mechanism of 5'LTR suppression and 3'LTR promoter activation of *c-myc* in avian B-cell lymphomas. J. Virol., 66: 4818-4823

Colicelli J and Goff SP (1987) Isolation of a recombinant murine leukemia virus utilizing a new primer tRNA. J. Virol., 57: 37-45

Franz T, Hilberg F, Seliger B, Stocking C and Ostertag W (1986) Retroviral mutants efficiently expressed in embryonal carcinoma cells. Proc. Natl. Acad. Sci. USA, 83: 3292-3296

Gottesman MM and Pastan I (1993) Biochemistry of multidrug resistance mediated by the multidrug transporter. Annu. Rev. Biochem., 62:385-427

Grez M, Akgün E, Hilberg F, and Ostertag W (1990) Embryonic stem cell virus, a recombinant murine retrovirus with expression in embryonic cells. Proc. Natl. Acad. Sci. USA, 87: 9202-9206

Hawley RG, Fong AZC, Brns BF and Hawley TS (1992) Transplantable myeloproliferative disease in mice induced by an interleukin 6 retrovirus. J. Exp. Med., 176: 1149-1164

Hawley RG Fong AZC, Ngan BY, de Lanux VM, Clark SC and Hawley TS (1993) Progenitor cell hyperplasia with rare development of myeloid leukemia in interleukin 11 bone marrow chimeras J. Exp. Med., 178: 1175-1188

Hilberg F, Stocking C, Ostertag W and Grez M (1987) Functional analysis of a retroviral host-range mutant: altered long terminal repeat sequences allow expression in embryonal carcinoma cells. Proc. Natl. Acad. Sci. USA, 84: 5232-5236

Kempler G, Freitag B, Berwin B, Nanassy O and Barklis E (1993) Characterization of the Moloney murine leukemia virus stem cell-specific repressor binding site. Virology, 193: 690-699

Mann R, Mulligan RC and Baltimore D (1983) Construction of a retrovirus packaging mutant and its use to produce helper-free defective retrovirus. Cell, 33: 153-159

Markowitz D, Goff S and Bank A (1988a) A safe packaging line for gene transfer: seperating viral genes on two different plasmids. J. Virol., 62: 1120-1124

Markowitz D, Goff S and Bank A (1988b) Construction and use of a safe and efficient amphotropic packaging cell line. Virology, 167: 400-406

Miller AD (1992a) Retroviral vectors. Curr. Top. Microbiol. Immunol., 158: 1-24

Miller AD (1992b) Human gene therapy comes of age. Nature (London), 357: 455-460

Miller AD and Rosman GJ (1989) Improved retroviral vectors for gene transfer and expression. BioTechniques, 7: 980-990

Moritz T and Williams DA (1994) Gene transfer into the hematopoietic system. Curr. Opin. Hematol., 1: 423-428

Stocking C, Grez M and Ostertag W (1993) Regulation of retrovirus infection and expression in embryonic and hematopoietic stem cells. in: Virus Strategies. Molecular biology and pathogenesis. pp 433-455. Doerfler W and Böhm P (eds). VCH Verlagsgesellschaft mbH, Weinheim, Federal Republic of Germany

Weiher H, Barklis E, Ostertag W, and Jaenisch R (1987) Two distinct sequence elements mediate retroviral gene expression in embryonal carcinoma cells. J. Virol., 61: 2742-2746

Site-directed Mutagenesis of the cGMP Phosphodiesterase Inhibitory γ Subunit from Bovine Rods: New Mechanisms of Catalytic Subunits Inhibition and Activation

Valery M. Lipkin, Andrey M. Alekseev, Vladimir A. Bondarenko,
Khakim G. Muradov, Vasily E. Zagranichny*
Branch of Shemyakin & Ovchinnikov Institute of Bioorganic Chemistry,
Russian Academy of Sciences, Pushchino, Moscow Region, Russia

ABSTRACT

The recombinant and 30 mutant phosphodiesterase (PDE) γ subunit (PDEγ) genes were expressed by sequential transcription and translation *in vitro*. Inhibitory properties of these mutants and their interactions with PDE catalytic and transducin α subunits were studied. To explain the properties of PDEγ mutants we have proposed a new mechanism of PDE functioning: i) There are two sites on PDE catalytic complex (PDEαβ) for PDEγ binding (A and B). PDEγ sites interacting with sites A and B are structurally different. The site on PDEγ that interacts with the B-site partly overlaps with the binding

---

* **Address correspondence to:** V.Zagranichny, Branch S&O Institute of Bioorganic Chemistry, Pushchino, Moscow Region, 142292, RUSSIA.

The abbreviations used are: PDE, cGMP phosphodiesterase; PDEα, PDEβ, and PDEγ, cGMP phosphodiesterase α, β, and γ subunits, respectively; PDEαβ, phosphodiesterase catalytic dimer consisting of the α and β subunits; Tα, transducin α subunit; Tα•GTP and Tα•GDP, transducin α subunit in a complex with GTP and GDP, respectively; R24E, point mutation in PDEγ with arginine at position 24 substituted by glutamic acid (the other point substitutions in PDEγ are named similarly); wtPDEγ, wild-type PDEγ; ROS, rod outer segments; Tα•GTPγS, transducin α subunit in a complex with GTP non-hydrolysable analog, guanosine-5'-O-(3-thiotriphosphate); tPDE, trypsin activated PDE.

NATO ASI Series, Vol. H 94
Gene Technology
Edited by A. R. Zander, W. Ostertag,
B. V. Afanasiev, F. Grosveld

site for transducin α subunit complexed with GTP (Tα•GTP). ii) The PDEγ bound to the B-site provides the main contribution to inhibition of the PDE catalytic activity. iii) Tα•GTP first interacts with the PDEγ bound to the A-site and removes it in a PDEγ•(Tα•GTP) complex. iv) After removal of PDEγ from A-site, another Tα•GTP molecule is enabled to interact both with PDEαβ and with PDEγ bound to the B-site. This interaction results in the formation of a membrane-bound fully catalytically active triple complex, PDEαβ•PDEγ•(Tα•GTP).

INTRODUCTION

In vertebrate retinal rods, light-activated rhodopsin catalyses the formation of the transducin α subunit complex with GTP (Tα•GTP) (Stryer, 1986). In turn, this complex activates cGMP phosphodiesterase (PDE) (Fung *et al.*, 1981). PDE is a peripheral membrane protein consisting of three types of subunits: α (98 kDa) (Ovchinnikov *et al.*, 1987), β (98 kDa) (Lipkin *et al.*, 1990), and two identical γ ones (each of 10 kDa) (Ovchinnikov *et al.*, 1986). The latter (PDEγ) are responsible for non-competitive inhibition of the catalytic heterodimer consisting of α and β moieties (PDEαβ) (Hurley and Stryer, 1982). The degree of PDE activation depends reciprocally on the concentration of PDEγ and Tα•GTP (Wensel and Stryer, 1986). PDEγ is the basic polypeptide consisting of 87 amino acid residues. 10 of the 13 basic amino acid residues are situated between positions 24 and 45 of the structure. Earlier, using monoclonal antibodies, we demonstrated the importance of this region in the interaction between PDEγ and transducin α subunits (Lipkin *et al.*, 1988). Furthermore, it was shown (Lipkin *et al.*, 1988; Cunnick *et al.*, 1990) that PDEγ C-terminus takes part in PDE catalytic subunits inhibition.

To study the role of individual amino acid residues in the PDEγ structure, we used oligonucleotide-directed mutagenesis and expression *in vitro*, which allowed us to obtain radioactively labeled functionally active PDEγ (Muradov *at al.*, 1990). A number of PDEγ point mutants was obtained (Fig. 1 and Table 1), as well as mutants with the deletions of seven C-terminal (Δ7C) and six N-terminal (Δ2-7N) amino acid residues and two double point mutants (P27,28L and M17,57L). We determined the following properties of PDEγ mutants: (1) incorporation into rod outer segment (ROS) membranes; (2) inhibition of trypsin-activated PDE (tPDE); (3) capability of the transducin α subunit charged with GTPγS (guanosine 5'-O-(3-thiotriphosphate)), a non-hydrolysable GTP analogue, to activate tPDE inhibited by PDEγ mutants.

EXPERIMENTAL PROCEDURES

Purification of PDE and Tα•GTPγS. Isolation of ROS membranes, elution of PDE by a hypotonic buffer, its purification on DEAE-cellulose were carried out as described previously (Lipkin *et al.*, 1993). tPDE preparation was carried out as follows: purified PDE was treated with trypsin at enzyme:substrate ratio 1:50 for 5 min on ice, the reaction was stopped by adding of 5-fold excess of soybean trypsin inhibitor. Then tPDE was isolated by gel permeation chromatography on Superose 6 column HR 10/30 (Pharmacia, Sweden) in buffer A (30 mM Tris-HCl, pH 7.5, 300 mM NaCl, 5 mM $MgCl_2$, 5 mM 2-mercaptoethanol), concentrated with Centriflo C 25 and stored at -20°C in 50% glycerol. tPDE specific activity was of 400 nmol of cGMP/min/μg of protein. Tα•GTPγS was eluted from PDE-depleted ROS membranes with 30 μM GTPγS in buffer B (10 mM Tris-HCl, pH 7.5, 2 mM $MgCl_2$, 2 mM 2-mercaptoethanol). The hypotonic supernatant containing Tα•GTPγS was applied on DEAE-cellulose and eluted with the NaCl gradient (from 0.1 to 0.4 M) in buffer B. The use of Centriflo C 25 allowed us to concentrate Tα•GTPγS containing protein peak. Subsequent purification was performed on a Superose 12 column HR 10/30 (Pharmacia, Sweden) in buffer A. Tα•GTPγS was stored at -20°C in 50% glycerol.

Cloning, mutagenesis and expression of PDEγ and its mutant forms. Site-directed mutagenesis of the bovine PDEγ gene was accomplished as described previously (Lipkin *et al.*, 1993). Oligodeoxyribonucleotides used to introduce point substitutions or deletions into PDEγ coding sequence are listed in Table 1. PDEγ or mutant cDNAs were inserted into the pGEM2 plasmid treated with BamHI and PstI.

Table 1. List of PDEγ mutants obtained by site-directed mutagenesis and their phenotypes.

No No	Mutant	Phenotype wt or !mut	Mutation position in bovine PDEγ	Amino acid residue in bovine PDEγ	Replaced with	Oligonucleotide(s) used for mutagenesis [a]
1	Δ2-7N	wt	2-7		del.	5'-GGTTTTAAATAT***GG***CAGAGATCCGGT [b]
2	M17L	wt	17	Met	Leu	5'-GTCGGCCACACGCGTGCTGGGGGAC
3	P23L	wt	23	Pro	Leu	5'-ACCCGTCACTCTGAGGAAAGGGCC
4	R24G	!mut	24	Arg	Gly	5'-CCCGTCACCCCGGGAAAAGGGCC
5	R24H	wt	24	Arg	His	5'-CCGTCACTCCCCACAAAGGGCCCCC
6	R24K	wt	24	Arg	Lys	5'-CCGTCACTCCCAACAAAGGGCCCCC
7	R24E	!mut	24	Arg	Glu	5'-CCGTCACTCCGGAGAAAGGGCCC
8	R24L	wt	24	Arg	Leu	5'-GGGGGACCGGTCACCCCCCTGAAAGGGCCC
9	K25H	wt	25	Lys	His	5'-TCACTCCCAGGCACGGGCCCCCGAA
10	K25I	wt	25	Lys	Ile	5'-TCACTCCCAGGATCGGGCCCCCGAA
11	K25R	wt	25	Lys	Arg	5'-TCACTCCCAGGCGCGGGCCCCCGAA
12	K25S	wt	25	Lys	Ser	5'-TCACTCCCAGGAGCGGGCCCCCGAA
13	K25T	wt	25	Lys	Thr	5'-CACTCCCCGTACTGGGCCCCCG
14	P27L	wt	27	Pro	Leu	5'-CAGGAAAGGGCTGCCCAAATTTAAGC
15	P28L	wt	28	Pro	Leu	5'-CAGGAAAGGGCCGCTCAAATTTAAGC
16	P27,28L	wt	27,28	Pro	Leu	5'-CAGGAAAGGGCTGCTCAAATTTAAGC
17	K29T	!mut	29	Lys	Thr	5'-GGGCCCCCGACTTTTAAACAGC
18	K31R	!mut	31	Lys	Arg	5'-CCCGAAATTTAGACAGCGGCAAAC
19	K31T	!mut	31	Lys	Thr	5'-CCCGAAATTTACACAGCGGCAAAC
20	R33G	!mut	33	Arg	Gly	5'-AAATTTAAACAAGGTCAAACCAGG
21	K45T	wt	45	Lys	Thr	5'-CCCCCCAAGACCGGGGTCCAAG
22	P55L	!mut	55	Pro	Leu	5'-TGATGACATCCTGGGAATGGAAGG
23	M57L	wt	57	Met	Leu	5'-CATCCCTGGACTCGAGGGCCTGGGA
24	M17,57L	wt	17,57	Met	Leu	5'-GTCGGCCACACGCGTGCTGGGGGAC 5'-CATCCCTGGACTCGAGGGCCTGGGA
25	C68S	wt	68	Cys	Ser	5'-CACCGTCATCTCACCGTGGGAGGC
26	P69L	!mut	69	Pro	Leu	5'-CGTCATCTGCCTTTGGGAGGCCTT
27	E77G	wt	77	Glu	Gly	5'-AACCACCTGGGTTTACACGAGCT
28	H79L	!mut	79	His	Leu	5'-CCTGAGCTCTTAGAGCTGGCC
29	Y84A	!mut	84	Tyr	Ala	5'-GCTGGCCCAGGCAAGGCATCATCT
30	Δ7C	!mut	81-87		del.	5'-AACCACCTGGA***GT***AGCCCTGGACC [c]

[a] Substituted nucleotides are underlined;

[b] PDEγ cDNA sequence between printed in italic letters ***GG***, 5'-AATTTAGAACCACCAAAA coding for 2nd-7th amino acid residues, was deleted;

[c] PDEγ cDNA sequence between printed in italic letters ***GT***, 5'-CTGGCCCAGTACGGCATCATC coding for 81st-87th amino acid residues, was deleted.

The plasmids containing PDEγ or mutant genes under control of the SP6 promoter were used for *in vitro* transcription and translation in the wheat germ extract (Lipkin *et al.*, 1993) in the presence of [^{14}C]-labeled amino acids: Arg, Glu, Ile, Leu, Phe, Pro, Thr, Val with the specific activity of 0.2 Ci/mmol. The translation levels were

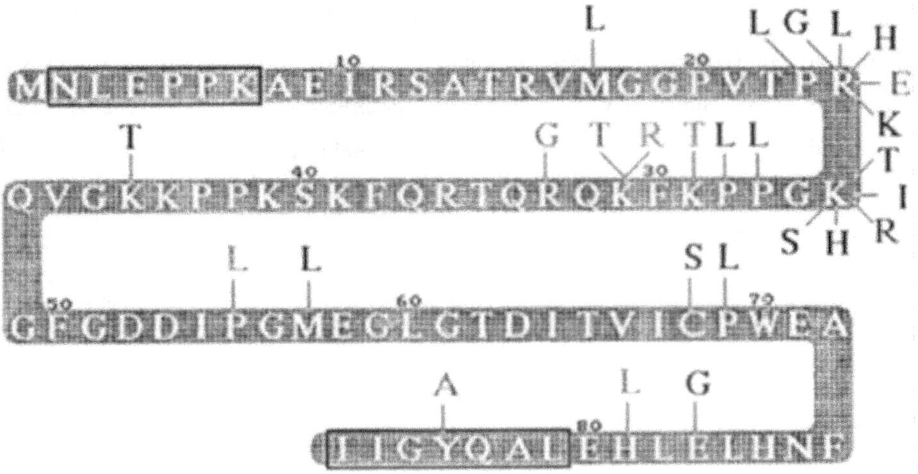

Fig. 1. The primary structure of bovine rod PDEγ and its mutants

determined by counting the radioactivity bound to Whatman glass microfiber filters in buffer B. Translation mixtures, containing PDEγ and its mutants, were used for the activity determination without additional purification.

Incorporation of PDEγ mutants into ROS membranes. ROS membranes containing about 5 μg of PDE were incubated at 4°C for 20 min with aliquots of the translation mixtures containing 4 pmol PDEγ or its mutants. Unbound radioactivity was washed out three times with 1 ml of 100 mM NaCl in buffer B at 10 000 g for 10 min. Pellets of ROS membranes containing bound PDEγ were counted. In these conditions about 20% of wild-type PDEγ (wtPDEγ), which makes approximately 2.5% of the endogenic γ subunit, was bound to the ROS membranes. This binding of wtPDEγ to ROS membranes was taken as 100%.

Determination of PDEγ inhibitory activity. Different amounts of PDEγ or its mutants were added to the mixture (100 μl) containing 80 mM Tris-HCl, pH 7.5, 4 mM $MgCl_2$, 0.01% bovine serum albumin and about 200 pM of tPDE. An addition of cGMP to 2 mM initiated the reaction. The mixture was incubated at 30°C for 10 min. The reaction was stopped by heating (90°C, 2 min). Then 20 μl (0.1 unit) of bacterial alkaline phosphatase were added and incubation was continued at 30°C for 15 min. The inorganic phosphate content was determined using $(NH_4)_6Mo_7O_{24}$ according to (Eibl and Lands, 1969). tPDE activity was plotted against the concentration of PDEγ or its mutant added. To determine the transducin influence on tPDE inhibition by PDEγ or its mutant forms, the assay was carried out in the presence of 2 μM Tα•GTPγS. tPDE specific activity of 400 nmol of cGMP/min/μg of protein was taken as 100%. The results were the mean values of three or more independent measurements with a standard deviation of less than 5%.

RESULTS AND DISCUSSION

The γ subunit incorporation into ROS membranes is easy to measure, because PDEγ expressed *in vitro* is the only radioactively labeled protein. Incorporation of the PDEγ mutants into ROS membranes characterizes their affinity to catalytic subunits. Earlier, we showed that after incorporation, the radioactively labeled PDEγ was cross-linked only with α or β subunits of PDE (Muradov *at al.*, 1990). Moreover, washing of ROS membranes with a hypotonic buffer led to co-extraction of PDE and recombinant PDEγ. PDE-depleted ROS membranes are bound to the recombinant γ subunit much more poorly.

PDEγ mutant with Arg at position 24 substituted with Gly (R24G, (Lipkin *et al.*, 1993)) poorly binds to ROS membranes and practically does not inhibit tPDE (data not shown). At the same time, other point substitutions (R24H and R24K (Lipkin *et al.*, 1993), R24L) of Arg24 residue in PDEγ (except for negatively charged glutamate in case of R24E, see below) influence neither PDEγ incorporation into ROS membranes nor catalytic subunits inhibition (data not shown). This leads to the conclusion that the positive charge at PDEγ position 24 is not essential for its function. It could be suggested that R24G mutation entails deterioration of the PDEγ tertiary structure, thus preventing its interaction with PDE catalytic subunits. Either positively charged or even uncharged amino acid residue with a long side chain at this position does not affect PDEγ wild-type phenotype. Surprisingly, proline substitutions near the position 24 (P23L (Lipkin *et al.*, 1993), P27L, P28L, and P27,28L) have no effect on any PDEγ

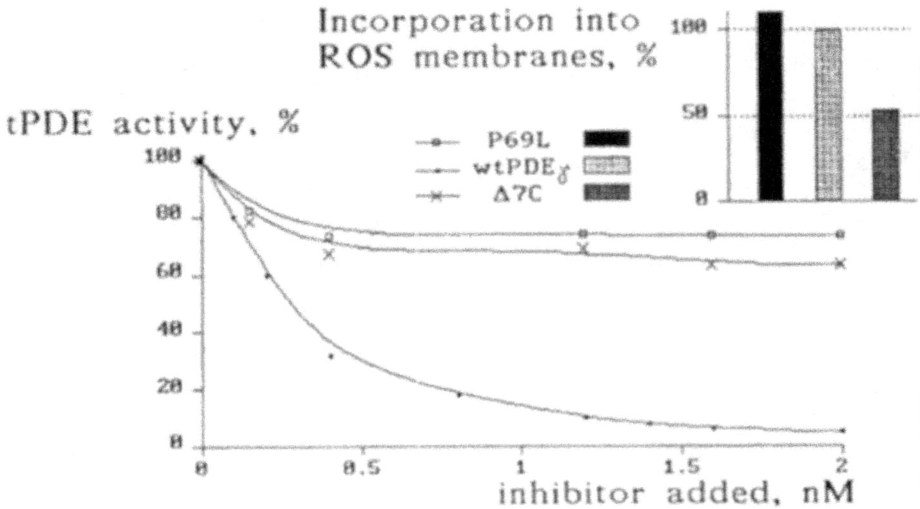

Fig. 2. Incorporation into ROS membranes and tPDE inhibition by wtPDEγ and mutants P69L and Δ7C.

property studied (data not shown), although they should cause serious conformational changes in its polypeptide chain. Together with an unusual effect of the negative charge introduced into position 24 (see below) this shows the importance of this residue in PDEγ interaction with catalytic subunits.

Other mutants under study could be divided into four groups. The first group includes 12 mutants: Δ2-7N, M17L, K25H, K25I, K25R, K25S, K25T, K45T, M57L, C68S, E77G, and M17,57L. At all concentrations, the mutants demonstrate inhibitory property analogous to that of the recombinant γ subunit (data not shown). The second group includes two PDEγ mutants, P69L and Δ7C. The removal of 7 C-terminal amino acid residues from the PDEγ polypeptide chain only slightly lowers its affinity to the catalytic subunits (probably mainly due to the absence of the Tyr84 residue, see below) and at the same time drastically decreases the inhibitory activity of this PDEγ deletion mutant (Fig. 2). The mutant P69L demonstrated an affinity to the PDEαβ identical to that of wtPDEγ, but its inhibitory activity was extremely low. In our opinion this is because the P69L mutant lacks the β-turn of its polypeptide chain in the 68-71 region. The data allow us to suggest that PDEγ interaction with catalytic subunits consists of two

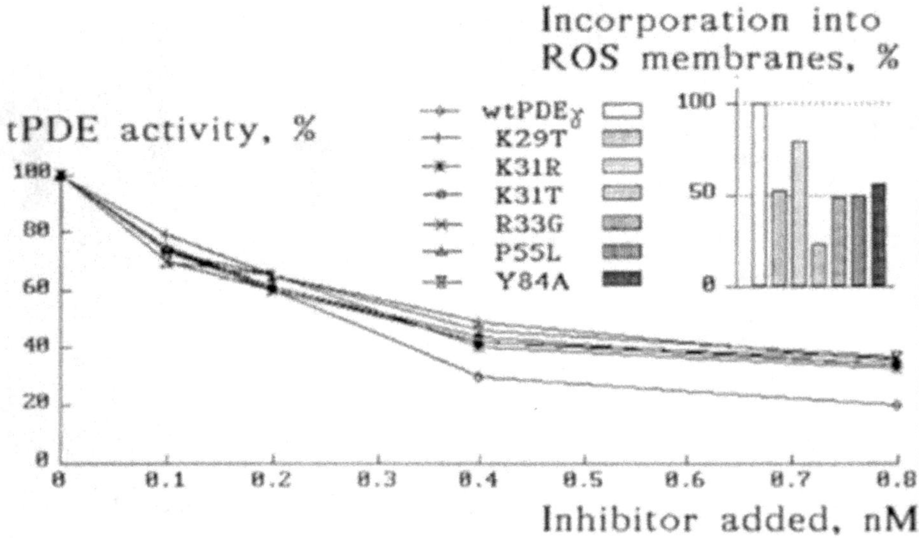

Fig. 3. Incorporation into ROS membranes and tPDE inhibition by wtPDEγ and mutants K29T, K31R, K31T, R33G, P55L and Y84A.

steps - primary binding and sequential PDEαβ inhibition, and the C-terminus plays a key role in PDEαβ inhibition. An appropriate spatial orientation of the C-terminus is required for inhibition (Lipkin *et al.*, 1993).

The third group of PDEγ mutants includes six ones (K29T, K31R, K31T, R33G, P55L, and Y84A (Lipkin *et al.*, 1993)). Each mutant has a lower affinity to PDE catalytic subunits. Their inhibitory properties are similar to that of the recombinant γ subunit at low concentrations, but are seriously deteriorated at higher concentrations (Fig. 3). This may occur because the mutants are bound to the first center on PDEαβ with normal affinity and to the second one with lower affinity.

Last two PDEγ mutants, R24E and H79L from the forth group, demonstrate some unusual features distinct from those of others studied by us. All other mutants which differed from wtPDEγ in affinity to the PDE catalytic subunits changed also their inhibitory activity. R24E with a lower affinity to rod outer segments (ROS) membranes (*i.e.* to PDE catalytic subunits) and H79L with a higher affinity are virtually indistinguishable from wtPDEγ on the PDE inhibitory plot (Fig. 4).

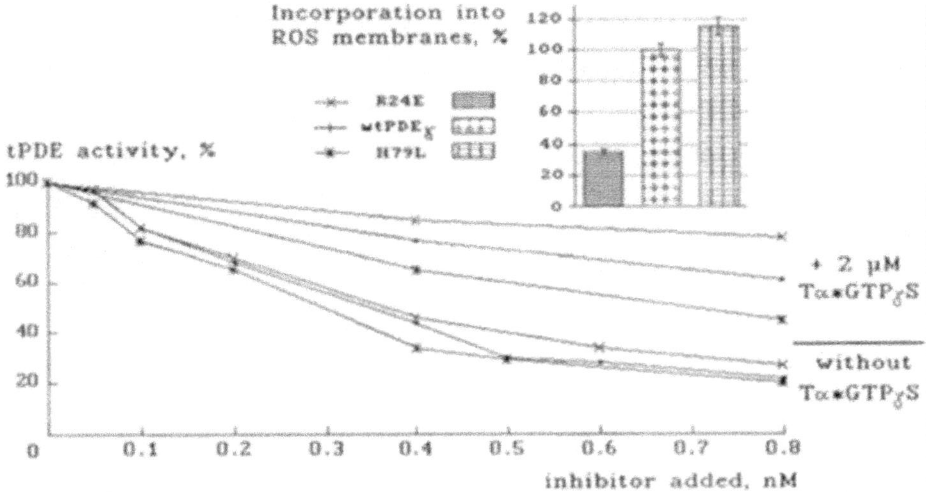

Fig. 4. Incorporation into ROS membranes and tPDE inhibition by wtPDEγ and mutants R24E and H79L in the presence and in the absence of 2 μM Tα•GTPγS.

Another feature of the R24E and H79L mutants is their ability, in a manner differing from that of wtPDEγ, to affect tPDE activation with the transducin α subunit in a complex with the GTP non-hydrolysable analog (Tα•GTPγS). Here, a reverse correlation is observed between the affinity of the mutants to the PDE catalytic subunits and the ability of Tα•GTPγS to activate the tPDE preinhibited by mutant PDEγ (Fig. 4). Surprisingly, none of other PDEγ mutant of the large number studied by us differed from the wtPDEγ by its effect on tPDE activation by transducin (data not shown).

None of the previously suggested mechanisms of PDE functioning explain these new findings. A new model of PDE functioning should answer the following questions: 1) Why two distinct groups of PDEγ mutants differing from wtPDEγ in their affinities to PDEαβ affect PDE inhibition and activation in a different manner? 2) Why R24E and H79L substitutions affecting the mutant PDEγ interaction with PDEαβ do not influence tPDE inhibition in the absence of Tα•GTPγS whereas substitutions K29T, K31R, K31T, R33G, P55L and Y84A demonstrate their effects on either tPDE inhibition or PDEαβ binding? 3) Why do R24E and H79L, at the same time, affect the PDE activation with Tα•GTPγS and why do these effects are in a reverse correlation with the

R24E and H79L affinities to PDEαβ? 4) Why the other studied PDEγ mutations do not affect PDE activation with Tα•GTPγS?

Fig. 5 represents the proposed mechanisms of the inhibition by the PDEγ and the activation by the Tα•GTP of mammalian ROS PDE. This essentially refined scheme fits well with all the properties of the PDEγ mutants studied by us and other authors and in its general features is consistent with the scheme suggested by Whalen and Bitensky (Whalen and Bitensky, 1989) and further modified by Takemoto and Cunnick (Takemoto and Cunnick, 1990).

According to Whalen and Bitensky (Whalen and Bitensky, 1989), PDEαβ from mammalian rods has two centers for PDEγ binding, each with a different affinity to the inhibitor. The authors (Whalen and Bitensky, 1989) were unable to calculate the binding constants. It is likely that the difference between these constants is insignificant. We presume that the mutations R24E and H79L affect PDEγ binding to one of these centers while the mutations K29T, K31R, K31T, R33G, P55L, and Y84A to the other. Let us name these centers the A-site and the B-site, respectively. Since the latter six mutations influence tPDE inhibition only at high concentrations, it is tempting to speculate that these mutants have a lower affinity to the low affinity center of PDEγ binding on PDEαβ. However, at present we do not define each of these two centers as the "high" or "low" affinity ones and we do not identify which of the two PDE catalytic subunits (α or β) forms each of the centers. The removal of one PDEγ from the virtually inactive PDEαβγ_2 complex in bovine ROS membranes results only in 17% of maximal PDE activation (Whalen and Bitensky, 1989). Thus, the contribution of each of the two PDEγ molecules bound to different centers on PDEαβ to the inhibition of the PDE catalytic activity must be different. We assume that the PDEγ bound to the B-site imparts the major contribution to the inhibition of PDE catalytic activity. Correspondingly, the PDEγ mutants with altered parameters of interaction with the B-site noticeably differ from wtPDEγ on the inhibitory plots. These mutants include K29T, K31R, K31T, R33G, P55L, and Y84A. At the same time, mutations affecting PDEγ interaction with the other center (A-site) may not demonstrate any substantial effect on the tPDE inhibitory plot, as it is in the case of R24E and H79L.

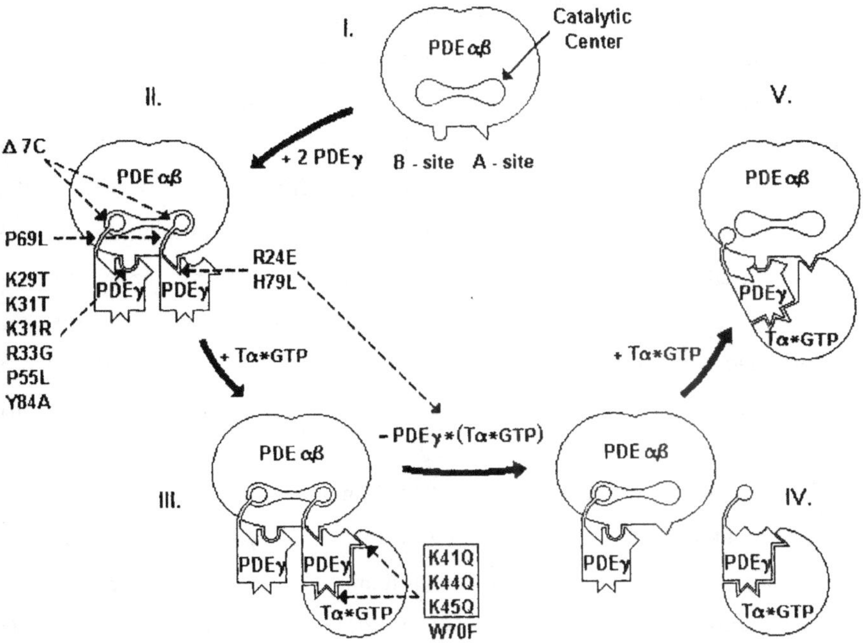

Fig. 5. The proposed mechanism of PDE functioning:

I. There are two structurally different sites on PDEαβ for PDEγ binding, the A-site and the B-site. **II.** Sites on PDEγ, binding to the A- and the B-sites, are also structurally different. The site on PDEγ interacting with the B-site partly overlaps with the binding site for the Tα•GTP formed by two regions of the PDEγ molecule. The PDEγ bound to the B-site provides the main contribution to inhibition of the PDE catalytic activity. The PDEγ C-terminus plays the key role in inhibition. The spatial orientation of the PDEγ C-terminus is of great importance here. **III. and IV.** Tα•GTP first interacts with the PDEγ bound to the A-site on the PDE and removes it in a PDEγ•(Tα•GTP) complex. This results in a slight activation of the PDE catalytic activity. **V.** After removal of PDEγ from the A-site, another Tα•GTP molecule is enabled to interact both with PDEαβ and with PDEγ bound to the B-site. This interaction results in the formation of a membrane-bound fully catalytically active triple complex, PDEαβ•PDEγ•(Tα•GTP).

PDEγ mutations K29T, K31T, K31R, R33G, P55L, and Y84A affect PDEγ binding to PDEαβ in the B-site. Substitutions in PDEγ R24E and H79L affect the PDEγ interaction with PDEαβ in the A-site and the PDEγ dissociation from the A-site in the complex with transducin. PDEγ mutations Δ7C (the deletion of 7 C-terminal amino acid residues) and substitution P69L affect PDEαβ inhibition. PDEγ mutations K41Q,K44Q,K45Q (Brown, 1992) and W70F (Otto-Bruc *et al.*, 1993) prevent PDEγ binding to transducin.

Thus, site-directed mutagenesis data support the assumption that the sites on PDEγ interacting with the A-site and the B-site differ structurally. Consequently, the A-site and the B-site themselves must be different in structure. This is consistent with the results of Takemoto *et al.* (Oppert *et al.*, 1991; Oppert and Takemoto, 1991) on the binding of PDEγ to synthetic peptides corresponding to different regions of the PDEα and PDEβ polypeptide chains. The authors postulated that the PDEγ binding centers on PDEα and PDEβ (demonstrating a significant homology of their amino acid sequences, 72% in case of bovine rod PDE (Lipkin *et al.*, 1990)) are situated in the non-homologous regions and are composed of PDEα fragments 16-30 and 78-90 (Oppert *et al.*, 1991) and of PDEβ fragments 15-34, 91-110, and 211-230 (Oppert and Takemoto, 1991).

It has been shown (Artemyev *et al.*, 1992; Cunnick *et al.*, 1994) that PDEγ has two Tα•GTP binding regions with coordinates 24-45 and 70-76, that the center of PDEαβ binding is also located in the 24-45 region and that this center can partly overlap with the binding site for the Tα. Our results on PDEγ site-directed mutagenesis indicate that the PDEγ site of binding to the B-site includes residues K29, K31 and R33 and, consequently, namely this site of the PDEγ can partly overlap with the transducin binding one. If so, then the only PDEγ bound to the A-site in the holoenzyme complex PDEαβγ_2 is accessible for the transducin α subunit to interact.

PDE activation by transducin, according to the suggested PDE functioning scheme, in this part analogous to those of Bitensky and Takemoto (Whalen and Bitensky, 1989; Takemoto and Cunnick, 1990; Whalen *et al.*, 1990), begins with the dissociation of one PDEγ molecule in the PDEγ•(Tα•GTP) complex from the holoenzyme PDEαβγ_2 with the subsequent formation of a weak catalytically active complex PDEαβγ. We presume that here the PDEγ molecule bound to A-site dissociates. It is likely that Tα•GTP interaction with the PDEγ molecule allosterically favors this dissociation. It seems apparent that mutations R24E and H79L, affecting the process of PDEγ interaction with the A-site in a different manner, should have a reverse effect on the PDEγ dissociation from the holoenzyme complex. The analogous action of these mutations on the process of PDE activation by transducin can also be explained. We assume that the main role in this activation belongs to the second Tα•GTP molecule. This Tα•GTP binds to PDEγ (in the PDEαβγ complex) at one of the two Tα-binding sites (probably 70-76, since the other one, 24-45, is unavailable as a result of the interaction with the B-site) and interacts also with PDEαβ. This should hamper dissociation of the PDEγ bound to the B-site. As a result of such interaction the PDEγ, remaining in the B-site as part of the PDEαβ•PDEγ•(Tα•GTP) complex, ceases inhibiting. It is likely that the second Tα•GTP molecule is able to bind to PDE only after PDEγ dissociation from the A-site and that the Tα-binding site and A-site on PDEαβ overlap. Accordingly, if the dissociation of the first PDEγ is hampered, as in the case of H79L mutation, PDE activation by transducin is hampered also. On the contrary, if the PDEγ mutant affinity to the A-site is weakened, as in the case of R24E mutation, the PDEγ dissociation from the A-site is facilitated and, hence, transducin activation is higher than in the case of wtPDEγ.

We have shown earlier (Lipkin *et al.*, 1988) that the PDEγ sites of primary binding to PDEαβ and of inhibitory action on PDEαβ are spatially separated, and the exact spatial interlocation of the PDEγ C-terminal region and the PDEαβ catalytic (or allosteric) center is of great importance for the process of inhibition. The changes of the spatial arrangement of P69L mutant PDEγ polypeptide chain C-terminus hinder its interaction with the enzyme catalytic (or allosteric) center. An analogous process apparently takes place at transducin activation of PDE. The second Tα•GTP molecule binding simultaneously with the PDEγ (bound to the B-site) and with the PDEαβ (probably to the A-site) and acting as a lever, shifts the location of the PDEγ C-terminal region relative to the PDEαβ catalytic (or allosteric) center, thus terminating its inhibitory action.

Two PDEγ mutants have been reported recently to hinder transducin activation of PDEαβ after their binding with the PDE catalytic complex: these are the triple mutant K41Q,K44Q,K45Q (Brown, 1992) and the point mutant W70F (Otto-Bruc *et al.*, 1993). Both mutants bind to PDEαβ and inhibit the tPDE catalytic activity in the same way as the recombinant wtPDEγ. Consequently, we reason that both these mutations affect only the site of PDEγ interaction with transducin and have no effect on their interaction with PDEαβ in the A-site or in the B-site.

The proposed mechanism is also consistent with the data of Cherione *et al.* (Kroll *et al.*, 1989; Phillips *et al.*, 1989) on the probability of direct interaction of the transducin α subunit with PDE catalytic ones during activation and with the results of

Clerc and Bennett (Clerc and Bennett, 1992) who showed that transducin remains bound to the activated by it PDE catalytic complex in ROS membranes. It can be speculated that such a relatively complicated mechanism of PDE activation in mammalian rods facilitates the process of enzyme reinhibition after GTP hydrolysis by the transducin α subunit and subsequent dissociation of the Tα•GDP from the catalytically active multisubunit aggregate. In this case, 80-90% of PDE inhibition could be achieved immediately after Tα•GDP dissociation since the PDEγ still bound to the B-site restores its inhibitory action.

* * *

Acknowledgments - We thank L.M.Ginodman, V.T.Ivanov, M.Yu.Natochin and A.N.Obukhov for critical reading the manuscript. The research described in this publication was made possible in part by Grant No MU5000 from the International Science Foundation. This research has also been supported by Russian Fundamental Research Fund (project code 93-04-7310).

REFERENCES

Artemyev NO, Rarick HM, Mills JS, Skiba NP and Hamm HE (1992) Sites of interaction between rod G-protein α-subunit and cGMP-phosphodiesterase γ-subunit. J Biol Chem 267:25067-25072.

Brown RL (1992) Functional regions of the inhibitory subunit of retinal rod cGMP phosphodiesterase identified by site-specific mutagenesis and fluorescent spectroscopy. Biochemistry 31:5918-5925.

Clerc A and Bennett N (1992) Activated cGMP phosphodiesterase of retinal rods. J Biol Chem 267:6620-6627.

Cunnick JM, Hurt D, Oppert B, Sakamoto K and Takemoto DJ (1990) Binding of the γ-subunit of retinal rod-outer-segment phosphodiesterase with both transducin and the catalytic subunits of phosphodiesterase. Biochem J 271:721-727.

Cunnick J, Twamley C, Udovichenko I, Gonzales K and Takemoto DJ (1994) Identification of a binding site on retinal transducin α for the phosphodiesterase inhibitory γ subunit. Biochem J 297:87-91.

Eibl H and Lands WEM (1969) Sensitive determination of phosphate. Anal Biochem 30:51-57.

Fung BK-K, Hurley JB and Stryer L (1981) Flow of information in the light-triggered cyclic nucleotide cascade of vision. Proc Natl Acad Sci USA 78:152-156.

Hurley JB and Stryer L (1982) Purification and characterization of the γ regulatory subunit of the cyclic GMP phosphodiesterase from retinal rod outer segments. J Biol Chem 257: 11094-11099.

Kroll S, Phillips WJ and Cerione RA (1989) The regulation of the cyclic GMP phosphodiesterase by the GDP-bound form of the α subunit of transducin. J Biol Chem 264:4490-4497.

Lipkin VM, Dumler IL, Muradov KG, Artemyev NO and Etingof RN (1988) Active sites of the cyclic GMP phosphodiesterase γ subunit of retinal rod outer segments. FEBS Lett 234:287-290.

Lipkin VM, Khramtsov NV, Vasilevskaya IA, Atabekova NV, Muradov KG, Gubanov VV, Li T, Johnston JP, Volpp KJ and Applebury ML (1990) The β-subunit of bovine rod photoreceptor cGMP phosphodiesterase: comparison with the phosphodiesterase family. J Biol Chem 265:12955-12959.

Lipkin VM, Bondarenko VA, Zagranichny VE, Dobrynina LN, Muradov KhG and Natochin MYu (1993) Site-directed mutagenesis of the cGMP phosphodiesterase γ subunit from bovine rod outer segments: role of separate amino acid residues in the interaction with catalytic subunits and transducin α subunit. Biochim et Biophys Acta 1176:250-256.

Muradov KhG, Natochin MYu, Bondarenko VA, Skiba NP and Lipkin VM (1990) Interactions between bovine retinal rod cGMP phosphodiesterase subunits. Biol Membr 7:565-572.

Oppert B, Cunnick JM, Hurt D and Takemoto D (1991) Identification of the retinal cyclic GMP phosphodiesterase inhibitory γ-subunit interaction sites on the catalytic α-subunit. J Biol Chem 266:16607-16613.

Oppert B and Takemoto D (1991) Identification of the γ-subunit interaction sites in the retinal cyclic-GMP phosphodiesterase β-subunit. Biochim Biophys Res Comm 178:474-479.

Otto-Bruc A, Antonny B, Vuong TM, Chardin P and Chabre M (1993) Interaction between the retinal cyclic GMP phosphodiesterase inhibitor and transducin. Kinetics and affinity studies. Biochemistry 32:8636-8645.

Ovchinnikov YuA, Lipkin VM, Kumarev VP, Gubanov VV, Khramtsov NV, Akhmedov NB, Zagranichny VE and Muradov KG (1986) Cyclic GMP phosphodiesterase from cattle retina. Amino acid sequence of the γ-subunit and nucleotide sequence of the corresponding cDNA. FEBS Lett 204:288-292.

Ovchinnikov YuA, Gubanov VV, Khramtsov NV, Ischenko KA, Zagranichny VE, Muradov KG, Shuvaeva TM and Lipkin VM (1987) Cyclic GMP phosphodiesterase from bovine retina. Amino acid sequence of the α subunit and nucleotide sequence of the corresponding cDNA. FEBS Lett 223:169-173.

Phillips WJ, Trukawinski S and Cerione RA (1989) An antibody-induced enhancement of the transducin-stimulated cyclic GMP phosphodiesterase activity. J Biol Chem 264:16679-16688.

Stryer L (1986) Cyclic GMP cascade in vision. Ann Rev Neurosci 9:87-119.

Takemoto DJ and Cunnick JM (1990) Visual transduction in rod outer segments. Cellular Signalling 2:99-104.

Wensel TG and Stryer L (1986) Reciprocal control of retinal rod cyclic GMP phosphodiesterase by its γ subunit and transducin. PROTEINS: Struct Funct Genet 1:90-99.

Whalen MM and Bitensky MW (1989) Comparison of the phosphodiesterase inhibitory subunit interactions of frog and bovine rod outer segments. Biochem J 259:13-19.

Whalen MM, Bitensky MW and Takemoto DJ (1990) The effect of the γ-subunit of the cyclic GMP phosphodiesterase of bovine and frog (Rana catesbiana) retinal rod outer segments on the kinetic parameters of the enzyme. Biochem J 265:655-658.

Part 4

Oncogenes and Antioncogenes as Targets for Gene Therapy

Genetic Approaches to Defining Signaling by the CML-Associated Tyrosine kinase BCR-ABL[1]

Daniel E.H. Afar, Andrei Goga, Lucie Cohen, Charles L. Sawyers, Jami McLaughlin, Randolph N. Mohr and Owen N. Witte

Department of Microbiology and Molecular Genetics, Molecular Biology Institute and Howard Hughes Medical Institute, University of California-Los Angeles.
5-748 MacDonald Building
10833 Le Conte Ave.
Los Angeles, CA 90024-1662
U.S.A.

Human chronic myelogenous leukemia (CML) is a cancer that is associated with a specific cytogenetic marker known as the Philadelphia (Ph[1]) chromosome (Nowell and Hungerford, 1960; Rowley, 1973; reviewed in Kurzrock et al., 1988). Cells of multiple lineages carry Ph[1], suggesting that the disease originates in a pluripotent stem cell. Ph[1] is the result of a reciprocal translocation of chromosomes 9 and 22, which results in the fusion of two cellular genes encoding BCR and c-ABL (Shtivelman et al., 1985; Clark et al., 1988). The resulting fusion gene (see Figure 1) encodes BCR-ABL, an activated protein tyrosine kinase. The tyrosine kinase activity of BCR-ABL is deregulated when compared to c-ABL (Konopka and Witte, 1985) and is the most essential feature for its transforming activity. In nature two fusion products are detected. P210 BCR-ABL is associated with CML and P185 BCR-ABL is associated with the more aggressive acute lymphocytic leukemia (ALL). The tyrosine kinase activity of P185 is more potent than P210 kinase activity, which also correlates with their respective transforming activities (Lugo et al., 1990).

[1]Originally published in: Cold Spring Harbour Symposia on Quantitative Biology, Volume LVIX. © 1994 Cold Spring Harbour Press.

NATO ASI Series, Vol. H 94
Gene Technology
Edited by A. R. Zander, W. Ostertag,
B. V. Afanasiev, F. Grosveld

The association of BCR-ABL with CML suggests that BCR-ABL plays a role in the initiation of the disease. However, cells expressing BCR-ABL during the chronic phase of CML, which may endure for 3-5 years, are still capable of differentiation.

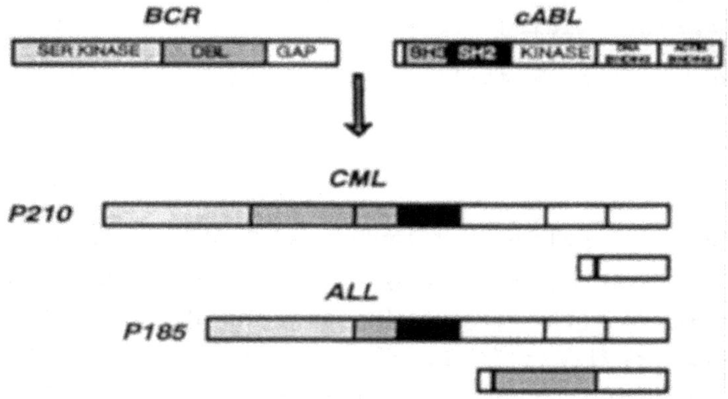

Figure 1: Schematic representation of BCR, c-ABL and the fusion product BCR-ABL. BCR exhibits a serine kinase domain at the amino-terminus (Maru and Witte, 1991), a DBL homology domain that encodes a putative GTP/GDP exchange factor (Ron et al., 1991) and a carboxyl-terminal GTPase activating domain (GAP) (Diekmann et al., 1991). The c-ABL tyrosine kinase contains SH3 and SH2 domains at the N-terminus, a tyrosine kinase domain and DNA (Kipreos and Wang, 1992) and actin binding sequences (McWhirter and Wang, 1991). Both BCR-ABL fusion products, CML-associated P210 and ALL-associated P185, contain BCR sequences fused upstream of the SH3 domain of c-ABL. P210 includes the serine kinase and DBL regions from BCR. P185 only contains the serine kinase region from BCR.

The terminal phase of CML is an acute leukemia, characterized by expansion of predominantly myeloid or lymphoid precursor cells (reviewed by Kantarijan et al., 1993). This time course of events suggests that additional genetic alterations are necessary to trigger blast crisis.

In an attempt to recapitulate human CML several experimental animal systems have been developed. These involved generating transgenic mice with BCR-ABL (Heisterkamp et al., 1990) or reconstituting lethally irradiated mice with bone marrow infected with BCR-ABL containing retrovirus (Daley et al., 1990; Elefanty et al., 1990; Kelliher et al., 1990, Kelliher et al., 1991; Scott et al., 1991, Gishizky et al., 1993). In both systems mice expressing BCR-ABL in hematopoietic cells developed various malignancies, including CML-like syndromes. In most cases the disease developed within 20 weeks and could not be propagated when transplanted into syngeneic recipient mice. This suggested that the experimentally induced syndrome was due to stimulation of the proliferative capacity of committed myeloid progenitor cells by BCR-ABL. In cases where the disease developed with a longer latency period (>20 weeks), however, the leukemia was transplantable (Gishizky et al., 1993). Molecular and cellular analysis demonstrated that this CML-syndrome was a consequence of the disease originating in a multipotent stem cell that gave rise to cells of various lineages (Gishizky et al., 1993). This situation more closely reflected the etiology and pathology of human CML.

While providing valuable information on the development of CML, animal experiments have contributed little insight on the genes that cooperate with BCR-ABL in the disease. Earlier work with in vitro systems showed that BCR-ABL is sufficient for transforming primary bone marrow and fibroblasts (McLaughlin et al., 1987; Lugo and Witte, 1989). The in vivo setting differs from this situation in that BCR-ABL alone is not capable of inducing blast crisis, for which additional genetic lesions are required. However, the in vitro systems provide us with a valuable tool in understanding BCR-ABL signaling. Using three approaches we developed genetic tests to identify components and pathways involved in BCR-ABL signaling. These include: (1) the use of dominant negative forms of genes believed to cooperate with BCR-ABL in transformation; (2) a complementation strategy to rescue signaling defective mutants of BCR-ABL for transformation; (3) an approach to identifying new genes associated with BCR-ABL induced malignancy. These strategies provide us with knowledge on genes acting downstream of BCR-ABL signaling, which are potentially important in accelerating the disease.

Dominant Negative Forms of MYC Block BCR-ABL Mediated Transformation

Dominant negative genes have been extensively used to demonstrate the involvement of specific genes in certain signaling pathways (Feig and Cooper, 1988; Dang et al., 1989; DeClue et al., 1991; Stacey et al., 1991a; Stacey et al., 1991b; Thomas et al., 1992, Wood et al., 1992). Their modes of action vary depending on the gene in question. In general, dominant negative molecules form non-functional (hetero or homo-dimeric) complexes with binding partners, effector molecules or regulatory molecules. Regardless of the mechanism, they function to downregulate the activity of the wild-type protein.

The protooncogene c-myc encodes a nuclear transcription factor that synergizes with BCR-ABL in transforming rat-1 fibroblasts (Lugo and Witte, 1989). The activity of MYC depends on its ability to heterodimerize with its binding partner MAX (Blackwood and Eisenman 1991; Prendergast et al., 1991). MYC competes for MAX binding with MAD, another MAX binding protein, which opposes the action of MYC (Ayer et al., 1993). The MYC-MAX complex formation is mediated by helix-loop-helix and leucine zipper domains in both binding partners. The complex is capable of binding to specific DNA sequences via a C-terminal basic region in MYC. Transcriptional activation is mediated by an activation domain at the N-terminus (Kato et al., 1990).

To analyze the importance of MYC expression during BCR-ABL mediated transformation, we used molecules with a deletion within the activation domain (D106-143) and molecules with a linker insertion adjacent to the DNA-binding basic region (ln373) (Sawyers et al., 1992). These MYC mutants retain the capacity to interact with its binding partner MAX, but are deficient in transcriptional activation. Therefore, expression of dominant negative MYC should compete with wild-type MYC for interaction with MAX and block a MYC dependent signal.

To test the effect of dominant negative MYC on the transforming potential of BCR-ABL, rat-1 indicator cell lines expressing the different MYC mutants were generated. These cell lines expressed about 10 times more dominant negative MYC over endogenous wild-type MYC (Sawyers et al., 1992). Retroviruses containing BCR-ABL were then used to infect the different cell lines. The transforming activity was assessed by plating infected cells into soft agar. The growth of cellular colonies in soft agar signifies a transformed phenotype. Using this assay, we determined that BCR-ABL mediated transformation of fibroblasts was reduced by a factor of 10 in the presence of dominant negative MYC (Sawyers et al., 1992).

Similar results were obtained with primary mouse bone marrow infected with BCR-ABL retrovirus. In this case, the retroviruses contained both BCR-ABL as well as the different forms of MYC. The advantage of having both genes in the same retrovirus is that both proteins will be expressed in the same infected cells. Using this strategy, BCR-ABL with dominant negative MYC was unable to efficiently transform hematopoietic cells (Sawyers et al., 1992). Together with the rat-1 data, these results demonstrated that c-MYC is essential for a BCR-ABL generated signal critical for transformation.

Complementation of Signaling Defective Mutants of BCR-ABL with c-MYC

To genetically define the minimum number of signals needed for oncogenesis we developed a genetic complementation strategy (Afar et al., 1994) (see Figure 2). Point mutations were generated in domains of the gene of interest that might be expected to play a role in connecting to downstream effector molecules. To define its signaling pathways, these inactive point mutants are assayed in the presence of a gene known to function downstream of a signal generated by the oncogene. If the downstream gene is unable to rescue any of the mutants, the experiment becomes uninformative. If all mutants are rescued by the downstream gene, then the same signaling pathway is inactivated by the different mutations. Complementation of one point mutant, but not the other, would signify that each mutation blocks a different pathway.

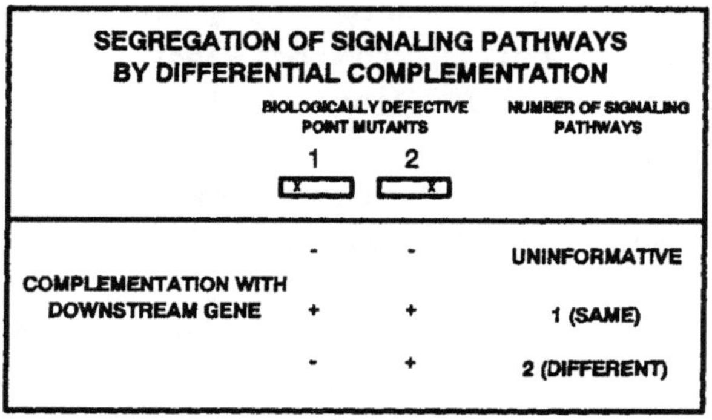

Figure 2: Complementation strategy. Signaling defective point mutants of an oncogene are tested for restoration of transformation in the presence of a downstream gene. Information about the signaling pathways can be gained with the differential rescue of the point mutants (see text).

Point mutation were generated in the src-homology 2 (SH2) region of BCR-ABL, the major tyrosine autophosphorylation site of the kinase domain, and a tyrosine phosphorylation site in the BCR region known to interact with Grb-2, an adaptor molecule involved in Ras activation (Pendergast et al, 1993a; Pendergast et al., 1993b; Afar et al., 1994). The SH2 region is a molecular domain that mediates protein-protein interactions by binding specifically to tyrosine phosphorylated residues (reviewed in Koch et al., 1991; Pawson and Gish, 1992). The residue targeted for mutagenesis in the SH2 domain was the Arg in the FLVRES motif, which is the single most important amino acid for phosphotyrosine binding (Waksman et al, 1992).

All BCR-ABL point mutants retained kinase activity, but were incapable of transforming rat-1 cells as determined by soft agar growth (Pendergast et al., 1993a; Pendergast et al., 1993b; Afar et al., 1994). Since c-MYC cooperates with BCR-ABL in transformation and since dominant negative MYC blocks BCR-ABL signaling, we decided to test the ability of c-MYC to complement the BCR-ABL point mutants for transformation. Rat-1 cell lines were generated that stably expressed the various BCR-ABL forms. These cells were then superinfected with retrovirus containing c-MYC and plated into soft agar. The results showed that only cells expressing the SH2 mutant of BCR-ABL were rescued for transformation with c-MYC hyperexpression (Afar et al., 1994) (summarized in Figure 3). The number and size of colonies generated by the SH2 mutant with c-MYC overexpression was comparable to those generated by wild-type BCR-ABL.

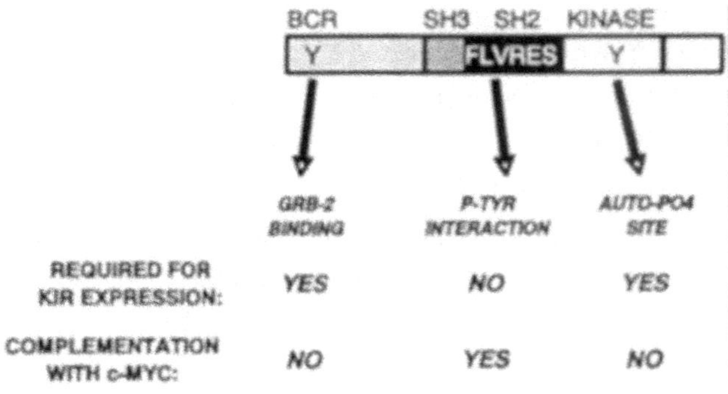

Figure 3: Summary of the signaling pathways activated by BCR-ABL. A signal dependent on the SH2 domain functions upstream of MYC, but is not required for induction of kir expression. Signals dependent on Grb-2 binding and autophosphorylation of the kinase domain are required for kir expression, but function independent of MYC.

These results support the concept that BCR-ABL generates at least two signals for transformation. MYC complementation of the FLVRES mutant, but not the other mutants, segregates these pathways. While the Grb-2 binding mutant is defective in signaling to Ras (Pendergast et al., 1993b), the defect in signaling by the autophosphorylation mutant is unknown. The possibility that the signals generated by Grb-2 binding and autophosphorylation ultimately feed into the same pathway, however, cannot be ruled out.

Rescue of transformation by MYC further supports the concept that c-MYC functions downstream of a BCR-ABL signal. Our data provides genetic evidence that the FLVRES motif in the SH2 domain is responsible for activating the BCR-ABL signal via the MYC pathway. Since Grb-2 binding was shown to link tyrosine kinases to ras activation (reviewed in Pawson and Gish, 1992; Bogurski and McCormick, 1993), this data also provides functional evidence to segregate ras activation from MYC transcription.

Identification of a Novel Ras-Related Gene Associated with BCR-ABL-Induced Malignancy

An important issue in cancer biology is knowledge about specific genes that are expressed in response to an oncogenic signal. This question seems particularly critical in the case of BCR-ABL-induced leukemia, where knowledge of genetic changes that precipitate blast crisis would be very valuable in understanding and managing the disease. To address this issue, we developed an approach to identify genes that are associated with BCR-ABL-mediated transformation of hematopoietic cells.

Previous work has shown that wild-type BCR-ABL as well as the autophosphorylation mutant of BCR-ABL can induce pre-B cell lines to proliferate in the absence of growth factor (Pendergast et al., 1993a). When tested in vivo, however, only cells expressing wild-type BCR-ABL formed highly metastatic and lethal leukemias (Pendergast et al., 1993a). Since cells expressing wild-type BCR-ABL and mutant BCR-ABL would be expected to express similar genes associated with growth in culture, we anticipated a difference in the expression of genes associated with malignant transformation. Therefore, both cell lines were screened using the method of differential display of mRNA by polymerase chain reaction, first described by Liang and Pardee (1992).

Using this technique, five cDNA clones were identified that displayed differential expression by Northern analysis. The clone that showed the highest quantitative difference, was chosen for sequence analysis. Sequencing of the full length cDNA revealed a 40-75% homology at the amino acid level with the ras superfamily of genes (Cohen et al., submitted). This novel gene was named **kir**, for **k**inase-**i**nducible **r**as. Using S1 nuclease protection of isolated RNA, kir expression is induced in pre-B cells transformed by wild-type BCR-ABL and v-ABL. Its expression is absent in the parental cell line and in cells expressing the autophosphorylation mutant of BCR-ABL.

Kir displayed the highest homology with rad, another new member of the ras-superfamily that is induced in skeletal muscle of type II diabetes (Reynet and Kahn, 1993). Interestingly, rad is a gene that is expressed in cells which are defective in a tyrosine kinase signal, while kir is induced in cells with a hyperactive tyrosine kinase signal.

How does BCR-ABL induce the expression of kir? We began to address this issue by examination of the DAGM myeloid cell line which was transformed with the different BCR-ABL point mutants. Expression of kir was monitored by S1 nuclease protection (Cohen et al., submitted) and was shown to be induced in cells expressing only wild-type BCR-ABL and the SH2 mutant of BCR-ABL (summarized in Figure 3). The Grb-2 binding mutant and the autophosphorylation mutant showed undetectable or highly reduced levels of kir expression. This result reinforces our model of multiple signaling pathways generated by BCR-ABL and once again segregates a signal dependent on the SH2 domain from signals dependent on Grb-2 binding and autophosphorylation. These results also suggest that signals dependent on autophosphorylation and Grb-2 binding ultimately converge at a point upstream of transcriptional activation of kir.

Alternative Modes of Signaling by ABL

The disruption of c-ABL in the genesis of BCR-ABL may contribute to the abnormal cell physiology in CML. Knowledge of normal c-ABL function will add to our understanding of the transformation process. c-ABL exhibits some homology to the SRC-family of tyrosine kinases. Similar to SRC, c-ABL contains a myristylation site at the amino-terminus, an SH3 domain, an SH2 region and a tyrosine kinase domain. It differs from the SRC-family by exhibiting a long carboxyl terminus with a nuclear localization signal and DNA and actin binding capabilities (see Figure 1). Unlike SRC, which is cytoplasmic, c-ABL is predominantly nuclear in localization (Van Etten et al., 1989) and is postulated to play a role in cell cycle regulation (Kipreos and Wang, 1992; Welch and Wang, 1993, Sawyers et al., 1994). A direct interaction was recently demonstrated between c-ABL and the retinoblastoma gene product (RB) (Welch and Wang, 1993), a growth suppressor acting in the cell cycle. The binding of RB to c-ABL is postulated to control c-ABL tyrosine kinase activity. More recently, we have shown that c-ABL itself functions to suppress cell growth by arresting cells in G1 (Sawyers et al., 1994). This means that in the case of BCR-ABL, a disruption of the normal c-abl gene may have a dual effect: (i) the generation of an activated oncogene and (ii) the loss of a growth suppressor. Whether the loss of a single copy of c-abl contributes to the malignancy induced by BCR-ABL remains to be tested.

To further understand the biological function of c-ABL, a random mutagenesis scheme was developed to isolate transformation active mutants of c-ABL (Goga et al., 1993). An in-frame deletion mutant was identified, which was capable of transforming fibroblasts, but not lymphoid cells in culture (Goga et al., 1993). This mutant was named ΔLX, because the deletion occurred in the last exon of c-ABL, which includes the DNA binding region. There are several features which distinguish ΔLX from other activating ABL mutants, such as BCR-ABL and v-ABL. BCR-ABL and v-ABL are the result of mutations at the 5′ end of c-ABL sequences. They are cytoplasmic in localization, exhibit elevated autophosphorylation activity and are capable of transforming lymphoid cells (reviewed in Rosenberg and Witte, 1988). In contrast, ΔLX is both nuclear and cytoplasmic in localization, does not exhibit detectable autophosphorylation activity and is incapable of transforming lymphoid cells (Goga et al., 1993). These data suggest that ΔLX transforms cells via a different mechanism than BCR-ABL and v-ABL (see Figure 4).

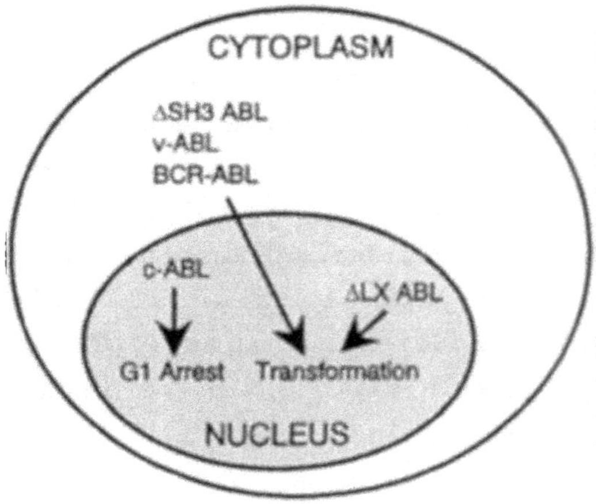

Figure 4: Positive and negative growth effects of the ABL tyrosine kinase. The nuclear c-ABL proto-oncogene is believed to control cell growth in the cell cycle. Activated mutants of ABL, including ΔSH3 ABL, v-ABL and BCR-ABL signal via cytoplasmic pathways to cause transformation. In contrast, the ΔLX mutant of ABL initiates a nuclear signal for transformation.

Removal of the nuclear localization signal in ΔLX blocks its transformation activity (A. Goga and O.N. Witte, unpublished observation), suggesting that the mechanism of action involves a nuclear signal. The details of ΔLX signaling are not clear, but may involve a disruption of normal c-ABL activity in growth suppression.

Conclusion and Future Prospects

Our studies with dominant negative target genes, complementation of signaling defective point mutants, identification of new target genes and elucidation of c-ABL biology provide us with important functional data on the complexity of BCR-ABL signaling. With the identification of more pathways and downstream genes involved in BCR-ABL signaling, we hope to gain an understanding of oncogenic signaling that is crucial to the development of therapeutic intervention. These strategies should be useful for delineating signaling pathways generated by other oncogenes and signal transduction molecules.

Acknowledgements

This work was supported by the NCI Physician Scientist Award CA01551 (C.L.S.), Jonsson Comprehensive Cancer Center seed fund (C.L.S.), grant CA 53867 from the NIH (O.N.W.) and fellowship support from the Medical Research Council of Canada (D.E.H.A.), the UCLA Medical Scientist Training Program (A.G.), Human Frontier Science Program (L.C.) , and a USPHS NRSA award #GM-07104 (R.N.M.). O.N.W. is an Investigator of the Howard Hughes Medical Institute.

References

Afar, D.E.H., A. Goga, J. McLaughlin, O. Witte, and C.L. Sawyers. 1994. Differential rescue of Bcr-Abl point mutants with c-Myc. *Science* **264**:424-426

Ayer, D.E., L. Kretzner, and R.N. Eisenman. 1993. Mad: A heterodimeric partner for max that antagonizes Myc transcriptional activity. *Cell* **72**:211-222

Blackwood, E.M., and R.N. Eisenman. 1991. Max: A helix-loop-helix zipper protein that forms a sequence-specific DNA-binding complex with myc. *Science* **251**:1211-1213

Boguski, M., and F. McCormick. 1993. Proteins regulating Ras and its relatives. *Nature* **366**:643-654

Clark, S.C., J. McLaughlin, M. Timmons, A.M. Pendergast, Y. Ben-Neriah, L.W. Dow, W. Crist, G. Rovera, S.D. Smith, and O.N. Witte. 1988. Expression of a distinctive BCR-ABL oncogene in Ph1-positive acute lymphocytic leukemia (ALL). *Science* **239**:775-777

Cohen, L., R. Mohr, Y.-Y. Chen, M. Huang, R. Kato, D. Dorin, F. Tamanoi, A. Goga, D. Afar, N. Rosenberg, and O. Witte. 1994. transcriptional activation of a novel ras-like gene (kir) by oncogenic tyrosine kinases. *Proc. Natl. Acad. Sci. USA* **91**:12448-12452.

Daley, G.Q., R.A. Van Etten, and D. Baltimore. 1990. Induction of chronic myelogenous leukemia in mice by the $P210^{bcr/abl}$ gene of the Philadelphia chromosome. *Science* **247**:824-830

Dang, C.V., M. McGuire, M. Buckmire, and W.M.F. Lee. 1989. Involvement of the 'leucine zipper' region in the oligomerization and transforming activity of human c-myc protein. *Nature* **337**:664-666

DeClue, J.E., K. Zhang, P. Redford, W.C. Vass, and D.R. Lowy. 1991. Suppression of *src* transformation by overexpression of full-length GTPase-activating protein (GAP) or of the GAP C terminus. *Mol. Cell. Biol.* **11**:2819-2825

Diekmann, D., S. Brill, M.D. Garrett, N. Totty, J. Hsuan, C. Monfries, C. Hall, L. Lim, and A. Hall. 1991. *Bcr* encodes a GTPase activating protein for $p21^{rac}$. *Nature* **351**:400-402

Elefanty, A.G., I.K. Hariharan, and S. Cory. 1990. *bcr-abl*, the hallmark of chronic myeloid leukaemia in man, induces multiple haemopoietic neoplasms in mice. *EMBO J.* **9**:1069-1078

Feig, L.A., and G.M. Cooper. 1988. Inhibition of NIH 3T3 cell proliferation by a mutant ras protein with preferential affinity for GDP. *Mol. Cell. Biol.* **8**:3235-3243

Gishizky, M.L., J. Johnson-White, and O.N. Witte. 1993. Efficient transplantation of BCR/ABL induced CML like syndrome in mice. *Proc. Natl. Acad. Sci. USA* **90**:3755-3759

Goga, A., J. McLaughlin, A.M. Pendergast, K. Parmar, A. Muller, N. Rosenberg, and O.N. Witte. 1993. Oncogenic activation of c-ABL by mutation within its last exon. *Mol. Cell. Biol.* **13**:4967-4975

Heisterkamp, N., G. Jenster, J. ten Hoeve, D. Zovich, P.K. Pattengale, and J. Groffen. 1990. Acute leukaemia in *bcr/abl* transgenic mice. *Nature* **344**:251-253

Kantarijan, H.M., A. Deisseroth, R. Kurzrock, Z. Estrov, and M. Talpaz. 1993. Chronic myelogenous leukemia: a concise update. *Blood* **82**:691-703

Kato, G.J., J. Barrett, M. Villa-Garcia, and C.V. Dang. 1990. An amino-terminal c-Myc domain required for neoplastic transformation activates transcription. *Mol. Cell. Biol.*

10:5914-5920

Kelliher, M., A. Knott, J. McLaughlin, O.N. Witte, and N. Rosenberg. 1991. Differences in Oncogenic Potency but Not Target Cell Specificity Distinguish the Two Forms of the *BCR/ABL* Oncogene. *Mol. Cell. Biol.* **11**:4710-4716

Kelliher, M.A., J. McLaughlin, O.N. Witte, and N. Rosenberg. 1990. Induction of a chronic myelogenous leukemia-like syndrome in mice with v-*abl* and *BCR/ABL*. *Proc. Natl. Acad. Sci. USA* **87**:6649-6653

Kipreos, E.T., and J.Y.J. Wang. 1992. Cell cycle-regulated binding of c-Abl tyrosine kinase to DNA. *Science* **256**:382-385

Koch, C.A., D. Anderson, M.F. Moran, C. Ellis, and T. Pawson. 1991. SH2 and SH3 Domains: Elements That Control Interactions of Cytoplasmic Signaling Proteins. *Science* **252**:668-674

Konopka, J.B., and O.N. Witte. 1985. Detection of *c-abl* tyrosine kinase activity in vitro permits direct comparison of normal and altered *abl* gene products. *Mol. Cell. Biol.* **5**:3116-3123

Kurzrock, R., J.U. Gutterman, and M. Talpaz. 1988. The molecular genetics of Philadelphia chromosome-positive leukemias. *N. Engl. J. Med.* **319**:990-998

Liang, P., and A.B. Pardee. 1992. Differential display of eukaryotic messenger RNA by means of the polymerase chain reaction. *Science* **257**:967-971

Lugo, T., and O.N. Witte. 1989. The BCR/ABL oncogene transforms Rat-1 cells and cooperates with v-myc. *Mol. Cell. Biol.* **9**:1263-1270

Lugo, T.G., A. Pendergast, A.J. Muller, and O.N. Witte. 1990. Tyrosine kinase activity and transformation potency of bcr-abl oncogene products. *Science* **247**:1079-1082

Maru, Y.M., and O.N. Witte. 1991. The BCR gene encodes a novel serine/threonine kinase activity within a single exon. *Cell* **67**:459-468

McLaughlin, J., E. Chianese, and O.N. Witte. 1987. *In vitro* transformation of immature hematopoietic cells by the P210 *BCR/ABL* oncogene product of the Philadelphia chromosome. *Proc. Natl. Acad. Sci. USA* **84**:6558-6562

McWhirter, J.R., and J.Y.J. Wang. 1991. Activation of Tyrosine Kinase and Microfilament-Binding Functions of *c-abl* by *bcr* Sequences in *bcr/abl* Fusion Proteins. *Mol. Cell. Biol.* **11**:1553-1565

Nowell, P.C., and D.A. Hungerford. 1960. A minute chromosome in human chronic granulocytic leukemia. *Science* **132**:p. 1497

Pawson, T., and G.D. Gish. 1992. SH2 and SH3 Domains: from structure to function. *Cell* **71**:359-362

Pendergast, A.M., M.L. Gishizky, M.H. Havlik, and O.N. Witte. 1993a. SH1 domain autophosphorylation of P210 BCR/ABL is required for transformation but not growth factor-independence. *Mol. Cell. Biol.* **13**:1728-1736

Pendergast, A.M., L.A. Quilliam, L.D. Cripe, C.H. Bassing, Z. Dai, N. Li, A. Batzer, K.M. Rabun, C.J. Der, J. Schlessinger, and M.L. Gishizky. 1993b. BCR-ABL-induced oncogenesis is mediated by direct interaction with the SH2 domain of the GRB-2 adaptor protein. *Cell* **75**:175-185

Prendergast, G.C., D. Lawe, and E.B. Ziff. 1991. Association of Myn, the Murine Homolog of Max, with c-Myc Stimulates Methylation-Sensitive DNA Binding and Ras Cotransformation. *Cell* **65**:395-407

Reynet, C., and C.R. Kahn. 1993. Rad: A Member of the Ras Family Overexpressed in Muscle of Type II Diabetic Humans. *Science* **262**:1441-1444

Ron, D., M. Zannini, M. Levis, R.B. Wickner, L.T. Hunt, G. Graziani, S.R. Tronick, S.A. Aaronson, and A. Eva. 1991. A region of proto-dbl essential for its transforming activity shows sequence similarity to a yeast cell cycle gene, CDC 24, and the human breakpoint cluster gene, bcr. *New Biolog*. **3**:372-379

Rosenberg, N., and O.N. Witte. 1988. The viral and cellular forms of the Abelson (*abl*) oncogene. *Adv. in Virus Res.* **39**:39-81

Rowley, J.D. 1973. A new consistent chromosomal abnormality in chronic myelogenous leukemia identified by quinacrine fluorescence and giemsa staining. *Nature* **243**:290-293

Sawyers, C.L., W. Callahan, and O.N. Witte. 1992. Dominant negative myc blocks transformation by ABL oncogenes. *Cell* **70**:901-910

Sawyers, C.L., J. McLaughlin, A. Goga, M. Havlik, and O. Witte. 1994. The nuclear tyrosine kinase c-Abl negatively regulates cell growth. *Cell* **77**:121-131

Scott, M.L., R.A. Van Etten, G.Q. Daley, and D. Baltimore. 1991. v-*abl* causes hematopoietic disease distinct from that caused by *bcr-abl*. *Proc. Natl. Acad. Sci. USA* **88**:6506-6510

Shtivelman, E., B. Lifshitz, R.P. Gale, and E. Canaani. 1985. Fused transcript of *abl* and *bcr* genes in chronic myelogenous leukaemia. *Nature* **315**:550-554

Stacey, D.W., L.A. Feig, and J.B. Gibbs. 1991a. Dominant inhibitory Ras mutants selectively inhibit the activity of either cellular or oncogenic Ras. *Mol. Cell. Biol.* **11**:4053-4064

Stacey, D.W., M. Roudebush, R. Day, S.D. Mosser, J.B. Gibbs, and L.A. Feig. 1991b. Dominant inhibitory Ras mutants demonstrate the requirement for Ras activity in the action of tyrosine kinase oncogenes. *Oncogene* **6**:2297-2304

Thomas, S.M., M. DeMarco, G. D'Arcangelo, S. Halegoua, and J.S. Brugge. 1992. Ras is essential for nerve growth factor- and phorbol ester-induced tyrosine phosphorylation of MAP kinases. *Cell* **68**:1031-1040

Van Etten, R.A., P. Jackson, and D. Baltimore. 1989. The mouse type IV *c-abl* gene product is a nuclear protein, and activation of transforming ability is associated with cytoplasmic localization. *Cell* **58**:669-678

Waksman, G., D. Kominos, S.C. Robertson, N. Pant, D. Baltimore, R.B. Birge, D. Cowburn, H. Hanafusa, B.J. Mayer, M. Overduin, M.D. Resh, C.B. Rios, L. Silverman, and J. Kuriyan. 1992. Crystal structure of the phosphotyrosine recognition domain SH2 of v-*src* complexed with tyrosine-phosphorylated peptides. *Nature* **358**:646-653

Welch, P.J., and J.Y.J. Wang. 1993. A C-terminal protein-binding domain in the retinoblastoma protein regulates nuclear c-Abl tyrosine kinase in the cell cycle. *Cell* **75**:779-790

Wood, K.W., C. Sarnecki, T.M. Roberts, and J. Blenis. 1992. Ras mediates nerve growth factor receptor modulation of three signal-transducing protein kinases: Map kinase, Raf-1 and RSK. *Cell* **68**:1041-1050

Analysis of Oncosuppressor Gene Function in the Mouse by Gene Targeting

Martin L. Hooper
Cancer Research Campaign Laboratories
Department of Pathology
University of Edinburgh
Teviot Place
Edinburgh EH8 9AG
United Kingdom

When functioning normally, oncosuppressor genes, also known as tumour suppressor genes or antioncogenes, prevent the development of one or more types of cancer (reviewed by Knudson, 1993). A powerful way of investigating the mechanism of action of such genes is to examine the consequences of inactivating them in an experimental animal. This can be done via gene targeting in embryonal stem (ES) cells (reviewed by Hooper, 1992). These cells, established in culture from the pluripotent inner cell mass cells of the peri-implantation embryo, retain the capacity to colonise the somatic tissues and germ line of chimaeras produced by injecting them into blastocyst-stage embryos. Tailored mutations can be introduced into chosen genes by introducing into the cultured cells a gene targeting DNA vector and isolating cells in which the vector has undergone homologous recombination with the endogenous gene. These mutations can then be introduced via chimaeras into the mouse germ line. Here I review the study of the *Rb-1* and *p53* oncosuppressor genes using this approach.

The oncosuppressor gene paradigm is the retinoblastoma susceptibility gene *RB-1*. Retinoblastomas are childhood retinal tumours of which about 40% of all cases are familial. Knudson (1971) proposed that these tumours arise as a result of two mutational

NATO ASI Series, Vol. H 94
Gene Technology
Edited by A. R. Zander, W. Ostertag,
B. V. Afanasiev, F. Grosveld

events, of which one is present in the germline in familial cases. Recent molecular analysis has confirmed this hypothesis, and demonstrated that both events involve modification or loss of the same gene, *RB-1*. More than 90% of individuals constitutively heterozygous for an *RB-1* mutation develop retinoblastoma as a result of a somatic event occurring in one or more cells of the retina, or of its precursor tissues, that eliminates the function of the wild-type allele, usually by allele loss (reviewed by Knudson, 1993). There are several candidates for the exact cell type of origin of retinoblastoma including primitive neurotubular cells, glial cells and photoreceptor cells, the last possibility being most easily reconcilable with biochemical, immunocytochemical and morphological properties (Rootman et al., 1987). Photoreceptor-like ultrastructural features are particularly marked in benign tumours, termed retinomas or retinocytomas, which develop in a small proportion of *RB-1* heterozygotes and are thought to result from loss of the residual wild-type allele in a more mature stage of the retinoblastoma precursor cell type (Gallie et al., 1982). These tumours have been described as cone-like (Margo et al., 1983), although this should not be over-interpreted since the organisational features that distinguish rods from cones would probably be obscured in a tumour. In addition to retinal tumours, about 15% of *RB-1* heterozygotes develop osteosarcomas. *RB-1* allele loss is also seen in sporadic lung, breast, prostate and bladder carcinomas, although no increase in incidence of these tumours is seen in germline heterozygotes (see Knudson, 1993).

The *RB-1* gene codes for a 105kD protein present in most cell types that undergoes cell cycle-dependent changes in phosphorylation as a result of the action of cyclin-dependent kinase and phosphoprotein phosphatase activity (see Hollingsworth et al., 1993). The hypophosphorylated protein associates with the cell nucleus and binds to transcription factors such as E2F, with effects that depend upon the transcription factor and in some cases also on the cell type. Transcription of genes containing an E2F recognition site is inhibited by the binding of the *RB-1* gene product; such genes include several whose products are required during S phase. Depending upon the phase of the cell cycle, E2F may associate with the *RB-1* gene product or with one of two other related proteins, p107 or p130 (Cao et al., 1992; Shirodkar et al., 1992; Li et al., 1993; Hannon et al., 1993; Cobrinik et al., 1993) and it is likely that this provides a further level at which the expression of E2F-dependent genes can be regulated. A different control element appears to mediate the stimulation of *c-fos* expression by the *RB-1* gene product (Robbins et al., 1990). Further levels of complexity may be present in the form of feedback control loops, one involving *RB-1* and TGF-β1 implicated in the regulation of *c-myc* expression (see Hollingsworth et al., 1993) and another that involves *RB-1* and a cyclin-dependent kinase inhibitor, $p16^{INK4}$ (Serrano et al., 1993). Numerous other proteins also bind to the *RB-1* gene product, implicating it in mechanisms unrelated to the control of gene expression. The importance of *RB-1* in cell cycle regulation is underlined by the fact that several tumour viruses have independently evolved proteins that inactivate the hypophosphorylated gene product by binding to it. Examples are SV40 T antigen, adenovirus E1A protein and human papillomavirus E7 protein (see Hollingsworth et al., 1993).

The corresponding murine gene, designated *Rb-1*, which has 91% sequence homology to the human *RB-1* gene, is widely expressed in mouse tissues (Bernards et al., 1989; Szekely et al., 1992). In collaboration with the laboratory of Dr. Anton Berns at the Netherlands Cancer Institute, we have introduced mutations into the mouse *Rb-1* gene by gene targeting (Clarke et al., 1992), as have two other groups (Lee et al., 1992; Jacks et al., 1992). All three groups, with minor variations, have reported similar results, which I will summarise with particular reference to our own data. We used two targeting vectors to generate the mutant *Rb-1* alleles, designed to insert respectively the selectable marker genes *neo* and *hyg*, in each case driven by the *pgk-1* promoter, into exon 19 of the *Rb-1* gene (Clarke et al., 1992). Both insertions carry translation termination codons in all three reading frames and polyadenylation signals, resulting in truncation of the *Rb-1* reading frame in the targeted allele upstream of the coding sequences for SV40 T antigen-binding domain number 2, which is essential for the function of the gene product (Hu et al., 1990). We therefore consider both targeted alleles as null alleles.

In contrast to the situation in humans, no retinoblastomas have to date been detected by any of the three groups in mice heterozygous for a null allele of *Rb-1*. We have considered a number of possible explanations for this. First, because of their smaller size and shorter lifespan, mice may simply have a lower probability of a somatic mutation occurring in the retina during the susceptible period, so that only a very small proportion of heterozygotes develop retinoblastomas. However, as the aggregate number of heterozygous animals examined increases, this hypothesis becomes progressively less tenable as the sole explanation. Second, in view of the relationship to photoreceptor cells discussed above, it may be significant that the photoreceptors of the predominantly nocturnal mouse are almost exclusively rods, while humans also possess cones of three different spectral sensitivities (see Bowmaker, 1991). Third, mice may differ from humans in the level of exposure of their retinas to some required environmental agent. As laboratory mice are maintained under conditions of artificial light and are active principally in the hours of darkness, one candidate for such an environmental agent is sunlight. This has prompted us to undertake an analysis of published incidences of human retinoblastoma. Intriguingly, these increase significantly with decreasing geographical latitude (Hooper, 1994), indicating that this hypothesis is worthy of further investigation. This is currently being explored by controlled exposure of heterozygous mice to light. Fourth, there may be a need in the mouse for an additional genetic event or events. Two observations suggest that this idea should be taken seriously: (i) retinoblastomas are uncommon if not non-existent in veterinary practice (Hogan and Albert, 1991), suggesting that humans may be unusually susceptible; (ii) it is possible to induce retinoblastoma in the mouse by expressing SV40 T antigen in the retina, thereby inactivating both the *Rb-1* and *p53* gene products (Windle et al., 1990). This effect requires expression of T antigen in developing photoreceptors (Al-Ubaidi et al., 1992a; Howes et al., 1994a), and its expression in rods after withdrawal from the cell cycle leads only to photoreceptor degeneration (Al-Ubaidi et al., 1992b) A role for *p53* inactivation in the genesis of these tumours has recently been established by Howes et al. (1994b) who showed that expression of the human papilloma virus (HPV) E7 oncogene in the developing photoreceptor cells of transgenic mice, thereby inactivating the *Rb-1* gene product but not the *p53* gene product, led not to retinoblastoma but to retinal degeneration by photoreceptor apoptosis. However, if the transgene was crossed on to a *p53*-deficient genetic background, retinal tumours developed. Nevertheless, crossing into our *Rb-1* mutant stock the null allele of *p53* described below has not led to the development of any retinoblastomas (Harrison et al., 1994) — although Jacks and co-workers report that $Rb\text{-}1^{+/-}$ $p53^{-/-}$ mice develop retinal dysplasia (Williams et al., 1994a) — and it is possible that more than one of the above factors is important.

Although retinoblastomas have not been seen in *Rb-1* heterozygous mice, pituitary adenocarcinomas develop in most if not all of these animals (Jacks et al., 1992; Hu et al., 1994; Harrison et al., 1994). These tumours express pro-opiomelanocortin mRNA, which is normally found in the corticotroph cells of the anterior pituitary and in the melanotroph cells that make up the intermediate lobe, the prohormone itself being processed differently in these two cell types with the formation of adrenocorticotrophic hormone (ACTH) in both but α-melanocyte stimulating hormone (α-MSH) only in melanotrophs (Harrison et al., 1994). That it is the melanotrophs that are the precursor cells of the tumours was established by Hu et al. (1994), who showed that the tumours express α-MSH and develop from foci in the intermediate lobe of the pituitary, which is present only in vestigial form in adult humans. Interestingly, they also show patchy expression of the photoreceptor S-antigen (Hu et al., 1994). The time of onset of overt signs of disease in mice known to have inherited their mutant allele paternally is significantly shifted to the lower end of the spectrum, suggesting that the development of the pituitary tumours is influenced by gametic imprinting (Harrison et al., 1994). In humans, gametic imprinting at the *RB-1* locus appears to play a role in the development of sporadic osteosarcomas but not of retinoblastomas (see Marshall, 1991).

A high incidence of medullary thyroid carcinoma has been reported by Williams et al. (1994a) in mice heterozygous for the targeted *Rb-1* allele produced by Jacks et al. (1992). In our mice the thyroid was macroscopically normal in every case, but detailed examination of seven serially sectioned thyroids revealed the presence in three animals of focal hyperplasia of the C cell population associated with some atrophy. There was, however, no evidence of malignancy. The difference between the observations in the two strains may be a consequence of differences in the targeted allele, in the genetic background or in environmental exposure.

Our pituitary tumour-bearing mice showed a number of other pathological changes, some of which may be secondary to systemic effects of the tumour while others are less likely to be so (Harrison et al., 1994). Seven of eighty heterozygotes had severe steatosis in the liver, probably reflecting an effect of ACTH secretion by the tumour. There was an increased level of iron, in the form of haemosiderin, in the spleen of almost all the heterozygotes, suggesting an abnormality in the turnover and homeostasis of erythrocytes which may be related to the abnormalities of haematopoiesis seen in homozygous embryos (see below). Four of the mice showed enlarged, hyperplastic islets of Langerhans in the pancreas, which may represent an intermediate stage in the genesis of islet cell tumours that have been reported in mice carrying mutations in both *Rb-1* and *p53* (Williams et al., 1994a).

No case of constitutional homozygosity for a null *RB-1* allele has ever been documented in humans, and so the production of *Rb-1* mutant mice allowed for the first time a study of the consequences of germline homozygosity at this locus. When heterozygous mice were intercrossed, homozygous embryos failed to survive to term, and at midgestation showed abnormalities in the haematopoietic and nervous systems, in both cases involving increased levels of apoptotic cell death and overabundant or ectopic mitosis (Clarke et al., 1992; Lee et al., 1992; Jacks et al., 1992). The tissues affected normally show high-level *Rb-1* expression (Szekely et al., 1992) and are those in which a dividing precursor cell population matures to a postmitotic differentiated cell earliest in embryonic development, suggesting a generalised role for *Rb-1* in the maturation of precursor cells. The presence of parallel effects on cell division and cell death is consistent with the hypothesis that *Rb-1* functions to maintain cells in a quiescent state characterised by reduced levels of both mitosis and apoptosis (Dive and Wyllie, 1993). In the nervous system, many of the ectopically dividing cells die by apoptosis shortly after entry into S phase (Lee et al., 1994). Staining for differentiation markers revealed that neuronal cells failed to mature properly. In particular, there was reduced expression of the neurotrophin receptors TrkA, TrkB and $p75^{LNGFR}$, but not of their ligands. Consistent with this finding, primary cultures of dorsal root and trigeminal ganglion cells showed reduced neuronal cell survival and neurite outgrowth even in the presence of nerve growth factor (Lee et al., 1994). Abnormalities of development due to inappropriate cell proliferation and apoptosis are also present in the lens of homozygous embryos (Morgenbesser et al., 1994; Robanus-Maandag et al., 1994). The apoptosis is *p53*-dependent, since it is almost completely suppressed in embryos doubly homozygous for *Rb-1* and *p53*-deficiency (Morgenbesser et al., 1994). Similar effects can be seen in adult transgenic mice in which Rb-1 function in the lens is abolished by expressing the HPV E7 oncogene from the α_A-crystallin promoter. In this case the simultaneous inactivation of p53 by expressing HPV E6 from the same promoter not only suppressed apoptosis but led to lens tumour formation (Pan and Griep, 1994). There are obvious parallels here with the effects seen in the retina (see above).

The death of *Rb-1* homozygous embryos in the second half of gestation prevents their being used to examine the effects of homozygosity on events occurring later in development. This problem can be to some extent overcome by examination of chimaeras formed by injecting ES cells in which both *Rb-1* alleles have been inactivated by sequential targeting into wild-type blastocysts. The *Rb-1*-null cells contribute to all adult tissues examined, including mature erythrocytes and retina (Williams et al., 1994b; Robanus-Maandag et al., 1994), although the contribution to the adult retina is much reduced as a result of substantial cell degeneration in association with ectopic mitoses during late gestation (Robanus-Maandag et al., 1994). It is not clear whether this reduction affects uniformly all cell types in the retina or whether there is a preferential, perhaps absolute, exclusion from certain lineages. No retinoblastomas were seen in these chimaeras, which may mean either that *Rb-1*-null cells fail to contribute to the precursor cell type or that, consistent with some of the hypotheses already discussed, loss of *Rb-1* function is not in itself sufficient for retinoblastoma formation in the mouse. Pituitary tumours similar to those seen in heterozygotes developed from the *Rb-1*-null cells, but with an earlier time of onset consistent with the lack of requirement for somatic allele loss. Other abormalities reported were cataracts, hyperplasia of the adrenal medulla and the presence of enlarged cells in cerebellum and liver (Williams et al., 1994b). The contribution of the mutant cells to mature erythrocytes indicates that the effects of the mutation on haematopoiesis are not cell-autonomous, a conclusion foreshadowed by the earlier observation that *Rb-1*-deficient embryonal stem cells can give rise to phenotypically normal mature B and T lymphocytes in chimaeras produced by injecting them into blastocysts homozygous for a mutation in the *Rag-2* gene (Chen et al., 1993). A further non-cell-autonomous effect on haematopoiesis in highly chimaeric embryos was reported by Robanus-Maandag et al. (1994), namely an increase in the number of nucleated red cells during fetal liver erythropoiesis that involved both mutant and wild-type cells. It would appear, therefore, that the effect of *Rb-1* homozygosity on the turnover of haematopoietic cells is mediated through the microenvironment.

The *p53* gene was originally discovered by virtue of its ability to form a complex with SV40 T antigen (Lane and Crawford, 1979), and was at first thought to be an oncogene. It was subsequently realised, however, that the allele initially cloned and studied was, in fact, a mutant allele, and the wild-type gene is now known to function as an oncosuppressor gene (see Knudson, 1993). Allele losses in human malignancy are more common at the *p53* locus than at any other, and humans heterozygous for a mutant allele in their germline have Li-

Fraumeni syndrome, a predisposition to a spectrum of tumours, most commonly affecting breast and brain (reviewed by Knudson, 1993). The gene codes for a sequence-specific DNA-binding protein that is capable of activating the expression of a reporter gene adjacent to the binding site, and has an N-terminal transcription activation domain, a central DNA-binding core domain and a C-terminal oligomerisation domain (reviewed by Vogelstein and Kinzler, 1994). The core domain has been studied as a DNA complex by X-ray crystallography and shown to have a structure in the form of a sandwich that forms a scaffold for two large loops and a loop-sheet-helix motif that form the DNA-binding surface. Most of the mutations identified in tumours lie at the DNA-binding surface and others involve residues important for stable folding of the core domain (Cho et al., 1994). The oligomerisation domain has been studied by nuclear magnetic resonance spectroscopy and shown to form a symmetric tetramer made up of a dimer of dimers (Clore et al., 1994). The protein is phosphorylated in a cell cycle-dependent manner, and its levels rise following DNA damage as a result of post-translational stabilisation, leading to G1 growth arrest, mediated by induction of the expression of a 21kD cyclin-dependent kinase inhibitor, variously designated WAF-1, Cip-1 or sdi-1 (reviewed by Nasmyth and Hunt, 1993). Loss of wild-type *p53* is accompanied by genomic instability, and it has been proposed that the p53 protein functions to prevent cells entering S phase prior to repair of DNA damage (Lane, 1992). Studies using cultured cell lines have also implicated *p53* in the regulation of apoptosis (Yonish-Rouach et al., 1991; Shaw et al., 1992). Some mutant alleles of *p53* exert a dominant negative effect and, in the presence of wild-type *p53*, can immortalise normal early-passage cultured cells and cooperate with *ras* to cause transformation of embryo fibroblasts. This effect results from the ability of the mutant protein to sequester wild-type protein into inactive oligomers. Such alleles lack the ability of the wild-type allele to suppress embryo fibroblast transformation by other oncogenes, which is itself not dependent on oligomerisation (see Slingerland et al., 1993).

Mice with a targeted mutation in the *p53* gene were first described by Donehower et al. (1992). They found that heterozygous animals developed tumours at low frequency and that homozygotes developed normally to birth but all developed tumours, of various types, in the first few months of life. Although the targeted allele carried by these mice was designed as a null allele, it was theoretically possible that an active polypeptide fragment could be produced from the 3' end of the modified gene. However, we have used gene targeting to produce mice with a *p53* allele in which exons 2 to 6 are replaced by an insert which, as described above for *Rb-1,* truncates the reading frame, producing an unambiguously null allele (Clarke et al., 1993), and these mice have a similar phenotype. Homozygotes develop tumours, predominantly T cell lymphomas, during the first six months of life; soft tissue sarcomas are relatively more important in heterozygotes, which develop tumours later (Purdie et al., 1994). Some mice developed pathological signs in the absence of obvious tumour development, mostly involving reactive lymphoid hyperplasia, suggesting a perturbation of the immune system. A similar tumour spectrum is shown by a third *p53* mutant strain (Jacks et al., 1994). Small inter-strain differences in tumour spectrum are probably a consequence of differences in genetic background (Harvey et al., 1993a). Most tumours in our *p53* heterozygotes had suffered loss of the wild-type *p53* allele, indicating classical oncosuppressor gene behaviour. In contrast to the pituitary tumours that develop in *Rb-1* heterozygotes, most of the tumours in *p53* heterozygous and homozygous mice contained an aneuploid or polyploid cell population detectable by flow cytometry (Purdie et al., 1994), consistent with the idea that, if p53 is non-functional, cells enter division without having repaired DNA damage, and this leads to abnormalities in chromosome segregation (Lane, 1992). As well as having an elevated spontaneous tumour incidence, p53-deficient mice have recently been shown to have an increased susceptibility to induction of tumours by γ-radiation (Kemp et al., 1994).

One of the mechanisms underlying the tumour suppressor activity of *p53* has been elucidated by studies of thymocyte apoptosis. Wild-type thymocytes set up in culture in the absence of serum undergo spontaneous apoptosis whose rate can be stimulated by a variety of stimuli including physiological agents such as glucocorticoids and DNA-damaging agents such as γ-radiation. Thymocytes from homozygous *p53*-deficient mice showed normal response to agents of the former class, but were resistant to induction of apoptosis by DNA-damaging agents, with heterozygous cells showing an intermediate response (Clarke et al., 1993; Lowe et al., 1993b). This suggested that stimuli can activate the apoptosis pathway via different routes, and placed *p53* in a branch route linking DNA damage to the main apoptosis pathway.

In order to investigate whether this behaviour is peculiar to thymocytes, we have extended the analysis to other tissues. In collaboration with Dr. Stephen Griffiths and Prof. Mel Greaves (Institute of Cancer Research, London) we have shown that *p53* is also required for induction of apoptosis in B lymphocyte precursors by X-irradiation. Unlike thymocytes, these cells can be grown clonally in culture, and this has made it possible to demonstrate that the *p53*-deficient pre-B cells surviving irradiation have a considerably increased frequency of mutation, assayed at the *Hprt* locus (Griffiths, S.D., Clarke, A.R., Ford, A.M., Hooper, M.L., Wyllie, A.H. and Greaves, M.F., manuscript in preparation). We have also shown that in peripheral lymphocytes, induction of apoptosis by γ-irradiation is p53-dependent, while its induction by antibody to CD4 (Howie et al., 1994) is p53-independent (Coutts et al., 1994). To extend the observations to an epithelial tissue undergoing rapid turnover from a stem cell population, we have examined the response of intestinal crypts to γ-irradiation *in vivo*. Four hours after irradiation, crypt epithelial cells of both small intestine and colon in wild-type mice showed abundant apoptosis, but corresponding cells in *p53* homozygotes were completely resistant to its induction. Again, cells in heterozygotes showed an intermediate response. In contrast, two other early responses of wild-type crypts to γ-irradiation, i.e. reduction in the proportion of cells in S phase and growth arrest in the G2 phase of the cell cycle (reviewed by Potten, 1990) were both largely intact in *p53*-deficient animals (Clarke et al., 1994). Merritt et al. (1994) also observed resistance of intestinal epithelial cells in homozygous *p53*-deficient mice to induction of apoptosis by γ-irradiation, but did not observe a reduced response in heterozygotes, perhaps because they gave a higher radiation dose. In both studies the low background level of intestinal apoptosis appeared to be *p53*-independent, indicating that, as with thymocytes, *p53* is required to link apoptosis to DNA damage rather than for apoptosis itself. This is consistent with the finding that homozygous *p53*-deficient mice develop normally to birth, implying no anomalies in developmental programmed cell death (Donehower et al., 1992), and with the lack of requirement for *p53* in the apoptosis that occurs in prostatic glandular cells following androgen ablation (Berges et al., 1993). However, the simple picture in which the requirement for *p53* is associated with instances of apoptosis induction by DNA damage is not consistent with all observations. Apoptosis induced by *Rb-1*-deficiency (see above) or by activating the *c-myc* proto-oncogene in serum-starved fibroblasts (Hermeking and Eick, 1994) is not known to involve DNA damage but is *p53*-dependent; in contrast, cycling T-lymphoma cells and mitogenically activated T lymphocytes undergo apoptosis by a *p53*-independent pathway after DNA damage (Strasser et al., 1994).

This leaves the question of how cells decide between different *p53*-mediated responses to DNA damage, i.e. between entry into growth arrest, thereby possibly allowing DNA repair, and entry into apoptosis. Relevant here is recent work of Lowe et al. (1993a), who showed using fibroblasts of different *p53* genotypes that the apoptotic response is seen in these cells only when G1 growth arrest is circumvented by engineering the expression of adenovirus E1A protein in the cells. This suggests that in the general case the choice of response depends upon the other genes being expressed by the cell. Fibroblasts from *p53* mutant mice have also been used by Harvey et al. (1993b) to confirm that p53-deficiency promotes genetic instability and to demonstrate that it facilitates immortalisation. Enhanced proliferative potential has been noted in various cell types from *p53*-deficient mice (Tsukuda et al., 1993).

It seems reasonable to conclude that one mechanism contributing to the high incidence of tumours in *p53*-deficient mice is failure to eliminate damaged, potentially tumourigenic cells by the normal apoptotic pathway. This emphasises that the regulation of cell death can be as important as the regulation of cell division in protecting against tumour development. It also has implications in cancer therapy: some anti-cancer drugs exert their cytotoxic effect by induction of the endogenous apoptotic pathway (reviewed by Dive and Wyllie, 1993) and therefore loss of *p53* from human cancers may be an important mechanism of resistance to such drugs (cf. Lowe et al., 1994). Future years are likely to bring the advent of novel drugs specifically designed to treat cancer by activating the natural suicide pathway of the cell.

Acknowledgements: Work in my laboratory is supported by the Cancer Research Campaign and the Medical Research Council.

References

Al-Ubaidi, M.R., Font, R.L., Quiambo, A.B., Keener, M.J., Liou, G.I., Overbeek, P.A. and Baehr, W. (1992a). Bilateral retinal and brain tumours in transgenic mice expressing simian virus 40 large T antigen under the control of the human interphotoreceptor retinoid-binding protein promoter. *J. Cell Biol.* **119,** 1681-1687.

Al-Ubaidi, M.R., Hollyfield, J.G., Overbeek, P.A. and Baehr, W. (1992b). Photoreceptor degeneration induced by the expression of simian virus 40 large tumor antigen in the retina of transgenic mice. *Proc. Nat. Acad. Sci. USA* **89,** 1194-1198.

Berges, R. R., Furuya, Y., Remington, L., English, H. F., Jacks, T. and Isaacs, J. T. (1993). Cell proliferation, DNA repair, and p53 function are not required for programmed death of prostatic glandular cells induced by androgen ablation. *Proc. Nat. Acad. Sci. USA* **90,** 8910-8914.

Bernards, R., Schackleford, G. M., Gerber, M. R., Horowitz, J. M., Friend, S. H., Schartl, M., Bogenmann, E., Rapaport, J. M., McGee, T., Dryja, T. P. and Weinberg, R. A. (1989). Structure and expression of the murine retinoblastoma gene and characterization of its encoded protein. *Proc. Nat. Acad. Sci. USA* **86,** 6474-6478.

Bowmaker, J. K. (1991). The evolution of vertebrate visual pigments and photoreceptors. In *Vision and Visual Dysfunction* vol. 2 (ed. J. R. Cronly-Dillon and R. L. Gregory) pp. 63-81. London: Macmillan Press.

Cao, L., Faha, B., Dembski, M., Tsai, L.-H., Harlow, E. and Dyson, N. (1992). Independent binding of the retinoblastoma protein and p107 to the transcription factor E2F. *Nature* **355,** 176-179.

Chen, J., Gorman, J. R., Stewart, V., Williams, B., Jacks, T. and Alt, F. W. (1993). Generation of normal lymphocyte populations by Rb-deficient embryonic stem cells. *Curr. Biol.* **3,** 405-413.

Cho, Y., Gorina, S., Jeffrey, P.D. and Pavletich, N.P. (1994). Crystal structure of a p53 tumor-suppressor-DNA complex: Understanding tumorigenic mutations. *Science* **265,** 346-355.

Clarke, A. R., Robanus Maandag, E., van Roon, M., van der Lugt, N.M.T., van der Valk, M., Hooper, M. L., Berns, A. and te Riele, H. (1992). Requirement for a functional *Rb-1* gene in murine development. *Nature* **359,** 328-330.

Clarke, A. R., Purdie, C. A., Harrison, D. J., Morris, R. G., Bird, C. C., Hooper, M. L. and Wyllie, A. H. (1993). Thymocyte apoptosis induced by p53-dependent and independent pathways. *Nature* **362,** 849-852.

Clarke, A. R., Gledhill, S., Hooper, M. L., Bird, C. C. and Wyllie, A. H. (1994). p53-dependence of early apoptotic and proliferative responses within the mouse intestinal epithelium following γ-irradiation. *Oncogene,* **9,** 1767-1773.

Clore, G.M., Omichinski, J.G., Sakaguchi, K., Zambrano, N., Sakamoto, H., Appella, E. and Gronenborn, A.M. (1994). High-resolution structure of the oligomerization domain of p53 by multidimensional NMR. *Science* **265,** 386-391.

Cobrinik, D., Whyte, P., Peeper, D. S., Jacks, T. and Weinberg, R. A. (1993). Cell cycle-specific association of E2F with the p130 E1A-binding protein. *Genes Dev.* **7,** 2392-2404.

Coutts, S.B., Clarke, A.R., Howie, S.E.M. and Harrison, D.J. (1994). Cell death and p53 in mature T lymphocytes. J. Path. **173** Suppl, 162A.

Dive, C., and Wyllie, A. H. (1993). Apoptosis and Cancer Chemotherapy. In *Frontiers in Pharmacology: Cancer Chemotherapy* (eds. Hickman, J.A. and Tritton, T.T.) pp 21-56. Oxford: Blackwell Scientific.

Donehower, L. A., Harvey, M., Slagle, B. M., McArthur, M. J., Montgomery, C. A., Jr., Butel, J. S. and Bradley, A. (1992). Mice deficient for p53 are developmentally normal but susceptible to spontaneous tumours. *Nature* **356,** 215-221.

Gallie, B. L., Ellsworth, R. M., Abramson, D. H. and Phillips, R. A. (1982). Retinoma: spontaneous regression of retinoblastoma or benign manifestation of the mutation? *Brit. J. Cancer* **45,** 513-521.

Hannon, G. J., Demetrick, D. and Beach, D. (1993). Isolation of the Rb-related p130 through its interaction with CDK2 and cyclins. *Genes Dev.* **7,** 2378-2391.

Harrison, D.J., Hooper, M.L., Armstrong, J.F. and Clarke, A.R. (1994). Effects of heterozygosity for the *Rb-1t19neo* allele in the mouse. *Oncogene,* submitted..

Harvey, M., McArthur, M. J., Montgomery, C. A., Jr., Bradley, A. and Donehower, L. A. (1993a). Genetic background alters the spectrum of tumors that develop in p53-deficient mice. *FASEB J.* **7,** 938-943.

Harvey, M., Sands, A. T., Weiss, R. S., Hegi, M. E., Wiseman, R. W., Pantazis, P., Giovanella, B. C., Tainsky, M. A., Bradley, A. and Donehower, L. A. (1993b). *In vitro* growth characteristics of embryo fibroblasts isolated from p53-deficient mice. *Oncogene* **8,** 2457-2467.

Hermeking, H. and Eick, D. (1994). Mediation of *c-myc*-induced apoptosis by *p53*. *Science* **265,** 2091-2093.

Hogan, R. N. and Albert, D. M. (1991). Does retinoblastoma occur in animals? *Prog. Vet. Comp. Ophthalmol.* **1,** 73-82.

Hollingsworth, R. E., Jr., Hensey, C. E. and Lee, W.-H. (1993). Retinoblastoma protein and the cell cycle. *Curr. Opin. Genet. Dev.* **3,** 55-62.

Hooper, M. L. (1992). *Embryonal stem cells: introducing planned changes into the animal germline* (Harwood Academic Publishers gmbh, Chur, Switzerland).

Hooper, M. L. (1994). The role of the *p53* and *Rb-1* genes in cancer, development and apoptosis. *J. Cell Sci. Suppl.* **18**, in press.

Howes, K.A., Lasudry, J.G.H., Albert, D.M. and Windle, J.J. (1994a). Photoreceptor tumours in transgenic mice. *Invest. Ophthalmol. Visual Sci.*, **35**, 342-351.

Howes, K.A., Ransom, N., Papermaster, D.S., Lasudry, J.G.H., Albert, D.M. and Windle, J.J. (1994b). Apoptosis or retinoblastoma: Alternative fates of photoreceptors expressing the HPV-16 E7 gene in the presence or absence of p53. *Genes and Development* **8,** 1300-1310.

Howie, S.E.M., Sommerfield, A.J., Gray, E. and Harrison, D.J. (1994). Peripheral T lymphocyte depletion by apoptosis after CD4 ligation *in vivo*: selective loss of $CD44^-$ and 'activating' memory T cells. *Clin. Exp. Immunol.* **95,** 195-200.

Hu, N., Gutsmann, A., Herbert, D. C., Bradley, A., Lee, W.-H. and Lee, E. Y.-H. P. (1994). Heterozygous $Rb\text{-}1^{\Delta 20}/+$ mice are predisposed to tumours of the pituitary gland with a nearly complete penetrance. *Oncogene* **9,** 1021-1027.

Hu, Q., Dyson, N. and Harlow, E. (1990). The regions of the retinoblastoma protein needed for binding to adenovirus E1A or SV40 large T antigen are common sites for mutations. *EMBO J.* **9,** 1147-1155.

Jacks, T., Fazeli, A., Schmitt, E. M., Bronson, R. T., Goodell, M. A. and Weinberg, R. A. (1992). Effects of an Rb mutation in the mouse. *Nature* **359,** 295-300.

Jacks, T., Remington, L., Williams, B. O., Schmitt, E. M., Halachmi, S., Bronson, R. T. and Weinberg, R. A. (1994). Tumour spectrum analysis in p53 mutant mice. *Curr. Biol.* **4,** 1-7.

Kemp, C.J., Wheldon, T. and Balmain, A. (1994). p53-deficient mice are extremely susceptible to radiation-induced tumorigenesis. *Nature Genetics* **8,** 66-69.

Knudson, A. G. (1971). Mutation and cancer: statistical study of retinoblastoma. *Proc. Nat. Acad. Sci. USA* **68,** 820-823.

Knudson, A. G. (1993). Antioncogenes and human cancer. *Proc. Nat. Acad. Sci. USA* **90,** 10914-10921.

Lane, D. P., and Crawford, L. V. (1979). T antigen is bound to a host protein in SV40-transformed cells. *Nature* **278,** 261-263.

Lane, D. P. (1992). Cancer: p53, guardian of the genome. *Nature* **358,** 15-16.

Lee, E. Y.-H. P., Chang, C.-Y., Hu, N., Wang, Y.-C. J., Lai, C.-C., Herrup, K., Lee, W.-H. and Bradley, A. (1992). Mice deficient for Rb are nonviable and show defects in neurogenesis and haematopoiesis. *Nature* **359,** 288-294.

Lee, E. Y.-H. P., Hu, N., Yuan, S.-S. F., Cox, L.A., Bradley, A., Lee, W.-H. and Herrup, K. (1994). Dual roles of the retinoblastoma protein in cell cycle regulation and neuron differentiation. *Genes Dev.* **8,** 2008-2021.

Li, Y., Graham, C., Lacy, S., Duncan, A. M. V. and Whyte, P. (1993). The adenovirus E1A-associated 130-kD protein is encoded by a member of the retinoblastoma gene family and physically interacts with cyclins A and E. *Genes Dev.* **7,** 2366-2377.

Lowe, S. W., Ruley, H. E., Jacks, T. and Housman, D. E. (1993a). p53-dependent apoptosis modulates the cytotoxicity of anticancer agents. *Cell* **74,** 957-967.

Lowe, S. W., Schmitt, E. M., Smith, S. W., Osborne, B. A. and Jacks, T. (1993b). p53 is required for radiation-induced apoptosis in mouse thymocytes. *Nature* **362,** 847-849.

Lowe, S.W., Bodis, S., McClatchey, A., Remington, L., Ruley, H.E., Fisher, D.E., Housman, D.E. and Jacks, T. (1994). p53 status and the efficacy of cancer therapy in vivo. *Science* **266,** 807-810.

Margo, C., Hidayat, A., Kopelman, J. and Zimmerman, L. E. (1983). Retinocytoma: a benign variant of retinoblastoma. *Arch. Ophthalmol.* **101,** 1519-1531.

Marshall, C.J. (1991). Tumor suppressor genes. *Cell* **64,** 313-326.

Merritt, A. J., Potten, C. S., Kemp, C. J., Hickman, J. A., Balmain, A., Lane, D. P., and Hall, P. A. (1994). The role of p53 in spontaneous and radiation-induced apoptosis in the gastrointestinal tract of normal and p53-deficient mice. *Cancer Res.* **54,** 614-617.

Morgenbesser, S.D., Williams, B.O., Jacks, T., and DePinho, R.A. (1994). *p53*-dependent apoptosis produced by *Rb*-deficiency in the developing mouse lens. *Nature* **371,** 72-74.

Nasmyth, K. and Hunt, T. (1993). Cell cycle: Dams and sluices. *Nature* **366,** 634-635.

Pan, H. and Griep, A.E. (1994). Altered cell cycle regulation in the lens of HPV-16 E6 or E7 transgenic mice: Implication for tumour suppressor gene function in development. *Genes Dev.* **8,** 1285-1299.

Potten, C. S. (1990). A comprehensive study of the radiobiological response of the murine (BDF1) small intestine. *Int. J. Radiat. Biol.* **58,** 925-973.

Purdie, C. A., Harrison, D. J., Peter, A., Dobbie, L., White, S., Howie, S. E. M., Salter, D. M., Bird, C. C., Wyllie, A. H., Hooper, M. L. and Clarke, A. R. (1994). Tumour incidence, spectrum and ploidy in mice with a large deletion in the p53 gene. *Oncogene* **9,** 603-609.

Robanus-Maandag, E.C., van der Valk, M., Vlaar, M., Feltkamp, C., O'Brien, J., van Roon, M., van der Lugt, N., Berns, A. and te Riele, H. (1994). Developmental rescue of an embryonic-lethal mutation in the retinoblastoma gene in chimeric mice. *EMBO J.* **13,** 4260-4268.

Robbins, P.D., Horowitz, J.M. and Mulligan, R.C. (1990). Negative regulation of human *c-fos* expression by the retinoblastoma gene product. *Nature* **346,** 668-671.

Rootman, J., Carruthers, J. D. A. and Miller, R. R. (1987). Retinoblastoma. *Perspect. Pediatr. Pathol.* **10,** 208-258.

Serrano, M., Hannon, G. J. and Beach, D (1993). A new regulatory motif in cell-cycle control causing specific inhibition of cyclin D/CDK4. *Nature* **366,** 704-707.

Shaw, P., Bovey, R., Tardy, S., Sahli, R., Sordat, B. and Costa, J. (1992). Induction of apoptosis by wild-type p53 in a human colon tumour-derived cell line. *Proc. Nat. Acad. Sci. USA* **89,** 4495-4499.

Shirodkar, S., Ewen, M., DeCaprio, J. A., Morgan, J., Livingston, D. M. and Chittenden, T. (1992). The transcription factor E2F interacts with the retinoblastoma product and a p107-cyclin A complex in a cell cycle-regulated manner. *Cell* **68,** 157-166.

Slingerland, J.M., Jenkins, J.R. and Benchimol, S. (1993). The transforming and suppressor functions of p53 alleles: effects of mutations that disrupt phosphorylation, oligomerization and nuclear translocation. *EMBO J.*, **12,** 1029-1037.

Strasser, A., Harris, A.W., Jacks, T. and Cory, S. (1994). DNA damage can induce apoptosis in proliferating lymphoid cells by p53-independent mechanisms inhibitable by Bcl-2. *Cell* **79,** 329-339.

Szekely, L., Jiang, W.-Q., Bulic-Jakus, F., Rosen, A., Ringertz, N., Klein, G. and Wiman, K.G. (1992). Cell type and differentiation dependent heterogeneity in retinoblastoma protein expression in SCID mouse fetuses. *Cell Growth Diff.* **3,** 149-156.

Tsukuda, T., Tomooka, Y., Takai, S., Ueda, Y., Nishikawa, S., Yagi, T., Tokunaga, T., Takeda, N., Suda, Y., Abe, S., Matsuo, I., Ikawa, Y. and Aizawa, S. (1993). Enhanced proliferative potential in culture of cells from p53-deficient mice. *Oncogene* **8,** 3313-3322.

Vogelstein, B. and Kinzler, K. W. (1994). Tumour-suppressor genes: X-rays strike p53 again. *Nature* **370,** 174-175.

Williams, B.O., Remington, L., Albert, D.M., Mukai, S., Bronson, R.T. and Jacks, T. (1994a). Cooperative tumorigenic effects of germline mutations in p53. *Nature Genetics* **7,** 480-484.

Williams, B.O., Schmitt, E.M., Remington, L., Bronson, R.T., Albert, D.M., Weinberg, R.A. and Jacks, T. (1994b). Extensive contribution of Rb-deficient cells to adult chimeric mice with limited histopathological consequences. *EMBO J.* **13,** 4251-4259.

Windle, J. J., Albert, D. M., O'Brien, J. M., Marcus, D. M., Disteche, C. M., Bernards, R. and Mellon, P. L. (1990). Retinoblastoma in transgenic mice. *Nature* **343,** 665-669.

Yonish-Rouach, E., Resnitzky, D., Lotem, J., Sachs, L., Kimchi, A. and Oren, M. (1991). Wild-type p53 induces apoptosis of myeloid leukaemic cells that is inhibited by interleukin-6. *Nature* **352,** 345-347.

Part 5

Receptors and Their Ligands in Normal Development and Disease

Cloning of retroviral insertion sites possibly involved in growth-factor autonomy.

Boris Fehse, Panos Papadopoulos, Klaus Kühlcke, Jörg Bergemann, Ilka Ratz, Carol Stocking, Wolfram Ostertag, and Heinz Lother[1].
Heinrich-Pette-Institut für Experimentelle Virologie und Immunologie an der Universität Hamburg, Hamburg, Federal Republic of Germany

Summary

A protocol has been established which allows obtaining growth-factor-independent cell mutants by retroviral insertion mutagenesis *in vitro*. In order to identify gene alterations possibly leading to growth-factor independency, two techniques were established for the cloning of retroviral insertion sites. One technique makes use of the amplification of retroviral flanking fragments by inverse polymerase chain reaction (IPCR). The other strategy involves complementation of a truncated kanamycin gene present in a bacterial plasmid vector by a neomycin gene fragment originating from the retroviral vector, which allows direct selection for kanamycin resistant bacterial cell clones. Using these techniques flanking fragments with several putative genes have been obtained. One flanking fragment shows high homology to the rat ionotrophic glutamate receptor, i.e. GluR5.

Introduction

Omission of growth factor from proliferating cells leads to cell death unless oncogenes such as those encoding tyrosine kinase activity abrogate the factor requirement of factor-dependent cells (Pierce et al., 1985; Mathey et al., 1986; Wheeler et al., 1987; Cleveland et al., 1989; Katzav et al., 1989; Meeckling et al., 1992). In fact, in hematopoietic stem cells, the primary function of colony-stimulating factors appears to be suppression of programmed cell death (apoptosis, Williams et al., 1993), and this effect allows an

[1] Communicating author

NATO ASI Series, Vol. H 94
Gene Technology
Edited by A. R. Zander, W. Ostertag,
B. V. Afanasiev, F. Grosveld

intrinsically determined pathway of differentiation to be followed (Fairbairn et al., 1993).

Several different messenger systems have been associated with induction of apoptosis, and the final response varies with the cell type and the other signals being received. Consequently, in addition to genes exclusively involved in cell death, there are likely to be some that can influence other aspects of cell behavior as well. In fact, genes mediating or modulating apoptosis such as myc, ras, p53, bcl-2, bcl-x, bax etc. are critically involved in control of proliferation and differentiation (reviewed by Freeman et. al., 1993). Since the molecular control of proliferation and differentiation has been studied far more extensively than control of apoptosis, some proteins with long established roles in these areas could also be important in apoptosis.

Retroviral insertion mutagenesis has proved useful for identifying oncogenes in provirally induced tumors (Habets et al., 1994), and for identifying genes or genetic loci involved in growth factor independence (Stocking et al., 1988, Dorsers et al., 1993). We asked whether it is applicable to obtain growth-factor independent mutants of hematopoietic stem cells *in vitro*. We anticipated that retroviruses integrate preferentially into open chromosomal domains (reviewed by Breindl et. al. 1989), which may lead to: a) gene inactivation, b) gene truncation yielding an altered phenotype or c) constitutive gene expression due to an exchange of transcription control regions.

Results and discussion

Establishment of growth-factor independent mutants

The human, growth factor-dependent, hematopoietic precursor TF-1 cell line derived from a patient with erythroleukemia (Kitamura et al., 1989) was used for the retroviral insertion mutagenesis. TF-1 cells require IL-3 or GM-CSF for long term growth. Other factors such as stem cell factor (SCF) and erythropoietin (Epo) induce short term proliferative signals. TPA induces differentiation into macrophage like cells. Hemin and Epo induce hemoglobin synthesis. TF-1 mutants grow stroma cell dependent (Itoh, et al., 1994).

For infection with the retroviral vector TF-1 cells were cocultivated with irradiated pM3neo-supF producing fibroblasts for 8 or 16 rounds of cocultivation. Subsequent growth factor removal allowed the detection of growth-factor independent cell mutants. This regime (Fig. 1) led to the generation of 241 growth-factor independent cell lines (Stocking et al., 1993). The construction of the retroviral vector pM3neo-sup containing a *gag-neo*^{R} fusion gene and the bacterial tRNA suppressor gene (*supF*) has been described earlier. Helper cell lines were infected to establish stably transfected amphotropic virus-producing cell lines (Stocking et al., 1993).

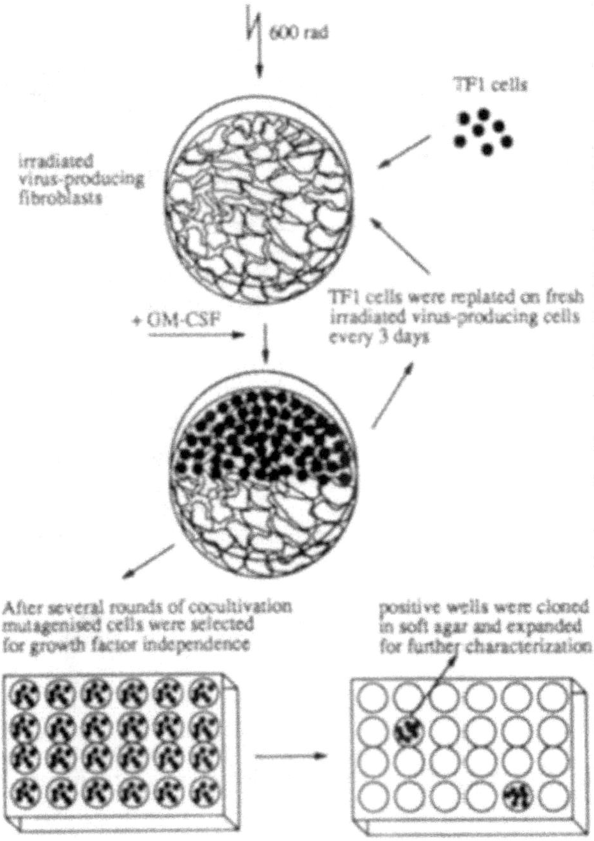

Fig. 1: Establishment of growth-factor independent TF-1 cell mutants.

Cloning of retroviral integration sites by inverse PCR

Since all our mutants contain more than one retroviral integration we intended to clone large numbers of integration sites in order to eliminate integration events not affecting genes involved in factor independency. To achieve this, we initially attempted to select for supF containing λ phages of genomic phage libraries. Due to the finding that the supF gene had suffered mutations (data not shown) probably during establishement of the packaging cell line, this technique could not be used. The first technique applied was therefore cloning by inverse PCR as is depicted in Figure 2. It involves cutting of genomic DNA by a restriction enzyme, ligation favouring intramolecular reaction (dilute solution), heating to 94°C for 30 min (nicking), and amplification of the circular genomic DNA using two primers which recognise retroviral sequences.

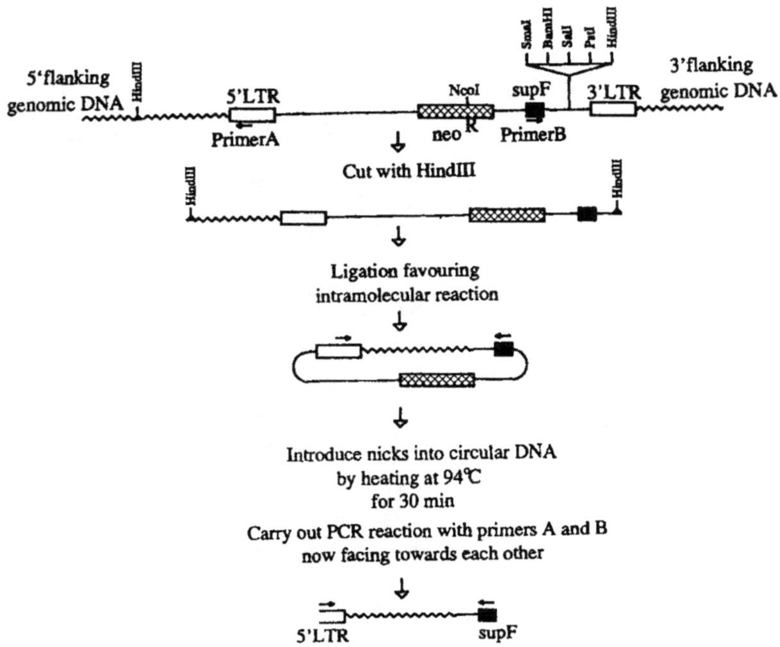

Fig. 2: Inverse PCR strategy for cloning retroviral integration sites.

Genomic fragments in the range of 0.15 kb to 1.4 kb were obtained. Southern-blot and sequence analysis confirmed that these fragments respresent retroviral flanking sequences (data not shown). However, Southern-blot analysis revealed as well, that several of the cloned fragments contained repeat elements. Most of these repeat elements were identified by sequence analysis as to be Alu elements. The estimated copy number of 700 000 Alu elements per haploid genome predicts a density of one Alu repeat every 4 kb of genomic DNA (reviewed by Makalowski et al., 1994). At the cytogenetic level, Alu repeats are concentrated in R bands, the most transcriptionally active areas of the genome. Alu elements are found in introns of most of known actively transcribed genes. It appears that integration of the retroviral vectors occurred with some preference close to repeat sequences, mainly Alu elements.

Cloning of retroviral flanking sequences by kanamycin gene complementation.

Mainly due to inheritent problems encountered by the IPCR cloning technique such as the presence of repeat elements within short fragments, which in part did not allow any further analysis, we established a second cloning protocol. We made use of the neomycin resistance gene, which confers kanamycin resistance to bacteria. Since no bacterial ribosomal binding site preceeded the neomycin gene, selecting for the complete proviral sequence including 3'- and 5'- flanking sequences could not be achieved. However, a cloning strategy for 3' flanking which involves cutting the genomic DNA by Nco1, ligation into a plasmid vector containing the 5' end of the kanamycin gene and a bacterial promoter depicted in Fig. 2 could be established. To obtain the highest possible transformation efficiency, the following parameters were optimized: bacterial strain, ligation conditions, transformation conditions, DNA concentration, and field strength (data to be shown elsewhere).

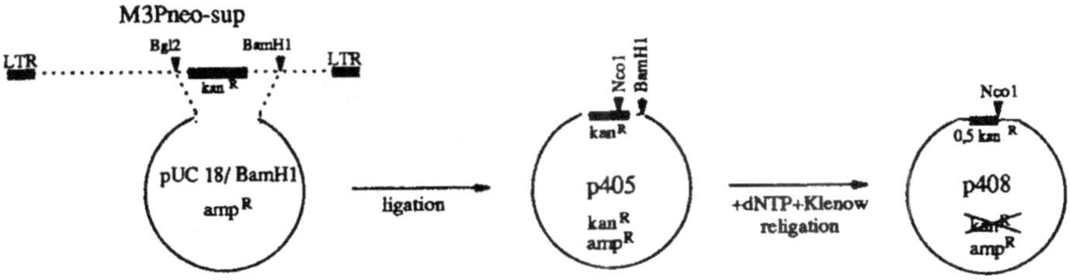

Fig. 3. Construction of the vector used for the cloning of 3'- flanking DNAs by kanamycin gene complementation.

Restriction of genomic DNA from TF-1 mutants by Nco1 and subsequent Southern-blot analysis using the Nco1/BamH1 3' fragment of the retroviral vector (see fig. 2) revealed that all identified fragments were larger than 2.5 kb, the size predicted by the vector part of the cloned fragment (data not shown). It is therefore suggested, that deletions or recombinations of the 3' part of the retroviral vector did occur with a rather low frequency if at all. Moreover, all fragments cloned by the kanamycin gene complementation approach did match in size with the corresponding genomic fragment seen by Southern-blot analyis (data not shown). Southern-blot analysis revealed as well, that approximately 50% of the cloned fragments do contain Alu repeats.

For further characterising the cloned fragments we started to sequence the first 300 bp of all probes using a LTR specific primer. In accordance with the data obtained by the inverse IPCR approach the integrations occurred mainly in non-coding regions. Despite the presence of repeat elements, the sequence analysis and the data base search revealed several interesting insertions:

1. One flanking sequence obtained from mutant 2B seems to contain an exon of an unknown transmembrane protein (data not shown), since a transmembrane spanning region could be predicted by various programs with high scores.

2. A database search of a flanking sequence obtained from mutant 19/1 suggests that the integration may have occurred into the partially sequenced region of the MHC II locus (Beck et al., 1992) between LMP2 and RING7 genes. A 3 kb and a 6 kb transcript were detected by Northern-blot analysis. Screening of a phage λ cDNA library allowed to detect several positive clones. However, these trancripts seem to define a gene not located in this chromosomal region. Due to the short sequence interval between LMP2 and RING7 genes (Trowsdale, personal communication) only a pseudogene corresponding to our cDNA clone may be mapped to the MHC II locus.

3. One integration of mutant 16/3 represents part of a transcript of about 8 kb which is expressed in the mutant cell line having this flanking sequence but not in the wild-type TF1 cell line or other insertion mutants. Several cDNA clones were obtained from a phage λ cDNA library.

Due to the large sizes of most of the cloned flanking regions and the high percent of repeat sequences within these fragments the exontrap Vector system (Mobitec) developed by Auch and Reth (1990) was adapted for identifying coding regions closed to the insertions (data to be shown elsewhere).

Using this approach, we could confirm and extend the initial sequence analysis. In particular, a retroviral integration cloned from mutant 29A allowed to identify an ionotrophic glutamate receptor. As was revealed by sequence analysis of the 5.5 kb flanking fragment, the human equivalent to the rat glutamate receptor, i.e. GluR5 (Bettler et al. 1990) was identified. The identified exon comprises 108 bp. It has 90% homology at the DNA level, and 100 % homology at the amino acid level to the rat GluR5 cDNA (Fig. 4). Using this exon as a probe a 900bp fragment was isolated from a human brain cDNA library, which bears 91% homology at DNA level and only a few conservative amino acid exchanges in comparision to the rat GluR5 (data not shown). No human GluR5 cDNA has been published to date. Its chromosomal localisation, however, has been assigned to 21q22 (Potier et al., 1993) in the vicinity of the gene for familial amyotrophic lateral sclerosis (Eubanks et al., 1993; Gregor et al., 1993).

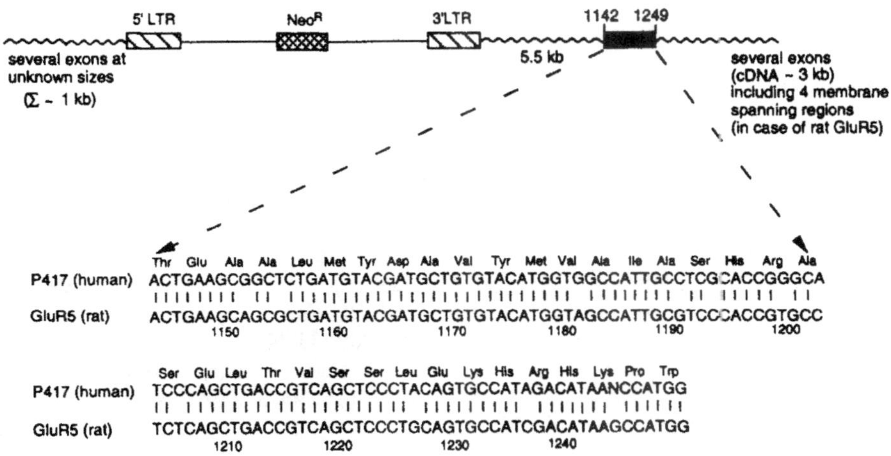

Fig. 4: Retroviral insertion into the Glutamate receptor GluR5.

It is generally agreed that receptors for the major excitatory neurotransmitter glutamate may play not only key roles in synaptic machanisms for distribution of information, and in neurodegeneration (reviewed by Barnard, 1992; Gasic and Heinemann, 1991; Sommer and Seeburg, 1992; Wisden and Seeburg, 1993) but also in mouse embryo developement as well, since the transcripts of GluR5 and GluR6 were detected by Northern analysis in mouse embryos of 11 days gestation (Gregor et al., 1993). This finding and the observation that GluR5 protein forms ion channels in Xenopus oocytes that are responsive to L-glutamate (Bettler, et al. 1990) allows speculating that the GluR5 may as well play a prominent role in the development of the hematopoietic system, in particular be involved in Ca2+ homeostasis. Based on these considerations we test at present whether a) the GluR5-R does play a role in differentiation and or proliferation of hematopoietic cells and b) whether common integration events can be identified.

Acknowledgements

This work was supported by a grant of the Deutsche Krebshilfe and AMGEN. The Heinrich-Pette-Institut is financially supported by Freie und Hansestadt Hamburg and Bundes ministerium für Gesundheit.

Literatur:

Auch, D., and Reth, M. (1990) Exon trap cloning: using PCR torapidly detect and clone exons from genomic DNA fragments. Nucl. Acid. Res. 18: 6743-6744.

Barnard, E.C. (1992) Receptor classes and the transmitter-gated ion channels. Tibs. 17: 368-374.

Beck, S., Kelly, A., Radley, E., Kurshid, F., Alderton, R.P., and Trowsdale J. (1992) DNA sequence analysis of 66 kb of the human MHC class II region encoding a cluster of genes for antigen processing. Mol. Biol. 228: 433-441.

Bettler, B., Boulter, J., Hermans-Borgmeyer, I., O'Shea-greenfiel, A., Deneris, E. S., Moll, C., Borgmeyer, U., Hollmann, M., and Heinemann S. (1990) Cloning of a novel glutamate receptor subunit, GLUR5: expression in the nervous system during development. Neuron. 5: 583-595.

Breindl, M., Scherdin, U., and Rhodes, K. (1989) Cellular target sequences for retroviral integration. In: Nato Asi Series, Vol H34. Vectors as tools for the study of normal and abnormal growth and differentiation. Lother, H., Dernick, R., and Ostertag, W. eds Springer-Verlag Berlin Heidelberg. pp. 77-88.

Cleveland, J.L., Dean, M., Rosenberg, N., Wang, J.Y.J. Rapp, U.R. (1989) Tyrosine kina se oncogenes abrogate intereleukin-3 dependence of murine myeloid cells through signaling pathways involving c-myc: Conditional regulation of c.myc transcription by temperture-sensitive v-abl. Mol. Cell. Biol. 9: 5685.

Dorsers, L.C.J., Vanagthoven, T., Dekker, A., and Vanagthoven, T.L.A. (1993). Induction of anti-estrogen resistance in human breast cancer cells by random insertional mutagenesis using defective retroviruses: identification of bcar-1, a common integration site. Mol. Endocrinol. 7: 870-878.

Eubanks, J.H., Puranam, R.S., Kleckner, N.W., Bettler, B., Heinemann, S.F., McNamara, J.O. (1993) The gene encoding the glutamate receptor subunit GluR5 is located on human chromosome 21q21.1-22.1 in the vicinity of the gene for familial amyotrophic lateral sclerosis. Proc.Natl.Acad.Sci.USA. 90: 178-82

Fairbairn, L.J., Cowling, G.J., Reipert, B.M., and Dexter, T.M. (1993). Suppression of apoptosis allows differentiation and development of a multipotent hemopoietic cell line in the absence of added growth factors. Cell. 74: 823-832.

Freeman, R.S., Estus; S., Horigome, K., and Johnson Jr., E.M. (1993) Cell death genes in invertebrates and (maybe) vertebrates. Current opinion in neurobiology. 3: 25-31.

Gasic, G.P. and Heinemann, S. (1991) Recptors coupled to ionic channels: the glutamate receptor family. Current Opinion in Neurobiology. 1: 20-26.

Gregor, P., Reeves, R.H., Jabs, E.W., Yang, X., Dackowski, W., Rochelle, J.M., Brown, R.H. Jr., Haines, J.L., O'Hara, B.F., Uhl, G.R., (1993) Chromosomal localizationof glutamate receptor genes: relationship to familial amyotrophic lateral sclerosis and other neurological disorders of mice and humans. Proc. Natl. Acad. Sci. USA. 90: 3053-7.

Gregor, P., O'Hara, B.F., Yang, X., Uhl, G.R. (1993) Expression and novel isoforms of glutamate receptor genes GluR5 and GluR6. Neuroreport. 4: 1343-1346.

Habets G. G. M., Scholtes E.H.M. Zuydgeests, D., van der Kammen, E.H.M., Stam, J.C., Berns, A., and Collard, J.G. (1994) Identification of an invasion-inducing gene, Tiam-1, that encodes a protein with homology to GDP-GTP exchangers for Rho-like proteins. Cell 77: 537-549.

Itoh, K., Klingler, K., Laker, C., Just, U., Mannweiler, K., Mori, K.J., Nishikawa, S.-I., Bittner, S., Nibbs, R., Harrison, P.R., and Ostertag, W., (1994) Soluble growth factors induce transient growth and clonal extinction of stroma cell dependent erythroleukemia cells. submitted for publication.

Katzav, S., Martin, Z.D., Barbacid, M., Hedge, A.M., Isfort, R., Ihle, J.N. (1989) The trk oncogene abrogates growth factor requirements and transforms hematopoietic cells. Oncogene. 4: 1129.

Kitamura, T., Tange, T., Terasawa, T., Chiba, S., Kuwaki, T., Miyagawa, K., Piao, Y.-F., Miyazono, K., Urabe, A., and Takaku, F. (1989) Establishment and characterization of a unique human cell line that proliferates dependently on GM-CSF, IL-3, or erythropoietin. J. Cell. Physiol. 140: 323-334.

Mathey, P.B., Nabel, G., Palacios, R., Baltimore, D. (1986). Abelson virus abrogation of interleukin-3 dependence in a lymphoid cell line. Mol. Cell Biol. 6: 4133.

Makalowski, W., Mitchell, G. A., and Labuda, D. (1994) Alu sequences in the coding region of mRNA: a source of protein variability. TIG. 10, 188-193.

Mecckling, G.K., Yee, S.P., Schrader, J.W., Pawson, T. (1992) A retrovirus encoding the v-fps protein-tyrosine kinase induces factor-independent growth and tumorigenicity in FDC-P1 cells. Biochim. Biophys. acta. 1137: 65.

Pierce, J.H., Di Fiori, PP., Aaronson, S.A., Potter, M., Pumphrey, J., Scott, A., Ihle, J.N. (1985). Neoplastic transformation of mast cells by Abelson-MuLV: Abrogation of IL-3 dependence by a nonautocrine mechanism. Cell. 41: 685.

Potier, M.C., Dutriaux, A., Lambolez, B., Bochet, P., Rossier, J. (1993) Assignment of the human glutamate receptor gene GLUR5 to 21q22 by screening a chromosome 21 YAC library. Genomics. 15: 696-7

Sommer, B., Seeburg, P.H. (1992) Glutamate receptor channels: Novel properties and new clones. Trends Pharmacol. Sci. 13: 291-296.

Stocking, C., Bergholz, U., Friel, J., Klingler, K., Wagner, T., Starke, C., Kitamura, T., Miyajima, A., and Ostertag, W., (1993) Distinct classes of factor-independent mutants can be isolated after retroviral mutagenisis of a human myeloid stem cell line. Growth factors, 8: 197-209.

Stocking, C., Löliger, C., Kawai, M., Suciu, S., Gough, N., and Ostertag, W. (1988). Identification of genes involved in growth autonomy of hematopoietic cells by analysis of factor-independent mutants. Cell 53, 869-879.

Wheeler, E.F., Askew, D., May, S., Ihle, J.N., Sherr, C.J. (1987). The v-fms oncogene induces factor-independent growth and transformation of the interleukin-3-dependent myeloid cell line FDC-P1. Mol. Cell Biol. 7: 1673.

Wisden, W., and Seeburg, P.H. (1993) Mammalian ionotropic glutamate receptors. Current Opinion in Neurobiology. 3: 291-298.

Williams, G.T., and Smith, C.A. (1993) Molecular regulation of apoptosis: Genetic controls of cell death. Cell. 74: 777-779.

Function of the IL-3 receptor system in hematopoiesis

Takahiko Hara and Atsushi Miyajima
Department of Cell Biology
DNAX Research Institute of Molecular and Cellular Biology
901 California Avenue
Palo Alto, CA 94304
USA

Introduction

Hematopoiesis is a complexed process of cell proliferation and differentiation that are regulated by a number of cytokines (Arai et al., 1990). Among such cytokines, interleukin-3 (IL-3) stimulates multipotential hematopoietic progenitors as well as lineage-committed cells including granulocytes, macrophages, megakaryocytes, erythroid cells, and mast cells, resulting in formation of the colonies of multiple lineages. Thus, IL-3 is also known as a multi-colony stimulating factor (multi-CSF). Granulocyte macrophage colony stimulating factor (GM-CSF) stimulates not only granulocytes and macrophages, but also various hematopoietic cells and exhibits activities similar to IL-3. In contrast, IL-5 has a narrow target cell specificity and mainly stimulates eosinophils and CD5-positive B cells. Interestingly, in factor-dependent cultured cells, these three cytokines are strong mitogens which induce almost identical

NATO ASI Series, Vol. H 94
Gene Technology
Edited by A. R. Zander, W. Ostertag,
B. V. Afanasiev, F. Grosveld

biochemical responses such as tyrosine phosphorylatin of intracellular proteins and activation of signaling pathways (Miyajima et al., 1993). Although IL-3, GM-CSF, and IL-5 show an overlapping biological activity on a certain type of cells (e.g. eosinophils), both GM-CSF and IL-5 do not possess multi-CSF activity on the primitive progenitor cells and mast cell stimulation activity.

Biological activities of IL-3, GM-CSF and IL-5 are mediated by their specific high affinity receptors which consist of α and β subunits, members of the class I cytokine receptor family (Miyajima et al., 1993). The α subunits are glycoproteins of 60-70 kilodalton (kD) and bind their specific ligand with low affinity. Reconstitution experiments have shown that the high affinity receptors for IL-3, GM-CSF, and IL-5 share the β subunit (β_C) that is a glycoprotein of 120-130 kD. The β_C does not bind any cytokine by itself, but forms high affinity receptors with any one of the three α subunits: i. e. IL-3Rα, GM-CSFRα, and IL-5Rα. Thus, the α subunits determine the cytokine specificity of the high affinity receptors. When cytokine and the α and β subunits form a complex on the cell surface, various intracellular signaling cascades are activated. Although intracellular domains of both α and β subunits are required for transmitting a proliferation signal (Takaki et al., 1994), the β subunit plays a major role in signaling including Ras/Raf activation, myc induction, and JAK2 kinase activation (Sato et al., 1993; Quelle et al., 1994). Thus the common biological responses are elicitted by IL-3, GM-CSF, and IL-5 through the common signal transducer, β_C. It should be noted that in mice there are two homologous β subunits, β_C and IL-3 specific β, β_{IL-3}, which binds IL-3 with low affinity and form a high affinity receptor with only IL-3Rα. The two high affinity IL-3Rs formed with either β_C or β_{IL-3} show any functinal differences (Takaki et al., 1991; Hara and Miyajima, 1992; Park et al., 1992).

Expression of α and β subunits of IL-3/GM-CSF/IL-5 receptors is mainly restricted in hematopoietic cells. The β subunits are expressed in myeloid progenitor cells, macrophages, mast cells, and CD5-positive B cells, but not in T cells and fibroblasts (Gorman et al., 1990). In contrast, expression of the α subunits is restricted to each cytokine-responsive cells (Hara and Miyajima, 1992). To better understand the function of each subunit in hematopoietic cell development, we have studied mice that show altered expression of IL-3Rα and IL-5Rα. In this article, we describe recent findings on the role of the IL-3R α subunit and the IL-3/IL-3R system in hematopoiesis.

Multi-CSF activity is mediated by the ectopically expresed functional IL-5R: IL-3Rα and IL-5Rα are functionally equivalent.

In the bone marrow, primitive hematopoietic progenitor cells are self-renewing and they are capable of differentiating into various lineage of mature hematopoietic cells upon stimulation with IL-3. Such a multipotent progenitor population can be isolated by using antibodies against several cell surface antigens (Heimfeld et al., 1991). As we expected, both α and β subunits (β_C and β_{IL-3}) of IL-3R are expressed in the multipotential progenitor cells (Fig. 1).

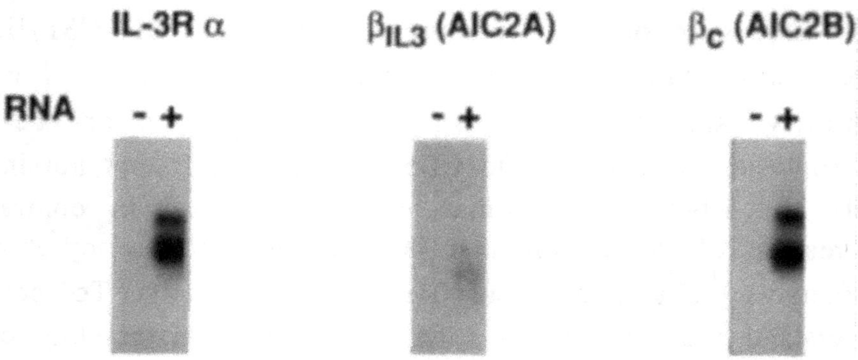

Fig. 1 Expression of the IL-3R subunits in sorted hematopoietic stem cells. Bone marrow cells were collected from mice and the stem cell population (Lin$^-$, Sca-1$^+$, and Thy-1^{lo}) was isolated according to the published procedure (Heimfeld et al., 1991). RNA expression of the α and two β subunits of mouse IL-3R was examined by reverse transcriptase-PCR by using specific primers. The PCR products were detected by hybridization using ^{32}P-labelled internal oligonucleotides as probes.

Since IL-5 does not stimulate these multipotential progenitor cells, we attempted to express IL-5Rα in these cells to examine whether IL-5 exhibits multi-CSF activity when IL-5Rα is expressed. For this purpose, we established the transgenic mouse expressing IL-5Rα cDNA which is driven by the PGK-1 promoter (Takagi et al., 1995). As shown in Fig. 2, bone marrow cells derived from the IL-5Rα transgenic mice gave rise to a various lineage of hematopoietic colonies in response to IL-5, which was the same as the colonies stimulated by IL-3. In the presence of erythropoietin, erythroid colonies also appeared in response to IL-5. Furthermore, when bone marrow cells from the transgenic mice were cultured in liquid medium containing IL-5 for a long time (> one month), IL-5-dependent mast cells started to grow (M. Takagi and A. M., unpublished data). Therefore we concluded that IL-5R has the same potential as IL-3R when the α subunit is ectopically expressed. In other words, identical signals which lead to cell differentiation can be delivered from both IL-3R and IL-5R. This result indicates that specific biological activities of IL-3 are due to the regulated expression of the IL-3Rα in certain populations of hematopoietic cells. Intriguingly, Longmore *et al.* has shown that erythropoietin acts as a multi-CSF on bone marrow cells infected with retrovirus carrying erythropoietin receptor (Longmore et al., 1994). Although further studies on other cytokine receptor systems are needed, hematopoietic cell differentiation signals could be transmitted by multiple signal transducing cytokine receptors when they are expressed in the immature precursor cells.

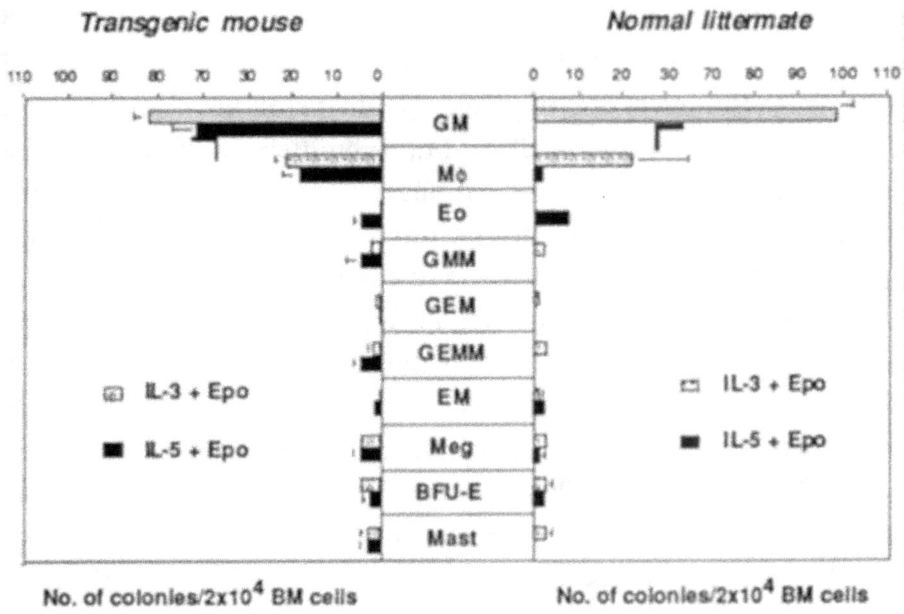

Fig. 2 Colony formation assays. Bone marrow cells (2 x 10^4) from IL-5Rα transgenic mice or normal littermates were cultured in the presence of FCS and various cytokines as indicated. Total number of various kinds of hematopoietic colonies was scored. Abbreviations of colony types are as follows: GM, granulocyte/macrophage; Mf, macrophage; Eo,

eosinophil; GMM, granulocyte/macrophage/megakaryocyte; GEM, granulocyte/erythrocyte/macrophage; GEMM, granulocyte/erythrocyte/macrophage/megakaryocyte; EM, erythrocyte/megakaryocyte; Meg, megakaryocyte; BFU-E, erythroid bursts; and Mast, mast cells.

Characterization of the IL-3 nonresponsive mice

So far as we know, IL-3 is the only cytokine which strongly stimulates multipotential hematopoietic progenitor cells. Hence it is of interest to know whether hematopoiesis is affected in mice lacking the functional IL-3R. Interestingly, IL-3 nonresponsive mice have been reported (Morris et al., 1990) and recently we found that expression of IL-3R in these mice are impaired (Ichihara et al., 1994; T. H., M. Ichihara, M. Takagi, and A. M., submitted). Bone marrow cells isolated from these mouse strains (A type) are nonresponsive to IL-3 in colony assays, however, they respond to GM-CSF normally, indicating that they lack an IL-3R specific component or signaling molecule. Detailed characterization has revealed that they are indeed deficient in expression of IL-3Rα (Fig. 3A). The IL-3-nonresponsive A type mouse strains possess a common mutation in intron 7 of the IL-3Rα gene. This mutaion, a 5 base-pairs deletion, disrupts the branchpoint consensus sequence for RNA splicing (Fig. 3B). Consequently, the alternatively spliced IL-3Rα transcript is produced in A type mice. Since the IL-3Rα protein produced from an alternatively spliced RNA lacks exon 8 encoding10 amino acid residues in the extracellular domain, it is discriminated from the normal product and is localized intracellularly but not translocated to the cell surface. The IL-3-nonresponsive mouse strains thus do not express a normal level of IL-3Rα nor high affinity IL-3Rs.

Remarkably, we have found 10 IL-3-nonresponsive A type mouse strains out of 27 inbred mouse strains (Table 1). The A type mouse strains do not apparently have a defect in constitutive hematopoiesis nor disadvantage in forming mouse colonies. Thus the A type IL-3Rα gene appears to be an allele which is not eliminated from the mice population in the laboratory settings. Interestingly, however, only 1 out of 21 wild-derived mouse strains we examined carry the A type IL-3Rα gene (T. H., M. Ichihara, M. Takagi, and A. M., submitted). This result may indicate some selective advantage of mice that possess the IL-3/IL-3R system in wild populations.

Table 1. Genotype of the IL-3Rα gene in inbred mouse strains

A type (IL-3 nonresponder)	A/J, A/WySnJ, A/HeJ, AKR/J, RF/J, C58/J, SM/J, BUB/BnJ, CE/J, NZB/B1NJ
B type (IL-3 responder)	C57BL/6J, C57L/J, BALB/cJ, CBA/J, DBA/1J, C3H/HeJ, 129/J, SWR/J, LP/J, YBR/Ei, BDP/J, SJL/J, PL/J, WC/Re Sl/+, ST/bJ, RIIIS/J, NZW/LacJ

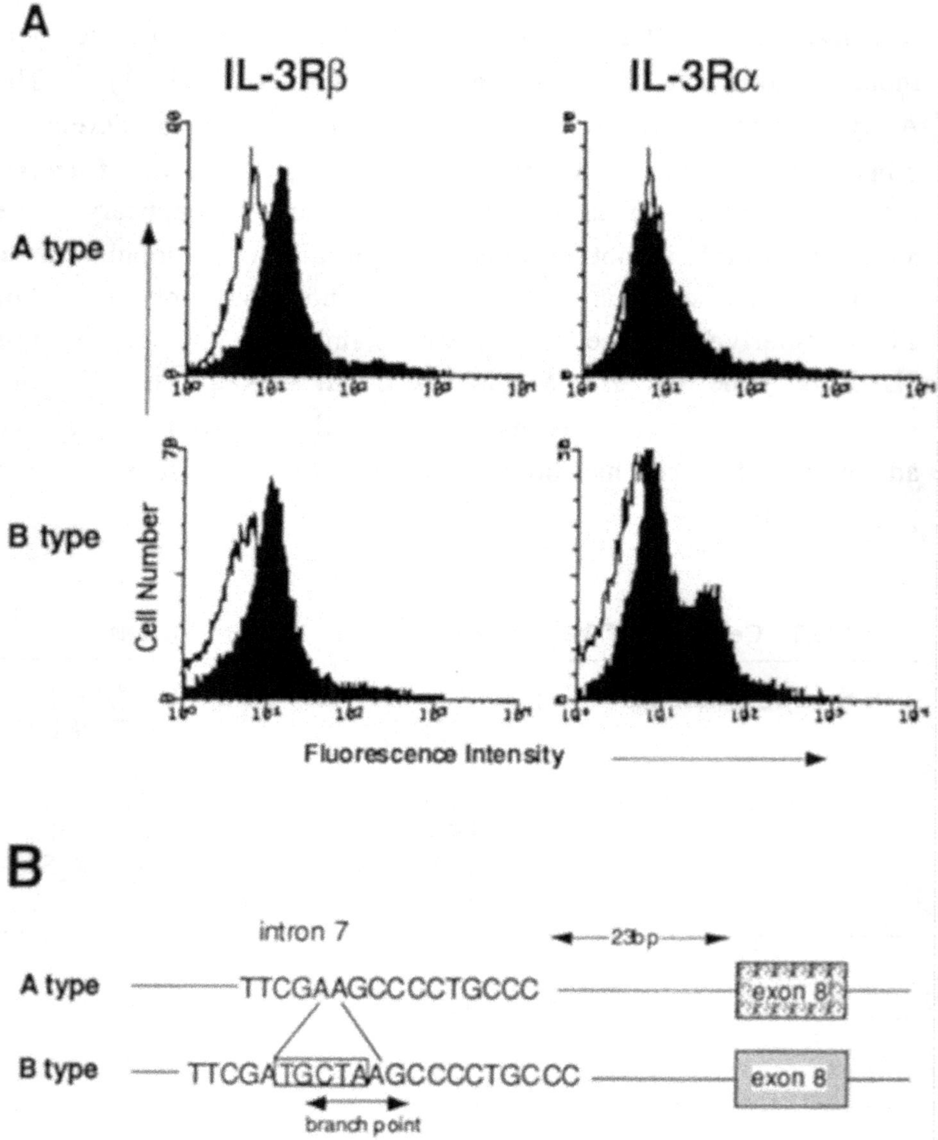

Fig. 3 Expression of the IL-3Rα protein and sequence comparison of the IL-3Rα gene in A and B type mouse strains. A. Bone marrow cells from A type mouse strain (A/J) or B type strain (C57BL/6J) were analyzed by flow cytometry using monoclonal antibodies against βIL-3 or IL-3Rα. Blank areas show staining profiles with an isotype control antibody.

B. Nucleotide sequence of intron 7 including a 5 base-pairs deletion (boxed) in A type mouse strains is shown. The consensus sequence for a branch point is also shown.

Many cytokines are coodinatedly produced and acting on a variety of cell types in the bone marrow where constitutive hematopoiesis is taking place. In the A type mouse strains, IL-3 alone does not stimulate the colony formation, however, IL-3 in combination with stem cell factor (SCF) stimulate multipotential primitive cells at a level similar to that of normal IL-3-responsive mice. This synergy is not caused by upregulation of the number of IL-3R by SCF. It is more likely that a small number of functional IL-3R, which is still present in the A type mice, delivers some signals that are not sufficient for triggerring cell division but can be complemented by other cytokine signals. The A type mice and cell lines derived from them would be useful tools to dissect signals and to find critical molecules for IL-3-mediated cell proliferation and differentiation.

We have described the role of IL-3R, particularly the α subunit, in hematopoiesis: 1) a functional similarity between the α subunits of IL-3, GM-CSF, and IL-5 receptors, and 2) normal hematopoiesis in IL-3Rα-deficient mice. To further understand the role of IL-3R system, characterization of β-knockout mice is underway in this laboratory.

Acknowledgements

The authors would like to thank Drs. S. Hudak, D. Rennick, M. Takagi, and M. Ichihara for their help and contributions to these studies. DNAX Research Institute of Molecular and Cellular Biology is supported by Schering-Plough Corporation.

References

Arai, K., Lee, F., Miyajima, A., Miyatake, S., Arai, N., and Yokota, T. (1990). Cytokines: coordinators of immune and inflamatory responses. Annu. Rev. Biochem. 59 : 783-836

Gorman, D. M., Itoh, N., Kitamura, T., Schreurs, J., Yonehara, S., Yahara, I., Arai, K., and Miyajima, A. (1990). Cloning and expression of a gene encoding an interleukin 3 receptor-like protein: identification of another member of the cytokine receptor gene family. Proc. Natl. Acad. Sci. USA 87: 5459-5463

Hara, T., and Miyajima, A. (1992). Two distinct functional high affinity receptors for mouse IL-3. EMBO J. 10: 1875-1884

Heimfeld, S., Hudak, S., Weissman, I., and Rennick, D. (1991). The in vitro response of phenotypically defined mouse stem cells and myeloerythroid progenitors to single or multiple growth factors. Proc. Natl. Acad. Sci. USA 88: 9902-9906

Ichihara, M., Hara, T., Takagi, M., Cho, L. C., Gorman, D. M., and Miyajima, A. (1994). Defective interleukin-3 (IL-3) response of the A/J mouse is caused by a branch point deletion in the IL-3 receptor α subunit gene. EMBO J. *in press*

Longmore, G. D., Pharr, P. N., and Lodish, H. F. (1994). A constitutively activated erythropoietin receptor stimulates proliferation and contributes to transformation of multipotent, committed nonerythroid and erythroid progenitor cells. Mol. Cell. Biol. 14: 2266-2277

Miyajima, A., Mui, A. L., Ogorochi, T., and Sakamaki, K. (1993). Receptors for granulocyte-macrophage colony-stimulating factor, interleukin-3, and interleukin-5. Blood 82: 1960-1974

Morris, C .F., Salisbury, J., Kobayashi, M., Townsend, P. V., and Hapel, A. J. (1990). Interleukin 3 alone does not support the proliferation of bone marrow cells from A/J mice: a novel system for studying the synergistic activities of IL-3. Br. J. Haematol. 74:131-137

Park, L. S., Martin, U., Sorensen, R., Luhr, S., Morrissey, P. J., Cosman, D., and Larsen, A. (1992). Cloning of the low-affinity murine granulocyte-macrophage colony-stimulating factor receptor and reconstitution of a high-affinity receptor complex. Proc. Natl. Acad. Sci. USA 89: 4295-4299

Quelle, F. W., Witthuhn, B. A., Inhorn, R., Ernst, T. J., Miyajima, A., Griffin, J. D., and Ihle, J. N. (1994). JAK2 associates with the β_c chain of the receptor for GM-CSF and its activation requires the membrane proximal region. Mol. Cell. Biol. 14:4335-4341

Sato, N., Sakamaki, K., Terada, N., Arai, K., and Miyajima, A. (1993). Signal transduction by the high-affinity GM-CSF receptor: two distinct cytoplasmic regions of the common beta subunit responsible for different signaling. EMBO J. 12:4181-4189

Takagi, M., Hara, T., Ichihara, M., Takatsu, K., and Miyajima, A. (1994). Multi-colony stimulating activity of interleukin-5 (IL-5) on hematopoietic progenitors from transgenic mice that express IL-5 receptor α subunit constitutively. J. Exp, Med. *in press*

Takaki, S., Kanazawa, H., Shiba, M., and Takatsu, K. (1994). A critical cytoplasmic domain of the interleukin-5 (IL-5) receptor α chain and its function in IL-5-mediated growth signal transduction. Mol. Cell. Biol. 14:7404-7413

Takaki, S., Mita, S., Kitamura, T., Yonehara, S., Yamaguchi, N., Tominaga, A., Miyajima, A., and Takatsu, K. (1991). Identification of the second subunit of the murine interleukin-5 receptor: interleukin-3 receptror-like protein, AIC2B is a component of the high-affinity interleukin-5 receptor. EMBO J. 10: 2833-2838

Altered Growth Factor Response in Myeloid Progenitor Cell Mutants Derived after Retroviral Insertional Mutagenesis

Christine Laker, Jutta Friel, Marie-Joseé Franz, Manuela Baccarini[1], Wolfram Ostertag, and Carol Stocking

Heinrich-Pette-Institut für Experimentelle Virologie und Immunologie an der Universität Hamburg
Martinistr. 52
D-20251 Hamburg, Germany

Summary

Retroviral insertional mutagenesis *in vivo* has proven to be an exquisite tool to identify previously unknown genes that are involved in tumorigenesis. Limitations of an *in vivo* system, *e.g.* selection for only one particular phenotype, can be overcome by *in vitro* insertional mutagenesis, which allows a variety of selection procedures for different phenotypes. In this study, we show that insertional mutagenesis can be used to define genes involved in receptor expression. For this purpose we used the murine hematopoietic cell line FDC-P1M that requires IL-3 for growth but does not normally respond to GM-CSF. Following retroviral insertional mutagenesis, mutants were selected that were either factor independent or had acquired the capacity to proliferate in response to GM-CSF. We describe here the phenotype of the GM-CSF responsive mutants and demonstrate the activation of the GM-CSF receptor α-chain responsible for the altered growth factor response.

Introduction

Retroviruses or transposable elements have successfully been used *in vivo* in mammals to identify genes that are causally involved in tumorigenesis (reviewed in van Lohuizen and Berns, 1990; Dührsen et al., 1990). This approach has also been used to define different oncogenic alterations in genes that cooperate during the successive stages of increased malignancy in multistep oncogenesis (van Lohuizen et al., 1991; Haupt et al., 1991). *In vivo* selection, however, has the disadvantage that selection is only possible for the final phenotype, *e.g.* for tumorigenesis. Therefore, an understanding of the different steps of oncogenesis and the type of selective forces opera-

[1]Frauenhofer Institut für Toxikologie und Molekulare Biologie, Nikolai-Fuchs-Str. 1, 3000 Hanover 51, Germany

NATO ASI Series, Vol. H 94
Gene Technology
Edited by A. R. Zander, W. Ostertag,
B. V. Afanasiev, F. Grosveld

ting becomes difficult. *In vitro* insertional mutagenesis, using defined tissue culture cells, has the advantage that mutagenesis and selection can be better controlled. Moreover, *in vitro* insertional mutagenesis allows selection for a variety of phenotypes depending on the test system used.

Work by our group has demonstrated that mutants can be isolated from hematopoietic cell lines that acquire factor-independent growth by different mechanisms, such as autocrine stimulation or constitutive activation of a growth factor receptor (Stocking et al., 1988; Heberlein et al. 1990; Hannemann et al. in press). In more recent experiments, in which cells of a factor-dependent human cell line were subjected to retrovirus infection, we could generate growth-factor independent mutants at a higher frequency, demonstrating that insertional mutagenesis can indeed be used to isolate a large number of mutants *in vitro* (Stocking et al, 1993).

We further wanted to test whether we could use selection protocols other than factor-independent growth to isolate mutants. Of special interest was a selection protocol based on responsiveness to novel growth factors, as such a selection would provide an alternative approach to identify receptors for orphan ligands, identify transcription factors that control receptor expression, or to identify subunits important for receptor signalling. For this purpose we chose to use cells of the murine hematopoetic precursor cell line FDC-P1M, clonally related to FDC-P1 cells (Dexter et al., 1980), but which are strictly dependent on IL-3 for long-term proliferation. Following insertional mutagenesis, cells were subjected to selection for proliferation in presence of GM-CSF. We have previously shown that in contrast to FDC-P1 cells, FDC-P1M cells do not express the α-subunit of the GM-CSF receptor, important for ligand binding and interaction with the β_c-subunit (Friel and Stocking, unpublished results). However, FDC-P1M cells do express the β_c-subunit, common to the receptor complex of IL-3, IL-5, and GM-CSF and important for transduction of the proliferation signal (reviewed in Miyajima et al., 1993), thus the number of potential target genes for insertional mutagenesis after GM-CSF selection protocol should be limited. GM-CSF-responsive mutants could show either (cis-)activation of the GM-CSF receptor α-subunit gene or activation of genes involved in regulation of this gene (trans-activation). A third subset of mutants may also be expected in which a "masterswitch" differentiation gene is activated, leading to the up-regulation of differentiation-specific genes, including the GM-CSF receptor, as has been described in mammalian myogenesis or insect morphogenesis (Davis et al., 1987; Gehring 1987).

Results

Increased mutation frequency of FDC-P1M cells by retroviral infection.

FDC-P1M is a myeloid precursor cell line that depends exclusively on IL-3 for long-term proliferation. FDC-P1M cells were subcloned in presence of IL-3 by standard agar cloning procedures. Subclones derived were checked for spontaneous mutation frequency to factor-independent growth by colony formation when plated in agar medium in absence of IL-3. FDC-P1M#4 showed a spontaneous mutation frequency lower than 1×10^{-8} and was used for further experiments. Infection of FDC-P1M#4 cells with an MPSV-based retroviral construct containing the procaryotic sequences for Tn5 phosphotransferase (neomycin resistance gene; *neoR*) and the tRNA suppressor gene (*supF*) (Stocking et al. 1993) was performed by cocultivation of hematopoietic cells on irradiated virus-producing PA317(neo^{R}supF) or GA12(neo^{R}supF) fibroblasts in the presence of IL-3. In parallel, a mock infection was performed in which cells were co-cultured with the irradiated packaging cell lines PA317 or GA12 which did not contain the neoRsupF retroviral vector. Because virions released from PA317 and GA12 packaging cell lines contain only retroviral vector RNA and not replication-competent viral RNA (Miller and Buttimore, 1986; Markowitz et al., 1988), multiple infections of the target cell are possible and the vector, with its selectable marker genes, is the only relevant insertional mutagen released. In four independent experiments a total of 4×10^8 cells were infected.

After repeated cocultures (8-12 cycles, depending on virus titers), each well (containing 5×10^5 cells) was split and subjected to growth in medium either without exogenous growth factor or containing GM-CSF conditioned medium (CM), to select for factor-independent or GM-CSF responsive mutants, respectively. Selection plates were cross-checked for double positives: factor-independent mutants should grow in medium either with or without GM-CSF CM, whereas GM-CSF responsive mutants survive only in presence of GM-CSF CM. The mutants were tentatively classified as GM or A (autonomous or factor-independent) mutants. Southern blot analysis proved clonality of mutants derived from individual wells (data not shown).

In four independent experiments performed, a 4.8-fold increase in the mutation frequency in retroviral infected cultures over mock-infected cultures was obtained. The number of proviral integration sites in independent mutants was determined by

Southern blot analysis (average of 1.7 copies/mutant) and compared to the number of mutants generated. The number of mutants per insertion was similar in all experiments performed (range from 1.3×10^{-7} to 4.6×10^{-7}) and thus strongly supports the causal relationship of retroviral infection and generation of mutants.

Isolated mutants from GM-CSF selection show a heterogeneous growth pattern. To ascertain the tentative classification of generated mutants, mutant cell lines were expanded under wild type conditions (IL-3 CM) and replated either in presence of IL-3 or GM-CSF CM or in absence of exogenous growth factor. This protocol allows discrimination between true GM-CSF-responsive mutants and factor-independent mutants, both expected to arise during GM-CSF selection.

As shown in Table 1 mutants derived from GM-CSF selection displayed different growth patterns. Eight out of 15 mutants tested clearly demonstrated a colony-forming ability in response to GM-CSF (Type I), seven mutants acquired factor independence (Type II). Typical cloning patterns for the two types of of GM-CSF responsive mutants are shown in Fig. 1.

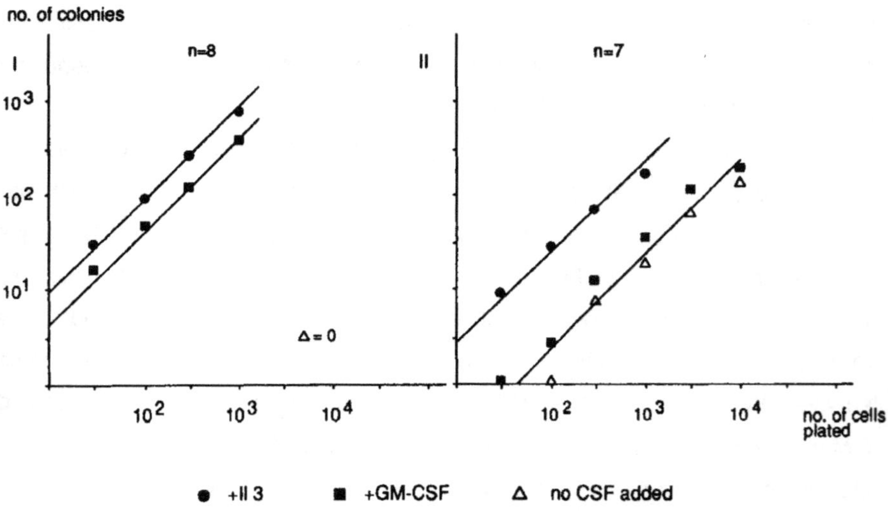

Figure 1. Cloning efficiency as a function of cell-density with or without factor defines two types of mutants after GM-CSF selection. Cells were cloned in methylcellulose either in the presence of GM-CSF or IL-3 or with no factor at different cell densities and the number of colonies was determined.

Table 1.
Growth Pattern of Cell Mutants

Cell line	Phenotype Classification	Cloning efficiency (%) in presence of IL3	GM-CSF	no factor
FDC-P1	wt	34	22	0
FDC-P1M#4		62	0	0
2GM#46†	I	21	19	0
2GM#1		21	16	0
2A317#3†		17	8	0
1GM#1		77	30	0
1GM#3†		25	9	0
1GM#4		84	37	0
2GM#20		25	10	0
2GM#23		34	3	0
2GM#4	II	15	2	0.7
2GM#7		27	3	1.5
2GM#10		15	2	0.7
2GM#52†		45	6	6.0
1GM#5		26	21	20.0
2GM#2		26	0.015*	0.015*
2GM#13†		62	0.6*	0.6*

*Density-dependent, non-linear colony formation; colony efficiency is calculated from wells were colony formation started.
†Cells produce a mitogenic activity able to support FDC-P1 cells.

Within the class Type I of GM-CSF responsive mutants no colony formation was detectable in absence of exogeneous growth factor even when cells were plated at high cell densities ($3x10^5$ cells/ml). All mutant cell lines showed colony-forming ability in presence of GM-CSF CM. There is, however, a variation seen in the efficiency of the GM-CSF response. In two mutants tested, the proliferation response to selection condition (GM-CSF CM) and wild-type condition (IL-3 CM) is interchangeable, whereas the response (as measured by colony formation) to GM-CSF CM as compared to IL-3 CM in all other mutants is lower (2-10 fold). Conditioned media from all mutants were prepared and tested on different target cell lines. CM from three GM-CSF responsive mutants (2GM#46, 2A317#3, 1GM#3) were able to stimulate GM-CSF responsive FDC-P1 cells but not parental FDC-P1M#4 cells. The stimulatory activity of 2GM#46 CM was determined to be GM-CSF (PCR analysis, data not shown), whereas the nature of the activity of the other conditioned media could not be identified. Whether the activation of growth factors is related to the activation of the GM-CSF response pathway remains unclear.

Type II class of mutants from GM-CSF selection comprises a group of seven mutants that acquired factor-independence. Five out of seven mutants tested showed a linear cloning pattern in absence of exogenous growth factor, indicating acquisition of factor-independence by activation of the signal transduction pathway downstream of the receptor (autonomous growth) whereas two mutants displayed a density dependent colony formation pattern in absence of exogenous growth factor (autocrine stimulation). Indeed, one of the two mutants of the latter subset secreted a stimulatory activity able to support the parental cell line FDC-P1M#4. PCR-analysis (data not shown) revealed the presence of IL-3 and SCF transcripts. IL-3 expression would be in line with density-dependent colony formation. Whether the newly transcribed SCF gene is responsible for the generation of the mutant remains unclear. Kit receptor transcripts in the mutant shows no alteration of expression in comparison to wild type FDC-P1M#4 cells, as judged by PCR-analysis. The presence of kit receptor transcripts, however, did not correlate with biological function; neither parental cells nor mutant cells could be stimulated by recombinant SCF.

A second subset of factor-independent mutants comprises a group of three mutants that show a slight increase of cloning efficiency in presence of GM-CSF CM as compared to absence of growth factor. The response to the GM-CSF CM was shown not to be a genuine response to GM-CSF, since use of recombinant GM-CSF did not improve colony formation above minus growth factor condition. GM-CSF CM was harvested from fibroblasts infected with a GM-CSF-expressing retroviral vector; fibroblasts are known to produce a variety of cytokine *e.g.* IL6, IL1, CSF-1. The mutant cell lines tested showed no additional proliferative response when exposed to recombinant cytokines, *e.g.* IL6, CSF-1, SCF and G-CSF. Thus, the 2-3-fold increase in cloning efficiency in presence of GM-CSF CM remains unresolved.

Mutant cell lines acquire secondary alterations.

Once established, mutants of both classes I and II developed an autonomous subpopulation over time (Table 2) regardless of whether mutant cell lines were maintained under wild type (IL-3 CM) or selection condition (GM-CSF CM). This observation is in line with previous studies of factor-dependent FDC-P1 cells which, after acquisition of factor-independence via autocrine stimulation due to aberrant GM-CSF expression, acquired autonomous growth properties (*e.g.* density-independent cloning) that could not be blocked by GM-CSF antibody (Laker et al., 1987). The transition of factor-dependent cells to fully transformed cells with tumorigenic potential could reflect the acquisition of additional mutations. These mutations may only be phenotypically detectable after a primary mutation has impaired the normal control of proliferation, in this case, after altered growth-factor responsiveness or factor-independent growth (Laker et al., 1989).

Table 2.

Shift in Mutant Cell Population to Autonomous Growth with Time

Cell line	Phenotype Class	Days of culture	Cloning efficiency (%) in presence of IL-3	GM-CSF	no factor
FDC-P1	wt	-	34	22	0
FDC-P1M#4		-	62	0	0
2GM#46	I	8	21	19	0
		18	48	53	0.03
		48	51	57	46
2GM#23	I	7	34	3	0
		14	16	7	0.0012
2GM#52	II	17	22	2	0.1
		30	45	6	6
		50	60	46	48

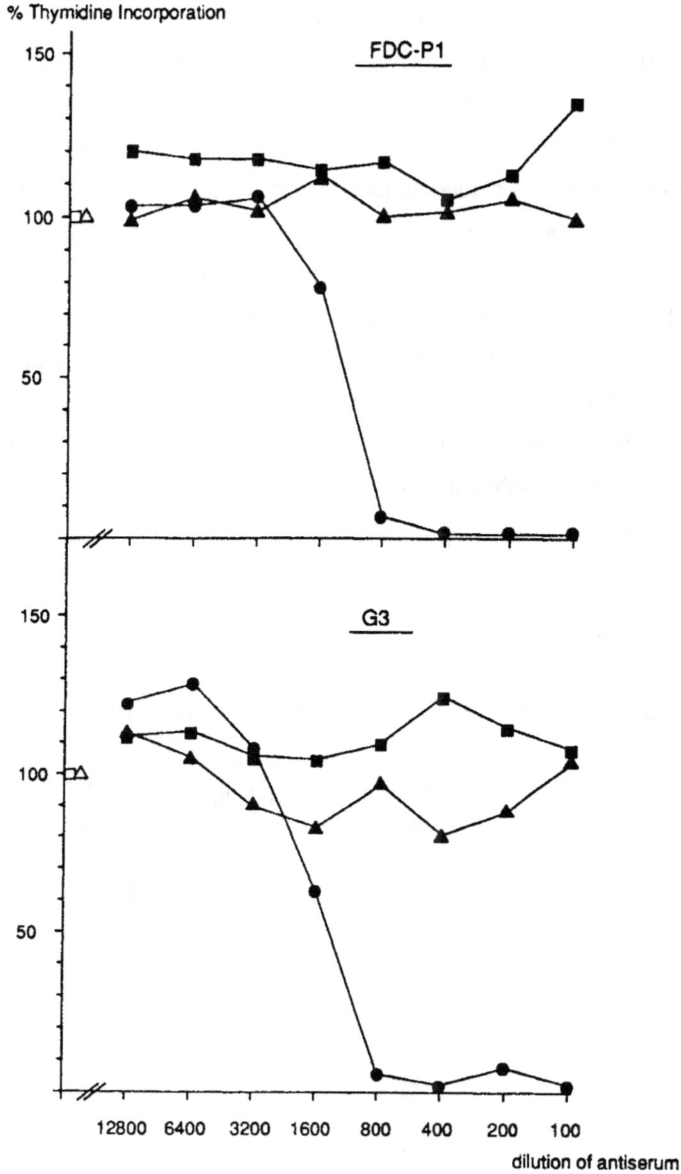

Figure 2. GM-CSF responsive mutants can be inhibited by α-GM-CSF antiserum. Proliferation assays of either control cells (FDC-P1) or mutant cells (1GM#3) were carried out in the presence of factor with or without antiserum. Cells (5000/well) were incubated in serial dilutions of antiserum for 72 h and pulsed with 0.5 μCi Thymidine / well for 6 h. Total cpm/minute obtained from cultures grown in IL-3 were taken as 100%.

(□) +IL-3 CM; (△) +rGM-CSF; (▲) +rGM-CSF plus titration with sheep preimmune serum; (■) +IL-3 CM plus α-GM-CSF serum; (●) +rGM-CSF plus α-GM-CSF serum.

Proliferation of GM-CSF responsive mutants can be inhibited by neutralizing GM-CSF antiserum.

To further analyse GM-CSF responsive mutants, mutant cell lines were tested for growth response to purified or recombinant factors. As shown in Table 2, the phenotype classification of GM-CSF responsive mutants was verified by agar cloning. In contrast to the parental cell line, all GM-CSF responsive mutants could be stimulated by recombinant GM-CSF as measured by ^{3}H-thymidine incorporation, *i.e.* DNA synthesis. This response could be specifically inhibited by use of neutralizing amounts of α-GM-CSF antiserum. Fig. 2 depicts the titration of α-GM-CSF antiserum of GM-responsive mutant 1GM#3. Factor-independent mutants of class II were used as controls; their growth could neither be stimulated by recombinant GM-CSF nor be inhibited by α-GM-CSF antiserum.

The GM-CSF α-receptor gene is transcritionally activated in GM-CSF responsive mutants.

To understand the molecular mechanism of newly acquired GM-CSF responsiveness, mutant cell lines were subjected to RT-PCR analysis. As shown in Fig.3, in all GM-CSF responsive class I mutants but not factor-independent class II mutants, a

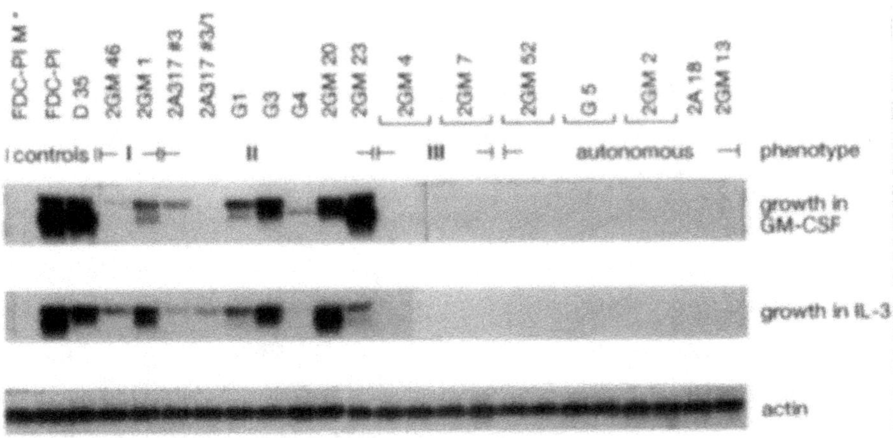

* FDC-PI M cells do not grow in GM-CSF

Figure 3. Transcriptional activation of the GM-CSF α receptor subunit was confirmed by RT-PCR. RNA from FDC-P1 (as positive control), the parental FDC-P1M#4, and GM-CSF responsive mutants was reversed transcribed and then amplified used primers specific for the GM-CSF α-subunit (Park et al., 1993) transcript of the GM-CSF receptor α-chain was detectable. Since all cell lines, including the parental cells, express the GM-CSF receptor common β-subunit (data not shown), the appearance of the newly transcribed α-chain mRNA is in agreement with the biological response to GM-CSF. Its expression allows the generation of a high affinity receptor capable of transducing a proliferation signal.

Discussion

The aim of our studies presented here was to establish an *in vitro* system that would allow identification of genes involved in growth control by use of retroviral vectors as insertional mutagens. The hematopoietic progenitor FDC-P1M#4 cell line used in our experiments allowed selection for distinct phenotypes such as factor-independent growth and alternate responsiveness to other growth factors *e.g.* GM-CSF. Retroviral insertional targeting of genes in an *in vitro* system allows the analysis of the molecular mechanism underlying subtle changes in growth behaviour, not detectable *in vivo* where tumorigenesis is the only selectable parameter.

The protocol described here gave rise to about 140 mutants following infection of parental cells with a MPSV-neo^{R}supF vector (Stocking et al., 1993). Biological classification of mutants confirmed the presence of the phenotypes expected, *i.e.* factor dependent mutants as well as GM-CSF responsive mutants. As shown by us and others, the acquisition of factor independence can be due to alteration in one of many possible genes in the signal transduction pathway (reviwed in Kung et al, 1991). The set of GM-CSF responsive mutants generated, however, provides the unique opportunity to gain insight into the control mechanism of the GM-CSF receptor gene expression. Since the common ß subunit is shared by IL-3 and GM-CSF receptors, we assumed that mutants of the IL-3 responsive parental cells would exhibit activation of the GM-CSF α-receptor gene. Here we show that retroviral infection of parental cells resulted in a 4.8 fold average increase in the number of mutants generated above the spontaneous background. Although this increase was not a high as the increase mutation frequency (up to 28-fold) obtained from similar experiments performed on a human erythroleukemia cell line selected for factor-independent growth (Stocking et al. 1993), the difference may reflect a number of parameters including infection frequency and inherent mutations in the cell lines occurring during the establishment that may synergize with *de novo* mutations. Significantly, the ratio of mutants established and integration events (2.2×10^{-7}) was similar with both cell lines.

Although a large number of genes may contribute to factor-independent growth, we speculated that a much smaller group of genes targeted by retroviral infection /insertion would lead to the tightly restricted phenotype of GM-CSF responsiveness. As expected, the mutants obtained from our selection protocol all showed activation of the GM-CSF α-receptor gene. However, preliminary results suggest that several mechanisms may be involved. No retroviral insertion has been detected in the proximity of the GM-CSF α-receptor gene, which could lead to cis-transcriptional activation. However, at least some of the GM-CSF-responsive mutants isolated may have resulted from cis-activation of a transactivator of the GM-CSF receptor gene. The recent success of several groups in identifying critical genes by characterizing common proviral integration sites following insertional mutagenesis both *in vivo* and *in vitro* (van Lohuizen and Berns 1990; Habets et al., 1994) supports the importance of identifying common sites in our independently isolated GM-CSF responsive. These common sites may harbor important regulatory elements of myeloid differentiation.

Acknowledgements

This work was supported by grants of the Deutsche Forschingsgemeinschaft to C.L. (Habilitations Stipend), W.O. (SFB 232), and C.S. (Sto 225/2-2). The Heinrich-Pette-Institut is financially supported by the Freie und Hansestadt Hamburg and the Bundesministerium für Gesundheit.

References

Davis RL, Weintraub H, Lassar AB (1987) Expression of a single transfected cDNA converts fibroblast to myoblasts. Cell 51:987-1000.

Dexter TM, Garland J, Scott D, Scolnick E, Metcalf D (1980) Growth of factor-dependent hemopoietic precursor cell lines. J. Exp. Med. 152:1036-1047.

Dührsen U, Stahl J, Gough NM (1990) *In vivo* transformation of factor-dependent hemopoietic cells: Role of intracisternal A-particle transposition for growth factor gene activation. EMBO J. 9:1087-1096.

Gehring WJ (1987) Homeo boxes in the study of development. Science 236:1245-1252.

Habets, GGM, Scholtes EHM, Zudygeest D, van der Kammen RA, Stam JC, Berns A, Collard JG (1994) Identification of an invacion-inducing gene, *Tiam-1*, that encodes a protein with homology to GDP-GTP exchangers for rho-like proteins. Cell 77:537-549.

Hannemann J, Hara T, Kawai M, Miyajima A, Ostertag W, Stocking C (1995) Sequential mutations in the IL3/GM-CSF/IL5 receptor β-subunit genes are necessary for the complete conversion to grwoth autonomy mediated by a truncated β_c-subunit. Mol. Cell. Biol., submitted.

Haupt Y, Alexander WS, Barri G, Klinken SP, Adams JM (1991) Novel zinc finger gene implacated as *myc* collaborator by retrovirally accelerated lymphomagenesis in Eμ-*myc* transgenic mice. Cell 65:753-763.

Heberlein C, Kawai M, Franz MJ, Beck-Engeser G, Daniel P, Ostertag W, Stocking C (1990) Retrotransposons as mutagens in the induction of growth autonomy in hematopoietic cells. Oncogene 5:1799-1807.

Kung HJ, Boerkoel C, Carter TH (1991) Retroviral mutagenesis of cellular oncogenes: A review with insights into the mechanism of insertional activation. In Current Topics in Microbio and Immnul. 171:1-25.

Laker C, Stocking C, Bergholz U, Hess N, DeLamarter JF, Ostertag W (1987) Autocrine stimulation after transfer of the granulocyte/macrophage colony-stimulating factor gene and autonmous growth are distinct but interdependent steps in the oncogenic pathway. Proc Natl Acad Sci 84:8458-8462.

Laker C, Kluge N, Stocking C, Just U, Franz MJ, Ostertag W, DeLamarter JF, Dexter M, Spooncer E (1989) Rates of mutation to growth factor autonmy and tumorigenicity differ in hematopoietic stem and precursor cells expressing the multilineage colony-stimulating factor gene. Mol Cell Biol 9:5746-5749.

Markowitz D, Goff S, Bank A (1988) Construction and use of a safe and efficient amphotropic cell line. Virology 167:400-406.

Miller AD, Buttimore C (1986) Redesign of retrovirus packaging cell lines to avoid recombination leading to helper virus production. Mol Cell Biol 6:2895-2903.

Miyajima A, Mui A, Ogorochi T, Sakamaki K. (1993) Receptors for granulocyte-macrophage colony-stimulating-factor, interleukin-3, and interleukin-5. Blood 82:1960-1974.

Park LS, Martin U, Sorense R, Luhr S, Morrissey PJ, Cosman D, Larsen A (1992) Cloning of the low-affinity murine granulocyte-macrophage colony-stimulating factor receptor and reconstitution of a high-affinity receptor complex. Proc Natl Acad Sci 89:4295-4299.

Stocking C, Bergholz U, Friel J, Klingler K, Wagener T, Starke C, Kitamura T, Miyajima A, Ostertag W (1993) Distinct classes of factor-independent mutants can be isolated after retroviral mutagenesis of a human myeloid stem cell line. Growth Factors 8:197-209.

Stocking C, Löliger C, Kawai M, Suciu S, Gough N, Ostertag W (1988) Identification of genes involved in growth autonomy of hematopoietic cells by analysis of factor independent mutants. Cell 53:869-879.

van Lohuizen M, Verbeek S, Scheijen, Wientjens E, van der Gulden H, Berns A (1991) Identification of cooperating oncogenes in Eμ-myc transgenic mice by provirus tagging. Cell 65:737-752.

van Lohuizen M and Berns AJM (1990) Tumorigenesis by slow-transforming retroviruses: An update. Biochim Biophysic Acta 1032:213-235.

Signaling by the Erythropoietin receptor in retrovirally transduced normal hematopoietic progenitors: A conserved proline-rich motif is required for association with JAK2.

Gregory D. Longmore[1,3,*], J. William Tanner[2,3], Grant A. McArthur[4], Pamela N. Pharr[5], Gregory R. Johnson[4], and Andrey S. Shaw[2,3]
[1]Departments of Medicine and Cell Biology, [2]Pathology, and [3]Center for Immunology, Washington University School of Medicine, St. Louis, MO 63110 USA
[4]The Walter and Eliza Hall Institute of Medical Research, 3050 Victoria, Australia
[5]Ralph H. Johnson Department of Veterans Affairs Medical Center, Charleston, SC 29401 USA
*corresponding author: Telephone 314-362-8800; FAX 314-362-8826;
E-mail longmore_g@wums.wustl.edu

Abstract

The receptor for erythropoietin is restricted to cells of mature erythroid and possibly megakaryocyte lineages. Studies in cell lines have suggested that cytokine receptors share a conserved signaling pathway for proliferation. We retrovirally transduced the erythropoietin receptor or a constitutively activated form of the EPO-R into normal hematopoietic progenitors, including blast cell colonies. The EPO-R was able to support the proliferation and differentiation of early erythroid, early megakaryocytic, and macrophage progenitors, but not granulocyte progenitors. Blast cell colonies transduced with the EPO-R proliferate in response to EPO but the development of erythroid cells was not favored over other lineages. These results with normal cells suggest that some but not all cytokine receptors exhibit shared signaling pathways, and that EPO signaling alone is not sufficient to drive erythroid development. The Janus family of cytosolic tyrosine kinases mediate cytokine initiated mitogenic signals. We have determined that the EPO-R box 1 cytoplasmic motif is required for the binding and activation of JAK2. However, sequences outside the box 1 domain most likely regulate the specificity of JAK kinase association.

NATO ASI Series, Vol. H 94
Gene Technology
Edited by A. R. Zander, W. Ostertag,
B. V. Afanasiev, F. Grosveld

Introduction

Erythropoietin (EPO) is a serum glycoprotein hormone required for the survival, proliferation and differentiation of committed erythroid progenitor cells, and is the principal hormone regulating the level of circulating red blood cells. The administration of recombinant human EPO to anemic patients suffering from chronic renal failure, AIDS, or bone marrow suppression due to chemotherapy has dramatically alleviated their need for blood transfusions.

In contrast to many other hematopoietic growth factors, EPO acts primarily on relatively mature erythroid progenitors, and to a significantly lesser extent megakaryocyte progenitors, within the fetal liver and adult bone marrow (reviewed in (Krantz, 1991). The receptor for EPO is normally restricted in its expression to relatively mature cells of the erythroid and megakaryocyte lineage(Youssoufian, Longmore et al., 1993), and has also been reported to be expressed in the placenta(Sawyer, Krantz et al., 1989), by embryonic stem cells(Keller, Kennedy et al., 1993), on endothelial cells(Anagnostou, Lee et al., 1990), and on some neuronal-like cell lines(Masuda, Nagao et al., 1993). The functional relevance of this developmentally diverse EPO-R gene expression is not clear, and quite apart from its function in hematopoiesis, EPO-Rs may play other roles in non-hematopoietic cells. In the case of the erythroid lineage, expression of the EPO-R is low or absent on immature progenitors such as BFU-E and is increased by the CFU-E stage of differentiation, before decreasing as the erythroid cells undergo terminal differentiation(Sawada, Krantz et al., 1990). Stimulation of adult BFU-E with EPO as a single stimulus fails to support the formation of erythroid colonies(Dai, Krantz et al., 1991). Whether EPO directly affects differentiation as well as proliferation is not clear.

A murine EPO-R cDNA, which stimulates EPO-dependent proliferation of several hematopoietic cell lines, was isolated by expression cloning and encodes a type I membrane-spanning protein of 66kD which is a member of the cytokine receptor superfamily(D'Andrea, Lodish et al., 1989). Studies have detected a single binding affinity for EPO (k_D = 300-800pM) in heterologous hematopoietic cells, fibroblasts, or COS cells transfected with the EPO-R cDNA. Binding studies on human and murine bone marrow and fetal erythroid progenitor cells have detected either one or two affinities for radioiodinated EPO (Youssoufian, Longmore et al., 1993). Two affinities for EPO may suggest the presence of other components which modulate the binding activity of the cloned EPO-R. However, it is unresolved whether there are indeed EPO-Rs of multiple affinities.

A constitutively active (cytokine-independent) form of the EPO-R has been isolated from a retroviral transduction system and found to contain a single point mutation, resulting in an Arg to Cys change at residue 129 of the exoplasmic domain

(Yoshimura, Longmore et al., 1990). The R129C receptors form disulfide-linked homodimers independent of hormone, yet retain the capacity to bind EPO with a single affinity (k_D= 700pM)(Watowich, Yoshimura et al., 1992). Since several members of the cytokine receptor family are active as ligand-induced homo- or heteromultimers, the disulfide-linked dimers most likely mimics the structure of the hormone-bound form of the EPO-R and thus transmit a constitutive proliferative signal. Chimeric c-kit/EPO-R receptors transmit mitogenic signals in response to added c-kit ligand (Ohashi, Maruyama et al., 1994). When an EPO-R lacking the entire cytoplasmic region is coexpressed, in excess, with wild type EPO-R EPO-induced mitogenic signals are blocked ("dominant negative" receptor form) (Watowich, Hilton et al., 1994). Taken together these observations strongly suggest that the functional cell surface form of the EPO-R is a multimeric protein complex consisting of, at least, a homodimer of the cloned cDNA.

Although the cytosolic domain of the EPO-R does not contain an obvious protein kinase domain, ligand induced multimerization of the EPO-R stimulates rapid and transient tyrosine phosphorylation of a number of cellular proteins, which are essential for EPO-induced proliferation (Miura, D'Andrea et al., 1991). Sequence analysis, and functional mutagenesis studies of members of the cytokine receptor superfamily of receptors have identified two small conserved motifs, box 1 and box 2, in the membrane proximal cytoplasmic domains, which are required for proliferative signals (Murakami, Narakaki et al., 1991). This membrane proximal domain has been shown to be necessary for binding and phosphorylation of Janus kinase family members (cytosolic, nonreceptor tyrosine kinases). There is some specificity for the JAK kinase member utilized by different receptors. For example, the EPO-R activates only JAK2, whereas the IL-6 receptor signal transducing component, gp130, activates either or both JAK1 and JAK2, depending upon the cell type expressing gp130 (Witthuhn, Quelle et al., 1993)(Stahl and Yancopoulos, 1993).

Results and Discussion

To identify features of the EPO-R that mediate its interaction with JAK2, we generated chimeric receptor proteins that contained the cytoplasmic domain of the EPO-R or gp130 fused to the extracellular and transmembrane domain of the VSV G protein. This modification allowed us to immunoprecipitate and immunoblot both EPO-R and gp130 receptors with the same monoclonal antibody and eliminated the variability of using different receptor antibodies for our comparison. Because the kinase activity of JAK2 is not detectable in the absence of cytokine stimulation we generated constitutively active forms of JAK1 and JAK2 by replacing the tyrosine kinase domain of the JAKs with an epitope tagged (myc) tyrosine kinase domain from p59*fyn* . Thus anti-myc monoclonal antibodies will immunoprecipitate the JAKs. The chimeric receptors and JAK kinases were transiently coexpressed in HeLa cells using the vaccinia-T7 expression system to determine if cytokine receptors and JAK kinases could form stable complexes.

The membrane proximal domain of the EPO-R was required for association with JAK2, and JAK1 did not associate with the EPO-R (Table 1). Specifically box 1, not box 2, was required for the association between the EPO-R and JAK2. Similarly the membrane proximal box 1 domain of gp130 was required for association with JAK2, but also associated with JAK1. This suggests that box 1 sequences are required for both JAK1 and JAK2 association with cytokine receptors but that sequences outside the box 1 domain regulate the specificity of JAK kinase association.

IL-6 and CNTF, cytokines whose receptors utilize gp130, can stimulate the phosphorylation of JAK1, JAK2, and tyk2 depending on the cell line examined(Stahl and Yancopoulos, 1993). In contrast only JAK2 has been shown to associate with the EPO-R, in vivo(Miura, Nakamura et al., 1994). Our results demonstrated that the EPO-R and gp130 behave differently with regards JAK activation, in a heterologous cell line. Thus, receptor interactions with JAK kinases could be regulated in two ways; receptors for cytokines such as EPO encode the specificity of association within receptor protein sequence whereas JAK kinase interaction with receptors like gp130 must be influenced by cell-specific factors. It will be important to determine what structural features of the receptors regulate the specific interaction of JAK2 with the EPO receptor protein.

Expression of exogenous EPO-R in some IL-3- and GM-CSF-dependent cell lines (BaF3, DA-1, 32D, and FDCP-1) confers upon these cells the capacity to proliferate in response to EPO(Youssoufian, Longmore et al., 1993). This is not the case for most IL-2-dependent cell lines (CTLL-2, HT-2)(Yamamura, Kageyama et al., 1992). Similarly, studies of the biochemical events following ligand binding to members of the cytokine receptor family have suggested conservation of signal transduction mechanisms between some members of the cytokine receptor family(Youssoufian, Longmore et al., 1993)(Ihle, Witthuhn et al., 1994). One possible interpretation for these observations is that specificity of response to a growth factor is obtained at the level of receptor expression rather than at the level of signal transduction. Given the normally limited expression of the EPO-R, we were interested to determine if the EPO-R was capable of generating a proliferative and differentiative signal in cells normally responsive to the ligands of other members of the cytokine receptor family. Retroviral vectors expressing EPO-R(wt) and EPO-R(R129C) were constructed, primary hematopoietic progenitor cells (d12.5 fetal liver and 5-fluorouracil treated adult bone marrow) were infected, and cultured in methylcellulose in the absence or presence of added EPO(McArthur, Longmore et al., 1995)(Pharr, Hankins et al., 1993).

Table 1. The association of JAK2 or JAK1 with the EPO-R or gp130

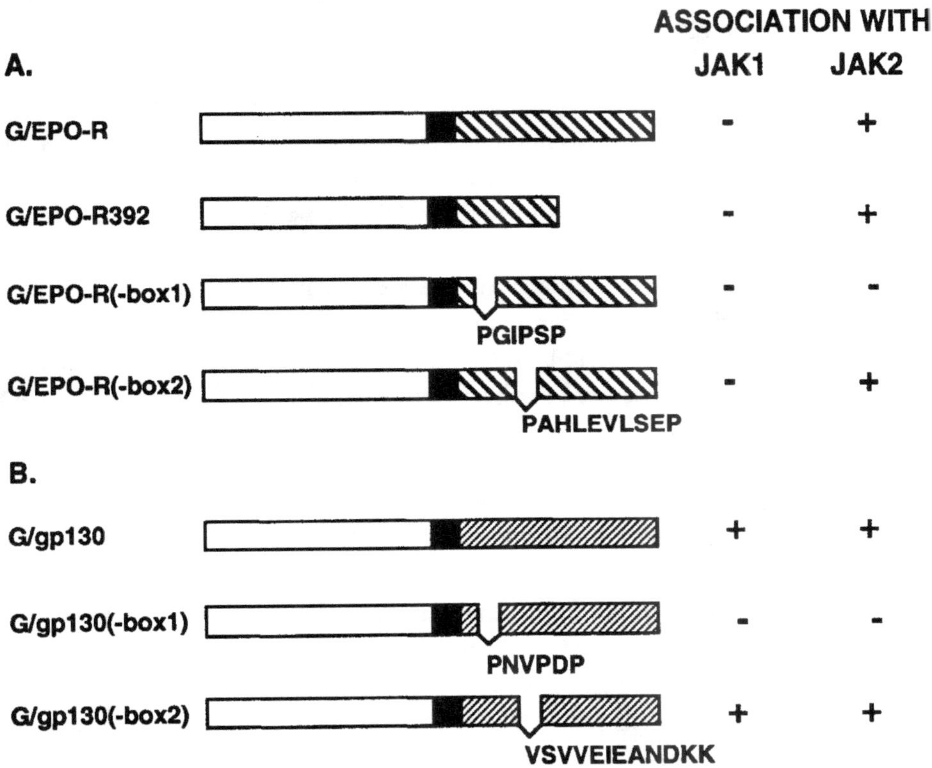

(A) Requirement of the membrane proximal box 1 of the EPO-R for JAK2 association. (B) Box1 but not box2 is critical for JAK association with the cytoplasmic domain of gp130. Open and closed boxes represent the extracellular and transmembrane domain of the VSV G protein, respectively. Hatched boxes represent the intracellular domain of the indicated cytokine receptor. HeLa cells were transfected with cDNAs corresponding to the indicated cytokine receptor chimeras and either JAK1 or JAK2. Cells were lysed in 1% Brij 96 and the detergent soluble extract was immunoprecipitated with anti-VSV G or anti-myc antibodies. Association of the cytokine receptor chimeras with the kinases was determined by in vitro kinase reactions, in the presence of ^{32}P-gATP, upon the immunoprecipitated pellets. The products of the in vitro kinase reactions were separated on a 7.5% acrlamide-SDS gel and analyzed by autoradiography. Expression of cytokine receptor chimeras was confirmed by anti-VSV G protein immunoblot, and JAK kinase by anti-myc IP kinase reactions.

Table 2 shows the effect of EPO upon erythroid progenitors transduced with wild type EPO-R or an activated form of the EPO-R, EPO-R(R129C). Cultures of cells infected with the EPO-R retrovirus demonstrated erythroid colony formation in response to EPO as a single stimulus when analyzed at day 2 or day 8. Normally stimulation of BFU-E by EPO does not support the development of day 8 colonies (Table 1 NeoR experiment). The finding of frequent day 8 erythroid colonies in EPO stimulated cultures of fetal liver cells transduced with EPO-R(wt) suggest that the signal transduction mechanism allowing fior EPO-induced cell proliferation and differentiation are present in some murine BFU-E.

Unstimulated cultures of fetal liver cells transduced with EPO-R(R129C) contained erythroid colonies when analyzed at day 2 but not day 8. However, the same cell population stimulated by EPO did contain day 8 erythroid colonies. This suggests that there may be some functional differences in signals generated by EPO-R(R129C) in the unstimulated versus the EPO-stimulated state. Analysis of cell surface EPO-Rs in BaF3 expressing EPO-R(R129C) detects a minority of receptor dimers as opposed to monomers(Watowich, Yoshimura et al., 1992). Thus EPO-stimulation may be able to generate additional EPO-R dimers leading to an enhanced signal.

When fetal liver or post 5-FU bone marrow cells transduced with EPO-R or EPO-R(R129C) were stimulated by EPO there was an increased number of megakaryocyte colonies (Table 3). This appeared to be due to the recruitment of additional progenitors cells rather than altered committment of existing progenitors as there was no change in non-megakaryocyte colony numbers. The recruitment of additional megakaryocyte clones was of both the BFU-Meg and CFU-Meg type. These observations suggest the presence of signal transduction mechanisms allowing EPO-induced proliferation and differentiation in cells of both erythroid and megakaryocyte lineages at stages of differentiation prior to the expression of significant numbers of EPO-Rs.

The current experiments also demonstrated a direct stimulation of individual clones of the macrophage lineage by EPO, when these cells are transduced with EPO-R or EPO-R(R129C) (Table 3). In contrast, the data did not demonstrate EPO-induced proliferation of isolated clones of the granulocyte lineage. Cells of the granulocyte lineage transduced with a c-fms (CSF-1R) also fail to proliferate in response to CSF-1(McArthur, Rohrschneider et al., 1994). These results, taken together, suggest that granulocyte progenitors may express a restricted range of signal transduction molecules influencing their response to specific cytokines.

Table 2. Erythroid colony formation in cultures of fetal liver cells induced to express EPO-R or EPO-R(R129C)

Gene transferred	stimulus	erythroid colonies day2	day8
NeoR	none	0	0
	EPO	++	0
	EPO + SCM	++	++
EPO-R	none	0	0
	EPO	+	+
	EPO + SCM	++	++
EPO-R(R129C)	none	+	0
	EPO	++	+
	EPO + SCM	++	++

Day 12.5 fetal liver cells were infected with retroviruses expressing NeoR, EPO-R/NeoR, or EPO-R(R129C)/NeoR by cocultivation with retroviral producing cells. Cultures contained 5 x 10^4 fetal liver cells. All cultures contained 1.5 mg/ml of G418. Results are from quadruplicate cultures and expressed as number of colonies relative to NeoR cultures. Cultures were stimulated by 2 U/ml of hEPO or 10% spleen conditioned media (SCM). Maximal responses are depicted in cultures of NeoR transduced cells containing EPO and SCM.

Hematopoietic cells at different stages in development are thought to have different complements of cytokine receptors. Whether the appearance of specific receptors initiates a particular developmental sequence is not known: Does the acquisition of a lineage-specific receptor induce differentiation? To address this question we utilized retroviral-mediated gene transfer to express the EPO-R in multilineage blast cell progenitors (Pharr, Ogawa et al., 1994) (Table 4).

By DNA PCR we demonstrated that pluripotent blast cell clones could be infected with EPO-R(R129C)-expressing retroviruses. Blast cells and their progeny, CFU-GM and CFU-mix, express retrovirally derived EPO-R(R129C) as determined by PCR of cDNA prepared from these colonies (not shown). We observed no evidence that blast cells transduced with EPO-R(R129C) could induce erythroid differentiation, even following the addition of EPO. These results are consistent with in vivo experiemtns in which EPO has been shown to regulate the rate at which committed erythrocyte progenitors become erythroblasts.

Table 3. Colony formation in cultures of post 5-FU bone marrow cells induced to express EPO-R or EPO-R(R129C).

		relative number of colonies			
Gene transferred	stimulus	G	GM	M	MEG
NeoR	none	0	0	0	0
	EPO	0	0	0	trace
	EPO + IL-3	++	++	++	++
EPO-R	none	0	trace	0	0
	EPO	trace	+	+	+++
	EPO + IL-3	+	++	++	+++
EPO-R(R129C)	none	0	trace	+	+++
	EPO	trace	+	+	+++
	EPO + IL-3	++	+++	+++	+++

Adult mice had their bone marrow harvested 4 days after treatment with 5-fluorouracil, 150 mg/kg. Cells were infected by cocultivation with retroviral producing cells. Cultures contained 5 x 10^4 post 5FU bone marrow cells. All cultures contained 2 mg/ml G418. Cultures were stimulated by 2 U/ml of hEPO, and 1000 U/ml of mIL-3. Results are from triplicate plates, scored after 7 days of culture, and reported as colony numbers relative to NeoR infected plates. G, granulocytes; GM, granulocyte-macrophage; M, macrophage; MEG, megakaryocyte.

The initiation of differentiation could be induced by an exogenous stimulus or could be a spontaneous random event with survival of committed cells dependent on the availability of a supportive milieu. Metcalf analyzed the progenitor content of developing blast cell colonies and found that GM-CSF or IL-3 with SCF increased the relative frequency of granulocytic progenitors. Others have demonstrated that primitive progenitors derived from human cord blood divide asymmetrically and that this was not affected by different combinations of cytokines. The results presented here are more consistent with a supportive role for EPO in the differentiation of red cells. Thus expression of lineage-specific receptors for EPO may be a consequence rather than a cause of differentiation.

Table 4. Effect of EPO-R(R129C) on the composition of mixed colonies derived from infected blast cells

Gene Transferred	growth factor	Cells/colony	Average % G	M	E	ME
none	+[a]	44,000	10	31	45	0.6
mEPO-R(R129C)	+[a]	102,000	13	28	43	0.8

Spleen cells were harvested from mice treated with 5-FU (150 mg/ml) 4 days earlier. These were cultured in methylcallulose. Blast cell colonies, identified on day 6 or 7, were picked, washed and resuspended in retroviral supernatant or control. Following infection samples of blast cells were plated at 50-100 per plate. Mixed colonies were identified, picked and cytospin preparations counted after Wrights staining. [a]cultures contained Stem Cell Factor (SCF) (3 U/ml), IL-3 (100 U/ml), and EPO (1 U/ml) and were scored on day 9. G, granulocytes; M, macrophage; E, erythroid; MEG, megakaryocyte.

References.

Anagnostou, A., E. S. Lee, N. Kessimian, R. Levinson and M. Steiner .1990. Erythropoietin has a mitogenic and positive chemotactic effect on endothelial cells. Proc. Natl. Acad. Sci. USA. **87**: 5978-5982.

D'Andrea, A. D., H. F. Lodish and G. G. Wong .1989. Expression cloning of the murine erythropoietin receptor. Cell. **57**: 277-285.

Dai, C. H., S. B. Krantz and K. M. Zsebo .1991. Human burst-forming units-erythroid need direct interaction with stem cell factor for further development. Blood. **78**: 2493-2499.

Ihle, J. N., B. A. Witthuhn, F. W. Quelle, K. Yamamoto, W. E. Thierfelder, B. Kreider and O. Silvennoinen .1994. Signaling by the cytokine receptor superfamily: JAKs and STATs. Trends in Biochem. Sci. **19**: 222-227.

Keller, G., M. Kennedy, T. Papayannopoulou and M. V. Wiles .1993. Hematopoietic commitment during embryonic stem cell differentiation in culture. Mol. Cell. Biol. **13**: 473-486.

Krantz, S. B. .1991. Erythropoietin. Blood. 77: 419-434.

Masuda, S., M. Nagao, K. Takahata, Y. Konishi, F. Gallyas, T. Tabira and R. Sasaki .1993. Functional erythropoietin receptor of the cells with neural characteristics. J. Biol. Chem. 268: 11208-11216.

McArthur, G. A., G. D. Longmore, K. Klinger and G. R. Johnson .1995. Lineage restricted recruitment of immature hematopoietic progenitor cells in response to erythropoietin after transfection of normal hematopoietic cells with the erythropoietin receptor. Exp.Hematol.,1995

McArthur, G. A., L. R. Rohrschneider and G. R. Johnson .1994. Induced expression of c-fms in normal hematopoietic cells shows evidence for both conservation and lineage restriction of signal transduction in response to M-CSF. Blood. 83: 972-980.

Miura, O., A. D. D'Andrea, D. Kabat and J. N. Ihle .1991. Induction of tyrosine phosphorylation by the erythropoietin receptor correlates with mitogenesis. Mol. Cell. Biol. 11: 4895-4902.

Miura, O., N. Nakamura, J. N. Ihle and A. Nobuo .1994. Erythropoietin-dependent association of phosphatidylinositol-s-kinase with tyrosine-phosphorylated erythropoietin receptor. J. Biol. Chem. 269: 614-620.

Murakami, M., M. Narakaki, M. Hibi, H. Yawata, K. Yasukawa, M. Hamaguchi, T. Taga and T. Kishimoto .1991. Critical cytoplasmic region of the interleukin 6 signal transducer gp130 is conserved in the cytokine receptor family. Proc. Natl. Acad. Sci. USA. 88: 11349-11353.

Ohashi, H., K. Maruyama, Y.-C. Liu and A. Yoshimura .1994. Ligand-induced activation of chimeric receptors between the erythropoietin receptor and receptor tyrosine kinases. Proc. Natl. Acad. Sci. (USA). 91: 158-162.

Pharr, P. N., D. Hankins, A. Hofbauer, H. F. Lodish and G. Longmore .1993. Expression of a Constitutively Active Erythropoietin Receptor in Primary Hematopoietic Progenitors Abrogates Erythropoietin Dependence and Enhances CFU-E, BFU-E, and GM Progenitor Growth. Proc. Natl. Acad. Sci. USA. 90: 938-942.

Pharr, P. N., M. Ogawa, A. Hofbauer and G. D. Longmore .1994. Expression of an activated erythropoietin receptor or a CSF-1 receptor by pluripotent progenitors enhances colony formation but does not induce differentiation. Proc. Natl. Acad. Sci. USA. 91: 7482-7486.

Sawada, K., S. B. Krantz, C. H. Dai, S. T. Koury, S. T. Horn, A. D. Glick and C. I. Civin .1990. Purification of human blood burst forming units-erythroid and demonstration of the evolution of the erythropoietin receptor. J. Cell. Physiol. 142: 219-230.

Sawyer, S. T., S. B. Krantz and K.-I. Sawada .1989. Receptors for erythropoietin in mouse and human erythroid cells and placenta. Blood. 74: 103-109.

Stahl, N. and G. D. Yancopoulos .1993. The alphas, betas, and kinases of cytokine receptor complexes. Cell. 74: 587-590.

Watowich, S. S., D. J. Hilton and H. F. Lodish .1994. Activation and inhibition of erythropoietin receptor function: role of receptor dimerization. Mol. Cell. Biol. 14: 3535-3549.

Watowich, S. S., A. Yoshimura, G. D. Longmore, D. Hilton, Y. Yoshimura and H. F. Lodish .1992. Homodimerization and constitutive activation of the erythropoietin receptor. Proc. Natl. Acad. Sci. USA. 89: 2140-2144.

Witthuhn, B. A., F. W. Quelle, O. Silvennoinen, T. Yi, B. Tang, O. Miura and J. N. Ihle .1993. JAK2 associates with the erythropoietin receptor and is tyrosine phosphorylated and activated following stimulation with erythropoietin. Cell. **74**: 227-236.

Yamamura, Y., Y. Kageyama, T. Matuzaki, M. Noda and Y. Ikawa .1992. Distinct downstream signaling mechanisms between erythropoietin receptor and interleukin-2 receptor." EMBO J. **11**: 4905-4915.

Yoshimura, A., G. Longmore and H. F. Lodish .1990. "Point mutation in the exoplasmic domain of the erythropoietin receptor resulting in hormone-independent activation and tumorigenicity." Nature. **348**: 647-649.

Youssoufian, H., G. Longmore, D. Neumann, A. Yoshimura and H. Lodish. 1993. Structure, Function, and Activation of the erythropoietin receptor. Blood. **81**(9): 2223-2236.

Spontaneous Knockout of CSF-1 Gene in the Mouse as a Model to Study the Organization of the Macrophage System

Wieslaw Wiktor-Jedrzejczak,
Andrzej Nowicki
Department of Immunology
Central Clinical Hospital
Military School of Medicine
PL00-909 Warsaw
Poland

Physiological role of many molecules and cells can be investigated by evaluating animals possessing genetically determined alterations in the production of those molecules and/or cells. While animals with either knockouts of genes for various molecules or transgenic for these molecules could be at present experimentally created, there is still not fully explored potential of natural mutants, the so called "experiments of nature" (Good, 1991). In particular, there are several natural mutants with osteopetrosis: a disorder of osteoclasts i.e. cells related to macrophages, and they all may have alterations in other parts of the macrophage system (Wiktor-Jedrzejczak et al., 1981, Marks 1987).

One of such models is the osteopetrotic *op/op* mouse (Marks and Lane, 1976). This mutant was found to possess very severe deficiency of macrophages, secondary to deficiency of a growth factor (Wiktor-Jedrzejczak et al.,1982). In particular, it allowed the identification of the total absence of a major macrophage growth factor: colony stimulating factor 1 (CSF-1 or M-CSF; Wiktor-Jedrzejczak et al.,1990, Felix et al.,1990) as the cause of cellular deficiences. In turn, CSF-1 absence was found to be due to inactivating mutation of the gene for that factor (Yoshida et al., 1990).

NATO ASI Series, Vol. H 94
Gene Technology
Edited by A. R. Zander, W. Ostertag,
B. V. Afanasiev, F. Grosveld

In consequence of these studies, the first mutant with severe congenital deficiency of macrophages became available for the investigation of the organization of the macrophage system, the function and diversity of cells belonging to that system and for verification of the evidence regarding this system obtained using other models (VanFurth et al., 1972, Nathan and Cohn, 1985, Gordon, 1986). Simultaneously, this model became molecularly defined.

Introduction to cells and regulatory molecules of the macrophage system

Macrophage lineage belongs to myelopoiesis but differs from other lines of myeloid differentiation in that:

- cells leaving the bone marrow, i.e. monocytes are not yet fully functionally mature, and in some species including mice they may even proliferate;
- there is a striking functional and phenotypic diversity among the end cells of the system, which include various forms of tissue macrophages as well as cells that are not phagocytes, such as dendritic cells and osteoclasts (Auger and Ross, 1992);
- there is an additional very large potential to increase this diversity of mature cells by their activation, with dramatic enhancement of some functions already expressed by resting cells, and the appearance of many new functions (Adams and Hamilton, 1992).

Classically, macrophage differentiation in the bone marrow was considered to begin at the level of hematopoietic stem cell, and then to proceed through the stage of bipotent neutrophil-macrophage progenitor to the first identified cell of the lineage namely the monoblast, and subsequently to monocytes, and macrophages, as well as to the other end cell of the system (VanFurth, 1993). However, in addition to these studies, also very primitive macrophage progenitor cells have been identified by Bradley and Hodgson (1979) suggesting that some macrophages may derive from progenitors unique to their line of differentiation and descending directly from the stem cells. Subsequent studies have suggested even greater heterogeneity of macrophage progenitors (Bertoncello et al., 1991, Suda et al., 1983).

There are three identified growth factors, that when individually added *in vitro* to macrophage progenitors stimulate their growth and maturation to macrophages

(Prystowsky et al., 1984, Koike et al., 1986, Falk and Vogel, 1988, Wiffeils et al., 1993). Such *primary macrophage growth factors* (Metcalf, 1991) include CSF-1, granulocyte-macrophage (GM)-CSF, and interleukin 3 (IL-3). CSF-1 is a large dimeric cytokine with homology to Steel Factor, and binding to a dimeric receptor of immunoglobulin superfamily with intrinsic tyrosine kinase activity: *c-fms*. Both GM-CSF and IL-3 belong to hematopoietin family and their receptors belong to the family of hematopoietin receptors.

The gene for CSF-1 is located on chromosome 3 at the *op* locus (Gisselbrecht et al.,1989), is composed of 10 exons, and is producing five different mRNA by alternative splicing (reviewed by Stanley, 1994). There are at least three different (N-terminus identical) protein forms of CSF-1:

- soluble proteoglycan with largest protein part of 522 aminoacids;
- soluble glycoprotein of 406 aminoacids, and:
- membrane-spanning glycoprotein of 224 aminoacids, which may be shed and also contribute to soluble CSF-1.

Genes for both GM-CSF and IL-3 are located on chromosome 11 and code for only one protein form of each factor (Gasson, 1991, Ihle and Weinstein, 1986). Receptors for GM-CSF and IL-3 have common beta subunit (responsible for signal transduction to the cell inside) and unique alpha subunits (responsible for cytokine binding). In the mouse there is an additional beta subunit exclusive for IL-3 receptor (Miyajima et al.,1992).

While IL-3 is only a paracrine factor acting at the vicinity of cells that produce it, GM-CSF is present in the circulation in small amounts (Cheers et al., 1988), suggesting that it may participate in steady-state regulation. However, the most of circulating macrophage colony stimulating activity is due to CSF-1, which is both endocrine (i.e. normally present in peripheral blood), paracrine, and cell contact molecule. Moreover, while both GM-CSF and IL-3 are considered to be mainly induced factors (Gasson, 1991, Ihle and Weinstein, 1986), CSF-1 has considerable level of constitutive expression, and therefore, may play a major role in steady-state regulation as opposed to stress regulation, which may be a predominant role for GM-CSF and IL-3 as well as for other molecules.

However, there is a considerable overlap between functions of CSF-1, GM-CSF and IL-3, and it is extremely difficult to dissect the actual role of each molecule in the regulation of the system using normal mouse models.

Basic biology of the *op/op* mouse model

The *op/op* mouse is deprived of all forms of CSF-1 (Wiktor-Jedrzejczak et al., 1990) as a consequence of the insertion of thymidine in position 262 of CSF-1 gene (Yoshida et al., 1990). This insertion shifts the reading frame and produces stop codon 21 bases downstream. Such defect should lead to the production of truncated protein, shorter than any known bioactive form of CSF-1 (Heard et al., 1987). Although theoretically possible, there is no evidence for the repair of this lesion in mutant mice, and the described partial resolution of osteopetrosis in old *op/op* mouse (Marks and Lane, 1976, Begg et al.,1993) is considered to be the effect of other compensatory mechanisms rather than being due to the leak in the defect (Wiktor-Jedrzejczak, 1993b).

Morever, the utilization of the *op/op* mouse model is facilitated by the availability of human recombinant CSF-1 (active in murine system), that can be used in reconstitution experiments (Ladner et al., 1987, Halenbeck et al., 1989). Consequently, the role of CSF-1 in the regulation of the macrophage system can be studied by a combination of the analysis of deficiences of cells of this system in mutant mice with analysis of mutant animals having reconstituted circulating level of CSF-1 by the administration of recombinant form of the factor.

Pattern of macrophage deficiencies in the *op/op* mouse

The macrophage deficiency in the *op/op* mouse is severe but not absolute. Functionally competent macrophages are present (although in reduced number) in mutant animals (Wiktor-Jedrzejczak et al.,1992b). Moreover, very profoud differences exist in the degree of affection of various local macrophage populations. This heterogeneity concerns not only different organs but also different specific locations within the same organ e.g. spleen.

There are tissues, where the number of resident macrophages is reduced to less than 5% of the normal and sometimes it is even negligible. They include peritoneal cavity (where the defect was originally identified: Wiktor-Jedrzejczak et al., 1982) pleural cavity, muscle, skin, periostium, synovium, uterus, kidney and peripheral blood monocytes (Naito et al., 1991, Wiktor-Jedrzejczak et al., 1992b, Witmer-Pack et al., 1993, Cecchini et al., 1994) as well as spleen metallophils (Witmer-Pack et al., 1993, Cecchini et al., 1994) and osteoclasts (Marks and Lane, 1976, Marks, 1982).

Organs, where the deficiency is less pronounced, and the number of macrophages is between 10 and 80% of the normal include the liver, other than metallophils populations of spleen macrophages, lung, intestine, salivary glands, adrenals, bladder as well as brain microglial cells (Naito et al., 1991, Wiktor-Jedrzejczak et al., 1992b, Witmer-Pack et al., 1993, Cecchini et al., 1994).

On the other hand, epidermal Langerhans cells, dendritic cells of lymphoid organs, as well as macrophages present in these organs are quantitatively normal in the *op/op* mice (Takahashi et al., 1992, Witmer-Pack et al., 1993, Takahashi et al., 1993, Cecchini et al., 1994), what suggests that they are completely CSF-1 independent.

Restoration of circulating CSF-1 by systemic administration of recombinant form of the factor corrected only some near- completely depleted macrophage-related populations including osteoclasts, monocytes, kidney macrophages, and spleen metallophils, as well as some partially depleted populations including bone marrow, spleen and liver macrophages (Wiktor-Jedrzejczak et al., 1991, Kodama et al., 1991, Cecchini et al., 1994). Such populations as peritoneal cavity macrophages could only be restored by local CSF-1 administration, and it was hypothetized that this is due to the existence of blood-tissue barrier for circulating CSF-1 and to the exclusively local control of macrophage populations in many tissues (Wiktor-Jedrzejczak et al., 1991). This hypothetical barrier was later found to be operative for almost all other local macrophage populations (Cecchini et al., 1994) in addition to the peritoneal and pleural cavities.

On the other hand, these observations may also be explained by the requirement for increasing CSF-1 gradient for monocyte migration to tissues. According to this concept, monocytes could only migrate from locations with lower CSF-1 concentration (blood) to higher CSF-1 concentration (tissues).

Whatever the answer, the reported combined data suggest that the local macrophage populations could be divided into:

- completely CSF-1 dependent;
- partially CSF-1 dependent, and:
- CSF-1 independent.

The first two subpopulations could be further subdivided into:

- dependent on systemic CSF-1, and;
- dependent on local CSF-1.

Moreover, there appears to be a certain logic in distinction between organs with considerable CSF-1 independent resident macrophage population, and organs without such population. The organs with considerable CSF-1 independent macrophage population are mainly those that possess large total macrophage populations such as the liver, spleen, lungs, intestine, and that are at high risk of exposure to microorganisms, and particulate materials such as cell debris and other. Almost all other organs appear to have only CSF-1 dependent resident macrophage population, and they are generally at low risk of exposure to stimuli that require macrophage action. However, also in these latter organs in the *op/op* mouse, it is easy to elicit macrophages at the absence of CSF-1 by for instance injection of endotoxin (Wiktor-Jedrzejczak et al., unpublished observations).

Furthermore, organs with normally large resident macrophage population such as the liver and spleen have their CSF-1 dependent macrophages under control of circulating, and not only of locally produced CSF-1. Therefore, it appears that the presence of resident macrophage population is these organs is assurred in multiple ways, that is by circulating CSF-1, local CSF-1, and by other primary macrophage growth factors. In contrast, in most other organs resident macrophages are almost exclusively dependent on local CSF-1 (Fig. 1). However, there are also organs such as the kidneys or specific subpopulations such as spleen metallophils, which are almost exclusively CSF-1 dependent, and which are dependent on circulating CSF-1 (Cecchini et al., 1994).

Additionally, it has to be pointed out that deficient macrophage populations in the *op/op* liver, spleen and other organs are maintained at the near absence of monocytes. Although diminished, these populations still constitute about 50% of the normal macrophage populations in these organs and are, therefore, of considerable size. This may suggest, that either only CSF-1 dependent resident macrophage populations require constant supply of monocytes or CSF-1 independent macrophages are replenished by a tiny monocyte subpopulation with very high turnover.

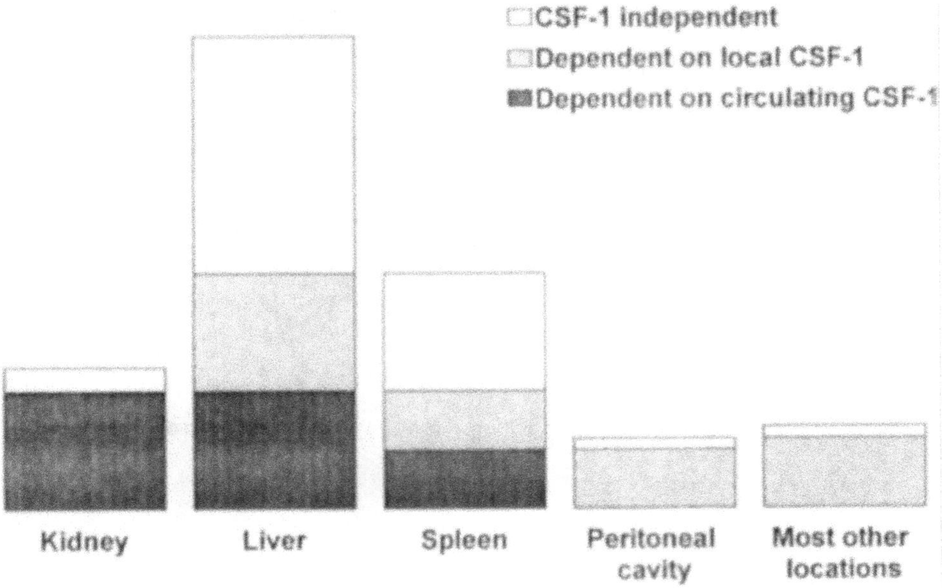

Fig. 1. Growth factor dependency of various local macrophage populations. Organs such as the liver and spleen possess both CSF-1 dependent and CSF-1 independent macrophage populations of considerable size. Some of their CSF-1 dependent macrophages are under control of circulating, and some are under the influence of local CSF-1. Organs such as kidneys possess almost exclusively macrophages dependent on circulating CSF-1. Locations such peritoneal cavity and vast majority of others have macrophages dependent almost exclusively on local CSF-1 with only a few CSF-1 independent macrophages.

<u>The range of CSF-1 action in macrophage maturation: medullary versus extramedullary role of CSF-1</u>

Hematopoietic organs of the *op/op* mice possess large numbers of macrophage progenitors, with their frequency being normal in the bone marrow and increased in the spleen (Wiktor-Jedrzejczak et al., 1992b). Only after calculation of the total mouse macrophage progenitor number it was possible to detect a deficiency in this parameter in the *op/op* mouse and this deficiency could be a secondary consequence of reduced marrow due to osteopetrosis rather than specifically related to CSF-1 absence. This suggests that CSF-1 is not necessary for the generation of macrophage progenitors from the hematopoietic stem cells. Macrophages could be generated *in vivo* in the *op/op* mice after administration of CSF-1 within only 48 hours (Wiktor-Jedrzejczak et al., in preparation for publication). This suggests that even quite late macrophage progenitors are formed at the absence of CSF-1, and that only then they reach block preventing further maturation.

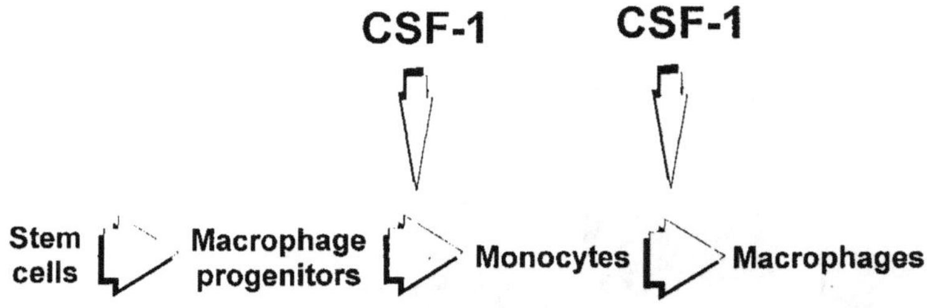

Fig. 2. Sites of CSF-1 action in macrophage formation and survival *in vivo*. In contrast to previous assumptions, most of the CSF-1 action seems to be exerted extramedullary in the tissues.

Consequently, only macrophage formation and not macrophage progenitor formation appears to be considerably CSF-1 dependent (Fig. 2). The process of macrophage formation may be further divided into formation of monocytes from macrophage progenitors and formation of macrophages from monocytes. Absence of more than 95% of monocytes in the *op/op* mouse and their restoration after CSF-1 treatment (Kodama et al.,1991) suggests that this process is near completely CSF-1 dependent. This agrees to some extent with published models of the development of macrophage system, where CSF-1 and other macrophage factors have been suggested to act mainly in the bone marrow effecting the production of monocytes (VanFurth, 1993, Johnson, 1993).

However, the very presence of profound macrophage deficiencies in the *op/op* mice suggests that the role of CSF-1 in transition of monocytes to the tissues, and in their further maturation to macrophages resident there, is much more important. Clearly, this suggests that the essential CSF-1 role in macrophage maturation is extramedullary, and that CSF-1 is not necessary in the bone marrow, but for most of the tissues (listed earlier) it is necessary in those tissues to maintain supply of monocytes, formation of macrophages, and local survival of these latter cells. This is supported also by the observation, that the CSF-1 treated *op/op* mouse has plenty of monocytes and still possess profound local macrophage deficiencies (Wiktor-Jedrzejczak et al., 1991, Cecchini et al., 1994).

The data from other models suggest that CSF-1 is not necessary for macrophage activation (Evans, 1991). Our studies suggest that it is clearly not necessary for the *in vivo* activation of those CSF-1 independent macrophages that are still present in the *op/op* mouse (Wiktor-Jedrzejczak et al., unpublished observations). However, the dependency of activation of CSF-1 dependent macrophages on CSF-1 has not been experimentally approached in the mutant mice. This is testable by restoring CSF-1 dependent macrophages in the *op/op* mouse with exogenous CSF-1, withdrawing this factor and testing macrophage activation at CSF-1 absence. Therefore, the results of such experiment should provide more definite answer to this question.

<u>The role of other primary macrophage growth factors in the regulation of the macrophage system in relation to CSF-1</u>

In addition to the *op/op* mouse, animals with knockouts of a gene for the 2nd major primary macrophage factor: GM-CSF have been created recently (Dranoff et al., 1994, E. Stanley et al.,1994). Moreover, also a double knockout: GM-CSF-knockout-*op/op* mice have been bred and partly characterized (Lieschke et al., 1994). These models complement very well the advantages of the *op/op* mouse. Similarly to the *op/op* mouse, these mutants possess large numbers of macrophage progenitors (E. Stanley et al.,1994, Lieschke et al., 1994), confirming that CSF-1 is not necessary for their formation, and suggesting that also GM-CSF is playing only a minor role in that process.

In agreement with a notion that most of monocytes are produced under CSF-1 influence, GM-CSF knockout mouse possess normal number of these cells (Dranoff et al., 1994, E. Stanley et al., 1994). However, also double knockout: GM-CSF knockout-*op/op* mouse still possess some monocytes (Lieschke et al., 1994), suggesting that some other factor at the absence of both GM-CSF and CSF-1 is capable of maintaining formation of a few monocytes, and that CSF-1 independent macrophages present in the *op/op* mouse are not exclusively GM-CSF dependent macrophages. One possible candidate molecule for the supporting the production of CSF-1 independent macrophages is IL-3. However, the fact that it is produced only by activated T cells (Ihle and Weinstein, 1986) may suggest that also some other molecules may play a role in that process.

The discussion of contribution of other than CSF-1 factors to macrophage presence in the tissues has to begin from the appreciation that considerable numbers of resident macrophages are present in the *op/op* mice in such organs as liver and spleen suggesting that they can reach full functional maturity without CSF-1. This may suggest, that in agreement with the model proposed earlier (Wiktor-Jedrzejczak, 1993b) other than CSF-1 primary macrophage growth factors also play a role in terminal maturation of macrophages, and contribute to the formation and maintenance of resident (as opposed to induced) cells.

Both published reports concerning GM-CSF knockout mice failed to identify numerical macrophage deficiencies and pointed out pulmonary proteinosis as the major abnormality present in these mice. Whether this proteinosis due to surfactant accumulation is secondary to the deficiency of a few alveolar macrophages dependent on GM-CSF is still an open question. Theoretically, it is possible that GM-CSF may control surfactant production also by a macrophage-independent mechanism. However, the analysis of macrophage deficiencies in these mice was not sufficiently detailed to exclude the presence of even quite significant local deficits.

Until such data are available, it is difficult to answer such basic question as: whether CSF-1 dependent and GM-CSF dependent macrophage subpopulations in organs such as liver are regulated independently of each other or whether the total liver macrophage population is maintained by the combined activity of all macrophage growth factors? However, the regular character of macrophage deficiencies in the *op/op* mouse may suggest that they are created by subtraction of an inherently regulated CSF-1-only- dependent subpopulation. In fact, data are available from other model that CSF-1 level and CSF-1 dependent macrophage subpopulation that is consuming CSF-1 form an integral regulatory feedback system (Bartocci et al., 1987).

Besides, the *op/op* mouse under stress conditions such as bacterial infection can produce large numbers of macrophages. For instance in the course of *E. coli* fecal peritonitis the mutant mice are capable of eliciting millions of macrophages in peritoneal cavity within 48 hours postinfection (Wiktor-Jedrzejczak et al., submitted for publication). This suggests that CSF-1 independent (i.e. dependent on other macrophage growth factors) formation of induced macrophages is efficient in the *op/op* mouse, in contrast to deficient formation of resident macrophages. Interestingly, the induced local expansion of macrophages may originate from local progenitors. This is supported by a recent observation that in the presence of inflammatory stimulus such as glucan, local (liver) macrophages can considerably expand, at the absence of CSF-1, and at the near absence of monocytes (Takahashi et al., 1994).

Macrophage growth factor dependency and functional diversity of macrophages

We have suggested earlier, that CSF-1 dependent macrophages (absent in the *op/op* mouse) play mainly regulatory role through release of cytokines, while CSF-1 independent macrophages (present in the *op/op* mouse) are responsible primarily for the immune functions of macrophages (Wiktor-Jedrzejczak et al.,1992a). It has to be understood that in order to be called macrophages, cells have to possess common set of functions including phagocytosis. Other studies comparing CSF-1 induced and GM-CSF induced macrophages *in vitro* have shown that GM-CSF induced macrophages expressed higher levels of class II MHC (Doherty et al.,1993) what is in agreement with suggestion on primarily immune function of these cells. However, the same authors have also suggested that GM-CSF induced macrophages are better producers of cytokines including IL-1, IL-6 and TNF-α than CSF-1 induced macrophages. This is in contrast to the postulated major role of these latter cells in that process. On the other hand, not only efficiency of production on the per cell basis counts *in vivo*, but also availability of cells. From that point of view, in most locations the vast majority of resident macrophages are dependent on CSF-1 and they have to provide the respective cytokines during early phases of tissues response to trauma or infection (Fig.3). This notion is supported by the presence of TNF-α, and IL-1α deficiencies in the *op/op* mouse (Wiktor-Jedrzejczak et al., 1992a, Szperl et al., 1995).

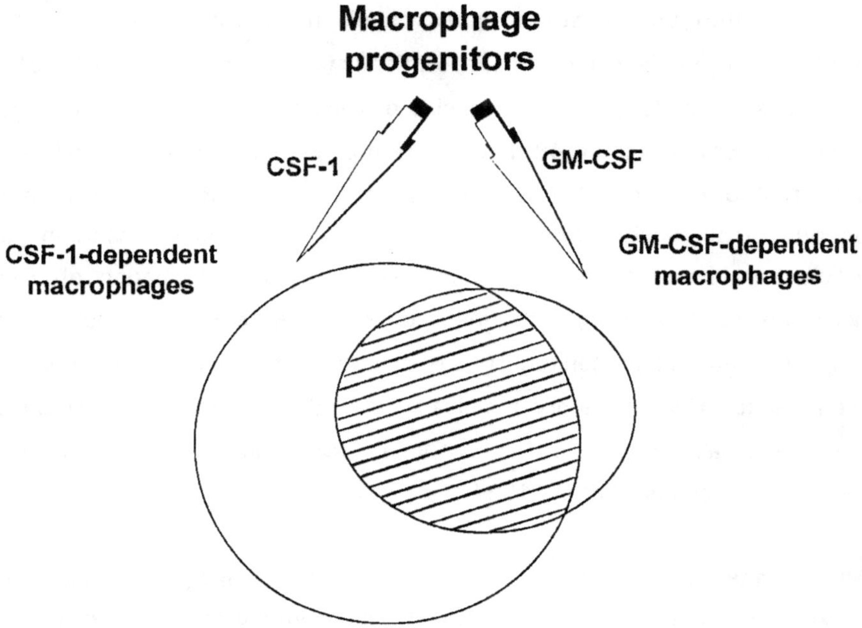

Fig. 3. Growth factor dependency of macrophage diversity. The diversity *in vivo* is a function of availability of cells, with much greater availability of CSF-1 dependent macrophages, and functional diversity. Most of the functions are shared by both subpopulations, which is indicated by shaded area, while some functions are unique to each subpopulation.

At present, there appears to be no data supporting the existence of separate CSF-1 dependent and CSF-1 independent macrophage progenitors. All macrophage progenitors seem to be able to respond to both CSF-1 and GM-CSF (Metcalf and Nicola, 1992). There is no synergy between these factors, what may suggest that they act on the same cell population.

However, after macrophages are formed, in addition to functions shared by macrophages formed under the influence of any factor, there appear to be functions that are exclusively dependent on CSF-1 or GM-CSF. Only CSF-1 dependent resident peritoneal macrophages are capable of recruiting lymphocytes to peritoneal cavity (Kalinski et al.,1993). On the other hand, eicosanoids production has been recently shown to be an exclusive function of GM-CSF and IL-3 produced macrophages, and not of CSF-1 induced macrophages (Shibata et al.,1994).

Concluding remarks

The studies employing the *op/op* mouse challenge several established views concerning the organization of the macrophage system and provide novel insights into regulation of macrophages at the tissue level. The model is far from being explored and awaites application to the studies reappraising various roles classically assigned to macrophages. Moreover, its possibilities may be increased by breeding mice combining CSF-1 deficiency with deficiencies of other earlier or later acting macrophage growth factors and macrophage activators such as interferon-γ. An example of this approach: breeding the *op/op* mouse possessing knockout of GM-CSF gene is already available (Lieschke et al., 1994). It is a strong belief of the present author that these models would allow final dissection of the organization and function of the macrophage system.

Acknowledgements

This study was supported in part by grants 4 1317 91 01 and 4 4381 91 02 from the Polish National Committee for Scientific Research to W.W-J.

References

Adams DO, Hamilton TA (1992) Molecular basis of macrophage activation: diversity and its origins. In: Lewis CE, McGee JOD (eds) The Natural Immune System: The Macrophage. IRL Press, Oxford, pp.75-114

Auger MJ, Ross JA (1992) The biology of the macrophage. In: Lewis CE, McGee JOD (eds) The Natural Immune System: The Macrophage. IRL Press, Oxford, pp.2-74

Bartocci A, Mastrogiannis DS, Migliorati G, Stockert RJ, Wolkoff AW, Stanley ER (1987) Macrophages specifically regulate the concentration of their own growth factor in the circulation. Proc Natl Acad Sci USA 84:6179-6183

Begg SK, Radley JM, Pollard JW, Chisholm OT, Stanley ER, Bertoncello I (1993) Delayed hematopoietic development in osteopetrotic (op/op) mice. J Exp Med 177:237-242

Bertoncello I, Bradley TR, Hodgson GS, Dumlop JM (1991) The resolution, enrichment, and organization of normal bone marrow high proliferative potential colony-forming cell subsets on the basis of rhodamine-123 fluorescence. Exp Hematol 19:174-178

Bradley TR, Hodgson GS (1979) Detection of primitive macrophage progenitor cells in bone marrow. Blood 54:1446-1450

Cecchini MG, Dominguez MG, Mocci S, Wetterwald A, Felix R, Fleisch H, Chisholm O, Hofstetter W, Pollard JW, Stanley ER (1994) Role of colony stimulating factor-1 in the establishment and regulation of tissue macrophages during postnatal development of the mouse. Development 120:1357-1372

Cheers C, Haigh AM, Kelso A, Metcalf D, Stanley ER, Young AM (1988) Production of colony-stimulating factors (CSFs) during infection: separate determinations of macrophage-, granulocyte-, granulocyte-macrophage-, and multi-CSFs. Infect Immun 56:247-251

Doherty TM, Kastelein R, Menon S, Andrade S, Coffman RL (1993) Modulation of murine macrophage function by IL-13. J Immunol 151:7151-7160

Dranoff G, Crawford AD, Sadelain M, Ream B, Rashid A, Bronson RT, Dickersin GR, Bachurski CJ, Mark EL, Whitsett JA, Mulligan RC (1994) Involvement of granulocyte-macrophage colony-stimulating factor in pulmonary homeostasis. Science 264:713-716

Evans R (1991) Clarification of the potential role of CSF-1 in activation of macrophages. J Leukocyte Biol 50:316

Falk LA, Vogel SN (1988) Comparison of bone marrow progenitors responsive to granulocyte-macrophage colony stimulating factor and macrophage colony stimulating factor-1. J Leukocyte Biol 43:148-157

Felix R, Cecchini MG, Hofstetter W, Elford PR, Stutzer A, Fleisch H (1990) Impairment of macrophage colony-stimulating factor production and lack of resident bone marrow macrophages in the osteopetrotic op/op mouse. J Bone Mineral Res 5:781-789

Gasson JC (1991) Molecular physiology of granulocyte-macrophage colony-stimulating factor. Blood 77:1131-1145

Gisselbrecht S, Sola B, Fichelson S, Bordereaux D, Tambourin P, Mattei M-G, Simon D, Guenet J-L (1989) The murine M-CSF gene is localized on chromosome 3. Blood 73:1742-1746

Good RA (1991) Experiments of nature in the development of modern immunology. Immunol Today 12:283-286

Gordon S (1986) Biology of the macrophage. J Cell Sci Suppl 4:267-286

Halenbeck R, Kawasaki E, Wrin J, Koths K (1989) Renaturation and purification of biologically active recombinat human macrophage colony-stimulating factor expressed in E. coli. Bio/Technology 7:710-715

Heard JM, Roussel MF, Rettenmier CW, Sherr CJ (1987) Synthesis, post-translational processing, and autocrine transforming activity of a carboxylterminal truncated form of colony stimulating factor-1. Oncogene Res 1:423-440

Ihle JN, Weinstein Y (1986) Immunological regulation of hematopoietic/lymphoid stem cell differentiation by interleukin 3. Adv Immunol 39:1-49

Johnson RB Jr (1993) Monocytes and macrophages. In Lachmann PJ, Peters K, Rosen FS, Walport MJ (eds) Clinical Aspects of Immunology. Blackwell, Boston,Vol.1,pp.467-479

Kalinski P, Urbanowska E, Kawiak J, Hoser G, Aukerman SL, Wiktor-Jedrzejczak W (1993) CSF-1 dependent but not GM-CSF dependent macrophages recruit lymphocytes to peritoneal cavity. Exp Hematol 21:1014

Kodama H, Yamasaki A, Nose M, Nijda S, Ohgame Y, Abe M, Kumegawa M, Suda T (1991) Congenital osteoclast deficiency in osteopetrotic (op/op) mice is cured by injections of macrophage colony-stimulating factor. J Exp Med 173:269-272

Koike K, Stanley ER, Ihle JN, Ogawa M (1986) Macrophage colony formation supported by purified CSF-1 and/or interleukin 3 in serum-free culture: evidence for hierarchical difference in macrophage colony-forming cells. Blood 67:859-864

Ladner MB, Martin GA, Noble JA, Nikoloff DM, Tal R, Kawasaki ES, White TJ (1987) Human CSF-1: gene structure and alternative splicing of mRNA precursors. EMBO J 6:2693-2698

Lieschke GJ, Stanley E, Grall D, Hodgson G, Sinickas V, Gall JAM, Sinclair RA, Dunn RA (1994) Mice lacking both macrophage- and granulocyte-macrophage colony stimulating factor have macrophages and coexistent osteopetrosis and severe lung disease. Blood 84:27-35

Marks SC Jr (1982) Morphological evidence of reduced bone resorption in osteopetrotic (op) mice. Am J Anat 163:157-167

Marks SC Jr (1987) Osteopetrosis: multiple pathways for the interception of osteoclast function. Appl Pathol 5:172-183

Marks SC Jr, Lane PW (1976) Osteopetrosis, a new recessive skeletal mutation on chromosome 12 of the mouse. J Hered 67:11-18

Metcalf D (1991) Control of granulocytes and macrophages: molecular, cellular, and clinical aspects. Science 254:529-533

Metcalf D, Nicola NA (1992) The clonal proliferation of normal mouse hematopoietic cells: enhancement and suppression by colony-stimulating factor combinations. Blood 79:2861-2866

Miyajima A, Kitamura T, Harada N, Yokota T, Arai K-i (1992) Cytokine receptors and signal transduction. Annu Rev Immunol 10:295-331

Naito M, Hayashi S-I, Yoshida H, Nishikawa S-I, Shultz LD, Takahashi K (1991) Abnormal differentiation of tissue macrophage populations in 'osteopetrosis' (op) mice defective in the production of macrophage colony-stimulating factor. Am J Pathol 139:657-667

Nathan C, Cohn ZA (1985) Cellular components of inflammation: monocytes and macrophages. In: Kelley WN, Harris ED Jr, Ruddy S, Sledge CB (eds) Textbook of Rheumatology. 2nd Ed. WB Saunders, Philadelphia, pp.144-169

Prystowsky IB, Otten G, Naujokas MF, Vardiman J, Ihle JN, Goldwasser E, Fitch FW (1984) Multiple hemopoietic lineages are found after stimulation of mouse bone marrow precursor cells with interleukin 3. Am J Pathol 117:171-179

Shibata Y, Bjorkman DR, Schidt M, Oghiso Y, Volkman A (1994) Macrophage colony-stimulating factor-induced bone marrow macrophages do not synthesize or release prostaglandin E_2. Blood 83:3316-3323

Stanley E, Lieschke GJ, Grail D, Metcalf D, Hodgson G, Gall JAM, Maher DW, Cebon J, Sinickas V, Dunn AR (1994) Granulocyte/macrophage colony-stimulating factor-deficient mice show no major perturbation of hematopoiesis but develop a characteristic pulmonary pathology. Proc Natl Acad Sci USA 91:5592-5596

Stanley ER (1994) Colony stimulating factor-1 (macrophage colony stimulating factor). In The Cytokine Handbook, Academic Press, New York, pp.387-418

Suda T, Suda J, Ogawa M (1983) Single-cell origin of mouse hemopoietic colonies expressing multiple lineages in variable combinations. Proc Natl Acad Sci USA 80:6689-6693

Szperl M, Urbanowska E, Ansari AA, Wiktor-Jedrzejczak W (1995) Increased resistance of macrophage-deficient, TNF-α-deficient, IL-1α-deficient op/op mouse to endotoxin. Ann NY Acad Sci (in press)

Takahashi K, Naito M, Shultz LD (1992) Differentiation of epidermal Langerhans cells in macrophage colony-stimulating-factor-deficient mice homozygous for the osteopetrosis (op) mutation. J Invest Dermatol 99:46S-47S

Takahashi K, Naito M, Shultz LD, Hayashi S-I, Nishikawa S-i (1993) Differentiation of dendritic cell populations in macrophage colony-stimulating factor deficient mice homozygous for the osteopetrosis (op) mutation. J Leukocyte Biol 53:19-28

Takahashi K, Naito M, Umeda S, Shultz LD (1994) The role of macrophage colony-stimulating factor in hepatic glucan-induced granuloma formation in the osteopetrosis mutant mouse defective in the production of macrophage colony-stimulating factor. Am J Pathol 144:1381-1392

VanFurth R (1993) Cell biology of mononuclear phagocytes. In VanFurth R (ed) Hemopoietic growth factors and mononuclear phagocytes. Karger, Basel, pp.1-9

VanFurth R, Cohn ZA, Hirsch JG, Humphrey JH, Spector WG, Langevoort HL (1972) The mononuclear system. A new classification of macrophages, monocytes, and their precursor cells. Bull WHO 46:845-852

Wiffeils JFAM, Derover Z, Kraal G, Beelen RHJ (1993) Macrophage phenotype regulation by colony-stimulating factors at bone marrow level. J Leukocyte Biol 53:249-255

Wiktor-Jedrzejczak W (1993a) Cure of osteopetrosis in op/op mice by bone grinding and good food? Exp Hematol 21:1314-1315

Wiktor-Jedrzejczak W (1993b) In vivo role of macrophage growth factors as delineated using CSF-1 deficient op/op mouse. Leukemia 7:S117-S121

Wiktor-Jedrzejczak W, Ahmed A, Szczylik C, Skelly RR (1982) Hematological characterization of congenital osteopetrosis in op/op mouse. Possible mechanism for abnormal macrophage differentiation. J Exp Med 156:1516-1527

Wiktor-Jedrzejczak W, Ansari AA, Szperl M, Urbanowska E (1992a) Distinct in vivo functions of two macrophage subpopulations as evidenced by studies using macrophage-deficient op/op mouse. Eur J Immunol 22:1951-1954

Wiktor-Jedrzejczak W, Bartocci A, Ferrante AW Jr, Ahmed-Ansari A, Sell KW, Pollard JW, Stanley ER (1990) Total absence of colony-stimulating factor 1 in the macrophage-deficient osteopetrotic (op/op) mouse. Proc Natl Acad Sci USA 87:4828-4832

Wiktor-Jedrzejczak W, Ratajczak MZ, Ptasznik A, Sell KW, Ahmed-Ansari A, Ostertag W (1992b) CSF-1 deficiency in the op/op mouse has differential effects on macrophage populations and differentiation stages. Exp Hematol 20:1004-1010

Wiktor-Jedrzejczak W, Skelly RR, Ahmed A (1981) Hematopoietic stem cell differentiation and its role in osteopetrosis. In: Gerschwin ME, Merchant B (eds) Immunologic defects in laboratory animals. Plenum Press, New York, vol.1:51-77

Wiktor-Jedrzejczak W, Urbanowska E, Aukerman SL, Pollard JW, Stanley ER, Ralph P, Ansari AA, Sell KW, Szperl M (1991) Correction by CSF-1 of defects in the osteopetrotic op/op mouse suggests local, developmental, and humoral requirements for this growth factor. Exp Hematol 19:1049-1054

Witmer-Pack MD, Hughes DA, Schuker G, Lawson L, McWilliam A, Inaba K, Steinman RM, Gordon S (1993) Identification of macrophages and dendritic cells in the osteopetrotic (op/op) mouse. J Cell Science 104:1021-1029

Yoshida H, Hayashi S-I, Kunisada T, Ogawa M, Nishikawa S, Okamura H, Sudo T, Shultz D, Nishikawa S-I (1990) The murine mutation osteopetrosis is in the coding region of the macrophage colony stimulating factor gene. Nature 345:442-444

Part 6

Immunology and Immunotherapy

Lifting a tip of the veil of human minor Histocompatibility antigens.

Els Goulmy, Dept. of Immunohaematology and Blood Bank, University Hospital, P.O. Box 9600, 2300 RC Leiden, The Netherlands.

Human bone marrow transplants performed as therapeutical treatment of severe aplastic anaemia, leukaemia and immune deficiency disease became available in the seventies (Bortin, 1970). In an artificial situation, such as organ transplantation, the major Histocompatibility (H) antigens function as a major transplantation barrier and thus play an important role in the survival of transplants and patients. Consequently, improved success in bone marrow transplantation was reported when matching for the HLA antigens was taken into account (Thomas, 1975). Between 1975 and the present day, the long-term results of allogeneic bone marrow transplantation (BMT) have greatly improved due to the use of HLA-matched siblings as marrow donors, advanced pretransplant chemoradiotherapy, the use of potent immunsuppressive drugs as Graft-versus-Host-Disease (GvHD) prophylaxis, better antibiotics and isolation procedures.

The results of clinical bone marrow (BM) transplantation reveal however that the selection of MHC identical donors/recipients is not a guarantee of avoidance of GvHD or disease free survival even when donor and recipient are closely related (Bortin, 1991). It is believed that disparities for minor Histocompatibility antigens (mHag) between donor and recipient constitute a potential risk for GvHD or graft failure (Martin, 1991; Beatty, 1989).

In man, the efforts of several investigators have led to the identification of a small number of mHag (Goulmy, 1977, 1983; Zier, 1983; Irlé, 1990; Van Els, 1992). Here, we will briefly focuss on their possible clinical relevance for BM transplantation in both the GvHD and the Graft-versus-Leukemia (GvL) reactivities. Furthermore, we provide preliminary evidence for the evolutionary conservation of human mHag.

Clinical relevance of anti-host cytotoxic T cells (CTLs) and helper T cells (Th) in the development of GvHD

Several reports demonstrated the presence of anti-host mHag specific CTL in patients suffering from GvHD after HLA genotypically identical BMT (Goulmy, 1983; Tsoi, 1980, 1983; Irlé, 1985; Van Els, 1990; Irscheck, 1992; Niederwieser, 1993). In our laboratory,

NATO ASI Series, Vol. H 94
Gene Technology
Edited by A. R. Zander, W. Ostertag,
B. V. Afanasiev, F. Grosveld

much effort was put into the further characterization of a (small) number of anti-host mHag specific CTLs. Hereto, CTL clones specific for host mHag were isolated from the peripheral blood (PBL) of patients suffering from severe GvHD. Subsequent immunogenetic analyses revealed that these CTL clones identified five non-sexlinked mHag, designated HA-1, -2, -3, -4, -5, which are recognized in a classical MHC restricted fashion (Van Els, 1992). mHag HA-3 is recognized in the presence of HLA-A1 and mHag HA-1, -2, -4 and 5 require the presence of HLA-A2. In order to document the effect of mH antigens in genotypically identical BMT on the occurrence of acute (grade ≥ 2) GvHD, we prospectively collected PBL from HLA-A1 and HLA-A2 positive patient/donor sibling pairs. This multi center study comprised 148 HLA genotypically identical BM donor/recipient combinations, adults as well as children, grafted between 1982 and 1990. The results of the mHag typing using the CTL clones specific for five well defined mHag HA-1 to HA-5 demonstrated (table 1) a significant correlation between mHag HA-1, -2, -4 and -5 mismatch and GvHD (Goulmy, 1994).

Table 1 Correlation of mHag HA-1, -2, -4 and -5, with the occurrence of GvHD

All patients		mHag	
		match	mismatch
GvHD	No	41	1
	Yes	40	12

P = 0.002

The last few years evidence has accumulated that in addition to CTLs mH antigen specific helper T cells (Th) could be relevant in the pathogenesis of GvHD. In vitro studies reporting on host directed Th cells have been described in patients having GvHD (Tsoi, 1980; Reinsmoen, 1984; Irlé, 1987). Van Els et al. reported on the long term kinetics of Th cells in response to host mH antigens in 16 patients and demonstrated that significant Th cell activity in vitro correlates with clinical acute GvHD (Van Els, 1990b). In a subsequent study, we demonstrated that these anti-host Th cells carry the CD4 phenotype and recognize mH antigens in the context of HLA-DR and -DP (Van Els, 1990c).

Most recent observations support the notion that mH antigen specific Th cells are by and large likely to play a role in the pathogenesis of acute GvHD (Theobald, 1992;

Schwarer, 1993). In both latter studies the primary in vitro putative mH antigen Th activities have been measured by IL-2 production of the responding cell population.

Possible involement of mHag in Graft versus Leukemia

The hypothesis that posttransplantation of bone marrow anti-host CTL activity may have a beneficial effect is based on the assumption of the postulated anti-leukaemic potential as a 'desired' side-effect of the post BMT complication GvH. (Bortin, 1973; Weiden, 1981a, 1981b). In search for anti-host CTL and Th cell activities post BMT, we observed earlier both absence and presence of anti-host CTL in patients without any clinical signs of GvHD (see table 2).

Table 2 Anti-host T cell activities after HLA identical BMT.

	Anti-host CTLs		Anti-host Th	
	+++	---	+++	---
no GvHD	5	4	2	5
acute GvHD	6	3	6	0
chronic GvHD	15	1	5	2
	N = 34		N = 20	

It was notable that in 3 out of 4 cases without anti-host CTL activity (see table 2), leukemia relapse was manifested. On the other hand, the presence of anti-host CTLs in the absence of GvHD (N-5) argues for the possible role of these CTLs in the anti-leukemic activity. In an attempt to study the post BMT anti-host CTL responses for their putative anti-leukemic activity in vitro, we observed "GvHD" related and "GvL' related acitivities. The latter type of CTL clones recognized patient's neoplastic cells only. The former type of CTL clones were reactive with ligands, like mH antigens, shared by host PBL and leukemic cells (Van Lochem, 1992). Another line of investigation support the notion that anti-host mHag specific CTL may play a role in the anti-leukemic effect of allogeneic BMT. Namely, mHag specific CTLs are capable of inhibiting in vitro outgrowth of clonogeneic leukemic precursor cells as well as lyse freshly obtained myeloid and lymphoid leukemic cells (Fakenburg, 1991; Van der Harst, 1994).

Human mHag are conserved in evolution

To substantiate the importance of the human mH antigenic systems, we investigated whether the mHag are conserved in evolution between man and chimpansee. Hereto, cells from chimpansees were transfected with the human HLA-A2.1 gene. Subsequent analyses with our human allo HLA-A2.1 and four mHag HLA-A2.1 restricted CTL clones revealed the presentation of chimpansees' allo and mHag peptides in the context of the transfected human HLA-A2.1 molecule by chimpansees' target cells (table 3). These results implicate that the chimpansee cell derived allo and mHag peptides investiged in this study are very similar to the human allo HLA-A2 and HLA-A2 restricted mHag peptides.

Table 3 Human mHag are evolutionary conserved

Chimp. target cells*		Human CTL clones				
		allo A2	A2HY	A2 HA-1	A2 HA-2	A2 HA-4
Theo	♂	76°	79	69	75	3
Japie	♂	48	58	45	54	0
Pearl	♀	37	2	35	40	2
Debbie	♀	42	1	37	48	1
Gwen	♀	23	0	38	32	1
Brigitte	♀	31	0	23	20	0
Sherry	♀	52	2	58	55	3

* chimp. target cells have been electroporated with the HLA-A2 gene.

° % specific cytotoxicity measured in a cell mediated lympholysis assay.

In conclusion, although lots of information was gathered during the past decades on the murine and human mHag, still many questions remain to be answered. Besides identification of the mHag and the genes they are encoded by, we must be able to dissect the majors from the minor minors. To achieve this, more information is needed on the Th and CTL defined human mHag repertoire, and to establish the immunodominant ones. To understand their biological role in bone marrow transplantation information on their cytokine secretion profile is essential.

Acknowledgements

I am indebted to Els Blokland, Ronald Bontrop, Cécile van Els, Fred Falkenburg, Ellen van Lochem and Jos Pool for their great contributions. Ingrid Curiël for typing the manuscript. This work was supported in part by grants from the Dutch Organisation for Scientific Research (NWO), the J.A. Cohen Institute for Radiopathology and Radiation Protection IRS) and the Dutch Cancer Society.

References

Beatty PG, Hervé P (1989) Immunogenetic factors relevant to acute GvHD. In: S.J. Burakoff, D.H.J. Deeg, S. Ferrara, K. Atkinson (eds): Graft-versus-Host-Disease, Immunology, Pathophysiology and Treatment. New York, Dekker, 415-23.

Bortin MM (1970) A compendium of reported human bone marrow transplants. Transplantation, 9: 571-587.

Bortin MM, Rimm AA, Salzstein EC, Rodey GE (1973) Graft versus leukemia. III. Apparent independent anti-host and anti-leukemic activity of transplanted immunocompetent cells. Transplantation 16: 182-188.

Bortin MM, Horowitz MM, Ursic M, Rimm AA and Sobocinskym KA (1991). Progress in bone marrow transplantation for leukemia: a preliminary report from the advisory committee of the international Bone Marrow Transplant Registry. Transplant Proc. 23: 61-62.

Falkenburg F, Goselink H, van der Harst D et al (1991). Growth inhibition of clonogenic leukemic precursor cells by minor histocompatibility antigen-specific cytotoxic T lymphocytes. J. Exp. Med. 174: 27-33.

Goulmy E, Schipper R, Pool J et al. Minor histocompatibility antigen mismatches influence the development of GvHD after HLA genotypically identical bone marrow transplantation. Manuscript subm. for publication 1994.

Goulmy E, Termijtelen A, Bradley BA, Van Rood JJ (1977). Y-antigen killing by T cells of women is restricted by HLA. Nature 266: 544-545.

Goulmy E, Gratama JW, Blokland E, Zwaan FE, van Rood JJ (1983) A Minor transplantation antigen detected by MHC restricted cytotoxic T lymphocytes during graft-versus-host-disease. Nature 302: 159-161.

Irlé C, Beatty PG, Mickelson E, Thomas ED, Hansen JA (1985) Alloreactive T cell responses between HLA identical siblings. Transplantation 40: 329-333.

Irlé C, Chapuis B, Jeannet M et al (1987) Transplant Proc. suppl. 1, 19: 2674.

Irscheck E, Hladik T, Niederwieser D et al (1992) Studies on the mechanism of tolerance or Graft-versus-Host Disease in allogeneic bone marrow recipients at the level of cytotoxic T cell precursor frequencies. Blood 79: 1622-1628.

Martin PJ (1991). Increased disparity for minor Histocompatibility antigens as a potential cause of increased GvHD risk in marrow transplantations from unrelated donors compared with related donors. Bone Marrow Transplantation. 8: 217-223.

Niederwieser D, Grassegger A, Auböck J, Herold M, Nachbaur D, Rosenmayr A, Gächter A, Nussbaumer W, Gaggl S, Ritter M and Huber C (1993) Correlation of minor histocompatibility antigen specific cytotoxic T lymphocytes with Graft-versus-Host Disease status and analyses of tissue distribution of their target antigens. Blood, 81: 2200-2208.

Reinsmoen NL, Kersey JH, Bach FH (1984). Detection of HLA restricted anti minor histocompatibility antigen(s) reactive cells from skin GvHD lesions. Human Immunol. 1, 11: 249-257.

Schwarer AP, Jiang JZ, Barrett JM et al (1993). Helper T-lymphocyte precursor (HTLp) frequency predicts the occurrence and severity of acute GvHD and survival after allogeneic BMT in both recepients of genotypically HLA-identical sibling (SIB) and phenotypically HLA-matched unrelated donor (MUD) marrow. Lancet 341: 203-205.

Theobald M, Nierle T, Bunjes D et al (1992). Host-specific interleukin-2-secreting donor T cell precursors as predictors of acute Graft-versus-Host Disease in bone marrow transplantation between HLA-identical siblings. N. Engl. J. Med. 327: 1613-1617.

Tsoi M-S, Storb R, Dobbs S, Medill I, Thomas ED (1980). Cell mediated immunity to non-HLA antigens of the host by donor lymphocytes in patients with chronic graft-vs-host disease. J. Immunol. 125: 2258-2262.

Tsoi M-S, Storb R, Santos E, Thomas ED (1983) Anti-host cytotoxic cells in patients with acute graft-versus-host disease after HLA identical marrow grafting. Transplant Proc. 15: 1484-1486.

Van der Harst D, Goulmy E, Falkenburg JHF et al (1994). Recognition of minor histocompatibility antigens on lymphocytic and myeloid leukemic cells by cytotoxic T-cell clones. Blood 83: 1060-1066.

Van Els C, Bakker A, Zwinderman AH, Zwaan FE, van Rood JJ, Goulmy E (1990a) Effector mechanisms in GvHD in response to minor Histocompatibility antigens. I. Absence of correlation with CTLs. Transplantation 50: 62-66.

Van Els CACM, Bakker A, Zwinderman AH, Zwaan FE, Van Rood JJ, Goulmy E (1990b) Effector mechanisms in GvHD in response to minor histocompatibility antigens. II: Evidence for a possible involvement of proliferative T cells. Transplantation 50: 67-71.

Van Els C, Zantvoort E, Jacobs N et al (1990c). Graft-versus-host disease associated T helper cell responses specific for minor histocompatibility antigens are mainly restricted by HLA-DR molecules. Bone Marrow Transplantation 5: 365-372.

Van Els C, D'Amaro J, Pool J, Bakker A, van den Elsen PJ et al (1992) Immunogenetics of human minor Histocompatibility antigens: their polymorphism and immunodominance. Immunogenetics 35: 161-165.

Van Lochem E, De Gast B and Goulmy E (1992). In vitro separation of host specific graft-versus-host and graft-versus-leukemia cytotoxic T cell activities. Bone Marrow Transplantation 10: 181-183.

Weiden PL, Flournoy N, Sanders JE, Sullivan KM, Thomas ED (1981a) Antileukemic effect of graft-versus-host disease contributes to improved survival after allogeneic marrow transplantation. Transplant. Proc. 18: 248-25'.

Weiden PL, Sullivan KM, Flournoy N, Storb R, Thomas ED (1981b) The Seattle marrow transplant team. Antileukemic effect of chronic graft-versus-host disease. N. Engl. J. Med. 304: 1529-1533.

Zier KS, Elkins WL, Pierson GR, Leo MM (1983) The use of cytotoxic T cell lines to detect the segregation of human minor alloantigen within families. Hum. Immunol. 7: 117-129.

Immune escape through T cell antagonism?

Paul Klenerman
David Lalloo
Rodney Phillips
Andrew McMichael

Molecular Immunology Group, Institute of Molecular Medicine, John Radcliffe Hospital, Oxford OX3 9DU, UK.

INTRODUCTION

A key question in HIV pathogenesis is how the virus survives in the face of active immunity. In other viral infections a critical mechanism in control and clearance of the virus is thought to be cellular immunity mediated by CD8 positive cytotoxic T lymphocytes (CTL). These cells recognise and destroy virally-infected cells - the recognition occurring through specific T cell receptors (TCRs). The latter interact with viral antigens presented to the immune system bound to MHC class I molecules on the cell surface (Townsend et al, 1984). Specific antiviral CTL activity is observed very early in HIV infection, prior to the appearance of neutralizing antibody, and the rise in CTL activity correlates with the initial control of viraemia (Koup et al, 1994). However, thereafter, although CTL levels remain elevated to an unusually high level, the virus persists and continues to replicate (Walker et al, 1987).

A feature of HIV which has attracted attention in this respect is its capacity for variation. This occurs at an extremely high rate as a result of the error-prone HIV reverse transcriptase, which lacks proof-reading capacity (Bebenek et al, 1989). Thus in an individual with an established infection the circulating virus is not monomorphic but instead is a swarm of closely-related viruses or "quasispecies". It has been proposed that some of these viruses may possess features which give them a relative advantage in the face of host immune defence. A process of evolution may therefore lead to the selection of 'escape mutants" (Phillips et al, 1991).

NATO ASI Series, Vol. H 94
Gene Technology
Edited by A. R. Zander, W. Ostertag,
B. V. Afanasiev, F. Grosveld

CTL IN HIV INFECTION

Antigen presentation in HIV

Mechanisms of immune escape in HIV are best understood in terms of the viral life-cycle and the antigen processing pathway. Virus enters the host target cell (CD4 positive lymphocyte or monocyte/macrophage) and the first stages thereafter involve reverse transcription of RNA into DNA, which must be transported into the nucleus for integration. Upon activation, virus transcription occurs, ultimately leading to synthesis of new viral structural proteins encoded in gag, pol and env which form new virions at the cell surface. Some nascent protein is side-tracked, however, into the cell's proteolytic pathways where they are broken down into small peptide fragments of 8-10 amino acids in length, representing CTL epitopes. These are passed through peptide transporters into the cell endoplasmic reticulum (ER) where they may bind MHC class I molecules. This binding is highly specific, dependent on the presence of "anchor" residues which can interact with the polymorphic antigen binding groove of MHC class I molecules. Stable complexes of peptide, MHC and beta-2-microglobulin are transported to the cell surface where it may be recognised by circulating CTL bearing the appropriate TCR.

Potential mechanisms of immune escape

Given the above scheme of antigen processing and presentation, virus could potentially escape CTL recognition by (see Fig 1):

a) remaining latent;

b) mutation in regions flanking an epitope therefore affecting proteolysis and generation of the appropriate peptide fragment (Koup, 1994);

c) mutation within the epitope affecting binding to the MHC molecule (ie the anchor residues) (Couillin I et al, 1994);

d) mutation within an epitope affecting TCR recognition (Phillips et al, 1991).

The latter could be of such an extreme degree that the TCR fails completely to interact with the MHC-peptide complex, and the epitope becomes effectively invisible to those CTL. Alternatively, the change induced may be more subtle, leading to a partial loss of recognition -

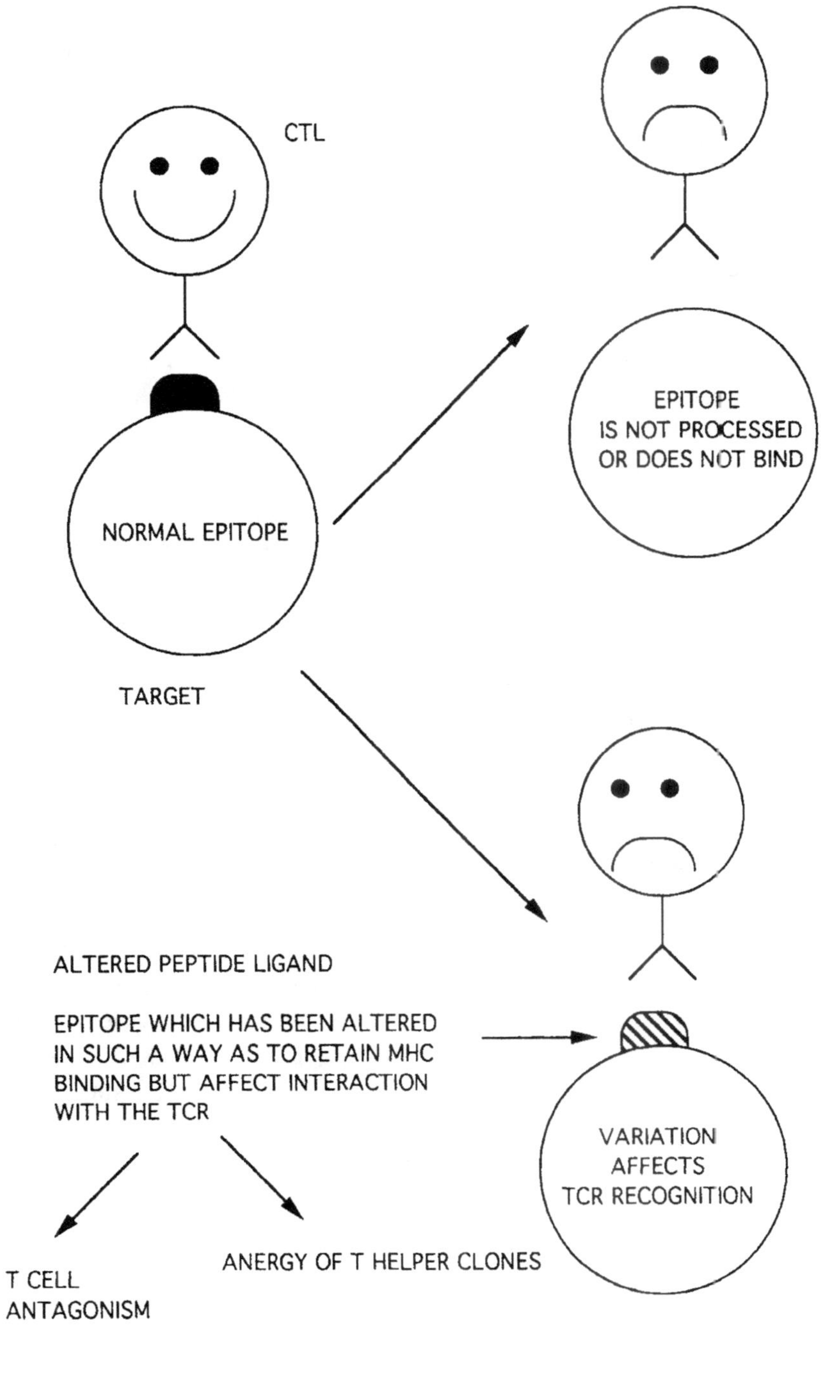

FIG 1 : MECHANISMS OF IMMUNE ESCAPE THROUGH VIRAL MUTATION

the new epitope representing a so-called "Altered Peptide Ligand" (Sette et al, 1994; Evavold et al, 1993). One important feature of APLs is their capacity to act as T cell antagonists.

Cytotoxic T Cell Antagonism in HIV Infection

T cell antagonism in context

The capacity of APLs to induce T cell antagonism was first observed in CD4 positive MHC class II restricted T cells where it was found that subtle changes in well-recognised epitopes led not only to loss of normal T cell proliferation but also to inhibition of normal T cell function in response to the original epitope (reviewed by Sette et al, 1994 and Evavold et al, 1993). This work was extended to CTL by Jameson and colleagues (1993) who documented inhibition of CTL responses to an epitope in ovalbumin when CTL were simultaneously exposed to mutated epitopes in the form of synthetic peptides. The working model for this observation was that TCRs may interact with the APL but that the signal received is either qualitatively or quantitatively different, so that the effector cell is only partially activated. In the case of CTL this may be insufficient to produce lysis of targets. Thus in the presence of mutant epitopes CTL TCRs may be occupied but fail to signal optimally, and what is observed is inhibition of cell lysis induced by the normal epitope.

T cell antagonism *in vitro*

Our investigations of the phenomenon of antagonism in HIV stem from the initial observation that mutations may arise and persist in HIV-1 gag CTL epitopes, and that these mutations may affect recognition by CTL (Phillips et al, 1991). For further work, we have concentrated particularly on the HLA-B8 restricted epitope in p17 gag (matrix) - GGKKKYKLK (p17-3). This has been also identified as a nuclear localisation signal for the virus (Bukrinsky et al, 1993).

Originally, it was not clear whether loss of CTL recognition reflected an effect on epitope binding MHC or TCR recognition, but two independent pieces of evidence suggested the latter;

a) in vitro assays of the capacity of variant p17-3 epitopes (observed in patients) to assemble complexes with MHC class I showed that 12/13 of these did so, all with approximately equal affinity (S McAdam et al, in press);

b) a panel of CTL clones raised against the epitope p17-3 showed cross-reactivity for some clones against all but one of the variants, confirming that the peptides must have the capacity to bind HLA B8.

Initial experiments using one of these variants GGKKKYRLK (p17-3 7**R**) indicated that exposure of CTL to this peptide inhibited lysis of cells bearing the normal epitope (Klenerman et al, 1994). This was not observed if an irrelevant peptide derived from influenza was used. However the effect was still observed if the peptide was presented to CTL on separate target cells to those bearing the normal epitope. This last result indicates that antagonism is not simply due to competition between peptides for the MHC molecules. It is also particularly relevant in HIV infection, where variant peptides are likely to be presented on adjacent cells rather than the same cell.

The next set of observations relate to the capacity of a patient's own set of variant epitopes to affect recognition by their own CTL. Virus was amplified from patient peripheral blood cells or circulating virions using PCR and sequenced. We have now studied several epitopes in HIV-1 gag and more recently in reverse transcriptase and used synthetic peptides representing these epitopes in assays of CTL antagonism in an epitope in p24 gag restricted by HLA B8 (P24-13), using a variant found in an infected donor. These results showed that the patients' own variants are potent antagonists of their own CTL, a result which has been paralleled in studies of CTL in Hepatitis B infection (Bertoletti et al, 1994).

T cell antagonism *in vivo* ?

To address the questions as to the potential *in vivo* relevance of this phenomenon, experiments were performed using naturally-processed epitopes as opposed to synthetic peptides. Target cells in these experiments were infected with either replicating HIV-1 (MN strain) or recombinant vaccinia viruses encoding p17 gag with a mutant p17-3, we observed that naturally processed mutant epitope may act as an antagonist. Finally, these results have been confirmed in freshly separated lymphocytes from infected donors, not subjected to culture or stimulation - this representing, presumably, the natural antiviral CTL repertoire of these individuals.

Although these results indicate that such a mechanism may be potentially relevant in vivo the question remains as to how important it might be, particularly as multiple epitopes may be simultaneously available for recognition. However, it does appear that certain epitopes may be immunodominant, at least for periods, in which case such activity may have relatively more influence (Nowak M et al, 1995).

Mechanisms

The mechanism of CTL antagonism is not understood. In CD4 positive T helper cells some APL lead to qualitative changes in T cell function including total anergy (Sloan-Lancaster et al, 1994). Such observations have led to the idea that a qualitative change in signalling occurs. Consistent with this is the observation that some naturally-occurring antagonists are active at low antagonist:agonist ratios. On the other hand a quantitative change in TCR signalling may account for some of the phenomena observed, although direct measurements of TCR affinities and binding kinetics are awaited. One feature of HIV APLs is their capacity to induce slow killing of targets, visible only in extended assays. This is consistent with the idea that T cells are receiving a positive stimulus, but that this is weaker than normal. Furthermore, anergy is not observed in CTL after exposure to HIV APLs; indeed they may act to stimulate CTL growth over the long term (P Klenerman et al, 1995). The effect of HIV APL on antiviral CD4 positive T helper cells in infected patients will prove an interesting area for future research.

CONCLUSIONS

The data presented at this meeting support the idea that CTL antagonism is potential mechanism of immune escape in HIV-1 infection. The potency of this mechanism is that if an antagonistic mutant were to arise, it could lead to reduced CTL lysis of all virally-infected cells not merely those infected with that variant. If variants were to induce long term anergy, as in the case of CD4 positive T helper clones the effects could be even more pronounced and sustained. The mechanism need not merely be restricted to viral infections such as HIV and Hepatitis B, but could be active in any infection where there is persistence of the pathogen, together with some capacity for strain variation. Cellular immunity directed against tumours may also be subject to such inhibition. Finally, we could learn a lesson from retroviruses, and the use of APLs to turn off immune responses in situations where they are harmful (eg in auto-immune diseases) clearly warrants further research.

References

Bebenek et al (1989) Specificity and mechanism of error-prone replication by HIV-1 reverse transcriptase. J Biol Chem **264:** 16948-56.

Bertoletti et al (1994) Natural variants of cytotoxic epitopes are TCR antagonists for antiviral CTL. Nature **369:** 407-11.

Bukrinsky M et al (1993) A nuclear localisation signal within HIV-1 matrix protein that governs infection of non-dividing cells. Nature **365:** 666-9.

Couillin I et al (1994) Imparied CTL recognition due to genetic variations in the main immunogenic region of the HIV-1 nef protein. J Exp Med **180:** 1129-34.

Evavold B et al (1993) Tickling the TCR. Immunol today **14:** 602-6.

Jameson SC et al (1993) Clone-specific TCR antagonists of MHC class-I restricted CTL. J Exp Med **177:** 1541.

Klenerman P et al (1994) CTL activity antagonised by Naturally-occurring HIV-1 gag variants. Nature **369:** 403-7.

Klenerman P et al. (1995) Effects of Natural APL on the whole blood response to HIV-1. **Eur J Immunol.** (In press)

Koup R et al (1994) Temporal association of cellular immune responses with control of viraemia in primary HIV-1 syndrome. J Virol **68:** 4650.

Koup R (1994) Virus escape from CTL recognition. J Exp Med **180:** 779-82.

Phillips RE et al (1991) HIV genetic variation that can escape CTL. Nature **354:** 453.

Sette A et al (1994) Antigen analogs/MHC complexes as specific T cell receptor antagonists. Ann Rev Immunol **12:** 413-31.

Sloan Lancaster J et al (1994) TH2 clonal anergy as a consequence of partial activation. J Exp Med **180:** 1195-1205.

Townsend A et al (1984 CTL recognition of the Influenza nucleoprotein and Haemaglutinin expressed in transfected mouse L cells. Cell **39:** 13-22.

Walker BD et al (1987) HIV specific CTL in seopositive individuals. Nature **328:** 345.

Acknowledgements

This work was supported by the Medical Research Council (UK) and the Wellcome Trust. We are grateful to Ute Meier, Sarah Rowland-Jones, Steve McAdam, Britta Köppe, Jon Edwards, Susan Daenke, William Rosenberg, Diana Boyd, Ann Edwards, Graz Luzzi and Paul Giangrande for their work towards this project.

The privilege of expressing CD44v - used by immune cells, misused by tumour cells.

J. Moll, M. Plug, H. Ponta and P. Herrlich
Institut für Genetik
Kernforschungszentrum Karlsruhe
Postfach 3640
76021 Karlsruhe
Germany

ABSTRACT

Transgenic mice were generated which express CD44v4-v7 on the surface of 20-80% of their T cells, depending on which lymphoid organ the T cells were derived from. Soluble CD44(v4-v7) is present in their serum, showing that, like CD44s, CD44v can be shedded from the cell surface. The incidence of chemically induced lymphomas in these animals was not increased as compared to control mice, nor were the tumours more aggressive. The importance of CD44v in malignancies and during early stages of hemopoietic differentiation are discussed.

INTRODUCTION

CD44 designates a family of surface glycoproteins believed to be involved in the cell-cell and cell-matrix adhesion. The latter can be mediated by the binding of CD44 to the extracellular matrix component hyaluronic acid (for review see Sherman et al., 1994). The smallest isoform (CD44s) is expressed in a wide variety of tissues and cell types. Variant CD44 isoforms (CD44v) are generated

NATO ASI Series, Vol. H 94
Gene Technology
Edited by A. R. Zander, W. Ostertag,
B. V. Afanasiev, F. Grosveld

by the insertion of additional sequences in the extracellular part of CD44s via alternative splicing of 10 variant exons in different combinations (for review see Herrlich et al., 1993; Lesley et al.,1993; Screaton et al.,1993). The expression of CD44v is restricted to a few cell types of the hemopoietic system and certain regenerating epithelia such as skin and intestine (Heider et al., 1993; Fox et al., 1994; Terpe et al., 1994). The isoform CD44v4-v7 was originally identified on the surface of a highly metastatic rat pancreatic adenocarcinoma cell line (BSp73ASML), but was not expressed on a locally growing cell line (BSp73AS) derived from the same tumour. The transfection of CD44v4-v7 cDNA into BSp73AS cells induced metastatic spreading and metastasis formation was abolished by the administration of monoclonal antibodies directed against an exon v6 encoded epitope (Günthert et al., 1991). A correlation between increased malignancy and expression of CD44v in human tumours was also found in non-Hodgkin´s lymphoma, mammary and colon carcinoma. In these cases, CD44v is hardly dectectable in normal tissue but appears on the surface of tumour cells. In the case of mammary tumours, the patient´s overall survival correlates inversely with the presence of CD44 variant encoded epitopes on the surface of tumour cells (Sinn et al., 1994; Sleeman et al., 1994). In non Hodgkin´s lymphoma the v6 expression is elevated in several agressive cases (Koopman et al., 1993). Additionally, a stage specific increase of CD44v6 in colorectal carcinomas was found (Heider et al., 1993). In a second type of CD44v expression pattern found in tumours, CD44v epitopes are present in normal tissue but are combined in a different way in neoplastic lesions. The latter is true in the case of cervical carcinomas, where new splice variants which combine CD44v exon v7 and v8 are expressed (Dall et al., 1994). These findings make CD44v a promising candidate as a tumour progression marker useful in diagnosis and therapy.

As CD44 possesses putative protease recognition sites within the extracellular part, the molecule can be shedded from the cellular surface (Bazil and Horejsi, 1992). This property may have some diagnostic value in the case of gastric and colon carcinomas (Guo et al., 1994), and certain malignant lymphomas (Ristamaki et al.,

1994) where high molecular weight soluble isoforms of CD44 were elevated in serum samples of patients with advanced stages of cancer, as compared to controls. Also, the appearance of exfoliated cells which express CD44v on their surface in the urine of patients suffering from bladder carcinoma could be useful as a diagnostic tool (Matsumura et al., 1994).

Functions for CD44 during embryogenesis as well as in the adult are likely. One example of a physiological function for CD44v is suggested by the observation that surface expression of CD44v6 is transiently upregulated on activated B cells, T cells and macrophages (Arch et al., 1992, Koopman et al., 1993). Activation could be inhibited by a monoclonal antibody recognizing a CD44v6 epitope, suggesting a direct requirement of CD44v6 for the establishment of an immune response. In order to study the functional role of CD44v in the immune system, we created several transgenic mice lines over expressing CD44v4-v7 on the surface of T cells. These animals were investigated with respect to the function of CD44v4-v7 in the immune system and its role during tumour progression. These animals will also be useful to test therapeutic regimes for cancer and autoimmune diseases.

RESULTS and DISCUSSION

To direct the expression of the rat CD44v4-v7 cDNA within the T cell lineage, the transgene was expressed under the control of the Thy-1 promoter and the immunoglobulin heavy chain enhancer which replaced the endogenous brain specific enhancer (Vidal et al., 1990; see Fig. 1). This combination resulted in the expression of the transgene in 20-80 % of T cells, depending on which lymphoid organ the T cells were derived from. In different lymphoid organs the relative proportion of T cells expressing the transgene was comparable to the expression of another unrelated transgene (CD23), expressed under the same promoter enhancer combination (Texido et al., 1994). Obviously, the expression of CD44v has no influence on lymphocyte recirculation. A role for CD44 during lymphocyte homing has been suggested by the finding that Hermes-1, a mab recognizing the hyaluronate binding site of CD44, is able to block lymphocyte extravasation in specific lymphoid organs such as Peyers Patches (Jalkanen et al., 1987).

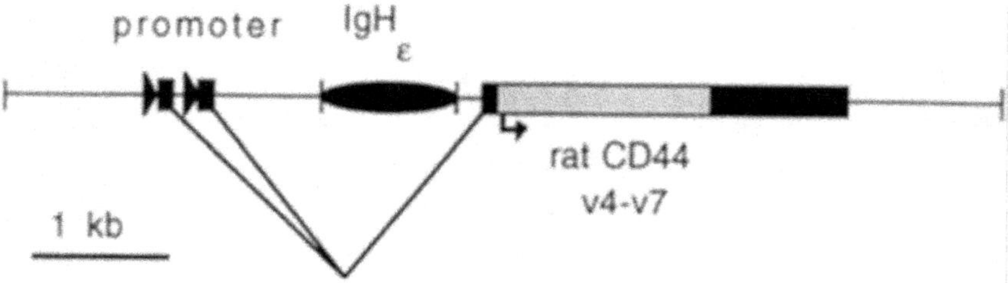

Fig 1: Expression of the transgene (rat CD44v4-v7) is mediated by the Thy-1 promoter in combination with the Ig heavy chain enhancer element.

The high expression of the transgene on the surface of T cells in transgenic animals provided us with an animal model to investigate shedding of CD44v. We therefore investigated whether shedded CD44v is present in serum from transgenic mice and, as a control, from non transgenic mice. Elisa assays using the monoclonal antibody 1.1 ASML, specific for rat CD44v6, showed that soluble CD44v is present in the serum of transgenic mice. Soluble CD44 was not detectable in the serum of non transgenic mice (Fig. 2). Estimations of the amount of soluble CD44 in the transgenic mice, using recombinant CD44 fusion protein as a standard, suggests a range of 2-9 µg/ml serum, with considerable variations between individual animals. Recently, a correlation of serum CD44 levels in with response to treatment in patients with malignant lymphomas could be shown (Ristamaki et al., 1994). The serum CD44 level was highly elevated in all patients with lymphoma and, isoforms containing v6 epitopes were also slightly increased. After treatment, patients with complete remission had similar CD44 serum levels as the controls. These results indicate that shedding of CD44 may be a useful diagnostic tool in monitoring both progression of certain tumours and the response to treatment of patients.

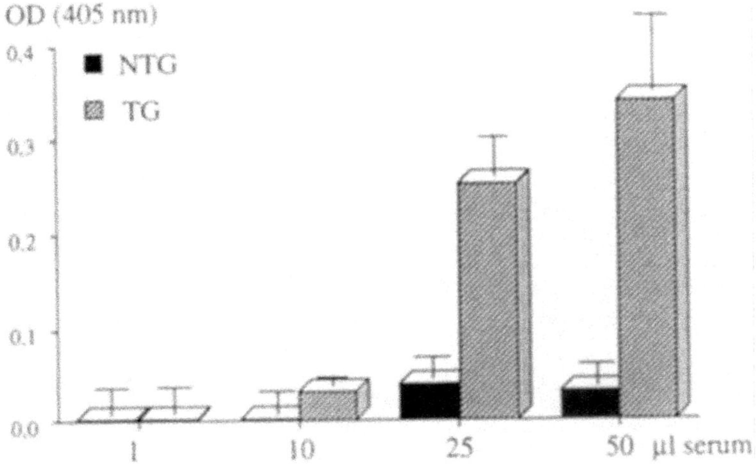

Fig 2: Detection of soluble CD44v4-v7 in the serum of transgenic (TG) and non transgenic (NTG) mice by Elisa. Plates were coated with OX 50, a monoclonal antibody recognising an epitope located within the standard portion of rat CD44. Bound CD44v4-v7 was detected using 1.1ASML, a monoclonal antibody against rat CD44v6. The values (measured OD minus background staining) represent the mean of three mice in duplicate. Three independent experiments gave similar results.

Phenotypic changes are observed upon immunological challenge of the transgenic animals as compared to their nontransgenic litter mates. In vitro as well as in vivo, the CD44(v4-v7)-transgenic T cells responded faster to immunological stimulation when activated by Concanavalin A or during an allogenic immune response (Moll et al., in preparation). This accelerated immune response, particularly during a primary stimulation, is a consequence of the presence of the transgene product on the surface of T cells. The presence of the monoclonal antibody 1.1ASML during immunological stimulation abolished the acceleration. Thus the transgene confers a proliferative advantage on T cells during early activation. Also, an accelerated repopulation of lymphoid organs after bone marrow transplantation could be observed. In this experiment, bone marrow derived from transgenic or control animals was grafted into lethally irradiated syngenic mice and the repopulation kinetics of both groups was compared. An explanation for the observed accelerated repopulation potential of bone marrow cells derived from transgenic animals could be the expression of low levels of the transgene in hemopoietic stem cells, resulting in an enhanced migration and/or production of progenitor cells within the bone marrow. Indeed, in the mouse, hemopoietic stem cells are known to express low levels of Thy-1 (Basch and Berman, 1982). Alternatively, the accelerated repopulation could also be explained by an enhanced proliferation of more mature lymphoid cells within the other lymphoid compartments during repopulation. In order to test these hypotheses we have started to examine immortalised hemopoietic stem cell lines, as will be discussed later.We also observed a reduced wound healing ability in older transgenic animals (our unpublished observations). This may be caused by an imbalanced T cell response which finally results in tissue destruction due to autoimmunity.

The proposed role of CD44v during tumour progression was tested by inducing thymic lymphomas after exposure to N-methyl-N-methylnitrosourea (MNU; Dumenco et al., 1993). Transgenic and non-transgenic mice were injected i.p. with a single dose of 50 mg/kg MNU. Most of the animals developed lymphomas. The majority of these lymphomas were of T cell origin resulting in a thymic hyperplasia in 70 % of injected animals. Approximately 50 % of the animals in each group developed extrathymic hyperplasias in lymph nodes and spleen and 25 % showed nodules in lung, kidney or liver. Most lymphomas were of thymic origin as was shown by the simultaneous expression of CD4 and CD8 surface markers detected by flow cytometry. No significant differences in the migratory behaviour of the tumour cells and in the overall survival of the mice were observed between transgenic mice and control siblings (Fig. 3).

These data argue against a role for CD44v during the progression of T cell lymphomas which originate in the thymus. In another T cell lymphoma animal model, T cell lymphomas with different leukemic potentials also showed no correlation of CD44 expression with spreading of the tumour cells (Dolcetti et al., 1994).

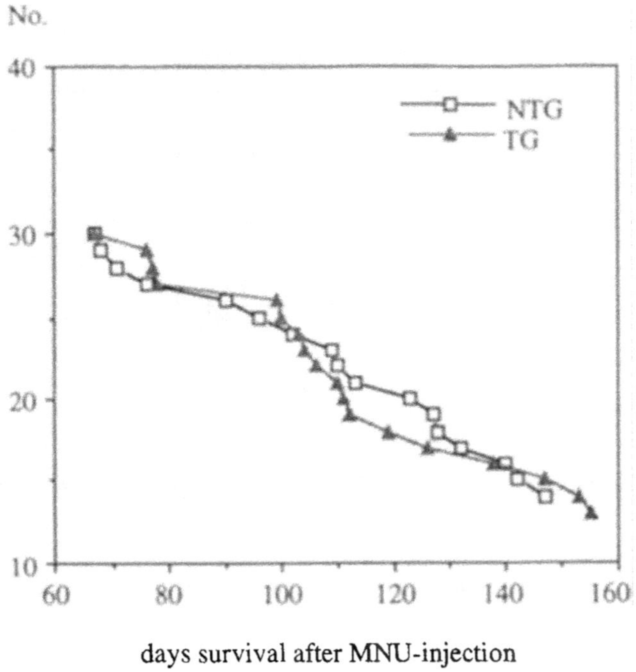

Fig. 3 Overall survival curve of TG and NTG after a single injection of MNU (50 mg/kg).

Interestingly, human leukemic cell lines which represent early stages in hemopoietic development display CD44v6 on their surface. These cell lines possess the capacity to differentiate into different hemopoietic lineages such as the erythroid, monocytic and megakaryocytic lineages upon treatment with TPA (12-O-tetradecanoylphorbol-13-acetate; Murate et al., 1993). The cell lines KG1 and TF1 also express the CD34 antigen on their surface which is a marker for primitive hemopoietic precursor cell types (Barrande et al., 1993; Simmons et al., 1992). The myelogeneous leukemia cell line K562 lacks CD34 expression on the surface, but has a translocation between chromosome 9 and 22 generating the

Philadelphia chromosome. Analysis of the expression of CD44 on the surface of KG1, TF1 and K562 cell lines showed differences in the expression patterns of CD44s and CD44v epitopes (Tab. 1).

cell line	CD44-epitope				
	st*	v5	v6	v7/8	v10
K562	+	nd	-	-	-
K562 + TPA	+++	nd	++	-	-
TF1	++	+	+	-	-
TF1 + TPA	+++	++#	+++#	-	-
KG1	++	nd	-	-	-
KG1 + TPA	+++	nd	+	-	-

15-30 % of cells * recognizes also CD44v

Table 1. Effect of TPA treatment on the expression of CD44 epitopes on the surface of the hemopoietic cell lines TF1, KG1 and K562 representing early differentiation stages. Data were obtained using flow cytometry and different monoclonal antibodies recognising human CD44v epitopes (Heider et al., 1993).

K562 cells express low levels of CD44s but no CD44v on their surface. Double fluorescence analysis, using fluorescence labeled hyaluronate and anti CD44 monoclonal antibodies, revealed that K562 cells are not able to bind hyaluronate, despite the presence of low levels of CD44s on their surface. This changed upon treatment of the K562 cells with TPA, which is known to induce differentiation. TPA treatment resulted in an approximately 20-fold induction of the CD44 expression. The majority of these CD44 molecules contain the CD44v6 epitope. In addition, K562 cells were activated to bind hyaluronate. The changed CD44 expression and hyaluronate binding properties could be induced by a short pulse (1h) of TPA and was detectable even several days after induction, although it was not detectable after 1 h . This observation suggests that the TPA-induced expression of CD44 is not directly mediated by effects of TPA on the CD44 promoter, but rather reflects a differentiation dependent event. Similar results were observed with the KG1 cell line, which constitutively expresses higher levels of CD44s in the uninduced state, compared to K562 cells. The TF1 cell line already expresses low levels of CD44v6 when uninduced, and only 15-30% of TF1 cells responded to TPA induction by upregulation of CD44 v5 and v6. We are currently checking the capacity of primary CD34 hemopoietic stem cells to express CD44v and to bind hyaluronate. A crucial role for CD44 during hemopoiesis is suggested by the observation that monoclonal antibodies recognizing CD44s are able to inhibit long term in vitro bone marrow cultures under conditions supporting lymphopoiesis and myelopoiesis (Miyake et al., 1990). We propose that during differentiation of hemopoietic stem cells, CD44 expression and the ability to bind hyaluronate is highly stage specific and mediates the interaction with stromal cells in the bone marrow. We are currently investigating this hypothesis by testing the adhesion and differentiation capacity of cell lines with hemopoietic precursor properties to bone marrow stromal cell lines.

REFERENCES

Arch R, Wirth K, Hofmann M, Ponta H, Matzku S, Herrlich P, Zöller M (1992) Participation in normal immune response of a metastasis-inducing splice variant of CD44. Science 257:682-685

Barrande C, Traore Y, Szekeres G, Hirn J (1993) Isolation and characterization of two new monoclonal antibodies against the CD34 molecule. Hybridoma 12:203-213

Basch RS and Berman JW (1982) Thy-1 determinants are present on many murine hematopoietic cells other than T cells. Eur J Immunol 12:359-364

Bazil V and Horejsi V (1992) Shedding of the CD44 adhesion molecule from leukocytes induced by anti-CD44 monoclonal antibody simulating the effect of a natural receptor ligand. J Immunol 149:747-753

Dall P, Heider KH, Hekele A, von Minckwitz G, Kaufmann M, Ponta H, Herrlich P (1994) Surface protein expression and messenger RNA-splicing analysis of CD44 in uterine cervical cancer and normal cervical epithelium. Cancer Res 54:3337-3341

Dolcetti R, Frisan T, Palmieri G, Rizzo S, Maestro R, Santoni A, Boiocchi M (1994) In vivo phenotypic characteristics of AKR T-cell lymphomas with different leukemic potential: possible role of alpha 4 beta 7 integrin in the progression towards the leukemic phenotype. Int J Cancer 56:560-567

Dumenco LL, Allay E, Norton K, Gerson SL (1993) The prevention of thymic lymphomas in transgenic mice by human O6-alkylguanine-DNA alkyltransferase. Science 259:219-222

Fox SB, Fawcett J, Jackson DG, Collins I, Gatter KC, Harris AL, Gearing A, Simmons DL (1994) Normal human tissue, in addition to some tumors, express multiple different CD44 isoforms. Cancer Res 54:4539-4546

Günthert U, Hofmann M, Rudy W, Reber S, Zöller M, Hausmann I, Matzku S, Wenzel A, Ponta H, Herrlich P (1991) A new variant of glycoprotein CD44 confers metastatic potential to rat carcinoma cells. Cell 65:13-24

Guo YJ, Liu G, Wang X, Jin D, Wu M, Ma J, Sy MS (1994) Potential use of soluble CD44 in serum as indicator of tumor burden and metastasis in patients with gastric or colon cancer. Cancer Res 54:422-426

Heider KH, Hofmann M, Horst E, Berg F van den, Ponta H, Herrlich P, Pals ST (1993) A human homologue of the rat metastasis-associated variant of CD44 is expressed in colorectal carcinomas and adenomatous polyps. J Cell Biol 120:227-233

Herrlich P, Zöller M, Pals ST, Ponta H (1993) CD44 splice variants: metastases meet lymphocytes Immunology Today 14:395-399

Jalkanen S, Bargatze RF, de los Toyos J, Butcher EC (1987) Lymphocyte recognition of high endothelium: antibodies to distinct epitopes of an 85-95-kD glycoprotein antigen differentially inhibit lymphocyte binding to lymph node, mucosal, or synovial endothelial cells. J Cell Biol 105:983-990

Koopman G, Heider KH, Horst E, Adolf GR, Berg F van den, Ponta H, Herrlich P, Pals ST (1993) Activated human lymphocytes and agressive non-Hodgkin lymphomas express a homologue of the rat metastasis-associated variant of CD44. J Exp Med 177:897-904

Lesley J, Hyman R, Kincade PW (1993) CD44 and its interaction with the extracellular matrix. Adv Immunol 54:271-335

Matsumura Y, Hanbury D, Smith J and Tarin D (1994) Non-invasive detection of malignancy by identification of unusual CD44 gene activity in exfoliated cancer cells. BMJ 308:619-624

Miyake K, Medina KL, Hayashi S, Ono S, Hamaoka T, Kincade PW (1990) Monoclonal antibodies to Pgp-1/CD44 block lympho-hemopoiesis in long-term bone marrow cultures. J Exp Med 11, 477-488

Murate T, Saga S, Hotta T, Asano H, Ito T, Kato K, Tsushita K, Kinoshita T, Ichikawa A, Yoshida S, Saito H (1993) The close relationship between DNA replication and the selection of differentiation lineages of human erythroleukemia cell lines K562, HEL, and TF1 into either erythroid or megakaryocytic lineages. Exp Cell Res 208:35-43

Ristamaki R, Joensuu H, Salmi M, Jalkanen S (1994) Serum CD44 in malignant lymphoma: an association with treatment response. Blood 84:238-243

Screaton GR, Bell MV, Bell JI, Jackson DG (1993) The identification of a new alternative exon with highly restricted tissue expression in transcripts encoding the mouse CD44 homing receptor. J Biol Chem 268:12235-12238

Sherman L, Sleeman J, Herrlich P, Ponta H (1994) Hyaluronate receptors: key players in growth, differentiation, migration and tumor progression. Current Opin Cell Biol 6:726-733
Sleeman JP, Moll J, Sherman L, Dall P, Pals ST, Ponta H, Herrlich P (1995) Cell adhesion and human disease. Wiley, Chichester (Ciba Foundation Symposium 189) 142-156
Simmons DL, Scatterthwaite AB, Tenen, DG and Seed B (1992) Molecular cloning of a cDNA encoding CD34, a sialomucin of human hematopoietic stem cells. J. Immunol. 148:267-271
Sinn H-P, Heider K-H, Skroch-Angel P, von Minckwitz G, Kaufmann M, Herrlich P, Ponta H (1994) Human mammary carcinomas express homologues of metastasis-associated variants of CD44. Breast Cancer Res Treat in press
Terpe HJ, Stark H, Prehm P, Günthert U (1994) CD44 variant isoforms are preferentially expressed in basal epithelia of non-malignant human fetal and adult tissue. Histochem 101:79-89
Texido G, Eibel H, Le Gros G, van der Putten H (1994) Transgene CD23 expression on lymphoid cells modulates IgE and IgG1 responses. J Immunol. 153:3028-3042
Vidal M, Morris R, Grosveld F, Spanopoulou E (1990) Tissue specific control elements of the Thy-1 gene. EMBO J. 9:833-840.

Transplantation Immunology: Peptides in the groove cause histoincompatibility

H.J. Stauss, C. Thomas, P. Nathan, E. Sadovnikova, O. Neth and L. Yin

Imperial Cancer Research Fund, Tumour Immunology Unit, 91 Riding House Street, London WC1 8BT, United Kingdom.

Summary:

Organ and bone marrow transplantation is frequently complicated by immune responses against major and minor histocompatibility antigens. This article will discuss current concepts of the molecular nature of histocompatibility antigens. The functional units that trigger immune responses by T lymphocytes are peptides bound to the peptide binding groove of major histocompatibility complex (MHC) class I and class II molecules. In MHC mismatched transplant situations, T lymphocytes recognise donor MHC molecules displaying donor derived peptides. In MHC matched transplant situations, T cell immunity is directed against donor-specific peptides bound to the groove of MHC molecules. These donor-specific peptides are derived from polymorphic cellular proteins, named minor histocompatibility (mH) antigens. We propose that immune responses against mismatched MHC antigens and mH antigens are probably directed against a small number of immunodominant peptides. In future, it may become possible to exploit such peptides for tolerance induction , which may lead to specific downregulation of unwanted T cell responses in organ and bone marrow transplantation.

Antigen Recognition by CD4+ T helper (Th) lymphocytes and CD8+ cytotoxic T lymphocytes (CTL):

T lymphocytes do not recognise protein antigens directly, but rather peptide fragments which are presented by MHC class I and class II molecules. Peptides presented by class I molecules are recognised by

NATO ASI Series, Vol. H 94
Gene Technology
Edited by A. R. Zander, W. Ostertag,
B. V. Afanasiev, F. Grosveld

CD8 expressing CTL. The antigen receptor (TCR) of CTL recognises peptides bound to the peptide binding groove of class I molecules, which is formed by the membrane distal α1 and α2 domains (Bjorkman, *et al.*, 1987a; Bjorkman, *et al.*, 1987b). Simultaneously, the CD8 molecule interacts with the membrane proximal class I α3 domain (Salter, *et al.*, 1989). Similarly, the TCR of Th cells recognises peptides bound to a groove formed by the membrane distal class II α1 and β1 domains (Brown, *et al.*, 1993), while the CD4 molecule interacts with the membrane proximal β2 domain. The mechanism of MHC restricted peptide recognition by CTL and Th lymphocytes applies for the majority of known immune responses except for superantigens, which bind to the side of MHC class II molecules and cross-link TCRs of CD4 Th cells expressing certain β-chain variable gene segments (Choi, *et al.*, 1990; Karp and Long, 1992). MHC unrestricted recognition by CTL has also been observed with antigens that contain repeated epitopes (Barnd, *et al.*, 1989; Jerome, *et al.*, 1991). It is possible that multivalent antigens sometimes can interact with several MHC molecules and cross-link TCRs in an unconventional fashion, leading to MHC unrestricted T cell recognition. Since there is currently no evidence of unconventional T cell recognition of allogeneic MHC and mH antigens, it is most likely that allo-recognition follows a conventional scheme involving at least four players: the TCR, the accessory molecules CD4 or CD8, the MHC molecules and the peptides bound to the MHC groove.

The peptides presented by MHC class I and class II molecules derive from distinct protein sources. Although this distinction is not absolute, class I molecules prefer peptides derived from cytosolic proteins, whereas class II molecules preferentially present peptides derived from secreted or transmembrane cell surface proteins. The MHC class I presentation pathway (Monaco, 1992) starts with cytosolic protein breakdown by the high molecular weight proteasomes consisting of multiple subunits, two of which are encoded by genes in the MHC region. Although there is currently no experimental evidence that other proteases generate peptides for MHC class I presentation, it is most likely that some proteasome independent pathway for peptide production does exists. Peptides produced in the cytosol are then transported into the lumen of the endoplasmic reticulum (ER) by two proteins called TAP1 and TAP2 (Deverson, *et al.*, 1990; Monaco, *et al.*,

1990; Trowsdale, *et al.*, 1990), which associate to form functional peptide transporter complexes. Inside the ER, peptides bind to newly synthesised MHC class I molecules which leads to the formation of stable MHC/peptide complexes which then travel to the cell surface.

In contrast, MHC class II molecules bind peptides in a recently identified class II compartment (Amigorena, *et al.*, 1994), but not in the ER. In the ER, newly synthesised MHC class II molecules bind to invariant chain which prevents binding of ER resident peptides, and which directs migration of class II molecules to the class II compartment. Here, invariant chain is cleaved and the MHC binding groove becomes accessible for peptide binding. The class II compartment also contains the recently identified DMA and DMB molecules (Sanderson, *et al.*, 1994), which are required for antigen presentation via the class II pathway. Several functions for the DM molecules have been proposed including removal of the invariant chain derived clip peptides, delivery of peptides which can bind to class II, and induction of conformational changes in class II molecules that allow peptide binding. The peptides which are available for class II binding are primarily produced by proteases in the endo-lysosomal compartment. Since this compartment preferentially contains endocytosed proteins derived from the cell surface or from the extracellular environment, most of the MHC class II presented peptides originate from these proteins.

In general, the pathways which produce MHC class I and class II presented peptides cannot discriminate between self and non-self proteins, although interferon-γ induced up-regulation of processing molecules during viral infections may lead to preferential presentation of viral peptides in infected cells. Under physiologic conditions, MHC class I and class II molecules are constitutively loaded with peptides derived from self proteins. Therefore, MHC molecules expressed by normal cells are not homogeneous, but contain large numbers of distinct peptides. For the immunobiology of tissue transplantation it is important that the T cell compartment has been rendered tolerant to complexes consisting of self MHC molecules containing self peptides. In a transplant situation immune responses can be triggered by donor MHC molecules containing donor peptides, and by syngeneic MHC molecules displaying donor peptides. Complexes of donor MHC plus donor peptides are encountered in MHC mismatched situations and trigger immune responses by allo-MHC-specific T lymphocytes. Syngeneic MHC plus donor peptides are encountered in MHC matched situations and trigger immune responses against mH antigens.

Immunobiology of allo-MHC antigens:

MHC molecules are encoded by genes of the MHC cluster which maps to chromosome 6 in man and chromosome 17 in mouse. MHC class I and class II genes are extremely polymorphic, and the proteins encoded by these genes stimulate strong immune responses by T lymphocytes in MHC mismatched tissue transplantation. Allogeneic MHC class I and class II antigens are extremely immunogenic as determined by in vivo and in vitro assays. In vivo models in mice show that allogeneic MHC molecules can cause rapid skin graft rejection within 10-14 days after transplantation. In vitro, T cell responses against allo-MHC antigens are readily detected in naive T cell populations, and limiting dilution experiments have shown T cell precursor frequencies as high as 1/100 to 1/1000. Two models have been proposed to explain the strong immunogenicity of allogeneic MHC antigens. The high ligand density model (Bevan, 1984) postulates that allogeneic class I and class II antigens stimulate strong T cell responses because they are expressed at much higher levels than 'conventional' antigens. The model assumes that high ligand density not only stimulates high affinity T lymphocytes but also T cells expressing low affinity TCRs, while in 'conventional' immune responses against antigens which are presented at low density, high affinity T lymphocytes are stimulated exclusively. Thus, in the high ligand density model the high T cell precursor frequencies against allogeneic MHC molecules are explained by the recruitment of T cells expressing high as well as low affinity TCRs.

Alternatively, it is possible that T cell responses against allogeneic MHC molecules are directed against a large number of distinct peptides bound to the groove of MHC molecules. Since the T cell compartment is tolerant only to complexes composed of self MHC plus self peptide, it is conceivable that all complexes composed of non-self MHC plus non-self peptide might stimulate T cell responses. Therefore, in the multiple ligand model (Matzinger and Bevan, 1977) it is postulated that allo-specific T cell responses are directed against a large number of distinct MHC/peptide complexes expressed on the surface of allogeneic cells. Recent studies have estimated that up to 10 000 distinct peptides derived from cellular proteins may be displayed on the cell surface by MHC class I molecules (Hunt, *et al.*, 1992). This observation is consistent with the suggestion that a large number of distinct MHC/peptide complexes are involved in T cell responses against allogeneic MHC molecules, which might account for the high precursor frequency.

Immunobiology of mH antigens:

Although loci encoding mH antigens have been identified and mapped to many different chromosomes, respective genes have not yet been isolated in man. Based on classical genetic studies of recombinant inbred mice, estimates of the number of mH antigens range from fifty to several hundred (Bailey and Mobraaten, 1969). The maternally transmitted antigen (Mta) and the myxovirus resistance protein (Mx) are the only mH antigens that have been molecularly identified in mouse. Mta is derived from the mitochondrially encoded ND1 protein (Loveland, *et al.*, 1990). Like in prokaryotes, proteins synthesised in mitochondria contain a N-formylmethionine at their amino-terminus. Peptides corresponding to the first 12 amino acids of ND1 are presented by non-classical H-2M3 class I molecules to CTL. These non-classical class I molecules have apparently evolved to present bacterial and mitochondrial antigens, since they preferentially bind peptides with formylated N-termini. The only non-formylated N-terminal amino acid that showed some binding to H-2M3 was glycine . The second molecularly identified murine mH antigen is encoded by a gene (Mx) that determines susceptibility or resistance to infection by myxovirus (Speiser, *et al.*, 1990). The Mx gene is not present in all mouse strains and it was shown that Mx negative mice rejected skin grafts from Mx positive donors, which correlated with the detection of Mx-specific CTL responses in vitro.

Skin transplantation experiments in mice have shown that graft rejection caused by mH antigens is slower than rejection caused by allo-

MHC antigens (Graff and Bailey, 1973). However, some mH antigens (e.g. the murine H-1 antigen) are nearly as immunogenic as MHC antigens as judged by the kinetics of graft rejection (Graff, *et al.*, 1966). Nevertheless, in general immune responses to mH antigens are weaker than responses to allo-MHC antigens. In vitro, T cell responses against mH antigens are usually not detectable in naive T cell populations, and detection of anti-mH responses requires prior in vivo immunisation.

In mice, mH antigens are defined as genetic loci which can cause skin transplant rejection when donor and recipient express the same MHC molecules. Although the composition of mH loci has not yet been fully dissected, it is likely that several genes are present in one locus. Since mH antigens usually induce responses by Th lymphocytes and by CTL, it has been suggested that distinct genes encode for antigens that are recognised by these T lymphocyte populations (Roopenian, 1992). One gene may encode a protein that is processed via the class II pathway to generate peptides that are presented by MHC class II molecules, while another gene may encode a protein that is channelled into the class I processing pathway to produce peptides that are presented by MHC class I molecules. Stimulation of both MHC class II restricted Th cells and MHC class I restricted CTL is probably required to cause graft rejection, while stimulation of only one T cell population may be insufficient. However, it does not necessarily follow that two distinct genes are required for stimulation of Th lymphocytes and CTL. For example, studies of T cell responses against human melanomas by Rosenberg's group have shown that both MHC class II and class I presented peptides were derived from tyrosinase (Topalian, *et al.*, 1994). It is therefore possible that a single gene can encode a protein that contains Th as well as CTL epitopes. Consequently, not all mH loci may contain multiple genes, although this has been shown to be the case for the murine mH loci H-3, H-4 and HY (Roopenian, 1992).

Immunodominance:

Immunodominance has been documented most extensively by E. Sercas and colleagues (Sercarz, *et al.*, 1993), who studied Th responses to hen egg lysozyme (HEL). Immunisation of mice with HEL stimulates strong Th responses against an immunodominant peptide, and responses to subdominant peptides are not detected. However, responses to these subdominant epitopes are readily detectable when immunodominant epitopes are absent during immunisation. Although immunodominance has not been demonstrated as rigorously for CTL responses, there is considerable evidence that it also applies. Firstly, CTL responses in mice infected with influenza virus, vesicular stomatitis virus, Sendai virus and lymphocytic choriomeningitis virus are directed against one or few viral peptides although these viruses express several proteins. Similarly, CTL from H-2^b mice immunised with ovalbumin always recognise an immunodominant 8mer peptide presented by H-2K^b class I molecules. A recent study has shown that a subdominant peptide can be detected when mice were immunised with high doses of ovalbumin. Although the mechanisms underlying immunodominance are not fully understood, the phenomena is most likely of importance for T cell responses to allogeneic MHC and mH antigens.

Immunodominance among mH antigens

The MHC matched mouse strains C57BL/6 and BALB/B differ by more than 29 mH loci (Bailey and Mobraaten, 1969). Early work by Wettenstein and colleagues has shown that C57BL/6 mice immunised with BALB/B spleen cells generated CTL responses against only two immunodominant mH loci, while the majority of mH loci were immunologically silent (Wettstein, 1986). Silence of these loci was not due to lack of immunogenicity but rather to the overriding effects of immunodominant loci. Omission of dominant mH antigens on the immunising cells revealed CTL responses against subdominant mH antigens that were previously undetectable (Wettstein, 1986).

The mechanisms of immunodominance among mH loci are currently not well understood. One possibility is that immunodominant loci contain multiple genes providing several peptide epitopes that are presented by MHC class II and class I molecules, which might lead to strong stimulation of Th cells and CTL. To investigate this possibility, we have analysed peptides that are involved in CTL responses of BALB/B mice against mH mismatched C57BL/6 stimulators (Yin, *et al.*, 1993).

Peptide purification by HPLC showed that only two HPLC fractions contained CTL recognised peptides. One of the fractions contained peptides that were presented by H-2K^b class I molecules, while the other HPLC fraction contained H-2D^b presented peptides. Further HPLC purification did not provide any evidence that more than one peptide was present in each of these HPLC fractions. The K^b presented peptides were found to be derived from the minor H-1 locus, while the D^b presented peptides were derived from another, unidentified mH locus. Together, these experiments provide evidence that immunodominance of these two mH loci was not caused by multiple CTL stimulating peptides. Only one, or a small number of closely related peptides, which are strongly immunogenic and account for the immunodominance of the analysed mH loci.

Peptide dominance in anti-allo MHC T cell responses?

There is now considerable evidence that a large proportion of CTL specific for allogeneic MHC class I molecules is peptide dependent. Immune responses to allogeneic MHC class II molecules have not been studied as extensively as anti-class I responses, but there is no reason to believe that there will be any fundamental difference. The first formal demonstration of peptide dependent allo-recognition came from experiments with human cells transfected with murine class I molecules. The human transfectants were not recognised by murine allo-specific CTL, but recognition was restored when the human cells were coated with peptides prepared by cyanobromide cleavage of proteins isolated from the cytosol of mouse cells (Heath, *et al.*, 1989). We have used limiting dilution analysis to show that the majority of CTL clones raised against the allogeneic H-2K^b class I molecule were peptide dependent (Aosai, *et al.*, 1991). Rotzschke et al. have documented that different CTL clones specific for a given allogeneic class I molecule recognised distinct HPLC purified peptide fractions (Rotzschke, *et al.*, 1991). Although these studies clearly indicate peptide dependent allo-recognition, it is not clear whether all peptides presented by allogeneic class I molecules are recognised by CTL, or whether a limited number of strongly immunogenic peptides dominate CTL responses. As mentioned above, immunodominance has been observed in many different situations, so that it would be surprising if peptide recognition in the context of allogeneic MHC molecules would be an exception.

Recent findings support the idea that immunodominant peptides are involved in CTL responses against non-self MHC class I molecules. For example, Henderson et al. have shown that five independently derived anti-HLA-A2 murine CTL clones recognised the same A2 presented peptide (Henderson, *et al.*, 1993). Also, Udaka et al. have sequenced a peptide (p2Ca) that was recognised by an allo-CTL clone specific for H-2L^d (Udaka, *et al.*, 1992). In an elegant study J. Connolly has analysed CTL responses of L^d negative dm2 mice stimulated with L^d expressing BALB/c cells (Connolly, 1994). In limiting dilution assays, approximately half of anti-L^d CTL precursors showed specificity for the p2Ca peptide. Interestingly, immunodominance of the p2Ca peptide was only observed in responder mice that expressed the Vβ8 gene segments of the TCR-β chain, but not in Vβ8 negative strains. The experiments by Connolly indicate that a single peptide can stimulate up to 50% of allo-reactive CTL, and that the immunodominant peptides may sometimes preferentially stimulates CTL precursors expressing certain TCR variable region gene segments.

Downregulation of allo-responses with immunodominant peptides?

The observation that immune responses to allogeneic MHC molecules and mH antigens may be dominated by strongly immunogenic peptides, raises the question whether such peptides can be exploited to downregulate T cell responses. Tolerance induction with synthetic peptides has been demonstrated with transgenic mice expressing the LCMV glycoprotein (GP) in pancreatic islet cells (Aichele, *et al.*, 1994). When these mice were repeatetly injected intraperitoneally with high doses of synthetic peptides corresponding to a CTL epitope in GP, they were subsequently resistant to T cell mediated induction of autoimmune diabetes. Oral antigen administration is another way of inducing antigen-specific unresponsiveness (Weiner, *et al.*, 1994). It seems that the mechanism of unresponsiveness might depend upon the dose of orally given antigen (Friedman and Weiner, 1994). Low dose antigen appears to preferentially induce TGF-β producing regulatory Th2 lymphocytes, which suppress responses by Th1 cells and CTL. In contrast, high dose antigen is less likely to induce regulatory T cells, and is more likely to cause deletion of antigen responsive T cells. A recent report by Sun et al. indicates that it may be possible to enhance induction of unresponsiveness by oral antigens (Sun, *et al.*, 1994). Cholera toxin (CT) consists of a central subunit A surrounded by five B subunits. Intact CT is a potent immune-stimulating oral adjuvant leading to stimulation of immune responses against oral antigens rather then tolerance induction. Sun et al. showed that recombinant subunit B has the opposite effect, since it can enhances induction of oral tolerance against proteins coupled to the B subunit. Ongoing clinical trials in multiple sclerosis patients and rheumatoid arthritis patients may benefit from this observation, since it seems possible to improve the design of tolerance inducing oral vaccines.

Peptides corresponding to sequences of MHC class I and class II molecules have been shown to inhibit T cell responses. One of the first studies showed that peptides derived from residues 98-113 of the α2 domain of HLA-A2 inhibited recognition by most A2-specific CTL (Perham, *et al.*, 1987). These HLA-A2 peptides probably inhibited CTL recognition by binding to the TCR. Subsequently, it was shown that peptides corresponding to the HLA-B7 residues 75-84 could induce tolerance in a rat heart allograft model (Nisco, *et al.*, 1994). A combination of HLA-B7 peptides together with low doses of cyclosporine A resulted in long-term heart allograft survival in most rats, while treatment with peptides alone or cyclosporine A alone had little effect. In this study, oral peptide feeding was as efficient as i.v. injection in inducing heart allograft survival. The mechanism by which HLA-B7 peptides can suppress immune responses against rat MHC molecules is currently unclear. Recently, it has been described that these peptides bind to two T cell molecules of 72kDa and 74kDa, and that binding is associated with an increase in intracellular Ca^{2+} (Krensky and Clayberger, 1994). These observations suggest that the immune-suppressing effect of these HLA-derived peptides may not be specific for immune responses against allo-MHC molecules. A similar lack of specificity is encountered with peptides corresponding to the conserved binding sites for CD8 and CD4 on MHC class I and class II molecules, respectively. In vitro, these peptides can inhibit differentiation of human CTL precursors and the mixed lymphocyte reaction of freshly isolated peripheral blood lymphocytes (Clayberger, *et al.*, 1994).

Sayegh et al. have described a HLA-DR2 derived peptide which inhibits proliferation in the mixed lymphocyte reaction, but not proliferation induced by mitogens or mumps (Krensky and Clayberger, 1994). This peptide corresponds to residues 182-94 of the transmembrane region of the DR $\alpha 2$ chain and shows binding to several HLA-DR molecules. Therefore, this peptide is probably involved in indirect allo-recognition, where peptides from allogeneic MHC molecules are presented to T cells in the context of self MHC products. Evidence that indirect allo-recognition can result in tissue transplant rejection comes from experiments with MHC class II knockout mice (Auchincloss, *et al.*, 1993). Normal mice receiving MHC class II negative skin transplants mounted a CD4 T cell responses which led to graft rejection. These CD4 T cells recognised donor derived peptides presented by host MHC class II molecules (Lee, *et al.*, 1994). In a rat model it was shown that indirect allo-recognition may be susceptible to peptide-specific tolerance induction. Peptides derived from rat MHC class II β chains RT1.B^u and RT1.D^u, which are probably involved in indirect allo-recognition, have been given orally to LEW rats which resulted in the down-regulation of allo-responses in vivo and in vitro (Sayegh, *et al.*, 1992).

The described results obtained with MHC derived peptides are encouraging and provide evidence that tolerance induction with peptides is feasible. The identification of strongly immunogenic peptides which dominate T cell responses against allogeneic MHC molecules and mH antigens may lead to their exploitation for tolerance induction. Synthetic peptides corresponding to mH antigens can be used directly for tolerance induction, since they can be presented by host MHC molecules. In contrast, peptides derived from the groove of allogeneic MHC molecules cannot be used directly, since they will not be presented properly by host MHC molecules. Therefore, complexes of donor MHC molecules plus donor peptide will be required for induction of donor-specific tolerance. Such complexes can be produced using recombinant MHC molecules generated in bacteria or yeast, which can be refolded with synthetic peptides to obtain homogeneous MHC/peptide complexes. An easier way to enrich for MHC molecules containing defined synthetic peptides is to acid treat intact cells in the presence of peptides, which leads to partial MHC denaturation, dissociation of endogenous peptides and binding of synthetic peptides. It has been shown that professional antigen presenting cells treated in this way were much more efficient in stimulating peptide-specific T cell responses than untreated APC (Langlade-Demoyen, *et al.*, 1994). Similarly, it is conceivable that peptide loading of non professional antigen presenting cells with immunodominant peptides may greatly enhance their ability to induce peptide-specific tolerance.

Conclusion:

Although immuno-suppressive drugs such as cyclosporine A can mediate acceptance of histoincompatible tissue transplants, their long-term use creates considerable problems. Induction of antigen-specific tolerance would be preferable over generalised immune-suppression. Animal models have provided evidence that synthetic peptides derived from sequences of MHC class I and class II molecules can prolong graft survival. The mode of action and the specificity of some of these MHC derived peptides is not yet clear. It is likely, that immunodominant peptides present in the peptide binding groove of MHC molecules can be identified and used for peptide-specific downregulation of T cell responses against allogeneic MHC molecules and against mH antigens. In future, an approach using sub-therapeutic doses of immuno-suppressive drugs in combination with tolerance inducing peptides derived from MHC sequences and from MHC bound peptides, may lead to long-term graft survival with minimal side effects.

References:

Aichele, P., Kyburz, D., Ohashi, P.S., Odermatt, B., Zinkernagel, R.M., Hengartner, H. and Pircher, H. (1994) Peptide-induced T-cell tolerance to prevent autoimmune diabetes in a transgenic mouse model. *Proc Natl Acad Sci U S A.* 91: 444-8.

Amigorena, S., Drake, J.R., Webster, P. and Mellman, I. (1994) Transient accumulation of new class II MHC molecules in a novel endocytic compartment in B lymphocytes [see comments]. *Nature.* 369: 113-20.

Aosai, F., Ohlen, C., Ljunggren, H.G., Hoglund, P., Franksson, L., Ploegh, H., Townsend, A., Karre, K. and Stauss, H.J. (1991) Different types of allospecific CTL clones identified by their ability to recognize peptide loading-defective target cells. *Eur J Immunol.* 21: 2767-74.

Auchincloss, H.J., Lee, R., Shea, S., Markowitz, J.S., Grusby, M.J. and Glimcher, L.H. (1993) The role of "indirect" recognition in initiating rejection of skin grafts from major histocompatibility complex class II-deficient mice. *Proc Natl Acad Sci U S A.* 90: 3373-7.

Bailey, D. and Mobraaten, L. (1969) Estimates of the number of loci contributing to the histoincompatiibility between C57BL/6 and BALB/c strains of mice. 7: 394-400.

Barnd, D.L., Lan, M.S., Metzgar, R.S. and Finn, O.J. (1989) Specific, major histocompatibility complex-unrestricted recognition of tumor-associated mucins by human cytotoxic T cells. *Proc Natl Acad Sci U S A.* 86: 7159-63.

Bevan, M.J. (1984) *Immunol. Today.* 5: 128.

Bjorkman, P.J., Saper, M.A., Samraoui, B., Bennett, W.S., Strominger, J.L. and Wiley, D.C. (1987a) The foreign antigen binding site and T cell recognition regions of class I histocompatibility antigens. *Nature.* 329: 512-8.

Bjorkman, P.J., Saper, M.A., Samraoui, B., Bennett, W.S., Strominger, J.L. and Wiley, D.C. (1987b) Structure of the human class I histocompatibility antigen, HLA-A2. *Nature.* 329: 506-12.

Brown, J.H., Jardetzky, T.S., Gorga, J.C., Stern, L.J., Urban, R.G., Strominger, J.L. and Wiley, D.C. (1993) Three-dimensional structure of the human class II histocompatibility antigen HLA-DR1 [see comments]. *Nature.* 364: 33-9.

Choi, Y.W., Herman, A., DiGiusto, D., Wade, T., Marrack, P. and Kappler, J. (1990) Residues of the variable region of the T-cell-receptor beta-chain that interact with S. aureus toxin superantigens. *Nature.* 346: 471-3.

Clayberger, C., Lyu, S.C., DeKruyff, R., Parham, P. and Krensky, A.M. (1994) Peptides corresponding to the CD8 and CD4 binding domains of HLA molecules block T lymphocyte immune responses in vitro. *J Immunol.* 153: 946-51.

Connolly, J.M. (1994) The peptide p2Ca is immunodominant in allorecognition of L^d by β-chain $V\beta8^+$ but not $V\beta8^-$ strains. *Proc. Natl. Acad. Sci.* 91: 11482-486.

Deverson, E.V., Gow, I.R., Coadwell, W.J., Monaco, J.J., Butcher, G.W. and Howard, J.C. (1990) MHC class II region encoding proteins related to the multidrug resistance family of transmembrane transporters [see comments]. *Nature.* 348: 738-41.

Friedman, A. and Weiner, H.L. (1994) Induction of anergy or active suppression following oral tolerance is determined by antigen dosage. *Proc Natl Acad Sci U S A.* 91: 6688-92.

Graff, R. and Bailey, D. (1973) The non-H-2 histocompatibility loci and their antigens. *Transplant Rev.* 15: 26-49.

Graff, R., Hildemann, W. and Snell, G. (1966) Histocompatibility genes of mice. VI. allografts in mice congenic at various non-H-2 histocompatibility loci. 4: 425-437.

Heath, W.R., Hurd, M.E., Carbone, F.R. and Sherman, L.A. (1989) Peptide-dependent recognition of H-2K^b by alloreactive cytotoxic T lymphocytes. *Nature.* 341: 749-752.

Henderson, R.A., Cox, A.L., Sakaguchi, K., Appella, E., Shabanowitz, J., Hunt, D.F. and Engelhard, V.H. (1993) Direct identification of an endogenous peptide recognized by multiple HLA-A2.1-specific cytotoxic T cells. *PNAS.* 90: 10275-10279.

Hunt, D.F., Henderson, R.A., Shabanowitz, J., Sakaguchi, K., Michel, H., Sevilir, N., Cox, A.L., Appella, E. and Engelhard, V.H. (1992) Characterization of peptides bound to class I MHC Molecule HLA-A2.1 by mass spectrophotometry. *Science.* 255: 1261-1263.

Jerome, K.R., Barnd, D.L., Bendt, K.M., Boyer, C.M., Taylor, P.J., McKenzie, I.F., Bast, R.J. and Finn, O.J. (1991) Cytotoxic T-lymphocytes derived from patients with breast adenocarcinoma recognize an epitope present on the protein core of a mucin molecule preferentially expressed by malignant cells. *Cancer Res.* 51: 2908-16.

Karp, D.R. and Long, E.O. (1992) Identification of HLA-DR1β chain residues critical for binding staphylococcal enterotoxins A and E. *J Exp Med.* 175: 415-424.

Krensky, A.M. and Clayberger, C. (1994) The induction of tolerance to alloantigens using HLA-based synthetic peptides. *Curr. Opin. Immunol.* 6: 791-796.

Langlade-Demoyen, P., Levraud, J.P., Kourilisky, P. and Abastado, J.P. (1994) Primary cytotoxic T lymphocyte induction using peptide-stripped autologous cells. *Intern. Immunol.* 6: 1759-66.

Lee, R.S., Grusby, M.J., Glimcher, L.H., Winn, H.J. and Auchincloss, H.J. (1994) Indirect recognition by helper cells can induce donor-specific cytotoxic T lymphocytes in vivo. *J Exp Med.* 179: 865-72.

Loveland, B., Wang, C.R., Yonekawa, H., Hermel, E. and Lindahl, K.F. (1990) Maternally transmitted histocompatibility antigen of mice: a hydrophobic peptide of a mitochondrially encoded protein. *Cell.* 60: 971-80.

Matzinger, P. and Bevan, M.J. (1977) *Cell. Immunol.* 29: 1.

Monaco, J.J. (1992) A molecular model of MHC class-I-restricted antigen processing. *Immunol Today.* 13: 173-9.

Monaco, J.J., Cho, S. and Attaya, M. (1990) Transport protein genes in the murine MHC: possible implications for antigen processing. *Science.* 250: 1723-6.

Nisco, S., Vriens, P., Hoyt, G., Lyu, S.C., Farfan, F., Pouletty, P., Krensky, A.M. and Clayberger, C. (1994) Induction of allograft tolerance in rats by an HLA class-I-derived peptide and cyclosporine A. *J Immunol.* 152: 3786-92.

Perham, P., Clayberger, C., Zorn, S.L., Ludwig, D.S., Schoolnik, G.K. and Krensky, A.M. (1987) Inhibition of alloreactive cytotoxic T lymphocytes by peptides from the α2 domain of HLA-A2. *Nature.* 325: 625-628.

Roopenian, D.C. (1992) What are minor histocompatibility loci? A new look at an old question. *Immunol Today.* 13: 7-10.

Rotzschke, O., Falk, K., Faath, S. and Rammensee, H.G. (1991) On the nature of peptides involved in T cell alloreactivity. *J Exp Med.* 174: 1059-71.

Salter, R.D., Norment, A.M., Chen, B.P., Clayberger, C., Krensky, A.M., Littman, D.R. and Parham, P. (1989) Polymorphism in the alpha 3 domain of HLA-A molecules affects binding to CD8. *Nature.* 338: 345-7.

Sanderson, F., Kleijmeer, M.J., Kelly, A., Verwoerd, D., Tulp, A., Neefjes, J.J., Geuze, H.J. and Trwsdale, J. (1994) Accumulation of HLA-DM, a regulator of antigen presentation, in MHC class II compartments. *Science.* 266: 1566-1569.

Sayegh, M.H., Khoury, S.J., Hancock, W.W., Weiner, H.L. and Carpenter, C.B. (1992) Induction of immunity and oral tolerance with polymorphic class II major histocompatibility complex allopeptides in the rat. *Proc Natl Acad Sci U S A.* 89: 7762-6.

Sercarz, E.E., Lehmann, P.V., Ametani, A., Benichou, G., Miller, A. and Moudgil, K. (1993) Dominance and crypticity of T cell antigenic determinants. *Annu Rev Immunol.* 11: 729-66.

Speiser, D.E., Zurcher, T., Ramseier, H., Hengartner, H., Staeheli, P., Haller, O. and Zinkernagel, R.M. (1990) Nuclear myxovirus-resistance protein Mx is a minor histocompatibility antigen. *Proc Natl Acad Sci U S A.* 87: 2021-5.

Sun, J.B., Holmgren, J. and Czerkinsky, C. (1994) Cholera toxin B subunit: An efficient transmucosal carrier-delivery system for induction of perigheral immunological tolerance. *Proc. Natl. Acad. Sci. USA.* 91: 10795-799.

Topalian, S.L., Rivoltini, L., Mancini, M., Markus, N.R., Robbins, P.F., Kawakami, Y. and Rosenberg, S.A. (1994) Human CD4+ T cells specifically recognise a shared melanoma-associated antigen encoded by the tyrosinase gene. *Proc. Natl. Acad. Sci. USA.* 91: 9461-9465.

Trowsdale, J., Hanson, I., Mockridge, I., Beck, S., Townsend, A. and Kelly, A. (1990) Sequences encoded in the class II region of the MHC related to the 'ABC' superfamily of transporters [see comments]. *Nature.* 348: 741-4.

Udaka, K., Tsomides, T.J. and Eisen, H.N. (1992) A naturally occurring peptide recognized by alloreactive CD8+ cytotoxic T lymphocytes in association with a class I MHC protein. *Cell.* 69: 989-98.

Weiner, H.L., Friedman, A., Miller, A., Khoury, S.J., al, S.A., Santos, L., Sayegh, M., Nussenblatt, R.B., Trentham, D.E. and Hafler, D.A. (1994) Oral tolerance: immunologic mechanisms and treatment of animal and human organ-specific autoimmune diseases by oral administration of autoantigens. *Annu Rev Immunol.* 12: 809-37.
Wettstein, P.J. (1986) Immunodominance in the T-cell response to multiple non-H-2 histocompatibility antigens. II. Observation of a hierarchy among dominant antigens. *Immunogenetics.* 24: 24-31.
Yin, L., Poirier, G., Neth, O., Hsuan, J.J., Totty, N.F. and Stauss, H.J. (1993) Few peptides dominate CTL responses to single and multiple minor histocompatibility antigens. *Int. Immunol.* 5: 1003-1009.

Part 7

Radiation Biology

COUPLING OF MEMBRANE PATHOLOGY AND DNA DAMAGE IN IRRADIATED LYMPHOID CELL POPULATIONS

Alexei B. Chukhlovin

Bone Marrow Transplantation Department,
N.N.Petrov Research
Institute of Oncology,
189646, St.-Petersburg, Russia

INTRODUCTION

Apoptotic variant of radiation-induced lymphoid cell death is an extensively studied pathology. It occurs in dose-dependent manner after ca. 0.5 to 10 Gy of X- or gamma-rays. Primary DNA damage is repaired during first hour after treatment for the most part. The secondary wave of massive DNA autolysis is the hallmark of apoptosis with involvement of APO-1/FAS membrane factors and several other modulatory gene products (for rev. see Marx, 1993; Wyllie, 1994).

A first increase in apoptotic cell numbers occurs following approximal 1 hr after radiation treatment. Apoptosis looks like "homogenization" of nuclear matter, combined with significant decrease of the average cell diameter. Several stages of nuclear "pycnosis" may be observed after the lysis of chromatin. DNA autolysis and concomitant shrinkage.,e.c. of thymic lymphoid cells, are proceeding very rapidly (Klassen et al., 1993). Apoptosis in lymphoid populations proceeds according to the "all-or-nothing" rule, thus allowing strict delineation of lethally affected cells when scoring irradiated lymphoid suspensions.

MEMBRANE FEATURES OF RADIATION-INDUCED APOPTOSIS

It is commonly accepted that the primary radiation damage to living cells consists of induction of DNA breaks, intermolecular cross-links along with peroxidation of membrane lipid molecules. Hence, the initial signal for DNA decay may originate from damaged chromatin (Cohen et al., 1992) or from affected membrane structures. The role of membrane damage is less clear (Ramakrishnan et al., 1993). Some examples of early membrane/surface alterations are listed in table 1.

NATO ASI Series, Vol. H 94
Gene Technology
Edited by A. R. Zander, W. Ostertag,
B. V. Afanasiev, F. Grosveld

Table 1.
Early membrane changes of irradiated lymphoid cells.

Alterations registered	Type of lymphoid cells	References
*Increase of cell volume	B-lymphocytes	Ashwell et al.(1986)
*Decreased physical adherence	Thymocytes	Sungurov (1988)
*Decreased surface electric charge	Thymocytes Thymocytes	Miyazava et al.(1979) Kubasova et al.(1993)
*Decreased activity of SOD	Thymocytes	Gudz et al. (1994)
*Induced protein kinase activities	Lymphoblasts Lymphocytes	Waddick et al. (1993) Ojeda et al. (1991)
*Increased Con A binding	Lymphocytes Thymocytes	Koteles (1986) Kubasova et al.(1993)
*Surface IgG loss	B-lymphocytes	Ojeda et al. (1991)
*Loss of receptors to autologous RBC	Thymocytes	Zherbin, Chukhlovin (1984)
*Loss of sheep RBC receptors	T-lymphocytes	Birkeland (1980); Szczylik, Viktor-Jedrzejczak (1981)
*Loss of Ig- and C'receptors	B-lymphocytes	Same workers
*Loss of CD 45 antigen	Peripheral lymphocytes	Carbonari et al. (1994)

This table showes that the membrane damage of lymphoid cell membranes may reflect general cell pathology (decreased electric charge, hypoadherence, increased ConA binding, alterations in cell volume) as well as expression of stress-inducible enzymes and alteration of lipid bilayers (increased SOD and protein kinases activities). All specific lymphoid receptors (T- and B-markers and receptors for autologous RBC on immature thymocytes) exhibited dose-dependent decrease. Intercellular contacts of such type are modelling some key interactions within the immune system, involving the CD 2-receptor on lymphoid cell and LFA-3 molecules on target cells (Springer, 1990). If such contacts are disrupted, critical disturbances in contact/humoral signalling between lymphocytes and stromal cell populations (e.c., macrophages and endothelium) that supply lymphoid cells with specific growth and survival factors may occur.

Hence, the severe alterations of lymphoid cell membranes seem to be critical to the ensuing DNA decay and "classical" apoptotic death. Direct connections between early surface damage and the ability to apoptosis were shown by Borisova et al. (1987). The numbers of cells with detectable apoptosis and with loss of autologous RBC receptors correlated well in different buoyant density fractions of irradiated thymus.

In general, malignant cells of lymphoid origin do not differ from normal lymphoid cells in their radiosensitivity (Thomson et al.,1985; Uckun et al., 1993). Their radiation death occurs mainly through apoptosis, either immediately or following several divisions (Radford et al., 1994). Postradiation DNA fragmentation in malignant lymphoblasts is also preceded by the very early surface alterations, involving the CD 19 antigen, activation of the lck protein kinase, and protein kinase C (Waddick et al., 1993). Hence, further studies of specific surface receptor changes in irradiated lymphoblasts will be of significant diagnostic and therapeutic value.

POSTRADIATION CHANGES OF LYMPHOID/STROMAL COMMUNICATIONS

During the first hours after a radiation insult, a local accumulation of dead cells occurs. Most of them are rapidly engulfed and resorbed, thus serving as functional stimuli for local macrophages (Zherbin and Chukhlovin, 1989; O'Brien-Ladner et al., 1993). However, increased IL1 release by them seems to occur after the main peak of apoptosis.

The stromal fibroblasts seem to be functionally radioresistant, at least after treatment with up to 10 Gy of X- or gamma-radiation. However, another stromal population, endothelial cells, have been shown to exhibit transient membrane damage and an increased incidence of apoptosis after irradiation in vitro and in vivo (Fuks et al. 1994).

Thymic epithelial cells, the "nursing" populations, are highly damageable, since they disappear from the thymus within the first days after radiation (Yarilin, Kharchenko, 1993).

Thus, radiation effect on the haemopoietic microenvironment may be suggested early after massive irradiation. Insufficient levels of IL 1, IL 2 and other growth factors may be critical for the survival of apoptosis-prone cells, e.c. lymphoid cells and radiosensitive haemopoietic precursors.

CONCLUSION

Multiple membrane changes and specific receptor loss from the surface of irradiated lymphoid cells may be combined with improper cytokine/growth factor compensation from the stromal microenvironment. Therefore, early "deafness" of irradiated lymphoid cells to their specific cytokines (e.c., IL1 and IL2) deserves further study, as a potential cause of radiation pathology. Apoptotic dath of normal lymphoid and immature blood precursors devoid of specific growth factors may serve as a useful model (Fesus et al., 1991; Raff, 1992). Thus, radiation-induced membrane pathology of lymphoid cells remains to be an informative model system for studying the surface/receptor damage of apoptosis-prone blood system cells (e.c., haemopoietic precursors of different lineages).

These future studies should include quantitation of cytokine and growth factor receptors on the surface of radiation-affected cells, and searching for the ways to prevent this type of acute haematological pathology.

REFERENCES

Ashwell JD, Schwartz RH, Mitchell JB, Russo A (1986) Effect of gamma radiation on resting B lymphocytes. J Immunol 136:3649-3656

Birkeland SA (1980) Influence of irradiation on the capacity of human T and B lymphocytes to form E and EAC rosettes. Transplantation 29: 23-24

Borisova EA, Chukhlovin AB, Seiliev AA, Zhivotovsky BD, Zherbin EA, Hanson KP (1987) Degree of chromatin fragmentation and frequency of nuclear pyknosis in Percoll-fractionated thymocytes of irradiated rats. Int J Radiat Biol 51: 421-428

Carbonari M, Cibati M, Cherchi M (1994) Detection and characterization of apoptotic peripheral blood lymphocytes in human immunodeficiency virus infection and cancer chemotherapy by novel flow cytometric method. Blood 83: 1268-1277

Cohen JJ, Duke RC, Fadok VA, Sellins K (1992) Apoptosis and programmed death in immunity. Ann Rev Immunol. 10:267-293

Fesus L, Davies PJA, Piacentini M (1991) Apoptosis: molecular mechanisms in programmed cell death. Eur J Cell Biol 56: 170-177

Fuks Z, Persaud RS, Alfieri A, McLoughlin M, Ehleiter D, Schwartz JL, Seddon A (1994) Basic fibroblast growth factor protects endothelial cells against radiation-induced programmed cell death in vivo and in vitro. Cancer Res 54:2582-2590

Gudz TI, Pandelova IG, Novgorodov SA (1994) Stimulation of respiration in rat thymocytes induced by ionizing radiation. Radiat Res 138:114-120

Klassen NV, Walker PR, Ross CK, Cygler J, Lach B (1993) Two-stage cell shrinkage and the OER of radiation-induced apoptosis of rat thymocytes. Int J Radiat Biol 64:571-581

Koteles GJ (1986) Cell membranes in radiation injury. Isotopenpraxis. 10:361-368

Kubasova T, Chukhlovin AB, Somosy Z, Ivanov SD, Koteles GJ, Zherbin EA, Hanson KP (1993) Detection of early membrane and nuclear alterations of thymocytes upon in vitro ionizing irradiation. Acta Physiol Hung. 81:277-288

Marx J (1993) Cell death studies yield cancer clues. Science 259: 760-761

Miyazawa T, Sato C, Kojima K (1979) Thymic phagocytosis and reduction in the negative surface charge of thymocytes after X-irradiation. Radiat Res 79: 622-629

O'Brien-Ladner A, Nelson ME, Kimler BF, Wesselius LJ (1993) Release of interleukin-1 by human alveolar macrophages after in vitro irradiation. Radiat Res 136:37-41

Ojeda F, Andrade J, Maldonado C, Guarda MI, Folch H (1991) Radiation-induced surface IgG modulation: protein kinase C involvement. Int J Radiat Biol 59:53-58

Radford IR, Murphy TK, Radley JM, Ellis SL (1994) Radiation response of mouse lymphoid and myeloid cell lines. Part II. Apoptotic death is shown by all cell lines examined. Int J Radiat Biol 65:217-227

Raff MC (1992) Social controls on cell survival and cell death. Nature 356: 397-400

Ramakrishnan N, McClain DE, Catravas GN (1993) Membranes as sensitive targets in thymocyte apoptosis. Int J Radiat Biol 63:693-701

Springer TA (1990) Adhesion receptors of the immune system. Nature 346:425-434

Sungurov AY (1988) Radiobiology of the cell surface (in Russian). Moscow, VINITI Publ

Szczylik C, Wiktor-Jedrzejczak W (1981) The effects of X-irradiation in vitro on subpopulations of human lymphocytes. Int J Radiat Biol 39: 253-263

Thomson AER, Vaughan-Smith S, Peel WE, Wetherley-Mein G (1985) The intrinsic radiosensitivity of lymphocytes in chronic lymphocytic leukemia, quantitatively determined independently of cell death rates factors. Int J Radiat Biol 48:943-961

Uckun FM, Chandan-Longlie M., Jaszcz W, Obuz V, Waddick K, Song CW (1993) Radiation damage repair capacity of primary clonogenic blasts in acute lymphoblastic leukemia. Cancer Res 53:1431-1436

Waddick KG, Chae HJP, Tuel-Ahlgren L, Jarvis LJ, Dibirdik I, Myers D, Uckun FM (1993) Engagement of CD 19 receptor on human B-lineage leukemia cells activates LCK tyrosine kinase and facilitates radiation-induced apoptosis. Radiat Res 136:313-319

Wyllie AH (1994) Death gets a brake. Nature 369:272-273

Yarilin AA, Kharchenko TY (1993) Effects of ionizing radiation upon thymic stroma of embryos and newborn mice (In Russian). Radiaz Biol Radioecol 33:542-548

Zherbin EA, Chukhlovin A.B. (1984) Possible association of membrane and nuclear changes in gamma-irradiated rat thymocytes. Int J Radiat Biol 45: 179-183.

Zherbin EA, Chukhlovin AB (1989) Radiation Haematology. (In Russian). Moscow, Medizina Publ.

EFFECTS OF RADIATION ON THE PLASMA MEMBRANE AND EXPRESSION OF MEMBRANE BOUND HEMATOPOIETIC CYTOKINES

Nicholas Dainiak and Joseph Albanese
Department of Medicine, Royal Victoria Hospital,
McGill University, Montreal, Quebec, Canada

Running Title: Radiation Effects on Plasma Membrane

Reprint Address: Nicholas Dainiak
Royal Victoria Hospital
687 Pine Avenue West, Room C6.82
Montreal, Quebec, Canada
H3A 1A1

ABSTRACT:

The plasma membrane is a dynamic organelle whose components are continually turned over. Proteins and lipids are synthesized and inserted into the plasma membrane where they reside for a finite time period, after which they are internalized or exfoliated from the cell surface. The release of plasma membrane constituents in association with vesicles exfoliated from the cell surface is reduced by exposure to ultraviolet radiation. Moreover, the expression of membrane-bound cytokines on the cell surface and on extracellular vesicles is also impaired by ultraviolet light treatment. These results are discussed in the context of radiation-induced alterations in the structure and function of plasma membrane components, and with respect to radiation-induced apoptotic cell death.

INTRODUCTION:

Radiation is well known to have direct effects on genomic DNA, resulting in its damage through the induction of DNA lesions, protein-DNA crosslinks, and single and double strand breaks; stimulation of abnormal DNA repair; and induction of oncogene expression. In addition to these nuclear effects, radiation may also alter other organelles such as the plasma membrane (Dainiak, 1995). Abnormalities in both the structure of plasma membrane components and functional expression of plasma membrane-associated molecules (such as those involved in intracellular communication and

NATO ASI Series, Vol. H 94
Gene Technology
Edited by A. R. Zander, W. Ostertag,
B. V. Afanasiev, F. Grosveld

signal transduction) have been described in irradiated cells (Krutzman *et al.*, 1990 and Matsui and DeLeo, 1990). Recently, the term "epigenetic" has been coined to represent alterations in the expression of genes at transcriptional, translational or post-translational levels, that lead to abnormalities in the structure and/or function of non-nuclear proteins (Trosko, 1995).

The effects of ionizing radiation on the lipid component of plasma membranes has been recently reviewed (Stark, 1991). The purpose of this short review is to address the effects of radiation exposure on structure and function of plasma membrane proteins that are involved in cell-cell signalling (i.e., intercellular communication). Discussion will focus on effects of ultraviolet (UV) radiation on exfoliation of regions of the plasma membrane as extracellular vesicles that bear signal molecules. The hypothesis that abnormal release of these molecules may mediate radiation-induced apoptosis will be presented.

DIRECT AND INDIRECT RADIATION EFFECTS ON THE PLASMA MEMBRANE:

While the primary cellular target of ionizing radiation is DNA, radiation exposure may both indirectly and directly affect the eukaryotic cell surface membrane. "Indirect" effects mediated by water radiolysis result in the generation of three reactive species (the aqueous electron, $OH^{\bullet}$, and $H^{\bullet}$) that in turn, react with plasma membrane constituents (Beir, 1990). A consequence of reactivity of these free radicals is the generation of superoxide and the chloride radical, respectively, both of which play an important role in killing of microbes by neutrophils (Babior, 1987). Although superoxide and the chloride radical play an essential role in immune defense, these species are also toxic to normal eukaryocytic cells, resulting in auto-oxidation and tissue damage. Among potentially toxic products resulting from the generation of highly reactive molecular species are the lipid hydroperoxides.

Lipid peroxidation results from the reaction of molecular oxygen with membrane lipids. Radiation induces lipid peroxidation that in turn, has profound effects on molecular interactions among plasma membrane constituents. Among the most important consequences of lipid peroxidation is induction of abnormalities in the permeability properties of lipid membranes resulting from an increase in the dielectric constant of the membrane interior and an increase in intrinsic plasma membrane microviscocity (Stark, 1991). These plasma membrane alterations may result in altered membrane stability and ultimately, local membrane lysis. It is of interest that at a constant absorbed dose, the radiation effect on structural stability and transport increase with decreasing dose rate. This "inverse dose rate" effect can be predicted mathematically where the kinetic chain length of the lipid is reduced at high dose rates (Stark, 1991).

Theoretically, it is also possible for radiation to initiate lipid peroxidation of membrane components directly (in the absence of the aqueous phase) (Stark, 1991). Experimentally, direct effects on electrical conductance of stacked monolayers of "artificial" membranes have been measured, using small-angle x-ray diffraction (Fatani and Rosen, 1992). While electrical conductance of Langmuir-Blodgette preparations of stearic acid were increased in a radiation dose-dependent fashion, dose rate had no effect on this parameter in artificial membranes. Provocatively, radiation-induced abnormalities in conductance were initially reversible but subsequent excitation resulted in irreversible damage to the multilayer preparations.

Using dihexadecylphosphatidylethanolamine vesicles, Cheng et al. observed complete transformation of a lamellar structure to a nonlamellar structure as analyzed by small-angle x-ray diffraction (Cheng *et al.*, 1993). These changes occurred at high radiation doses (2 x 10^{10} photons/s) (Cheng *et al.*, 1993). Utilizing Raman spectroscopy to evaluate the polyunsaturated phase transition of synthetic liposomes, Verma observed that low radiation doses (5-15 rads) also have a profound effect on membrane structure (Verma, 1986). These alterations correlate with increased membrane rigidity, indicating that lipid domains within the membrane may be directly affected by exposure to low level radiation (Verma, 1986). More recently, it has been suggested in studies of biological membranes that radiation triggers hydrolysis of plasma membrane sphigomyelin, resulting in production of ceramide and phosphocholine (Haimovitz-Freidman *et al.*, 1994). Ceramide may then function as a second messenger that mediates other cellular events such as radiation-induced programmed cell death (i.e., apoptosis).

UV-INDUCED MEMBRANE CHANGES:

UV-radiation of long wave lengths (UVA, 320-400 nm) and mid and short wave lengths (UVB, 290-320 nm; UVC, 100-290 nm) damages DNA directly (Freeman *et al.*, 1987 and Ronai *et al.*, 1988). Recently, UV-induced alterations in the plasma membrane have been carefully documented. Morliere et al. observed that UVA induces lipid peroxidation in long-term cultures of human fibroblasts (Morliere *et al.*, 1991). This effect is reversed by tocopherol, an antioxidant that protects normal cells from auto-oxidation (Lubin and Machlin, 1982). UVB and UVC have profound effects on the function of antigen-presenting cells (Krutzman *et al.*, 1990), and on the expression of cell surface molecules, including the intercellular adhesion molecule -1 (ICAM-1), interleukin-1 and basic fibroblast growth factor (Norris *et al.*, 1990 and Krammer *et al.*, 1993). Radiation dose-dependent inhibition of ICAM-1 expression at the cell surface was observed 24 hr after exposure to 10-100 mJ/cm^2 UV-B. On the other hand, augmentation of ICAM-1 expression by interferon-γ was observed up to 96 hr after radiation, indicating a biphasic response to UV-B. These changes correlated with the suppression of interferon-γ-stimulated keratinocyte proliferation in vitro (Norris *et al.*, 1990). More recently, Behrends and co-workers have demonstrated that ionizing radiation also inhibits expression of the gene for ICAM-1 (Behrends *et al.*, 1995).

It has been suggested that the production of soluble factors may be increased by UV radiation (Krammer *et al.*, 1993). Kramer and co-workers recently observed that UVC induces the synthesis and release of two molecules, interleukin-1 and basic fibroblast growth factor, that are known to be present in a plasma membrane bound form (Mignatti and Rifkin, 1991 and Kurt-Jones *et al.*, 1985). Provocatively, when cell-free culture media prepared from irradiated cells is removed and added to non-irradiated cells, biological effects reminiscent of those induced directly by UV irradiation are evident, including transcription of the collagenase I, HIV-1 and metallothionein IIA genes (Krammer *et al.*, 1993). Since membrane-anchored proteins may be exfoliated from the surface in association with extracellular, plasma membrane-derived vesicles (Dainiak, 1991), one may speculate that the release of factors such as interleukin-1 and basic fibroblast growth factor may result from direct effects on membrane exfoliation (see below). It is unknown whether such molecules are involved in cell-cell communication that may be important for induction of the radiation-induced transformed cell phenotype (Terzaghi-Howe, 1990), induction of clonal anergy among subpopulations of lymphocytes (Isobe *et al.*, 1992 and Kobata *et al.*, 1993) and other biological responses.

MEMBRANE AND NUCLEAR CHANGES IN IRRADIATED CHO CELLS:

We have previously observed that UVB induces profound changes at the lymphocyte surface, including loss of villi, blunting of villi, and abnormal exfoliation patterns (Dainiak *et al.*, 1993), (see Figure 1). To ascertain the time-course of changes in the plasma membrane, relative to those in the nucleus, Chinese hamster ovary (CHO) cells were irradiated for various times and distances from the UV source, and energy of 0-2800 JOULES/m^2 were delivered. We observed dose-dependent reduction in cell viability in irradiated CHO cells (see Figure 2). When a sub-lethal dose of UVB was delivered to CHO cells, changes consistent with apoptosis occurred, including shrinkage of the cell, and condensation and fragmentation of the nucleus. As shown in Figure 3, DNA fragmentation with a characteristic "step ladder" effect was observed after 6 hr. Together, these changes are characteristic of apoptotic cell death (Duvall and Wyllie, 1986).

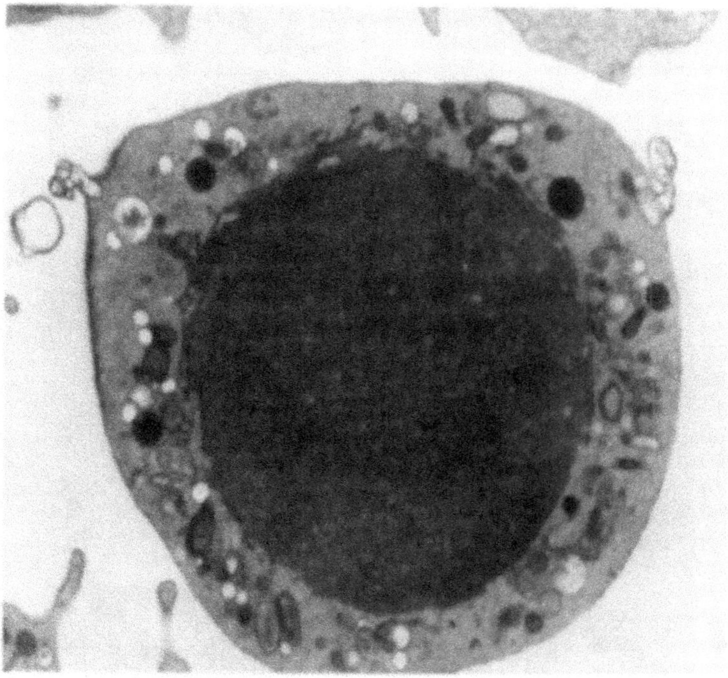

Figure 1: Effects of UVB on lymphocyte surface morphology. Lymphocytes treated with 1200 JOULES/m^2 have few villi and marked blunting of the villi. (Dainiak, reference #25)

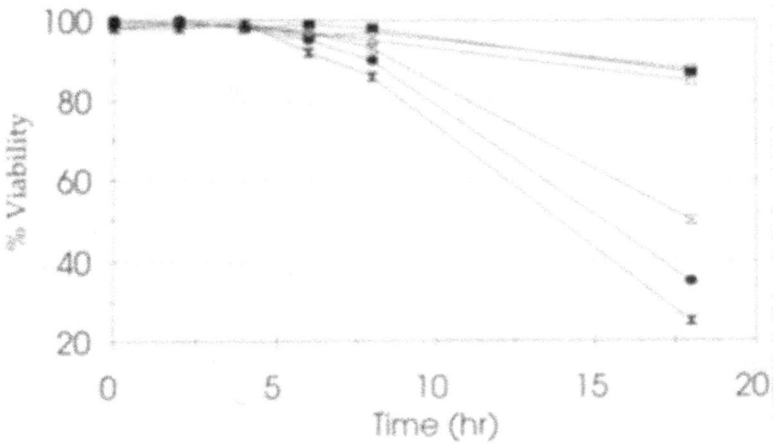

Figure 2: Effects of UVB treatment on viability of CHO cells. Shown are viabilities over time following radiation treatment with 0 (squares), 300 (open circles), 600 (triangles), 1200 (open inverted triangles), 1600 (shaded circles) or 2800 (closed inverted triangles) of UVB. Note a concentration-dependent reduction in cell viability over time.

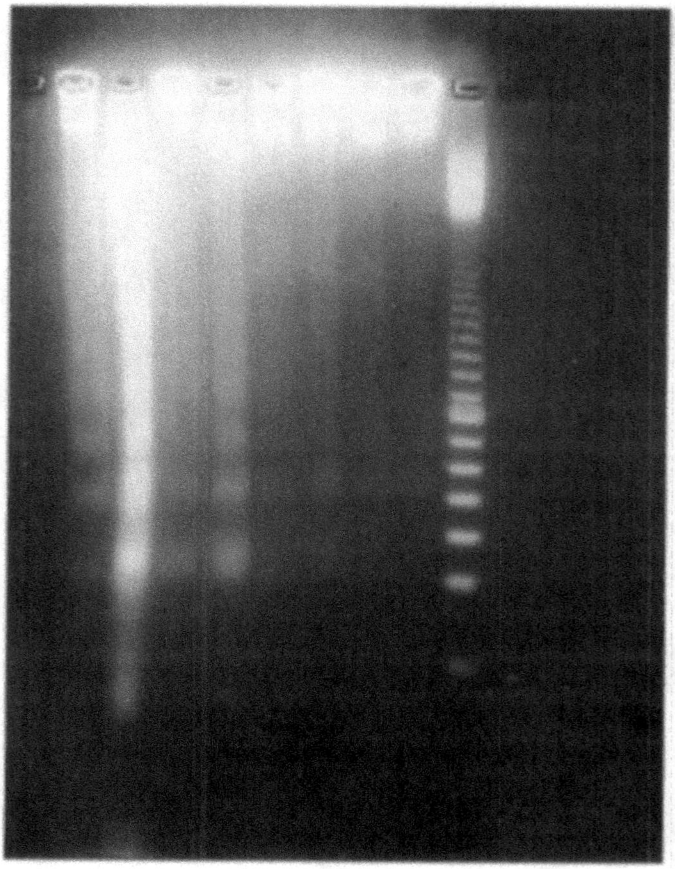

Figure 3: Agarose gel electrophoresis of DNA of CHO cells treated with the UVB (600 JOULES/m^2). Note the step ladder distribution of DNA that is characteristic of apoptosis in DNA isolated for fewer irradiated cells that were incubated for various times greater than 12 hr prior to study.

Radiation-induced cell death via necrosis or during mitosis is most likely due to DNA double strand brakes (Elkin, 1985). Inability to repair such damage is believed to be responsible for greatly enhanced sensitivity to radiation in cells from individuals with ataxia-telangiectasia syndrome (Cornforth and Bedford, 1985). Recently, the importance of cell death through apoptosis has been emphasized as a mechanism for radiation injury (Allan, 1992). Our results are consistent with those previously published indicating that low doses of UV radiation induce apoptosis in lymphocytes, germ cells and intestinal crypt cells (all of which turn over throughout the lifespan of an organism) (Allan, 1992). They also indicate that during apoptosis changes in the plasma membrane occur earlier than do changes in the cell nucleus. It is of interest that the surface of apoptotic cells may lose membrane phospholipid symmetry, thereby exposing phosphatidylserine on the outer leaflet that may trigger specific recognition and removal by macrophages (Fadok *et al.*, 1992). Whether specific DNA changes are also induced by UV radiation that in turn, trigger apoptosis is unknown. For example, it has been suggested that a functioning p53 tumor-suppressor gene must be present for at least some forms of apoptosis to take place (Lowe *et al.*, 1993), and that absence of wild-type p53 may alter sensitivity to radiation in cells from individuals with ataxia-telangiectasia (Kastan *et al.*, 1992). Nevertheless, p53 may not play a role in radiation sensitivity of other cell types (O'Connor *et al.*, 1993 and Slichenmyer *et al.*, 1993).

IMPAIRED RELEASE OF VESICLES FROM THE SURFACE OF IRRADIATED CHO CELLS:

The plasma membrane is a dynamic organelle that is continually turned over, being synthesized from within the cell from preexisting membranes, and at the same time being exfoliated or shed from the surface (Daniak, 1991, Quinn and Cherry, 1992 and Beaudoin and Grondin, 1991). Lipoproteins are transferred between the plasma membrane and the ambient plasma lipids as a consequence of lipid concentration, according to first-order kinetics (Johnson *et al.*, 1988). Work from this laboratory (Daniak, 1991 and Armstrong *et al.*, 1988) and others (Beaudoin and Grondin, 1991, Cone, 1988, Johnstone *et al.*, 1989 and Van Blitterswijk *et al.*, 1982) indicates that distinct areas of the plasma membrane are exfoliated in a directed (rather than random) fashion. Since UVB has a direct effect on the plasma membrane, we considered whether UVB alters the pattern of exfoliation of membrane components into serum-free, liquid culture medium.

The rates of exfoliation from UVB irradiated and control CHO cells that were dislodged and incubated in suspension culture were determined, as previously described (Dainiak and Sorba, 1991). As shown in Figure 4, the rate of exfoliation from UVB-treated cells after 2 hr, 37°C, is reduced by approximately 65%, relative to shedding from untreated, control cells. This effect after so short a time period is unexpected, and may represent an early event in radiation-induced apoptosis. Cumulative release of extracellular vesicles is likewise reduced (see Figure 5).

Whether release of extracellular vesicles is impaired after induction of apoptosis with differentiation factors or exposures other than radiation, remains to be determined. Furthermore, since anchorage-dependent cells undergo apoptosis when grown in suspension (Rouslahti and Reed, 1994), it is possible that cell surface-associated adhesion molecules such as ICAM-1 and other integrins may play a role in protecting cells from radiation-induced apoptosis. Additional studies are required to

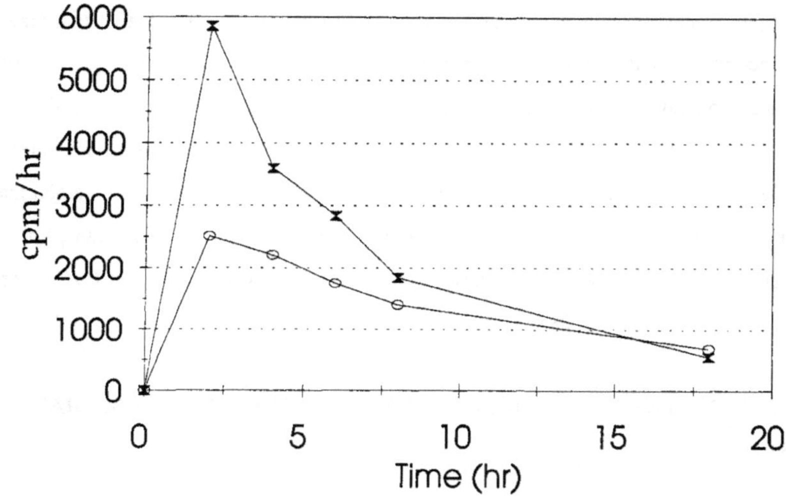

Figure 4: UVB treatment impairs the rate of vesiculation. Plasma membranes of intact CHO cells were labelled with ^{125}I, using iodobeads. The cells were washed and UVB-treated and control cells were resuspended in serum-free, liquid culture medium that was harvested at the indicated times of incubation. Rates of extracellular protein released in the form of membrane-derived vesicles were calculated as counts per minute (cpm) per hr. Note marked diminution in rate of release at 2 hr from irradiated cells (circles), relative to unirradiated cells (inverted triagles).

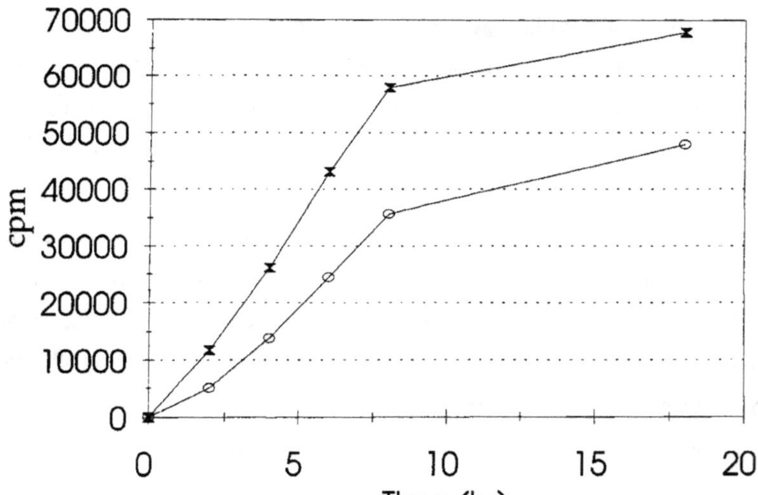

Figure 5: UVB treatment diminishes cumulative release of extracellular vesicles. Release of extracellular vesicles was quantified by measuring radioactivity (cpm) in pellets of serum-free medium conditioned by surface-radioiodinated CHO cells, as described in legend to Figure 4. Note that UVB treatment impairs cumulative release of extracellular vesicles into liquid culture medium. Irradiated cells: circles; unirradiated cells: inverted triangles.

investigate the possibility that dislodgement of CHO cells per se induces apoptosis in our test system. Moreover, contact among different cell types (such as lymphocytes and stromal cells) may also be important in regulating apoptosis (Kataoba *et al.*, 1993). Accordingly, regulation of radiation-induced apoptosis in vivo (where close cell-cell associations take place) may be highly complex, involving other cell types as well as extracellular matrix proteins (Rouslahti and Reed, 1994 and Frisch and Francis, 1994), particularly in the case of chronic, low-dose exposure.

IMPAIRED EXPRESSION OF MEMBRANE-BOUND CYTOKINES ON IRRADIATED CELLS:

Plasma membrane-derived vesicles exfoliated from cells that constitutively express membrane-bound cytokines or that have been transfected with cDNAs that encode membrane-bound cytokines, also express the biological activity (Dainiak and Sorba, 1991, Feldman *et al.*, 1987 and Tuck *et al.*, 1994). Relative to intact membranes, activity on extracellular vesicles is increased at a level of 2-3-fold higher per microgram of protein, suggesting that extracellular vesicles are enriched in selected membrane proteins. Therefore, exfoliation of plasma membrane components may serve as a mechanism by which signals are transferred from one cell to another (Dainiak, 1991 and Beaudoin and Grondin, 1991).

To determine whether UVB has an effect on the expression of cytokines on extracellular vesicles, growth factor-expressing cells were irradiated, and plasma membranes and extracellular vesicles shed from irradiated and control (unirradiated) cells were collected and assayed for biological activity. As shown in Figure 6, growth factor-bearing plasma membranes and vesicles shed from irradiated cells contain a reduced level of activity. This effect is related to radiation dose.

These results suggest that the function of hematopoietic cytokines at the cell surface can be abrogated by UVB. They have potential implications regarding the exposure of individuals to radiation in vivo. Several reports have indicated that accidental exposure to radiation resulting in defective hematopoiesis may be successfully treated with exogenous hematopoietic cytokines (Burtirini *et al.*, 1988 and Wu *et al.*, 1994). It is possible that defective expression of hematopoietic cytokines within the local microenvironment of developing hematopoietic cells may lead to cytopenias that are observed after radiation exposure. This possibility requires further study.

Interestingly, ultraviolet-treated lipoprotein vesicles induce changes in plasma membranes of intact cells that are reminiscent of direct ultraviolet treatment of those cells (Negre-Salvayre *et al.*, 1990).

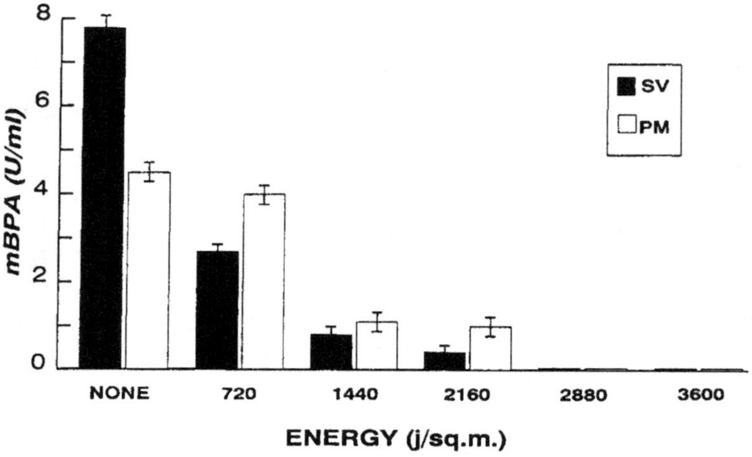

Figure 6: UVB treatment impairs hematopoietic cytokine expression. Levels of membrane bound erythroid burst-promoting activity (mBPA) in plasma membranes (PM) and shed vesicles (SV) from factor-producing cells were measured in bioassays (Dainiak and Sorba, reference #41). Note that vesicles shed from cells irradiated with 720-3600 JOULES/m^2 express dose-dependent reduced activity, relative to activity on vesicles shed from unirradiated cells (None). A similar reduction in activity was observed on intact plasma membranes prepared from irradiated cells (Dainiak, reference #1).

These changes include early blebbing of the plasma membrane of the unirradiated cells. Accordingly, radiation-induced defects in extracellular vesicles that are subsequently incorporated into the membranes of intact cells, may provide a mechanism by which radiation may impair cell-cell signalling within the microenvironment. Together with our results showing that radiation impairs exfoliation of membrane bound cytokines that are known to prevent apoptosis, the data are consistent with the hypothesis that radiation effects on the release and trafficking of extracellular vesicles may mediate programmed cell death of not only factor-producing but also factor-responsive cells. This hypothesis requires rigorous testing.

CONCLUSIONS:

In addition to altering genomic DNA, radiation has direct and indirect effects on the structure and function of biological membranes. Ultraviolet radiation induces structural changes in the plasma membrane and alters the expression of membrane-associated molecules, including those involved in adhesion and cell-cell signaling. The release of plasma membrane molecules is profoundly impaired by exposure to low dose UVB. Furthermore, expression of membrane bound cytokines at the cell surface is also reduced by UVB treatment. Chronic, low dose exposure to ultraviolet light may have significant effects on cells whose turnover is rapid. It is possible that exposure to ultraviolet radiation in outer space or in areas of the world wherein ultraviolet light radiation is increased, may be associated with alterations in homeostasis of the blood-forming tissues. Additional study is required to evaluate the possibility that chronic, low-level radiation exposure may have subtle effects on tissues that are continually replaced in man.

REFERENCES

Allan DJ. Radiation-induced apoptosis: its role in a MADCaT (mitosis-apoptosis-differentiation-calcium toxicity) scheme of cytotoxicity mechanisms. Int J Radiat Biol 62:145, 1992.

Armstrong MJ, Storch J, Dainiak N. Structurally distinct plasma membrane regions give rise to extracelluar membrane vesicles in normal and transformed lymphocytes. Biochim Biophys Acta 946:106, 1988.

Babior BM. The respiratory burst oxidase. Trends Biochem Sci 12:241, 1987.

Beaudoin AR, Grondin G. Shedding of vesicular material from the cell surface of eukaryotic cells different cellular phenomena. Biochim Biophys Acta 1071:203, 1991.

Behrends U, Peter R, Hintermeier-Knabe R, Eibner G, Holler E, Bornkamm GW, Caughman SW, Degitz K. Ionizing radiation induces human intercellular adhesion molecule-1 in vitro. J Invest Dermatol (in press) 1995.

Burtirini A, Gak RP, Lopes DM, et al. Use of recombinant granulocyte-macrophage colony-stimulating factor in the Brazil radiation accident. Lancet 2:471, 1988.

Cheng A, Hogan JL, Caffrey M. X-rays destroy the lamellar structure of model membranes. J Mol Biol 229:291, 1993.

Cone RE. Dynamic aspects of the lymphocyte surface: turnover and shedding of membrane proteins. In: JJ Marchalonis (ed) The Lymphocyte Structure and Function, Dekker, New York, 1988, p.55.

Cornforth MN, Bedford JS. On the nature of a defect in cells from individuals with ataxia-telangiectasia. Science 227:1589, 1985.

Dainiak N. Utility of biological membranes as indicators for radiation exposure: alterations in membrane structure and function over time. In: T.M. Fliedner, E.P. Cronkite (eds.) Biological Indicators of Exposure to Ionizing Radiation, Alpha Press, 1995 (in press).

Dainiak N. Surface membrane-associated regulation of cell assembly, differentiation and growth. Blood 78:264, 1991.

Dainiak N, Tan BJ, Guha A, DeLisa C, Hand A. Ultraviolet-B irradiation of human B cells blocks the expression of a membrane-bound cytokine and alters cell surface morphology. Blood 81:517, 1993.

Dainiak N, Sorba S. Intracellular regulation of the production and release of human erythroid-directed lymphokines. J Clin Invest 87:213, 1991.

Duvall E, Wyllie AH Death and the Cell. Immunol Today 7:115, 1986.

Elkin MM. DNA damage and cell killing: cause and effect? Cancer 56:2351, 1985.

Fadok VA, Voelker DR, Campbell PA, Cohen JJ, Bratton DL, Henson, PM. Exposure of phosphatidylserine on the surface of apoptotic lymphoctes triggers specific recognition and removal by macrophages. J Immunol 148: 2207, 1992.

Fatani W, Rosen D. Irradiation of Langmuir-Blodgett multilayer preparations of phospholipid and a fatty acid. 2. effect of x-radiation. Int J Radiat Biol 62:653, 1992.

Feldman L, Cohen CM, Riordan MA, Dainiak N. Purification of a membrane-derived human erythroid growth factor. Proc Natl Acad Sci USA 84:6775, 1987.

Freeman SE, Grange RW, Sutherland JC, Matzinger EA, Sutherland BM. Production of pyrimidine dimers in DNA of human skin exposed in situ to UVA radiation. J Invest Dermatol 88:430, 1987.

Frisch SM, Francis H. Disruption of epithelial cell-matrix interactions induces apoptosis. J Cell Biol 124:619, 1994.

Haimovitz-Freidman, Kan C-C, Ehleiter D, Persaud RS, McLoughlin M, Fuks Z, Kolesnick RN. Ionizing radiation acts on cellular membranes to generate ceramide and initiate apoptosis. J Exp Med 180:525, 1994.

Health Effects of Exposure to Low Levels of Ionizing Radiation. Beir V, National Academy Press, Washington, D.C., 1990.

Isobe M, Yagita H, Okumura K, Ihara A. Specific acceptance of cardiac allograft after treatment with antibodies to ICAM-1 and LFA-1. Science 255:1125, 1992.

Johnson WJ, Mahlberg FH, Chacko GK, Phillips MC, Rothblat GH. The influence of cellular and lipoprotein cholesterol contents on the flux of cholesterol between fibroblasts and high density lipoprotein. J Biol Chem 263:14099, 1988.

Johnstone RM, Bianchini A, Teng K. Reticulocyte maturation and exosome release: transferin receptor containing exosomes show multiple plasma membrane function. Blood 74:1844, 1989.

Kastan MB, Zhan Q, el Deiry WS, et al. A mammalian cell cycle checkpoint pathway utilizing p53 and GADD45 is defective in ataxia-telangiectasia. Cell 71:587, 1992.

Kataoka S, Naito M, Fujita N, Ishii H, Ishii S, Yamori T, Nakajima M, Tsuruo T. Control of apoptosis and growth of malignant T lymphoma cells by lymph node stromal cells. Exp Cell Res 207:271, 1993.

Kobata T, Ikeda H, Ohnishi Y, Urushibara N, Takahashi TA, Sekiguchi S. UV irradiation can induce in vitro clonal anergy in alloreactive cytotoxic T lymphocytes. Blood 82:176, 1993.

Kramer M, Sachsenmaier C, Herrlich P, Rahmsdorf HJ. UV irradiation-induced interleukin-1 and basic fibroblast growth factor synthesis and release mediate part of the UV response. J Biol Chem 68:6734, 1993.

Krutzman J, Khan IU, Wallis RS, Zhang F, Rich EA, Ellner JJ, Elmets CA. Cell membrane is a major locus for ultraviolet-B-induced alterations in accessory cells. J Clin Invest 85:1529, 1990.

Kurt-Jones EA, Beller DI, Mizel SB, Unanue ER. Identification of a membrane-associated interleukin-1 in macrophages. Proc Natl Acad Sci USA 82:1204, 1985.

Lowe SW, Schmitt EM, Smith SW, Osborne BA, Jacks T. p53 is required for radiation-induced apoptosis in m ouse thymocytes. Nature 362:847, 1993.

Lubin B, Machlin LJ. Vitamin E: biochemical, hematological and clinical aspects. Ann NY Acad Sci 393:1, 1982.

Matsui MS, DeLeo VA. Induction of protein kinase C activity by ultraviolet radiation. Carcinogenesis 11:229, 1990.

Mignatti P, Rifkin DB. Release of basic fibroblast growth factor, an angiogenic factor devoid of secretory signal sequence: a trivial phenomenon or novel secretion mechanism? J Cell Biochem 47:201, 1991.

Morliere P, Moysan A, Santus R, Huppe G, Maziere, J-C, Dubertret L. UV-A-induced lipid peroxidation in cultured human fibroblasts. Biochem Biophys Acta 1084:261, 1991.

Negre-Salvayre A, Lopez M, Levade T, Pieraggi M-T, Dousset N, Douste-Blazy L, Salvayare R. Ultraviolet-treated lipoproteins as a model system for the study of the biological effects of lipid peroxides on cultured cells, II. Uptake and cytotoxicity of ultraviolet-treated LDL on lymphoid cell lines. Biochim Biophys Acta 1045:224, 1990.

Norris DA, Lyons MB, Middleton MH, Yohn JJ, Kashihara-Sawami M. Ultraviolet radiation can either suppress or induce expression of intercellular adhesion molecule 1 (ICAM-1) on the surface of cultured human keratinocytes. J Invest Dermatol 95:132, 1990.

O'Connor PM, Jackman J, Jondle D, Bhatia K, Magrath I, Kohn KW. Role of the p53 tumor suppressor gene in cell cycle arrest and radiosensitivity of Burkitts's lymphoma cell lines. Cancer Res 53:4776, 1993.

Quinn PJ, Cherry RJ (eds). Structural and Dynamic Properties of Lipids and Membranes, Portland Press, London, 1992.

Ronai ZA, Okin E, Weinstein IB. Ultraviolet light induces the expression of oncogenes in rat fibroblast and human keratinocyte cells. Oncogene 2:201, 1988.

Ruoslahti E, Reed JC. Anchorage dependence, integrins, and apoptosis. Cell 77:477, 1994.

Slichenmyer WJ, Nelson WG, Slebos RJ, Kastan MB. Loss of a p53-associated G1 checkpoint does not decrease cell survival following DNA damage. Cancer Res 53:4164, 1993.

Stark G. The effect of ionizing radiation on lipid membranes. Biochim Biophys Acta 1071:103, 1991.

Terzaghi-Howe M. Interaction between cell populations influence expression of the transformed phenotype in irradiated rat tracheal epithelial cells. Radiat Res 121:242, 1990.

Trosko JE. Epigenetic biomarkers: potentials and limitations. In: ML Mendelsohn, JP Peeters, MJ Normandy (eds.) Biomarkers and Occupational Health: Progress and Perspectives, Joseph Henry Press, 1995, p.264.

Tuck D, Cerretti DP, Hand A, Guha A, Sorba S, Dainiak N. Human M-CSF is expressed and shed from the cell surface. Blood 84:2182, 1994.

Van Blitterswijk WJ, DeVeer G, Krol JH, Emmelot P. Comparative lipid analysis of purified and shed extracelluar membrane vesicles from normal murine thymocytes and leukemic GRSL cells. Biochim Biophys Acta 688:495, 1982.

Verma SP. Low levels of irradiation modify lipid domains in model membranes: a laser Raman study. Radiat Res 107:183, 1986.

Wu D, Bosch B, Keating A. Hematopoietic cytokines enhance survival of SCID mice undergoing high-dose irradiation. Exp Hematol 22:202, 1994.

Part 8

Clinical Leukemias

PROGNOSTIC VALUE OF DAY 8 PERIPHERAL BLOOD RESPONSE FOR CHILDREN WITH ACUTE LYMPHOCYTIC LEUKEMIA.

Silvia Regina Brandalise
Department of Pediatrics
University of Campinas
Caixa Postal 1170
13083-970 Campinas,São Paulo
Brazil

INTRODUCTION

Modern treatments for acute lymphocytic leukemia in children consistently produced in this last decade approximately 70 to 80% five year disease free survival for all risk groups. Intensified or "augmented" therapies failed to improve those results. Efforts have been done to identify during the first days of therapy, the group of patients known as slow early responders, who carries a significant poor prognosis. Perhaps for them new strategies based on different pharmacokinetics principles must be done in order to achieve better cure rates.

The prognostic significance of day 4 or day 7 marrow response,as well as, day 8 peripheral blast count reduction is well established by different investigators. However, considering by one side the known difficulties in the routine hematology laboratories to diagnose, after one week of intensive cytoreduction, what is a real lymphoblast and, by the other side, in order to avoid the discomfort involved in a centralized review, we proposed as an alternative to be determined, the WBC counts at day 8 of treatment, as a predictor for therapy failure

PATIENTS AND METHODS

Patient Eligibility

The study was done in the Boldrini's Children Research Hospital for Hematology and Oncology, at Campinas,São Paulo-Brazil. From May 1987 to May 1993, 273 consecutive new patients with acute lymphoblastic leukemia entered on study. Two hundred and sixty five were eligible for analysis. Four were lost of follow-up, two refused treatment and two children were transfered to other cities. To be eligible for the protocol patients had to be younger than 18 years of age at the time of the diagnosis and had no prior chemotherapy and/or corticosteroid treatment. B-ALL patients were excluded from the study.

NATO ASI Series, Vol. H 94
Gene Technology
Edited by A. R. Zander, W. Ostertag,
B. V. Afanasiev, F. Grosveld

Drug Administration Protocol

All patients received a modified ALL-BFM 83 protocol. They were exposed for one week to prednisone, before combination chemotherapy on day 8 has been started. At day 0 one additional dose of MTX was given intrathecal. Therapy for all patients with non-B ALL was the same for all risk groups. The patients received the Protocol I, Protocol M and Protocol II as defined by the BFM-83 study. Prophylactic cranial irradiation stratified according to age and CNS disease was given to all patients. Maintenance therapy with 6-MP and MTX was administered during 2 years.

RESULTS

Patient Characteristics

The patient's age ranged from one month to 17 and 8/12 years. Sixteen percent of the patients were over ten years of age. 15 children (5.5%) were below one year of age. The median age was 4 years and 5 months. There was a 1.07/1 male to female ratio. Ninety-one percent of the patients were caucasian, 8% black and 0.7% other races. In the 265 patients, mediastinal mass was present in 8 (3%). CNS disease at diagnosis was present in 4.5% of the patients. According to BFM risk factors and WBC count at diagnosis, the data are presented in Table 1. Only 12% of the patients belonged to the high risk group (RF ≥1.7) and 36 patients (13.5%) had more than 100,000 leukocytes/μl at diagnosis. E-rosette positivity (over 20%) and/or T-cell markers were positive in forty-five children (17%). Eighty-seven percent of the patients'lymphoblasts were CALLA positive.

Table 1 : Distribution of 265 patients according to the Risk Factor and WBC counts at diagnosis.

RISK FACTOR	Nº PTS.
< 0.8	62
≥ 0.8 < 1.2	72
≥ 1.2 < 1.7	99
≥ 1.7	32
INITIAL LEUKOCYTE COUNTS	
≤ 10.000	117
> 10.000 ≤ 50,000	82
> 50,000 ≤ 100.000	30
> 100.000	36

Survival

Four of 265 patients who completed induction did not achieve remission. There were 9 induction deaths. Remission induction rate was 95% (97%, 95% and 84% for standard risk (RF < 1.2), medium risk (RF ≥ 1.2 < 1.7) and high risk (RF >1.7), respectively).

Event-free survival for the entire study is 84% ± 2.1% at 7 years after diagnosis, with a median follow-up of 4 years (Figure 1). The EFS of males and females are similar on this study. Surface marker data are available in all the patients. There is no significant difference (p = 0.21) between T-ALL positive (EFS 78%) and non - T ALL survival rates (EFS = 69%). There is significant better EFS rates for the patients with CALLA positive (87%) in comparison with 65% for those CALLA negative (p = 0.01). Very few cytogenetic evaluation was performed in this study group.

Children over 10 years of age have an inferior EFS (60%) compared with those between 1 and 10 years (87. 2%). Eight infants out of 14 (≤12 months of age) are in CCR (57%) after a median follow-up of four years. Patients with over 100,000/mm3 WBC at diagnosis have a significantly reduced EFS (Figure 2).

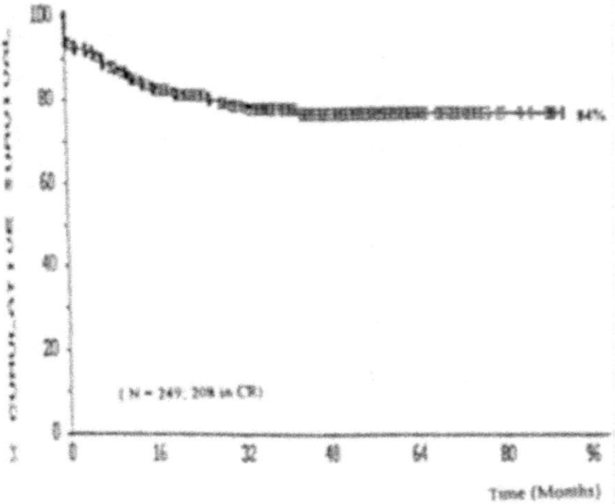

Figure 1 : Overall event free survival for the patients on study

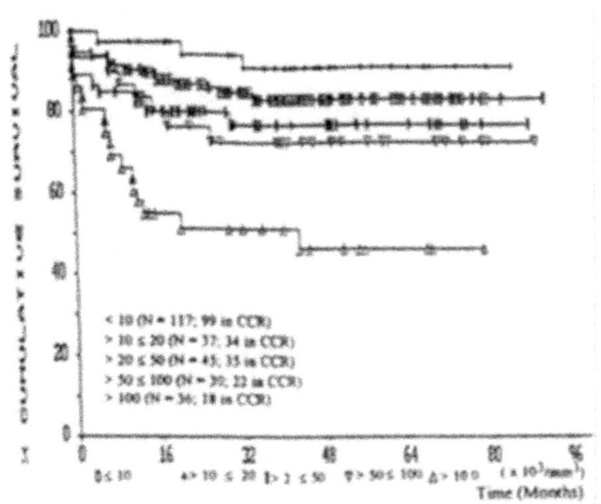

Figure 2 : Event free survival according to WBC counts at diagnosis

Table 2 contains the overall results of the 265 patients according to the risk group . Relapses are predominantely in the bone marrow. The isolated CNS relapse rate is higher in the high risk group (7.4%) in comparison with the other risk groups: 4.2% (RF≥1.2<1.7) and 3.0% (RF<1.2). Testicular relapse occured in only one patient. Nine children died in remission.

Table 2: Overall results of 265 ALL patients on study

PATIENTS AND EVENTS	N°	RISK FACTORS		
		SR	MR	HR
N° of Patients	265	134	99	32
% Total		50	38	12
Induction Deaths	9	3	3	3
Complete Remissions	252	131	94	27
% of Groups	95	37	95	84
Relapses				
BM Only	21	7	12	2
CNS Only	10	4	4	2
Testis	1	1	0	0
BM + CNS	9	2	6	1
Others	1	1		
Remission Deaths	9	5	2	2
Continous Remission	208	114	75	19
96 month EFS (%)	84	89	80	70

MEDIAN FOLLOW-UP = 4 years SR=RF<1,2 MR=RF≥1,2<1,7 HR=RF≥1,7

The overall EFS curves according to the Risk Factors is seen in Figure 3. The 7 year EFS rate for the group of good prognosis (RF < 1.2) is estimated in 89% , 80% for the intermediate risk group (RF ≥ 1.2 < 1.7) and 70% for the high risk patients (RF > 1.7).

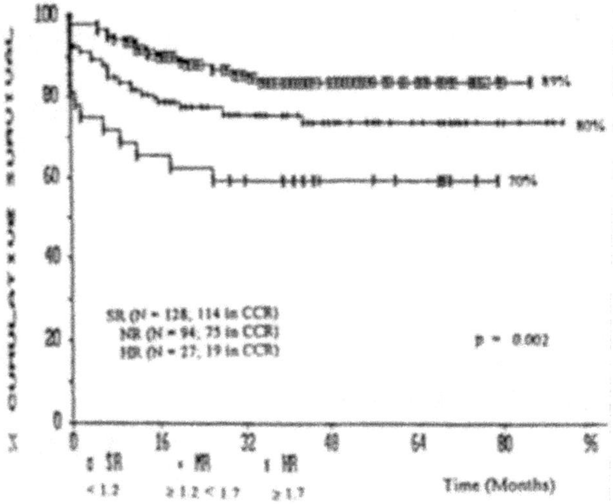

Figure 3:Event free survival according to the Risk Factors

Early Response

The day 8 peripheral blast response was evaluated in 265 patients. Twenty-five children (9%) had more than 1,000 leukemic blasts/mm^3 in the peripheral blood after one week of exposure to prednisone. The 96 month EFS was 81% and 56% for those below 1,000 and over 1,000 blasts/mm^3 on day 8, respectively (Figure 4). However the p-value was not significant ($p = 0.03$). For the BFM investigators the 3 year EFS is 75% and 43% for the group with < 1,000 and ≥ 1,000 blasts at day 8, respectively ($p < 0.0001$).

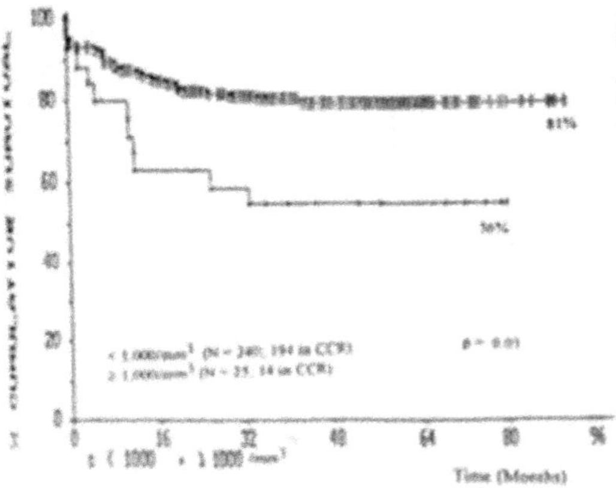

Figure 4: EFS according to the peripheral last counts on day 8

With a multivariate analysis including EFS, age and WBC at diagnosis and the peripheral blast count at day 8, the **r** value is 0.05 with a confidence interval between - 0.07 and + 0.17. A linear curve shows no correlation between EFS and day 8 peripheral blast count (Figure 5) and neither a correlation with WBC count at diagnosis and day 8 peripheral blast.

The early in vivo response of the patients on corticosteroid therapy measured by day 8 peripheral blasts is not an independent variable is this study.

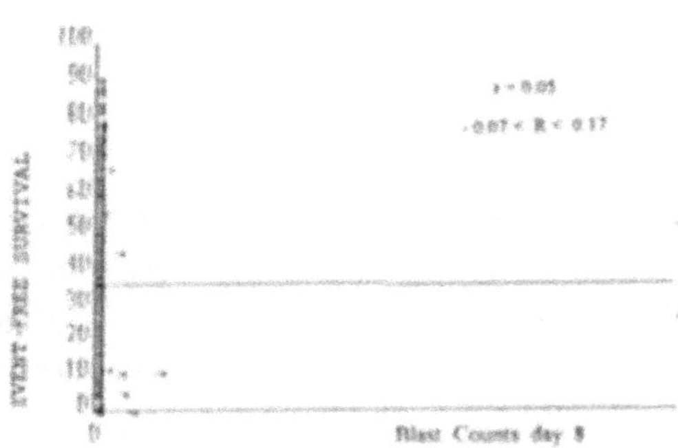

Figure 5: Linear regresion curve of EFS in accordance with blast counts (≥ 1000/mm3) at day 8.

However, the day 8 peripheral WBC count was highly predictive (Figure 6), with the 96 month EFS of 81% for those < 10,000 and only 31% for the patients with over 10,000 leukocytes/mm^3 ($p= 0.00000$). Interestingly the EFS rate for the group with severe leukopenia (< 1,000/mm^3) is 65% due to significant death rate during induction therapy (Figure 7).

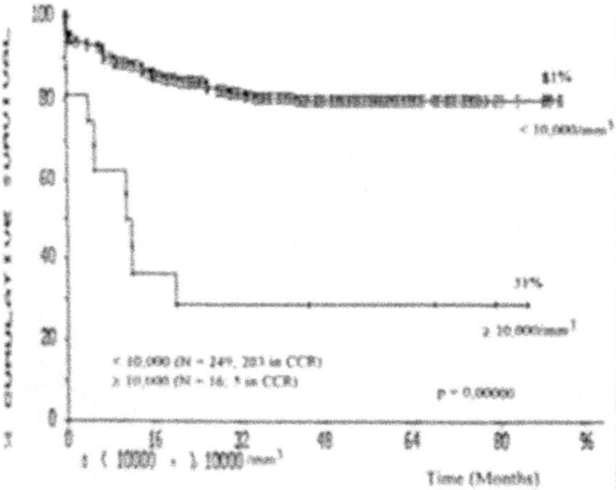

Figure 6 : EFS according to the peripheral WBC counts at day 8

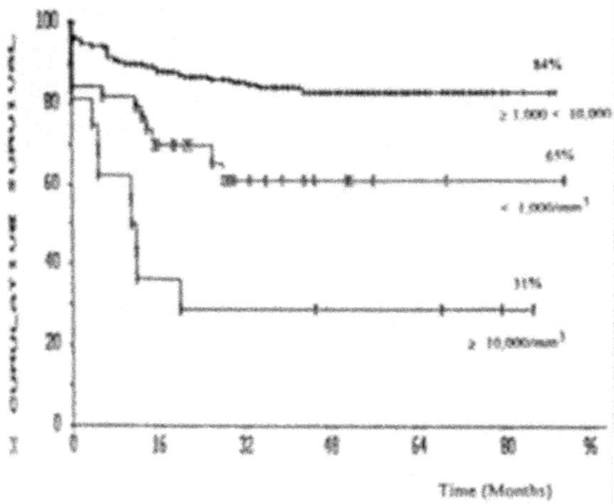

Figure 7: EFS according to the peripheral WBC counts at day 8

In the multivariate analysis considering EFS, age, initial WBC and the day 8 leukocyte counts, there was a significant correlation between the day 8 peripheral WBC counts and the EFS with **r** coeficient = - 0.15(- 0.27 < R < - 0.03). The greater day 8 WBC counts the shorter the EFS rates (Figure 8). Thus, the early in vivo corticosteroid response evaluated by the day 8 WBC counts is highly predictive of survival.

Even when we considered just those patients (N=15) with initial WBC over 200,000/mm^3 the multivariate analysis again shows a significant correlation with **r** = - 0.15 (- 0.27 < R < - 0.03). There is a negative relationship between day 8 WBC counts and EFS rates (Figure 9).

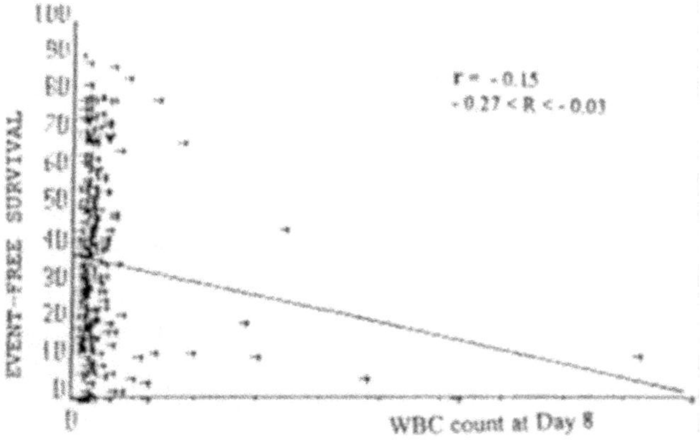

Figure 8 : Linear regression curve of EFS in accordance with peripheral WBC counts at Day 8.

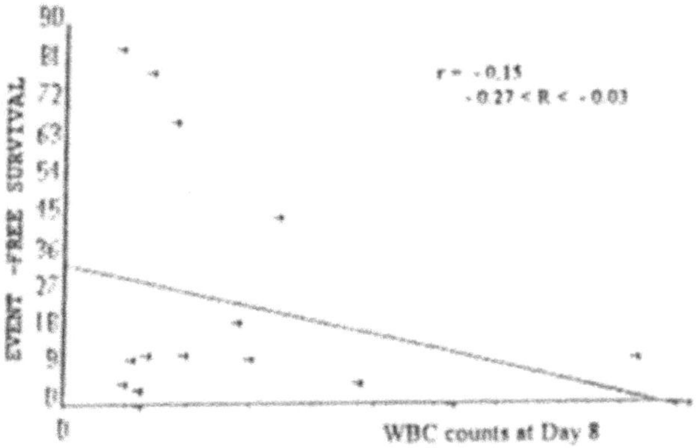

Figure 9 : Linear regression curve of EFS in accordance with day 8 WBC counts for the patients with initial WBC ≥ 200,000/mm3

CONCLUSIONS

1st - This modified BFM-83 protocol for treatment of childhood acute lymphocytic leukemia provides long term good results compared to those obtained in the developped countries.

2nd - In this study there is no correlation of day 8 peripheral blast counts neither with the EFS rates nor with the initial WBC counts.

3rd - The day 8 peripheral WBC counts is highly predictive of relapse. The higher leukocyte counts (>10,000/mm^3) are correlated with lower EFS rates. And finally,

4th - The in vivo corticosteroid response evaluated by the peripheral WBC count is a simple laboratory test, with high reprodutibility and without no subjective influence.

REFERENCES

Camitta B, Mahoney D, Leventhal SJ, et al 1994) Intensive Intravenous Methotrexate and Mercaptopurine Treatment of Higher-Risk Non-T, non-B Acute Lymphocytic Leukemia: A Pediatric Oncology Group Study.J.Clin. Oncol.12 (7) : 1383 - 1389

Evans WE, Rodman J, Relling MV, et al. (1991) Individualized dosages of chemotherapy as a strategy to improve response for acute lymphocytic leukemia. Semin. Hemat.28 : 15 - 21.

Land VJ, Shuster JJ, Crist WM, Ravindranath Y. et al (1994).Comparison of Two Schedules of Intermediate - Dose Methotrexate and Cytarabine Consolidation Therapy for Childhood B - Precursor Cell Acute Lymphoblastic Leukemia : A Pediatric Oncology Group Study. J.Clin.Oncol. 12 (9) : 1939 - 1945.

Niemeyer CM, Reiter A, Riehm H, et al (1993). Comparative results of two intensive treatment programs for childhood acute lymphoblastic leukemia: The Berlin - Frankfurt - Munster and Dana - Faber Cancer Institute protocols. Ann. Oncol. 2 : 745 - 749.

Van Eys J, Berry D, Crist W, et al. (1989). Treatment intensity and out come for children with acute lymphoblastic leukemia of standard risk. A Pediatric Oncology Group Study. Cancer 63 1466 - 1471

GM-CSF in Addition to Chemotherapy of ALL for Kinetics Based Protection of Stem Cells and Stimulation of Haemopoiesis. A randomized Study.

Jerzy Holowiecki
for the Polish Acute Leukemia Group*
Department of Haematology,
Silesian Medical Academy
Reymonta Str.8, 40029 Katowice
Poland

Intensive induction chemotherapy regimens has led to considerable improvement in the prognosis of adult acute lymphoblastic leukemia (ALL), but the results are still unsatisfactory and only about 35% of patients can be cured with current available therapy (7,10,11,14,). In conclusions from the studies of large series of adult ALL the complete remission (CR) rate ranges from 70 to 85% (2,6,7, 9,10, 11, 14,16,23,27,30). A further intensification of chemotherapy will increase CR rates but it is limited by hematopoietic toxicity and therefore the use of recombinant human growth factors in addition to induction chemotherapy has been attempted (13,18,19,22,23,25,29). The studies on bone marrow cells kinetics during GM-CSF treatment revealed , that 48 to 96 hours after GM-CSF discontinuation the proliferative activity of normal stem cells and progenitors decreases to levels lower than that observed prior to GM-CSF administration (1,5,12,28), so this represents a period of partial refractoriness of these cells to the cell cycle dependent cytostatics.

Based on the above data as well as on pharmacokinetics of anthracyclines and after succesful completing of the pilot study (12), an original cell kinetics based protocol using rhGM-CSF during induction and consolidation treatment of adult ALL has been designed and evaluated in a multicenter, randomized, prospective study. The aims of this study were as follows: 1) to evaluate the efficacy and tolerability of recombinant human GM-CSF in adult ALL patients, 2) to prove the concept that sequential administration of cell cycle specyfic cytostatics and GM-CSF during induction can enhance the myeloprotection by modulation of normal stem cell kinetics, 3) to evaluate the efficacy of GM-CSF for prevention of neutropenia after intensive consolidation therapy.

*Polish Acute Leukemia Study Group: Holowiecki J.,Cedrych I.,Krzemien S.,Rudzka E.,Holowiecka B.,Jagoda K., Krawczyk M., Hellmann A.,Maj S.,Konopka L.,Dmoszynska A.,Krykowski E.,Robak T.,Kuratowska Z., Skotnicki A.

NATO ASI Series, Vol. H 94
Gene Technology
Edited by A. R. Zander, W. Ostertag,
B. V. Afanasiev, F. Grosveld

Material and methods.

The multicenter, prospective, randomized trial, using GM-CSF in adult ALL patients has been carried out in 8 departments of haematology. Recombinant human GM-CSF (E.coli derived, Leucomax) had been supplied by Sandoz/Schering Plough.

Fifty two consecutive, previously untreated ALL adult patients were enrolled to the study (F=22,M=30,median age 29 (range16-59), immunological phenotypes: common=35, early pre B=7, pre T/T type=10, median WBC count=9,4 G\L, median Hb=9,35 g|l, median platelets count=59 G\L, hepatomegaly presented 63% of patients, splenomegaly-65%, lymphomegaly-71% of patients). Patients below 16 years and over 60 years old, FAB L3,ALL with "myeloid" suface antigens, history of anaphylaxis to human proteins, with severe heart, liver,lung or kidney impairment, were the exclusion criteria. After informed consent had been obtained, patients were stratified according to risk groups: the high risk one (age>35, WBC>30 G\L, "common" negative, phenotype Ph+) and the standard risk one (none of the above criteria). Within each group patients were randomized to receive either chemotherapy plus GM-CSF (GM arm) or chemotherapy alone (control arm). The patient's characteristics in each group is presented in the table 1.

[N=52 IN 8 CENTERS]

	HIGH RISK n=32		STANDARD RISK n=21	
Arm, n	A=15	B=17	A=10	B=10
sex F/M	4/11	9/8	3/7	6/4
age •	36	33	18	22
Common	9	6	10	10
early pre B	2	5		
T type	4	6		
Karnofsky sc.	80	72	70	70
Hb mg/dl•	8.7	8.6	10.25	9.2
WBC G/l •	80.2	41	3.35	5.9
Plt G/l •	42	75	85.5	85.5
Hepatomegaly	11/15	12/17	5/10	5/10
Splenomegaly	12/15	12/17	5/10	5/10
Lymphomegaly	12/15	11/17	6/10	8/10

• median values

tab. 1 Material

The patients treated according to the schedule presented in fig. 1.

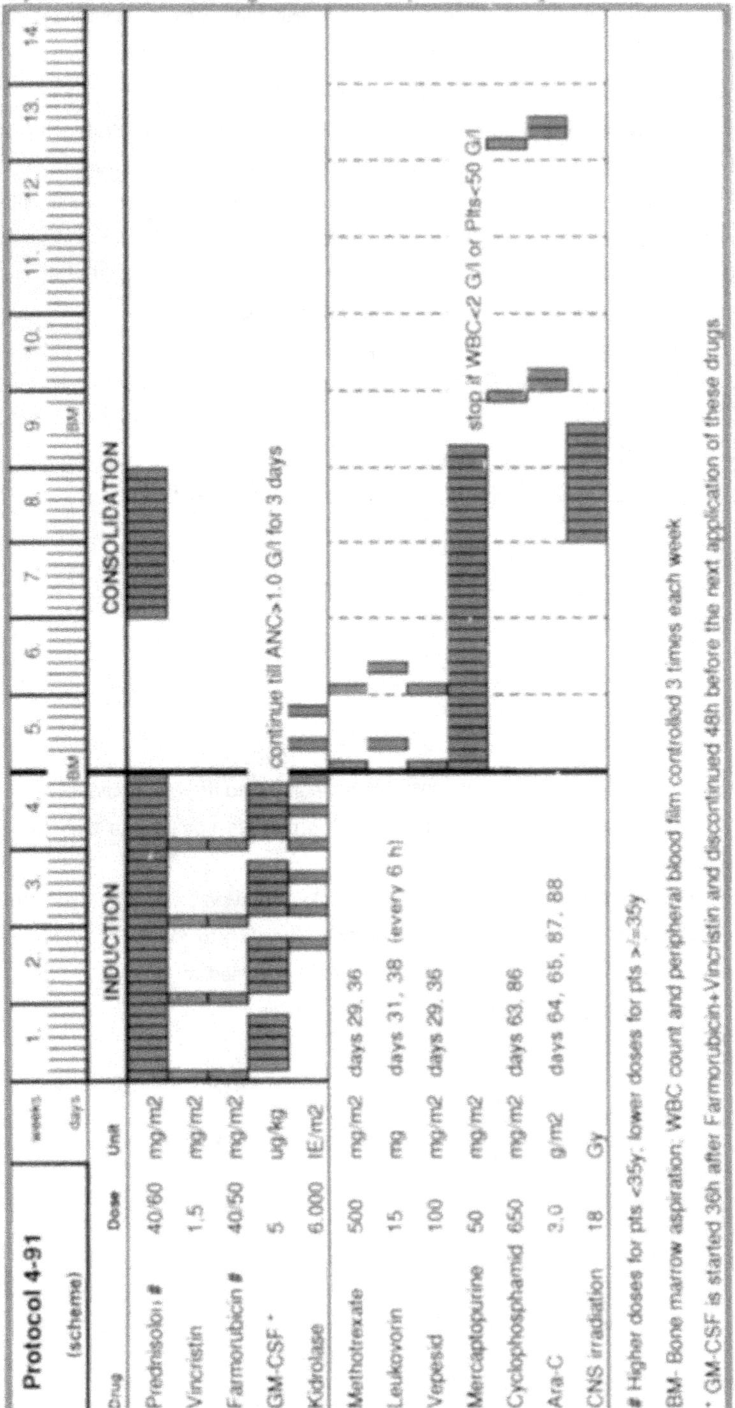

fig. 1 Scheme of the treatment protocol

Recombinant human GM-CSF (Leucomax), was administered subcutaneously , in daily dose of 5 ug/kg body weight, during induction, in five day cycles, each one was started 36 hours after Farmorubicin and Vincristin injection and discontinued 48 hours before the next injection of cytostatics. After completion of the induction, as well as during consolidation with Cyclophosphamide and HD ARA-C, GM-CSF was continued untill absolute neutrophil count (ANC) in peripheral blood reached 1 G\L during three consecutive days. The peripheral blood morphology with differential count, reticulocytes and platelets were monitored three times a week. In addition, the monitoring of peripheral blood CD34,CD38 positive cells was performed in 13 patients; 8 receiving GM-CSF and 5 treated in the control arm (each time on the last day of GM-CSF administration -days 3,20,27, and next after 48 hours, just before cytosta-tics dministration - days 1,8,15,22. Imunophenothyping of leukemic blasts at the start of induction and next peripheral blood active stem cells measurements during induction treatment were performed using flow cytometry (FacStar plus,Becton Dickinson) and immunoenzymatic APAAP staining.The statistal analysis of survival was performed using original database programme Leukos 7 and Biomedical Package Software BMDP.

Results

The overall CR rate equalled 73%, with 90% CR in the standard risk group and 62,5% in the high risk one. A higher CR rate of 80% has been obtained in patiente treated with GM-CSF (GM arm) as compared to the control group -66%.

All patients in the "GM arm" achieved CR after 1 induction cycle , whereas 33% of patients in the control arm received more than 1 cycle of induction. The time to reach CR in the "GM arm" was significantly shorter (median 29 days) than in the control one (median 41 days) $p<0.05$. The particular results obtained in each group of patients are schown in table 2.

	HIGH RISK		STANDARD RISK	
Arm	A	B	A	B
REMISSION INDUCTION				
CR	11/15•73,3%	9/17•52,9%	9/10•90%	9/10•90%
PR/NR	0/2	3/1	0	0/1
> 1 cycle to CR:	0	5•55%	0	1•10%
Death	2	4	1	0
POSTREMISSION				
Relapsed	5•33%	4•23%	3•33%	2•20%

No of evaluable pts 52,

tab. 2 Results of induction treatment

Analysis of bone marrow regeneration times revealed, that in the GM arm both the cytoreduction and the subsequent regeneration of particular haematopoetic lines occured significantly earlier than in the control arm (median of days to nadir granulocyte counts equaled 9 versus 15 days, $p<0.05$ and to nadir platelets 6.5 versus 12,5 days respectively, $p<0.05$). Times to granulocyte recovery and surprisingly also to platelets recovery were also significantly shorter in the "GM arm" than in the control one; median days to ANC >0.5 G\L 11 versus 28, to >1 G\L 21,5 versus 33,5, to >1,5 G\L 24 versus 34 days, $p<0.05$, median of days to reach platelet count >50 G\L 8.5 versus 12 days, >100 G/l 26,5 versus 33,5 days ($p<0.05$. The time to clearance of blasts from peripheral blood was also shorter in the GM arm (median 7 days) as compared to the control group (10 days).

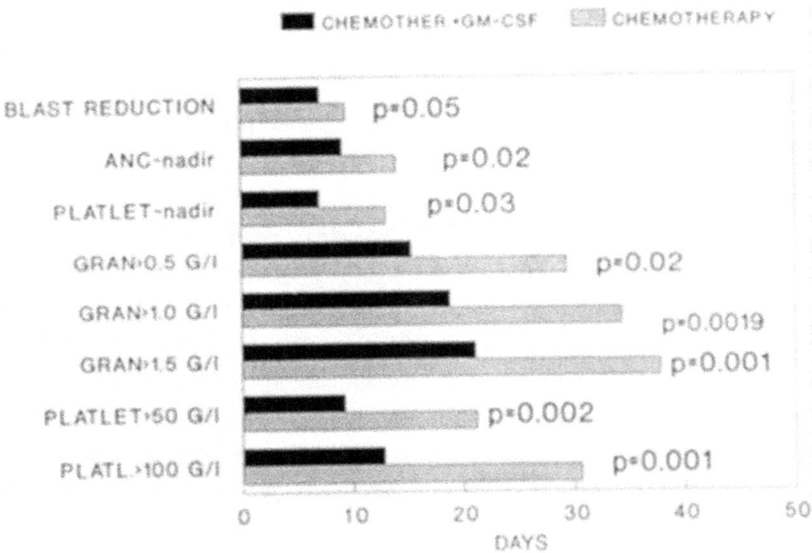

fig.2 Cytoreduction and regeneration times

The use of GM-CSF shortened significantly the neutropenic period during induction, the mean of days with ANC < 0.5 G\L in GM arm equaled 10,6 as compared to 21,4 in the control arm ($p<0.05$) and only 59% of GM-CSF treated patients displayed agranulocytosis, whereas 85% in the control group. The insidence and severity of infections during induction were reduced in the GM-CSF treated group (only 9% of patients in GM arm developed infection of WHO grade 4, compared to 23% in the control arm). As a consequence also the antibiotic treatment was significantly reduced in the GM arm ($p<0.05$) fig 3.

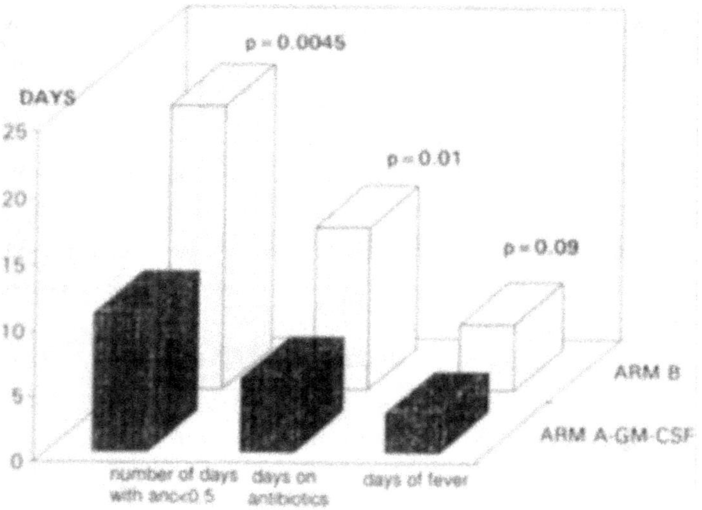

fig.3 Number of days of agranulocytosis, antibiotics treatment and fever compared in both groups

The need of RBC and platelets support during induction appeared to be lower in the GM arm than in the control one; median RBC liters equaled 1,5 versus 2,3, median platelets units 3 versus 3,8 respectively).

During consolidation, the GM-CSF treatment as an adjunct to high dose chemotherapy enabled a close adherence to the intensive therapy regimen (the average dalay of the subsequent cycle equaled 0.5 day in the MM arm versus 6.6 days in the control group, $p<0.5$). In the "GM arm" the frequency of infection, the fever rate, and the usage of antibiotics were also lower than in the control group.

The current follow up, reveales comparative results in both groups with not significant differences between Kaplan Meyer DFS curves (Wilcoxon Breslow-p=0.9,Mantel-Cox-p=0.6). The median CCR has not been reached at the mean observation time of 601 days in GM arm and 525 days in the control arm, although the tendency suggesting the better outcome in GM-CSF arm appears (75 th quantile has not been reached yet versus 298 days in control) - fig 4

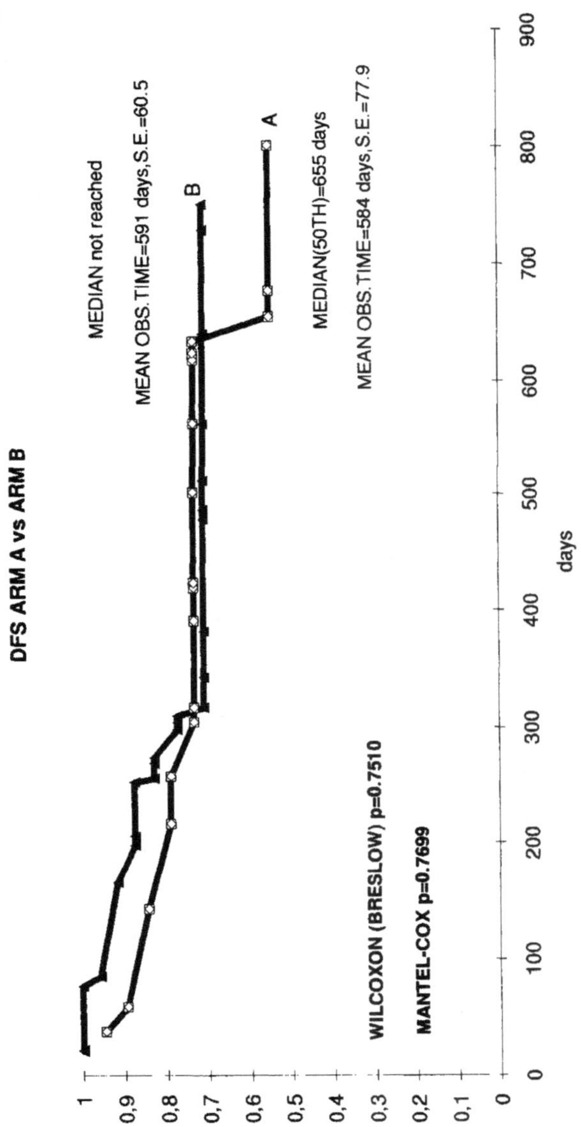

fig 4.Kaplan Meyer curves of disease free survival in both groups of patients

Repeated dual color staining for CD34CD38 flow cytometry analyses of peripheral blood cells demonstrated a significant increase in the absolute count and percent of these cells in the mononeulear and granulomonocytic fraction, during GM-CSF administration , followed by the rapid decrease 48 hours after the cessation of the cytokine. This phenomenon has been never observed in patients obtaining chemotherapy only, who displayed a distinct decrease of CD34CD38 positive cell count during the induction treatment. (fig.5)

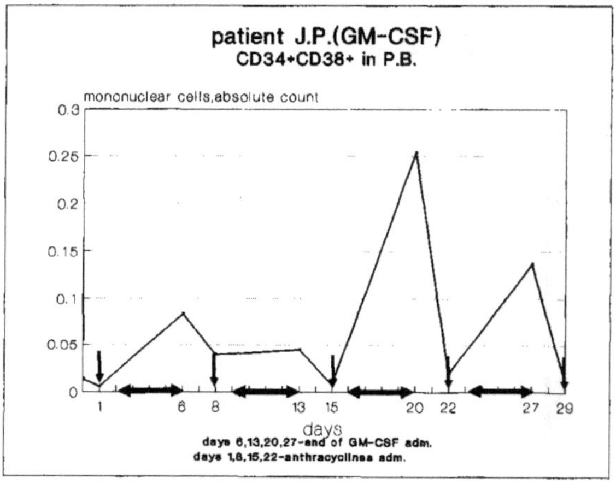

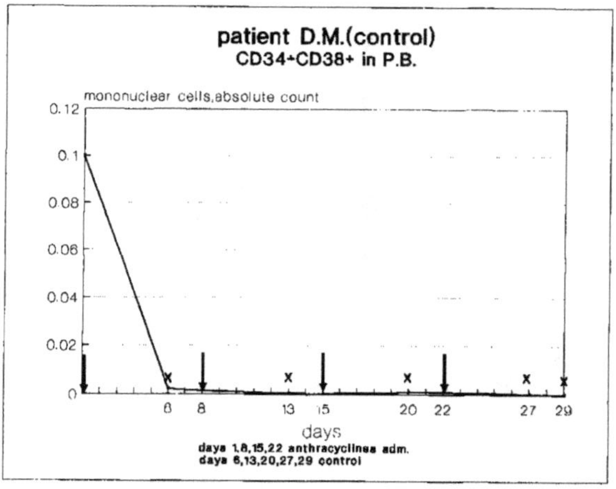

fig. 5.The kinetics of peripheral blood CD34+CD38+cells during GM-CSF and chemotherapy treatment compared to chemotherapy alone

Discussion

The current results of the study presented here support the idea of the beneficial use of growth factors in combination with chemotherapy in acute lymphoblastic leukemia. The results of multiple studies reflect the impact of the use intensive chemotherapy regimens in induction of ALL (2,6,7,9,10,11,13,14,16,18,22,23,25,27,30) to achieve rapid cytoreduction, but it is important to balance the antileukemic effect with the minimal myelotoxicity (14). In the German Multicentre Adult ALL Trial the granulocytopenia has appeared the major dose limiting toxicity, leading to delay or discontinuation of therapy and most investigators have demonstrated that an induction failure is equally due to the toxicity and to the refractoriness to chemotherapy (14). The use of recombinant growth factos such G-CSF and GM-CSF in the management of lymphoid malignancies have demonstrated their effectiveness in supportion of myelopoiesis, a good tolerability (4,8,11,13,18,24,25) with no evidence of malignant lymphoid clones stimulation (1,3,15,17,20,26). In addition to supportive function of growth factors in cytopenia, new perspectives have been offered based on prewious data suggesting that 48-96 hour period after discontinuation of GM-CSF administration represents a period of relative refractoriness of normal progenitors to cell cycle specyfic chemotherapy (1,3,5,12,26,28). Based on these findings we developed a novel remission induction protocol, and the data presented here confirm its efficacy. Our flow cytometric findings of kinetic changes of peripheral blood CD34 CD38 posivite cells during and after GM-CSF treatment appear to confirm that a period of relative refractoriness of normal stem and progenitor cells may be obtained using properly sequenced administration of rhGM-CSF and cytostatic drugs. Rapid granulopoietic reconstitution, as well as a rather unexpected shorter platelets recovery time , achieved in GM-CSF group should be considered as an another evidence of cytoprotection on multilineage stem cell level. Complete remission has been achiewed significantly more rapidly in patients receiving GM-CSF in addition to chemotherapy and this finding is of great importance. At first it confirs the value of the above discussed policy and secondly it has a prognostic value predicting for a longer remission duration. Data from large series of ALL patients demontrated that a short induction treatment duration , is an one of the most important parameters predicting long term disease free survival (9,10,11).

Further investigations are needed to explain the accelerated reduction of leukemic cells during GM-CSF administration. A priming effect of secondary secreted cytokines as well as activation of immunological mechanisms of tumor control should be taken into consideration. In summary, the protocol employing GM-CSF has following advantages in comparison to chemotherapy alone: shorter time to CR, ($p<0.05$), CR in all cases obtained after 1 chemotherapy cycle, accelerated blasts cytoreduction ($p=0.05$), accelerated hematopoietic reconstitution ($p<0.05$), better outcome of treatment with lower incidence and severity of infections. The use GM-CSF during high dose consolidation treatment permits close adherence to planned chemotherapy without delay.

Our clinical data supported by flow cytometric findings confirm the idea of a cell kinetics based protection of stem cells in ALL, using properly sequenced chemotherapy and GM-CSF.

The study protocol received the Local Ethics Commitee agreement.
This trial was subvenced in part by Polish Research Comitee, Grant No.406669101
Participating centers: Dept of Hematology Medical Academy Katowice, Dept.of Hematology M.A. Gdansk,Institute of Hematology Warsaw, Dept.of Hematology M.A. Lublin, Dept.of Hematology M.A. Lódz, .Dept.of Hematolgy M.A..Warsaw, Dept.of Hematology M.A. Kraków.

Bibliography

Aglietta M.Piacibello W.et al.Kinetics of Human Hematopoietic Cells After in Vivo Administration of Granulocyte Macrophage Colony Stimulating Factor J.Clin.Invest.1989.83.

Anglin P.Keating A.Adult Acute Lymphoblastic Leukemia Treated with a Pediatric High Risk Protocol : Single Centre Report of 50 Consecutive Cases.Blood.1993.83(10)Suppl 1

Bernstein I.D.,Andrews R.G.Recombinant Human Stem Cell Factor Enhances the Formation of Colonies by CD 34lin- Cells and the Generation of the Colony Forming Cell Progeny from CD 34+lin-Cells Cultured with Interleukin 3,Granulocyte Macrophage Colony Stimulating Factor or Granulocyte Colony Stimulating Factor .Blood.1991.77.

Blaise D,Vernant JP,Fiere D,Gluckman E,Reiffers J,et al.A Randomized Controlled Multicentre Trial of Filgrastim in Patients Treated by Bone Marrow Transplantation with Total Body Irradiation for Acute Lymphoblastic Leukemia or Lymphoblastic Lymphoma.Congress Abstract Review,34 th Annual Meeting of American Society of Hematology"Growth Factors in Hematology "December,1992

Broxmeyer HE,Benninger L,PatelS,Benjamin RS,Vadhan-Rai S,Differences in Kinetics Responses of Human Marrow Myeloid Progenitor Cells to in Vivo Treatment of G-CSF and GM-CSF.Congress Abstract Review,34 th Annual Meeting of American Society of Hematology"Growth Factors in Hematology "December,1992

Casssileth P.A.Andersen J.W.Bennett J.M.et al.Adult Acute Lymphocytic Leukemia:The Eastern Cooperative Oncology Group experience .Leukemia.1992. 6 Suppl 2

Gale R.P.Butturini A.Controversies in Therapy of Acute Lymphoblastic Leukemia.Leukemia.Leukemia.1992. 6,Suppl 2

Gerhartz H.H.,Englehard M,Meusers P,et al.Randomized Double Blind,Placebo Controlled Phase III Study of Recombinant Human Granulocyte Macrophage Colony Stimulating Factor as Adjunct to Induction Treatment of High Grade Malignant Non Hodkin Lymphomas.Blood,Vol 82,No 8,1993,pp.2329

Hoelzer D.Therapy of Newly Diagnosed Adult Acute Lymphoblastic Leukemia.Hematol-Oncol-Clin-North-Am.1993 Feb:7(1) 139-60

Hoelzer D.,Therapy of Acute Lymphoblastic Leukemia in Adults.Leukemia.1992.6,Suppl 2.

Hoelzer D.Thiel E.Ludwig WD.Loffler H.,Buchner TH.,Freud M.,et al.The German Multicentre Trials for Treatment of Acute Lymphoblastic Leukemia.Leukemia.1992 6.Suppl 2

Holowiecki J.,Cedrych I.et al.Efficacy of an Original Cell Kinetics Based Chemotherapy Regimen Using rHuGM-CSF for Induction and Consolidation Treatment of ALL.Preliminary Analysis.Acute Leukemias IV.Springer-Verlag Berlin Heidelberg 1993

Link H,Boogaerts MA,Carella AM,et al.A Controlled Trial of Recombinant Human Granulocyte-Macrophage Colony Stimulating Factor after Total Body Irradiation, High Dose Chemotherapy and Autologous Bone Marrow Transplantation for Acute Lymphoblastic Leukemia or Malignant Lymphoma Blood 1992,Nov 1,80(9) 2188-95

Linker C.A.Levitt L. J.et al.Treatment of Adult Acute Lymphoblastic Leukemia with Intensive Cyclical Chemotherapy:A Follow uo Report.Blood.1991.78.

Lord B.I.,Molineux G.et al.Myeloid Cell Kinetics in Mice Treated with Recombinant Interleukin 3,Granulocyte Macrophage Colony Stimulating Factor or Granulocyte Stimulating Factor.Blood.1991.77.

Mandelli F.Annino L.Vegna M.L.GIMEMA ALL 0288 : a Multicentric Study on Adult Acute Lymphoblastic Leukemia.Preliminary Results.Leukemia.1992 6.Suppl 2.

Ode D.L.Zhou M.et al.The Effect of Recombinant GM-CSF and G-CSF on the Bone Marrow Cells of Children with Acute Lymphoblastic Leukemia. Leukemia.1992.6(11)

Ottman OG,Hoelzer D,.Concomitant rmetHuG-CSF(Filgrastim) and Intensive Chemoradiotherapy as Induction Treatment in Adult ALL.A Randomized Multicenter Phase III trial.Blood.1993.83(10)Suppl 1

Reynoso E.E.Duenas S.Granulocyte Macrophage Colony Stimulating Factor (GM-CSF) to Atenuate the Hematological Toxicity of Phase II Induction of Adult Acute Lymphoblastic Leukemia.Blood.1993.83 (10) Suppl 1

Ridgway D.Borzy M.S.Differential Effect of Continous versus Cyclic Maintenance Chemotherapy on Cytokine Induced Myeloid Colony Growth.Hematol.Pathol.1990.4.87.

Ridgway D.Borzy M.S.Defective Production of Granulocyte Macrophage Colony Stimulating Factor and Interleukin 1 by Mononuclear Cells from Children Treated for Acute Lymphoblastic Leukemia.Leukemia 1992.6.

Scherrer R.Geissler K.et al.Granulocyte Colony Stimulating Factor as an Adjunct to Induction Chemotherapy of Adult Acute Lymphoblastic Leukemia.Ann.Hematol.1993.66.(6)

Schiffer C.A. Larson R.A. Bloomfield C.D.Cancer and Leukemia Study Group B (CALGB) Studies in Adult Acute Lymphocytic Leukemia.Leukemia.1992. 6,Suppl 2

Schulz G.Frish J.et al.New Therapeutic Modalities for the Clinical Use of thGM-CSF in Patients with Malignancies.Am.J.Clin.Oncol.1991.14.

Smedmyr B.Glycosylated rHuG-CSF as Supportive Treatment During Remission InductionTherapy for Adult Patients with Newly Diagnosed ALL\AUL.Blood.1993

Strauss L.C.Welsh S.B.et al.Initiation of DNA Synthesis by Colony Stimulating Factors in Subsets of Human CD34+Marrow Cells .Exp.Hematol.1991.19.

Stryckmans P.De Witte TH.et al.Therpay of Adult ALL:overview of 2 succesive EORTC studies :(ALL -2 & ALL-3).Leukemia.1992 6 Suppl 2

Vadhan Rai S.Hal. E. et al.Abrogating Chemotherapy Induced Myelosupression by Recombinant Granulocyte Macrophage Colony Stimulating Factor in Patients with Sarcoma.Protection at the Progenitor Cell Level.J.Clin.Oncol.1992.10.

Welte K.Mempel K.Reiter A,et al.A Randomized Phase III Study of Recombinant Human Granulocyte Colony Stimulating Factor (rmet Hu G-CSF) in Childchood High risk acute lymphoblastic leukemia.Blood 1993.83 (10) Suppl 1

Weiss M.Berman E.Rapid Remission Induction Therpay in Adult Acute Lymphoblastic Leukemia Without VCR or Prednisone.Blood 1993.83 (10) Suppl 1

CYTOGENETIC ABNORMALITIES AND THEIR CLINICAL AND PROGNOSTIC VALUE IN ACUTE LEUKEMIA

MARINETS O.V., MAMAEV N.N.*
DEPARTMENT OF MOLECULAR GENETICS
CLINICAL CENTER FOR ADVANCED MEDICAL TECHNOLOGIES
DYNAMO 3
197042 ST-PETERSBURG
*PAVLOV MEDICAL UNIVERSITY
L. TOLSTOY 6/8
197089 ST-PETERSBURG
RUSSIA

INTRODUCTION

Several chromosomal rearrangements are known to be associated with different types of acute leukemia (AL) (Fenauxet al., 1996; Kalwinsky et al., 1990; Pathak, 1990). Many studies have indicated an association between specific recurrent chromosomal changes and morphological, clinical and prognostic features of hematological diseases (Rowley, 1984; Heim, Mitelman, 1987; Secker-Walker, 1990).

The deletion or monosomy of chromosome 5 or 7 is very common cytogenetic markers in secondary AL. These chromosomal changes are associated with resistance to therapy and very poor prognosis (Keating et al., 1988; Bunin et al., 1991; Lewis et al., 1991).

Translocation t(8;21)(q22;q22) is highly correlated with a specific phenotype that is classified M2 in the FAB nomenclature (Fourth Internat. Workshop, 1984). Patients with translocation t(8;21) are characterized with good response to therapy and high complete remission (CR) rate (Groupe Franc. Cytogenet. Hematol., 1990; Kalwinsky et al., 1990).

However, in addition to these primary chromosomal markers secondary chromosomal changes are also often seen (Heim, Mitelman, 1986; Pathak, 1990). Secondary chromosomal aberrations arise in the course of tumor progression and seem to have little specificity to any AL type. Heim and Mitelman (1986) proposed that initial cytogenetic changes are important factors in determining the nature of subsequent chromosomal abnormalities, developing in the malignant clone. The clinical

NATO ASI Series, Vol. H 94
Gene Technology
Edited by A. R. Zander, W. Ostertag,
B. V. Afanasiev, F. Grosveld

significance of secondary chromosomal aberration at the present time remains unclear.

MATERIALS AND METHODS

Patients

The study included 94 AL patients who were examined between 1985 and 1992. There were 64 (66.9%) patients with acute nonlymphoblastic leukemia (ANLL) and 30 (33.1%) patients with acute lymphoblastic leukemia (ALL). Diagnosis was based on the criteria of the French-American-British (FAB) Cooperative Group (Bennet et al., 1976).

Procedure

Bone marrow samples obtained at diagnosis were processed according to the short-term unstimulated culture and stained with G-banding by a modification of the trypsin method of Seabright (Seabright et al., 1977; Verma, Babu, 1989).

Different cytogenetic markers were observed in 64 patients with AL. Chromosomal abnormalities were identified and classified according to the International System of Human Cytogenetic Nomenclature (ISCN, 1981).

All patients were divided into previously recognized clonal subgroups including monosomy or deletion of chromosome 5 or 7, n=13; t(8;21)(q22;q22) translocation, n=8. Other clonal chromosomal rearrangements and normal karyotypes were found in 73 cases and constituted the control group. Different additional chromosomal abnormalities were observed in 33 patients (51.6%).

The present study was performed using the computer program "HEMA" which has been developed at St-Petersburg Pavlov Medical Institute. Mean findings of all hematological and clinical data were calculated for each cytogenetic group. Comparison of the data was performed to define characteristics of cytogenetic variants of AL.

RESULTS AND DISCUSSION

The deletion or monosomy of chromosome 5 or 7 (fig. 1) was found in 13 patients with type M1 (30.1%), M2 (7.7%), M4 (23.1%), M5 ANLL (23.1%) and ALL (15.4%). This cytogenetic subgroup seems to have common clinical features (Table 1). Patients with -5/5q-;-7/7q- were characterized by older age. They had no CNS leukemia and fever at diagnosis. 15.4% of patients with these chromosomal changes had contact to mutagenes.

Table 1. Clinical features in AL with -5/5q-;-7/7q- compared with other cases AL

	-5/5q-;-7/7q- n=13	other AL n=81	P
median age (ys)	50.5(12.0-79.0)	37.5(4,0-81,0)	0.02
fever (%)	15.4	27.2	0.084
CNS leukemia (%)	0.0	6.1	0.001
mutagenes (%)	15.4	2.5	0.004
WBC count (x10 /l)	13.8(1.6-100.0)	28.7(1.0-350.0)	0.05
bands PB (%)	5.1(0-18.0)	2.5(0-33.0)	0.049
blasts BM (%)	47.2(5.6-83.8)	60.3(1.9-98.0)	0.062
bands BM (%)	4.3(0-11.2)	1.9(0-15.2)	0.002
CR rate (%)	7.7	45.2	0.005
resistance to ther.(%)	53.8	13.6	0.008
median CR duration (m)	4.0(-)	7.6(0.5-65.0)	0.001
median survival (m)	2.6(0.5-11.0)	10.0(0.5-67.0)	0.005

PB - peripheral blood BM - bone marrow

Patients with -5/5q-;-7/7q- had low WBC levels. It is of interest, that blast count level in peripheral blood (PB) was less but level of maturated cells was higher in these patients as compared with other AL patients. These phenomena may be connected to probable history of myelodisplasia (MDS) in these patients.

In accord with the literature, response to therapy in this subgroup was dismal. CR rate and medium survival were very short. The presence of certain clinical characteristics of patients with -5/5q-;-7/7q- was confirmed by comparison of response to therapy in a selected sub-group (table 2).

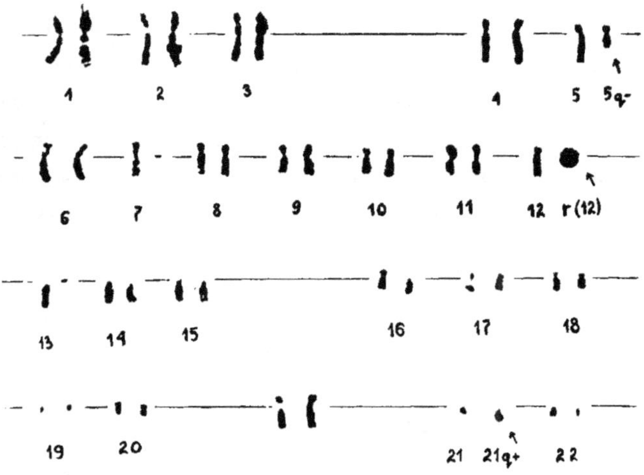

Fig.1. Karyotype of patient with M2 ANLL.
44,XX,del(5)(q13q31),-7,r(12)(p13q24),-13,t(12;21)(p13;q24)

Table 2. Response to therapy and survival in ANLL patients with different chromosomal abnormalities

	M1,M4,M5 ANLL -5/5q-;-7/7q- n=7	other M1,M4 M5 ANLL n=7	P
CR rate (%)	-	44.4	0.001
resistance to therapy (%)	63.6	33.3	0.05
median CR duration(m)	-	14.0(4.0-28.0)	0.001
median survival (m)	2.7(0.5-11.0)	11.2(0.5-30.0)	0.005

All patients with deletion or monosomy of chromosome 5 or 7 had different additional chromosomal aberrations. The appearance of double minute segments (DMS) and abnormalities of chromosome 4, 8, 12, 17, 21 were often found in this group of patients. We found that patients with abnormalities of chromosome 12, 21 and DMS had a very poor prognosis compared with other patients of the same group. They had also the worst medium survival (1.0-3.0 m.) and very often developed drug resistance (40.0-70.0% of patients).

The ANLL presenting the translocation t(8;21)(q22;q22) (fig. 2) was highly correlated with the M2 phenotype. We found that patients with this translocation were young, had hepatomegaly, low WBC and blast level count and good response to therapy (table 3).

Table 3. Clinical features in ANLL with t(8;21) compared with other cases ANLL

	t(8;21) n=8	other ANLL n=56	P
median age (ys)	37.0(4.0-52.0)	49.2(11.0-81.0)	0.05
hepatomegaly (%)	62.5	36.3	0.05
splenomegaly (%)	0.0	11.0	0.001
WBC count (x10 /l)	12.4(1.3-29.3)	39.5(1.0-187.0)	0.01
blasts PB (%)	25.9(0-50.0)	39.9(0-97.0)	0.01
CR rate (%)	62.5	30.0	0.05
resistance to therapy (%)	0.0	21.5	0.001

PB - peripheral blood

The investigation of clinical features in a selected group of patients with M2 ANLL revealed the importance of this chromosomal abnormalities (Table 4).

Table 4. Clinical features and response to therapy in M2 ANLL with t(8;21) compared with other cases M2 ANLL

	t(8;21) n=7	other M2 ANLL n=7	P
median age (ys)	34.3(4.0-52.0)	62.9(36.0-79.0)	0.001
CR rate (%)	71.4	28.5	0.001
resistance to therapy(%)	0	14.3	0.001

75.0% of patients with t(8;21) had additional chromosomal aberration. Loss one of one sex chromosome was a very common secondary chromosomal abnormality in these patients and was associated with young age (4.0-38.0) and good response to therapy in comparison with other patients with translocation t(8;21). All patients who had loss of either X, or Y chromosome responded to therapy and achieved a complete remission.

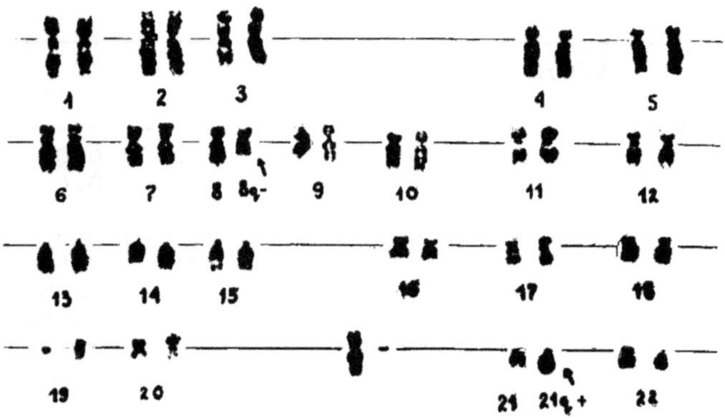

Fig.2. Karyotype of patient with M2 ANLL. 45,X,-X,t(8;21)(q22;q22)

REFERENCES

Bennet J., Catovsky D. (eds) (1976) Proposals for the classification of acute leukemia. Brit. J. Haematol. 33:451-459.

Bunin N., Nowell P. (eds) (1991) Chromosome 7 abnormalities in children with Down Syndrome and preleukemia. Cancer Genet. Cytogenet. 54: 119-126.

Fenaux P., Dube J.(eds) (1989) Cytogenetics and their prognostic value in de novo acute myeloid leukemia. Brit.J.Haematol.73:61-67

Fourth International Workshop on chromosome in leukemia-lymphoma (1984) A prospective study of acute nonlymphoblastic leukemia. Cancer Genet. Cytogenet. 11:249-360.

Groupe Francais de Cytogenetique Hematologique (1990) Acute myelogenous leukemia with an 8;21 translocation. Cancer Genet. Cytogenet. 44: 169-179.

Heim S., Mitelman F. (1986) Secondary chromosome aberration in the acute leukemia. Cancer Genet. Cytogenet. 22: 331-338.
Heim S., Mitelman F. (1987) Cancer cytogenetics. Alan. R. Liss. New-York.

ISCN(1985) International System for Human Cytogenetic Nomenclature.

Kalwinsky D., Raimondi S. (eds) (1990) Prognostic importance of cytogenetic subgroups in de novo pediatric ANLL. J. Clin. Oncol. 8: 75-83.

Keating M., Smith P, (eds) (1988) Cytogenetic pattern in acute myelogenous leukemia: a major reproducible determinant of outcome. Leukemia 2: 403-412.

Lewis S.,Solh H. (eds) (1991) Secondary acute nonlymphoblastic leukemia in childhood. Cancer Genet. Cytogenet. 55: 85-88.

Pathak S. (1990) Cytogenetic abnormalities in cancer. Cancer and metastasis Rev. 8: 299-318.

Rowley J.(1984) Biological Implications of consistent chromosome rearrangement in leukemia and lymphoma. Cancer Res.44:3159-68.

Seabright M.(1971) A rapid banding technique for human chromosomes. Lancet 2: 971.

Secker-Walker L. (1990) Prognostic and biological importance of chromosome findings in acute lymphoblastic leukemia. Cancer Genet. Cytogenet. 49: 1-13.

Verma R., Babu A. (1989) Human chromosomes: Manual of basic techniques. Pergamon Press. Inc. New-York.

PRODUCTION OF INTERLEUKIN-2 AND NATURAL KILLER (NK) ACTIVITY IN PATIENTS WITH MULTIPLE MYELOMA.

E.Podoltseva* , E.Morozova**.

* Bone Marrow Transplantation Department, St.Petersburg Center for Advanced Medical Technologes,197042, and ** Petrov Research Institute of Oncology, 189646, St.Petersburg, Russia.

INTRODUCTION.

Multiple myeloma (MM) is a clonal lymphoprolipherative disease. Survival ranges from a few months to many years (Kyle, 1988). At least, two factors determine the course of the disease: 1. the biology of malignant cells, 2. the interaction of malignant cells with the host organism,(Joshua, 1988). We have investigated the function of peripheral blood mononuclear cells (PBMC) in patients with different courses of MM in order to determine their prognostic significance in the progression of the disease.

MATERIALS AND METHODS.

48 patients with MM have been studied. Diagnosis was made according to previously described criteria (Chronic Leukemia -Myeloma, Task Forse, 1968).

We divided patients into three groups: 1. Indolent MM: patients with smoldering MM accordin to criteria of Kyle and Greipp (1980), patients who had been in remission for more than 5 year without chemotherapy. 2. Active MM: patients, responding to first-line conventiona chemotherapy. 3. Aggressive MM: patients with primary or secondary resistance to chemotherapy. Biological activity of interleukin-2 (IL-2) in conditioned media of PBMC was assayed accordin to the method reported by Bockman et al.(1981) and expressed as relative units. NK activity Effector cells were obtained from heparinized peripheral blood and cultured as described by Yod et al. (1988). K-562 cells were used as target cells. Cytotoxicity was measured in a standard 18-h H3-uridine microcytotoxicity assay, and expressed as percentage of controls.

Data were evaluated using Student's t-test. Survival curves were drawn by the Kaplan-Meier method.

RESULTS.

There was no significant difference in IL-2 production between normal controls and patients with indolent and active MM ($p>0.05$). A significant difference in IL-2 production was found between the patients with aggressive vs. active MM; between aggressive vs. indolent MM and normal controls ($p<0.01$) (Fig.1). The mean level of IL-2 production (M±SD) in normal donors (n=12) was 13.9±4.1 U in patients with indolent MM (n=10)- 12.9±3.7 U, in patients

NATO ASI Series, Vol. H 94
Gene Technology
Edited by A. R. Zander, W. Ostertag,
B. V. Afanasiev, F. Grosveld

with active MM (n=15)-11.9±2.6 U, in patients with aggressive MM (n=11) 8.6±1.9 U. There was no change in average NK activity of PBMC in MM patients as compared to normal controls. However, significant decrease of NK activity both in active and aggressive MM was observed in comparison to normal controls ($p < 0.05$) (Fig.2).

The mean level of NK activity (M+ SD) in indolent MM (n=9) was 36.6±20.0, ($p > 0.05$ with respect to normal controls); in active MM (n=5) 18.1±13.4; in aggressive MM (n=4) 13.8±6.35 ($p < 0.05$ with respect to normal controls).

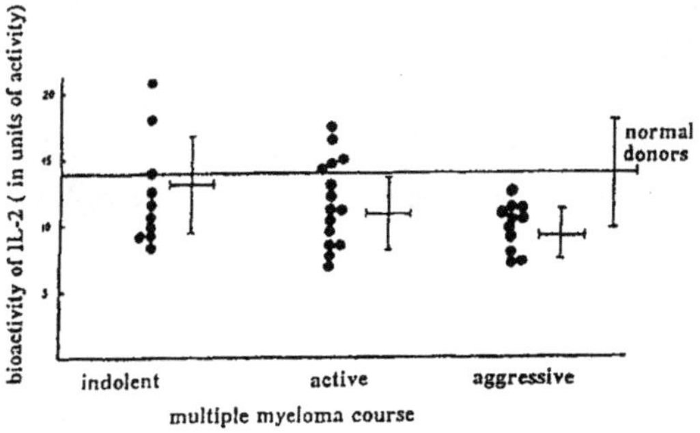

Fig.1. Production of IL-2 by PBMC in patients with different MM course.

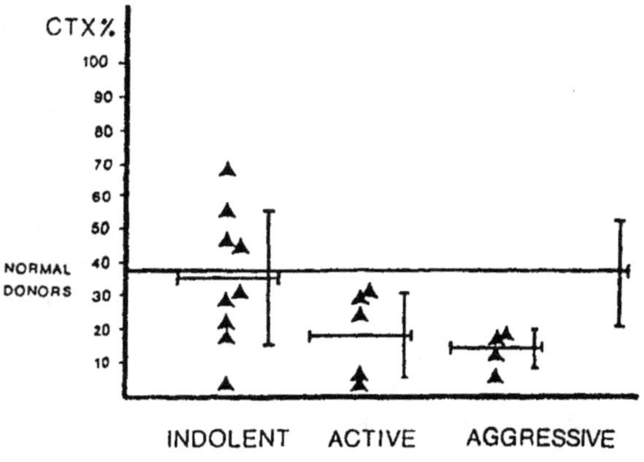

Fig.2. NK activity of PBMC in patients with different MM course.

Overall survival in MM patients with production of IL-2 >20 U was higher, than in patients with levels of IL-2< 20 U ($p < 0.001$).

Overall survival in MM patients with NK activity >60% was higher, than in patients with NK activity<60% ($p < 0.001$) (Fig.4).

DISCUSSION.

Our data suggest that IL-2 production by PBMC in indolent and active MM patients was similar to those in normal donors and, at the same time, it was significantly lower in aggressive MM. This result is in agreement with data reported by Commes et al. (1986), who have not revealed differences between the production of IL-2 in active MM patients and normal

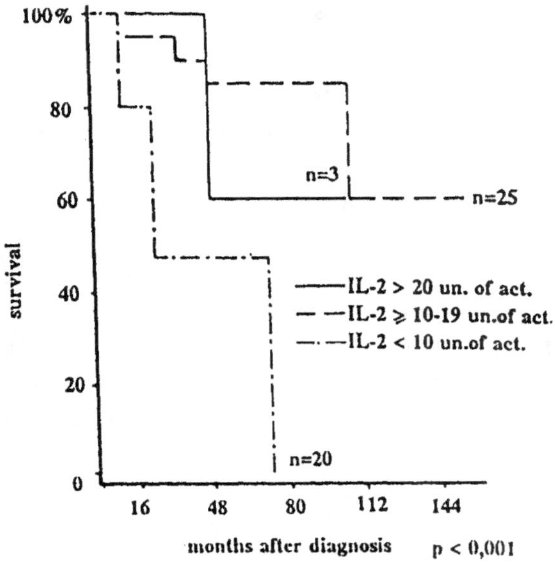

Fig.3. Overall survival in MM patients with different levels of IL-2 production ($p<0.001$).

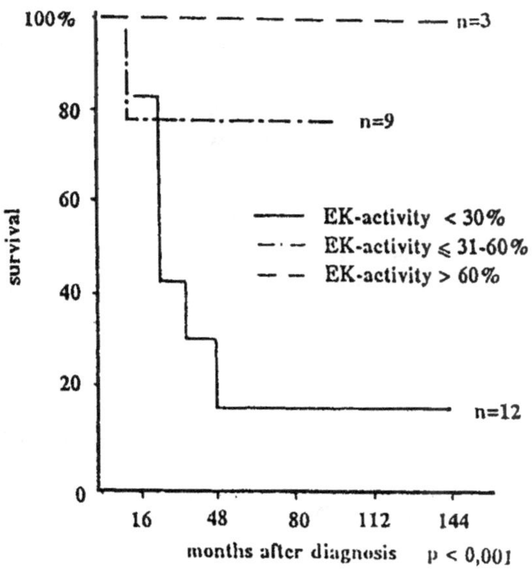

Fig.4. Overall survival in MM patients with different levels of NK activity ($p < 0,001$).

controls. However, the authors did not analyze separately the patients with aggressive course of MM.

Meanwhile, Gimino et al. (1989) have described an increase of serum IL-2 level in MM patients compared to normal controls. The elevated level of IL-2 has been revealed by Commes et al. (1986) in the plateau stage of the disease as well. Such controversy may result from the specificity of patient selection for the different studies.

Ambivalent data on the role of IL-2 for the pathogenesis of malignant blood disease exist: 1. Stimulation of tumor growth (Tsudo et al., 1984), 2. Increase of NK activity (Harst et al., 1988). No receptors for IL-2 were found in MM patients (Pizzolo et.al., 1987 ; Gimino et al., 1989). We believe that IL-2 is not a specific growth factor for myeloma cells. Hence, low level of IL-2 production by PBMC may reflect the depression of antitumor immune system in MM. These results do not interfere with our data about the worse survival of MM patients with low IL-2 production. Moreover, IL-2 activates cytotoxic lymphocytes directly, mediating in vitro destruction of myeloma cells (Shimazaki et al., 1988; Peest et al., 1989). It is known that the cytotoxicity of the lymphokine-activated killer (LAK) is provided mainly by IL-2 stimulated NK cells (Phillips et al., 1986). No difference was detected between NK activity of PBMC in MM patients and normal controls (Einhorn et al. 1986). There was no correlation between NK activity and the disease stage or the type of monoclonal protein (Einhorn et al., 1982). According to our data, NK activity of PBMC was sufficiently decreased in aggressive MM and, at the same time, it was rather high in indolent and active MM. These results agree with a significant decrease of IL-2 production in aggressive MM patients, this might be correlated with a depressed antitumor immune system.

Overall survival of MM patients with increased levels of NK activity was higher, than in those with low levels. Thus, the decrease of NK activity may be regarded as a prognostic test in MM. Applicability of IL-2 and LAK assays in the treatment and remission' maintenance in MM is under discussion now. Current data indicate, that IL-2 may be useful as a therapeutic approach to MM patients (Peest et al, 1994). Further studies are necessary to decide if immunotherapy can be widely used for MM treatment.

CONCLUSIONS.

1. Patients with aggressive MM had lower level of IL-2 production and NK activity compared to indolent, active MM and normal controls
2. Positive correlation was revealed between the survival of MM patients and the production of IL-2 and NK activity.
3. Our data suggest, that the level of IL-2 production and NK activity of PBMC can be useful as a prognostic test for MM course and the usefulness of the IL-2 and LAK - therapy in multiple myeloma.

REFERENCES.

Bockman R.S., Ropo M.A. (1981) Lymphokin-mediated bone resorption requires endogenous prostaglandin synthesis. J. Exp. Med. 154: 529-534

Commes T., Klein DB., Jourdan M. (1986) Production of interleukin-2 in multiple myeloma Clin. Exp.Immunol. 63: 533-540

Durie B.G.M., Salmon S.E. (1975) A clinical staging system of multiple myeloma: correlation of measured myeloma cell mass with presenting clinical features, response to treatment, and survival. Cancer 36: 842-848

Einhorn S., Ahre A., Blomoren H. et al. (1982) Interferon and natural killer activity in multiple myeloma. Lack of correlation between interferon-induced enhancement of natural killer activity and clinical response to human interferon. Cancer 30: 167-172

Gastl G.,Riederwieser D., Masth C. et al. (1984) Human large granular lymphocytes and their relationship to natural killer cell activity in varions disease states. Blood 74: 288-295

Gimino G., Awisati G., Amadori S. (1989) High serum interleukin-2 levels in patients with monoclonal gammopathy of undertermined significance (MGUS) and multiple myeloma. Nouv. Rev. Fr. Hematol. 31: 329-332

Harst D., Brand A., Luxemburg-Heys S.A.P. et. all. (1989). Lymphokin-activated killer cell functions in patients with leukemic B-lymphoproliferative diseases. Blood. 74: 2464-2470

Joshua D.E. (1988) Biology of multiple myeloma. Host-tumor interactions and immune regulation of disease activity. Hemat. Oncol. 6: 83-88.

Kyle R.A., Greipp P.R. (1980) Smoldering multiple myeloma. N. Engl. J. Med. 302: 1347-1349

Kyle R.A. (1988) Prognostic factors in multiple myeloma. Hematol.Oncol. 6: 125-130

Peest D., Vries J., Holscher K. et al., (1989) Effect of interleukin-2 on the ex vivo growth of human myeloma cells. Cancer. Immunol. Immunother. 30 : 227-232

Peest D., Leo R., Deicher H. (1994) Tumor-directed cytotoxicity in multiple myeloma - the basis for an experimental treatment approach with interleukin-2. Multiple myeloma. From biology to therapy. Current concepts. International conference. France. Abstract.

Phillips J.M., Lanier L.L. (1986) Dissection of the lymphokine-activated killer phenomenon. Relative contribution of peripheral blood natural killer cells and T-lymphocytes to cytolysis. J. Exp. Med. 184: 814-825

Pizzolo G., Chilosi M., Semenzato G. (1987) The soluble interleukin-2 receptor in hematological disorders. Br. J. Haem. 67: 377-380

Shimazaki C., Aizpodien J., Wisniewski D. (1988) Cell-mediated toxicity of interleukin-2 activated lymphocytes against autologous and allogenetic human myeloma cells. Acta Haemat. 80: 203-209

Tsudo M., Uchiyama T., Uchino M. (1984) Expression of TAC antigen on activated normal human B cells. J. Exp. Med. 160: 612-619

MULTIPLE MYELOMA ASSOCIATED WITH AMYLOIDOSIS

N.Podoltsev*, E.Podoltseva**, E.Morozova***.
* Internal Medicine Department N1, Pavlov Medical University, Leo Tolstoy str., 197089, St.Petersburg, Russia.

INTRODUCTION.

Survival data on patients with multiple myeloma (MM) associated with amyloidosis AL (AL) are variable: median survival varies from 5 months (Kyle, 1983) to 29 months (Pruzanski, Katz, 1976). According to Kyle et al. (1976) systemic AL with typical symptoms of organ and system damage is not often revealed among patients with apparent myeloma. Fielder and Durie (1980) suggest that MM associated with AL is more frequent than reported. The pathogenetic relationship between AL and plasma cell malignancy has not been elucidated. Retrospective analysis and prospective observation of patients with different MM activity and AL were carried out in order to determine the prognostic significance of AL for the disease course and survival. Progression of malignancy is determined by the biology of malignant cells, host organism and tumor mass interactions (Ioshua, 1988). Hence, we studied some functional characteristics of peripheral blood mononuclear cells (PBM) in MM patients with or without AL.

MATERIALS AND METHODS.

Thirty three MM patients with AL were studied: 8 with indolent, 15 with active and 10 with aggressive MM.

** Bone Marrow Transplantation Department, St.Petersburg Center for Advanced Medical Technologes, pr. Dinamo 3, 197042, St.Petersburg, Russia.
*** Bone Marrow Transplantation Department, Petrov Oncology Institute, Leningradskaya str. 68, 188646, Pesochny 2, Russia.

NATO ASI Series, Vol. H 94
Gene Technology
Edited by A. R. Zander, W. Ostertag, B. V. Afanasiev, F. Grosveld

1. Amyloid damage was justified by diagnosing typical pathology of skin biopsy specimens stained with Congo Red.

2. Our classification of MM patients, based on the disease activity, was used (Tab. 1).

Table 1. Clinical staging system of multiple myeloma based on the disease activity.

Type of course	Criteria	Median survival (months)
Indolent	Patients with smoldering MM according to criteria of Kyle Greipp (1980); patients who had been in remission for more than 5 years without chemotherapy	
Active	Patients responded to first line chemotherapy	72
Aggressive	Patients who had primary or secondary resistance to chemotherapy	24

3. The morphological type of MM was determined according to the classification of Carter et al., 1987.

4. The typing of light chains of immunoglobulins in serum and urine was performed by immunoenzyme assay with anti-kappa and anti-lambda antibodies.

5. Natural killer (NK) activity was studied in an H3-uridine radiometric assay (Rykova, 1981).

6. To measure interleukin-1 (IL-1) activity, PBM were cultivated for 24 hours with pyrogenal (40 mg/ml, Russia), in the thymocyte costimulation assay using human recombinant IL-1 as a standard (Gery, Misel 1972).

7. Data were estimated by means of BMDP Statistical Software (University of California, Los-Angeles, CA, 1981).

RESULTS.

A higher incidence of AL was seen in patients <60 years old (Tab. 2 , $p < 0.05$.)

Table 2. The frequency of AL in MM patients of different age.

	Age =< 60 years		Age > 60 years	
MM with AL	61.8 %	n=21	38.2 %	n=13
MM without AL	28.6 %	n=6	71.4 %	n=15

Most of the patients did not have clinical signs of AL (Tab.3)

Table 3. Clinical signs of AL in patients with MM.

Clinical signs of Amyloidosis	The type of MM course		
	Indolent	Active	Aggressive
Skin damage	8/8 (100%)	15/15 (100%)	10/10 (100%)
Tongue damage	1/8 (12.5%)	0/15 (0%)	0/10 (0%)
Peripheral nervous system damage	1/8 (12.5%)	0/15 (0%)	3/10 (30%)
Cardiac damage	1/8 (12.5%)	0/15 (0%)	0/10 (0%)
Kidneys damage	1/8 (12.5%)	2/15(13.3%)	4/10 (40%)

No correlation was found between the occurence of AL and the disease course, or the light chains type (Tab. 4, 5, $p>0.05$).

Table 4. AL frequency in patients with different MM course.

	Amyloidosis "+"		Amyloidosis "-"		Total
Indolent	n=8	57.2%	n=6	42.8%	n=14
Active	n=15	65.2%	n=8	34.8%	n=23
Aggressive	n=10	58.8%	n=7	41.2%	n=17

It was revealed, however, that the patients with active MM and AL have better survival than patients with active MM without AL (follow-up 8 to 96 months) (Fig. 1).

Table 5. Immunoglobulin light chain distribution in MM patients with and without AL.

	MM with AL	MM without AL
Non secretory MM	20%	15%
Kappa chains	60%	61%
lambda chains	20%	24%

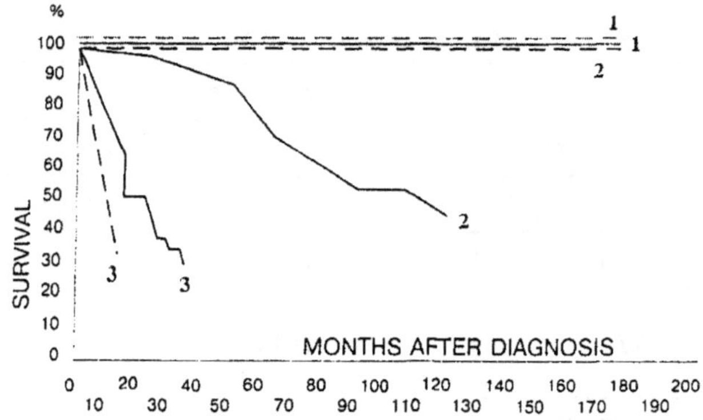

Fig. 1 Overall survival of MM patients with and without AL in different types of myeloma course: ___ - MM without AL , --- - MM with AL , 1 - indolent , 2 - active , 3 - aggressive.

Bone marrow plasma cells in active MM patients with AL as well as in all indolent MM patients had mature morphology. 27% of the myeloma cells were immature in active MM patients without AL. (Tab. 6).

Tab. 6 Biological features in MM patients with or without AL.

Parameters tested	MM with AL	MM without AL
Plasma cells maturation		
indolent MM	100% mature	100% mature
active MM	100% mature	27% immature 73% mature
aggressive MM	100% immature	6% plasmablastic 56% immature 38% mature

(continued Table 6)		
NK activity (M±SD) ($p<0.05$)	41.2±22.4 %	19.1±13.4 %
Pyrogenal-stimulated IL-1 production by PBM ($p<0.05$)		
>= 40 u/ml	100 % (n=6)	38.8 % (n=7)
< 40 u/ml	0 % (n=0)	61.2 % (n=11)

M=mean; SD, standard deviation

NK activity in MM patients with AL was significantly higher than in MM patients without AL (Tab. 6).
The activity of IL-1 produced by PBM in MM patients with AL was significantly higher than in MM patients without AL (Tab. 6).

CONCLUSIONS.

1. Amyloidosis (AL) was revealed with a higher frequency in MM patients < 60 years old.
2. No correlation was found between the incidence of AL MM patients and disease course or light chain type.
3. No difference in overall survival rates was detected between MM patients with or without AL.
4. Patients with active MM and AL had better survival rates than patients with active MM without AL.
5. The course of MM, either with or without AL, probably depends on myeloma cell maturity.
6. NK activity in MM patients with AL was higher than in AL-free MM patients.
7. The production of IL-1 by PBM in MM patients with AL was higher than in AL-free MM patients.

Further investigations are needed in order to determine the modifying role of AL on MM course.

REFERENCES

Fiedler K, Durie BGM, (1986) Primary amyloidosis associated with multiple myeloma. Predictors of successful therapy. Amer.J.Medicine 80: 413-418.

Gery J, Gessho RK, Waksman BH (1972) Potentiation of the T-lymphocyte response to mitogens. I. The responding cells. J.Exp.Medicine 136: 128-142.

Joshua DE, (1988) Biology of multiple Myeloma - host tumor interactions and immune regulation of disease activity. Hematol.Oncology 6: 83-86.

Kyle RA, Gripp PR, (1983) Amyloidosis clinical and laboratory features in 229 cases. Mayo Clin.Proc. 58: 665-683.

Kyle RA, Gripp PR, O'Fallon WM, (1986) Primary systemic amyloidosis: multivariante analysis for prognostic factors in 168 cases. Blood 68: 220-224.

Mizel SB, Oppenheim JJ, Rosenstreich DL, (1978) Characterization of lymphokine-activation factor produced by the macrophage cell line p388D1. I.Enhancement of LAF production by activated T-lymphocytes. J.Immunology 120: 1497-1503.

Pruzanski W, Katz A, (1976) Clinical and laboratory findings in primary generalized and multiple myeloma - related amyloidosis. Can.Med.Assoc.J. 114: 906-909.

Rykova MP, Spirande IV, Sedgenidse MS, Ljachov VV, Fux BV (1981) The new high-sensitive technique for natural killer cell activity testing. Immunobiology (Russia) 3: 88-91.

DISTRIBUTION OF HLA-DRB1 ALLELES IN PATIENTS WITH CHRONIC MYELOGENOUS LEUKEMIA DIFFERS FROM THAT IN HEALTHY DONORS.

A.Lyschov**, I.Komochkov*, S.Kapustin*, J.Aleksandrova*, E.Ilushik**, O.Frolova* and A.Zaritskey#

** Laboratory of Molecular Genetics, N.N. Petrov Research Institute of Oncology, St.Petersburg, 189646, Russia.
* BMT Center, Center for Advanced Medical Technologies Department of Haematology, St.Petersburg, 197042, Russia.
# Pavlov Medical University, St.Petersburg, 197089, Russia.

INTRODUCTION

Differences in distribution of HLA alleles in leukemic patients compared to healthy donors (HD) have been reported (Von Fliedner et al., 1980; 1981; Pawelec G. et al., 1988). These data imply a role of HLA alleles in susceptibility to leukemia (Bortin et al., 1987). However, at present there are no data on the distribution of HLA alleles in patients with chronic myelogenous leukemia (CML). Recent reports suggest that p210 bcr-abl (Ph-chromosome) may be a potential candidate for targeted immune intervention. Also, it is important to emphasize that HLA class II molecules appear to play an essential role in this process (Cheever et al., 1993). From this point of view, the study of the distribution of HLA class II alleles seems to be important.

In this report, we have analyzed the distribution of alleles of HLA-DRB1 in CML patients compared to normal donors. We show significant differences between these two groups.

MATERIALS AND METHODS

DNA of blood of 65 CML patients was tested. DNA of 118 HD were used as control group (Lyschov et al., 1993). DNA was extracted from peripheral blood lymphocytes of the CML patients using standard techniques (Bidwell and Bignon, 1991). The variable second exon of the HLA-DRB1 alleles was amplified by polymerase chain reaction. DRB1 typing was carried out using reverse dot blot method (Saiki et al., 1989). The nomenclature proposed by Bodmer et al. (1990) was used to define HLA-DRB1 alleles.
Statistical methods: Chi-square analysis was used.

NATO ASI Series, Vol. H 94
Gene Technology
Edited by A. R. Zander, W. Ostertag,
B. V. Afanasiev, F. Grosveld

RESULTS

Distribution of HLA-DRB1 alleles CML patients and healthy donors (HD) is shown in Table 1.

Table 1. Distribution of HLA-DRB1 alleles among CML patients and healthy donors.

HLA-DRB1 alleles	Healthy n	donors %	CML n	patients %
HLA-DRB1*01	25	(10.5)	17	(13.1)
HLA-DRB1*02	32	(13.6)	26	(20.0)
HLA-DRB1*03	32	(13.6)	11	(8.5)
HLA-DRB1*04	36	(15.3)	9	(6.9)
HLA-DRB1*11	24	(10.2)	23	(17.7)
HLA-DRB1*12	4	(1.7)	6	(4.6)
HLA-DRB1*13	27	(11.4)	14	(10.8)
HLA-DRB1*14	5	(2.1)	1	(0.8)
HLA-DRB1*07	34	(14.4)	17	(13.1)
HLA-DRB1*08	10	(4.2)	4	(3.1)
HLA-DRB1*09	3	(1.3)	2	(1.5)
HLA-DRB1*10	4	(1.7)	0	(0.0)
Total alleles	236	(100.0)	130	(100.0)
Total homozygotes	9	(7.6)	14	(21.5)
Total patients	118	(100.0)	65	(100.0)

P values <0.05 were considered to be statistically significant.

The comparative study of the distribution of HLA-DRB1 alleles in CML patients and healthy individuals revealed remarkable differences between these cohorts. There was a significantly higher frequency of HLA-DRB1 homozygotes in CML patients as compared to controls (21.5% and 7.6%; É = 0.008) (Table 1). The statistically significant increase in frequency of DRB1*11 allele was found in CML patients as compared to HD (17.7% and 10.2%; P = 0.04) (Table I). Also, there was significantly lower frequency of DRB1*04 (P = 0.03) (Table 1).

The distribution of HLA-DRB1 alleles in CML patients depending on the duration of the chronic phase (CP) is shown in Table 2. The frequency of DRB1*03 allele was significantly higher among CML patients with a more than average duration of CP (average duration of CP was 40 months) as compared to CML individuals with CP less than average (20.6% and 2.5%; P = 0.03). Moreover, the incidence of DRB1*13 allele was significantly higher in group of CML patients with duration of CP less than average, whereas no patient in the former group had the DRB1*13 allele (12.5% and 0.0%; P = 0.01).

Table 2. Distribution of HLA-DRB1 alleles among CML patients depending on the duration of the chronic phase.

	Duration of chronic phase			
HLA-DRB1 alleles	less than average		more than average	
HLA-DRB1*01	6	(15.0)	4	(11.8)
HLA-DRB1*02	10	(25.7)	7	(20.6)
HLA-DRB1*03	1	(2.5)	7	(20.6)
HLA-DRB1*04	4	(10.0)	3	(8.8)
HLA-DRB1*11	6	(15.0)	5	(14.7)
HLA-DRB1*12	0	(0.0)	2	(5.9)
HLA-DRB1*13	5	(12.5)	0	(0.0)
HLA-DRB1*14	0	(0.0)	0	(0.0)
HLA-DRB1*07	7	(17.5)	5	(14.7)
HLA-DRB1*08	1	(2.5)	0	(0.0)
HLA-DRB1*09	0	(0.0)	0	(0.0)
HLA-DRB1*10	0	(0.0)	0	(0.0)
Total alleles	40	(100.0)	34	(100.0)
Total homozygotes	6	(30.0)	0	(0.0)
Total patients	20		17	

P values <0,05 were considered to be statistically significant.

There was a significantly higher frequency of HLA-DRB1 homozygotes in CML patients with duration of CP less than 40 months as compared to more than 40 (30.0% and 0.0%; P = 0.03; Table 2).

The average duration of CP depending on distribution of HLA-DRB1 alleles in CML patients is shown on Table 3. As seen in Table, 3 DRB1*03-positive CML patients show a long average duration of CP whereas DRB1*13-positive CML individuals show short duration of CP.

Table 3. Average duration of chronic phase, depending on distribution of HLA-DRB1 alleles among CML patients.

HLA-DRB1 alleles	average duration of chronic phase (m)
HLA-DRB1*01	39
HLA-DRB1*02	42
HLA-DRB1*03	56
HLA-DRB1*04	40
HLA-DRB1*11	45
HLA-DRB1*12	51
HLA-DRB1*13	23
HLA-DRB1*14	nd
HLA-DRB1*07	38
HLA-DRB1*08	35
HLA-DRB1*09	nd
HLA-DRB1*10	nd
Total alleles	42

Note: nd - no data

DISCUSSION

HLA-DRB1 homozygous genotype was shown to be increased in CML patients compared to controls (Table 1). Also, a significantly increased incidence of HLA-DRB1 homozygotes was found in CML patients with a CP duration of less than average (Table 1). Thus, we suggest that a homozygous status is associated with high risk of occurrence of CML and with a poor prognosis of the disease as well.

It was shown that HLA antigens can vary between each other, regarding the spectrum of the capacity of oligopeptides binding (Braciale and Braciale, 1991). Therefore, homozygous combination leads to a reduction in the pattern of DR antigens which can be expressed by tumour cells and diminishes the spectrum of recognizable cellular protein abnormalities.

The other factor supposed to contribute to high risk for the development of CML is the presence of the DRB1*11 allele in the genotype (Table 1). The occurrence of this allele was increased in CML patients compared to HD patients. A strong association was shown between the presence of the DRB1*13 allele in the genotype of CML patients and a short duration of CP of the disease (Table 3). Presumably, HLA-DR molecules that are encoded by these "unfavorable" alleles present a narrower spectrum of protein changes associated with development of CML.

Interesting results were obtained on the incidence of the DRB1*04 allele among CML patients (Table 1). The frequency of this allele in CML patients was significantly lower than in the normal population. Probably, this observation might reflect the "protective" role of DRB1*04 allele from the development of CML. On the other hand, the remarkable difference in frequency of DRB1*03 allele between CML patients with a duration of CP more than average and less than average implies the possible protective role of this allele in the course of CML (Table 3).

Thus, based on our observations, we can conclude that DRB1*11, DRB1*04 might have a role in the development of CML. DRB1*03, DRB1*13 might determine the course of CML.

REFERENCES

Bidwell J., Bignon J. (1991) DNA-RFLP methods and interpretation scheme for HLA- and DQ typing. Eur. J. Immunogenetics 18:5-22.

Bodmer J.G., March S.G.E., Parham P. (eds.) (1990) Nomenclature for factors of the HLA system. Tissue Antigens. 35:1-8.

Bortini M.M., D'Amaro V., Bach F.H. (eds) (1987) HLA associated with leukemia. Blood. 70:227-232.

Braciale T.J. and Braciale V.L. (1991) Antigen presentation structural themes and functional variations. Immunol. Today. 12:124-129.

Cheever M.A., Chen W., Disis M.L. (eds) (1993) T-cell immunity to oncogenic proteins including mutated ras and chimeric bcr-abl. Ann N Y Acad Sci (Unated States). 690:101-112.

Lyschov A., Imyanitov E., Komochkov I., Chernitsa O. (1993) Specifics of distribution of alleles of HLA-DRB locus of II class major histocompatibility complex genes in tumor patients. Experimental oncology 1:43-49.

Pawelec G., Ehninger G., Muller C. (eds) (1988) Human leukocyte antigen-DP in leukemia. Cancer. 61:475-477.

Saiki R.K., Walsh P.S., Levenson C.H., Erlich H.A. (1989) Genetic analysis of amplified DNA immobilized with sequence specific oligonucleotide probe. Proc. Natl. Acad. Sci. USA. 86:6230-6234.

Von Fliedner V.E., Jolton Z., Jeannet M. (1980) HLA-w antigenes associated with acute leukemia. Tissue Antigene. 16:399-403.

Von Fliedner V.A., Sultan-Khan Z., Jeannet M. (1981) HLA-A and HLA-B antigenes in acute leukemia: A2-B12 phenotypes correlate with longer survival in acute myelogenous leukemia. Acta Haematol. (Basel). 65:73-75.

THE RELATIONSHIP OF MDR-1 GENE EXPRESSION AND CHEMOTHERAPY REGIMENS IN ACUTE LEUKEMIA.

T.Bykova, A.Sominskaya, I.Stuif, N.Anikina, A.Zaritskey
BMT Centre, Centre for Advanced Medical Technologies
Department of Haematology
Dynamo 3 197042, St.Petersburg
Pavlov Medical University,L.Tolstoy, 6/8
197089, St.Petersburg, Russia

SUMMARY

Data about induction of MDR-1 gene expression usually are derived from in vitro studies. Clinical data on the prognostic significance of MDR-1 gene expression in relation to the results of treatment are ambiguous (Savignano C. et al. 1993, Nooter K. et al. 1990, Ito J. et al. 1989).

Own data indicate that regimens containing vinca alcaloids (V), and/or epipodophyllin derivates (E) + anthracyclines (A) in comparison to those containing only A are more potent in induction of MDR-1 gene expression in acute leukemia. MDR-1 gene expression seems to be of less significance when treatment is based on high dose Ara-C regimens rather than on standard therapy.

Our data indicate that use of Vinca alkaloids (V)/epipodophyllin derivates (E) together with anthracyclines (A) results in higher frequency of MDR-1 gene induction than those containing only anthracyclines. Primary or induced MDR-1 related drug resistance can be overcome by high dose Ara-C in acute myelogenous leukemia (AML) and acute lymphoblastic leukemia (ALL).

INTRODUCTION.

MDR-1 gene expression is known to be responsible for the regulation of efflux of cell-toxic substances including chemotherapy agents such as vinca alkaloids, anthracyclines and epipodophyllin derivates (Hall A. et al. 1989, Tsuruo T. 1988).

In vitro studies show that the cell response to these types of cytostatic agents depends on the level of MDR-1 related protein (P-gly-180), (Natio M. et al. 1989, Nooter K. et al.

NATO ASI Series, Vol. H 94
Gene Technology
Edited by A. R. Zander, W. Ostertag,
B. V. Afanasiev, F. Grosveld

1990). Usually there is a correlation between MDR-1/mRNA level and related protein (Maslak P. et al. 1994, Hall A. et al. 1989, Gottesman M. et al. 1988). Clinical studies are more controversial (Ito J. et al. 1989, Nooter K. et al. 1990, Savignano C. et al. 1993). Recent studies have shown the expression of P-gly-180 and the correlation of its level and the function as drug efflux pump in both de novo and refractory/relapsed AL patients.

Modern trends of using high doses of Ara-C in AL (Buchner T. et al. 1991) make it difficult to analyze the impact of V/E + A vs A on drug resistance. Here we present data on prognostic significance and induction of expression of MDR-1 gene on standard and high dose therapy, as well as the impact of the one vs two drugs known to induce MDR-1 gene expression.

MATERIALS AND METHODS.

1.Patients. 21 de novo patients (5-ALL, 15-AML and 1-biphenotypic) and 21 pretreated patients (9-AML, 11-ALL, 1-biphenotypic) were included in the study.

2.The procedure is based on the method of in situ hybridization (Kanz L. et al. 1988) with slight modifications. Briefly, light density fractions of bone marrow mononuclear cells (d < 1,077 g/ml) were fixed with ethanol/acetic acid (3:1) on the slide circular areas followed by dehydration by increasing concentrations of ethanol (70,95%). Following steps were carried out according to Kanz et al. (1988).

3.Probe. The 1000-base pair MDR-1 specific DNA probe was labeled by standard, according to the protocol supplied with nick translation kit. We used P33 - dCTP.

4.Prehybridization of the wells was performed for 1 h and followed by hybridization for 16 to 24 h. Following posthybridization, the slides were covered by photographic emulsion 3M. Exposure of the slides was carried out 24 h at 4o°C in the dark. Thereafter the slides were Giemsa stained. The level of expression of studied gene was carried out according to the following criteria: 0-5 silver grains per cell were equal to background value or to the number of grains per peripheral blood cell of healthy donor; 6-15 grains per cell were considered as low value, this value was slightly more or equal to the number of grains per peripheral blood cell of healthy donor. These cells were considered as MDR-1-negative (MDR-). The expression exceeding 16-30 grains per cell was set as intermediate or high (> 30) and was considered as MDR-1-positive (MDR+).

RESULTS AND DISCUSSIONS.

In our department the overall complete remission (CR) rate in ALL is 78.2%, in AML - 77.2% and 3 year survival curves are comparable to the literature. This study is not representative for the overall treatment results because we studied more patients with persistent blast cell population in refractory/relapsed disease than in responsive disease.

We have analyzed primary and treated patients with blast cells in peripheral blood (Table 1). Not all pretreated patients could be called refractory. Pretreated patients received regimens containing A; A + V, or A + E in addition to other drugs.

Table 1. Comparison of MDR-1 gene expression in untreated and previously treated patients with AL (AML and ALL).

Patients	Expression of MDR-1 gene MDR+	MDR-	
de novo patients	5	16	$P<0.05$
pretreated patients	12	9	

It appeared that more previously treated patients expressed the MDR-1 gene in comparison to newly diagnosed patients. So, previous therapy including V, A and E can induce MDR-1 gene expression.

The number of primary leukemic patients (5/21) expressing MDR-1 gene is in accordance with the data of other authors (Maslak P. et all 1994, Savignano C. et al. 1993).

We tried to compare the impact of one or two drugs potentially inducing MDR-1 gene-related drug resistance: A vs A

+ V/E (Table 2). The former group of patients received regimens like: T-prolong, VAD-CVAD, VAPA, "7+3 + VP-16". The latter group of patients was treated with regimens :"7+3", TAD-HAM, sequential HAM.

Table 2. Comparison of MDR-1 gene expression in AL cells in patients receiving regimens containing A vs V / E + A.

Comparison of chemotherapy regimens	Expression of MDR-1 gene		
	MDR+	MDR-	
V/E + A	8	2	P<0.01
A	1	5	

Regimens containing two agents in comparison to one agent, known to be MDR-1 inducing, appeared to be more potent in induction of MDR-1 gene expression in AL cells. These data are interesting from both theoretical and practical points of view. Alternative explanation of the difference of MDR-1 gene induction in these groups of patients could be the total dose of A and its dose intensity. The total dose of A seems to be equal in both groups (data not shown). The intensity of dose of A will be studied.

The more potent impact of two drugs V/E+A than A in induction MDR-positivity can also be seen on normal bone marrow cells (Table 3).

Table 3. Comparison of MDR-1 gene expression in CR patients after consolidation or maintenance regimens containing A vs V/E+A.

Comparison of chemotherapy regimens	Expression of MDR-1 gene		
	MDR+	MDR-	
V/E + A	3	0	P<0.01
A	0	8	

Both groups of CR patients (MDR+ and MDR-) received the consolidation-maintenance regimens comparable according to the dose intensity and the total dose of A.

It appeared that MDR-1-positivity in CR BM cells was detected more often in patients previously treated with V / E + A. The high percentage of MDR-1-positive cells (10-30%) excludes that the positivity was due to residual leukemic population. The results indicate that two drugs (vincristine/E + A) appeare to be more potent in induction of MDR-1 gene expression than only A in normal bone marrow cells. These data seem to be of extreme interest for chemotherapeutic treatment of patients with solid tumors.

We have analyzed the response to therapy in relation to MDR-1 gene expression. (Table 4, 5, 6).

Table 4. Comparison of the results of standard and high dose treatment in relation to number of MDR-1-positivity cells.

Induction of remission	Expression of MDR-1 gene		
	MDR+	MDR-	
CR+	2	11	$P < 0.05$
Failed	8	5	

It appeared that more patients with MDR-1-negative cells entered CR as compared to MDR-1-positive patients (Table 4). The same results were obtained in the cohort of patients treated with conventional standard regimens (Table 5).

Table 5. Comparison of the results of standard dose treatment in relation to MDR-1-positivity.

Induction of remission	Expression of MDR-1 gene		
	MDR+	MDR-	
CR+	1	6	$P < 0.05$
Failed	6	3	

Table 6. Comparison of the results of high dose treatment (TAD or more intensive) in relation to MDR-1-positivity.

Induction of remission	Expression of MDR-1 gene		
	MDR+	MDR-	
CR+	1	6	$P > 0.05$

Failed	2	2

MDR-1-positivity had no significant impact on CR rate in the group of patients receiving high dose Ara-C chemotherapy.

CONCLUSIONS.

The high frequency of MDR-1 gene expression in de novo patients is one of the features of the biological resistance of the neoplastic clone. At the same time, the increase of the MDR-1 gene expression frequency in treated patients appears to reflect the reactivity of the leukemic cell as well as of normal cells to the chemotherapy agents.

Composition of two widely used chemotherapy agents known as MDR-1 gene inductive seems to be more potent in the MDR-1 induction in both leukemic blast cells and normal cells as compared to regimens containing one MDR-1 inductive drug.

According to our clinical data, the resistance of primary and drug-related MDR-1 positive leukemic clones results in poorer response to standard treatment and may be avoided by regimens containing high dose of Ara-C.

REFERENCES.

Buchner T., Hiddemann W., Loffler G. (eds) (1991) Improved cure rate by very early intensification combined with prolonged maintenance chemotherapy in patients with acute myelogenous leukemia: data from the AML cooperative group. Seminars in Haematology 28:76-79.

Gottesman M., Pastan I. (1988) The multidrug transporter, a double-edge sword. J.Biol.Chem. 263:12163-12166.

Hall A., Catton A., Proctor S. (1989) Mechanisms of drug resistance in acute leukemia. Cancer Res. 13:351-356.

Ito J., Tanimoto M., Ohno R. (1989) Increased P-gly expression and MDR gene amplification are infrequently found in fresh AL cells. Cancer 63:1534-1538.

Kanz L., Mielke R., Lohr G., Fauser A. (1988) Detection of messenger RNAs within single hemopoietic cells by in situ hybridization on small slid areas. Exp. Hematol. 16:394-399.

Maslak P., Hegewisch-Becker S., Godfrey L., Andreeff M (1994) Flow cytometric determination of the multidrug-resistant phenotype in acute leukemia. Cytometry 17:84-93.

Natio M., Tsuruo T. (1989) Competitive inhibition by verapamil of ATP-dependent high affinity vincristine binding to the plasma membrane of multidrug-resistant K-562 cells without calcium ion involvement. Cancer research 49:1452-1455.

Nooter K., Sonneveld P., Oostrum R., (eds) (1990) Overexpression of the MDR-1 gene in blast cells from patients with AML is associated with decreased anthracycline accumulation that can be restored by cyclosporin A. Int. J. Cancer 45:263-268

Savignano C., Geromin A., Michielli M., (eds) (1993) The expression of multidrug resistance related glycoprotein in adult acute lymphoblastic leukemia. Haematologica 78:261-263.

Tsuruo T. (1988) Mechanisms of multidrug resistance and application for therapy, Jpn J. Cancer Res. 79:285-296.

Immunophenotyping of blast crisis during the chronic myelogenous leukemia.

A.Ju.Baryshnikov*, A.G.Turkina**, I.N.Michailova**, N.P.Sedyakhina*, Ju.V.Shishkin*, A.M.Korolyova*, T.N.Palkina*, P.K.Ivanov*, E.A.Frolova*, R.A.Mokeyeva**, N.D.Khoroshko**.

* Cancer Research Center of RAMS, Kashirskoye shosse 24 115478 Moscow Russia

** Science Research Center of Hematology of RAMS

Introduction

Chronic myelogenous leukemia (CML) is a clonal myeloproliferative disorder of hematopoietic stem cells (P.G. Fialkov, 1981; R. P. Gale, 1991). CML initially presents chronic phase (CP), than it progresses into an accelerated phase (AP) and finishes as a rule by blast crisis (BC). The patients with BC have a lower response rate to the modern chemotherapy. However, the survival very various among different patients (N.M.Kantarjean, 1989). So this makes some prerequisites for the investigation of clinical, cytological and immunophenotypical characteristics and their influences on prognosis. It was shown that the blast cells are very heterogeneous (Janossy G et al, 1976; Baryshnikov AYu et al, 1989; Valeron O et al, 1989). It involves the myeloid, erythroid, megakaryocytic and lymphoid precursors. As blast cells have some differences, there were some attempts to describe clinical variants of blast crisis (Janossy G. et al., 1979). These investigations were not very successful because of lack of monoclonal antibodies, which could permit characterization of human hematopoietic progenitor cells. But now owing to hybridoma technology is a number of monoclonal antibodies by means which it is possible to characterize surface antigen markers of hematopoietic cells and identify their belonging to some of the lines of these cells. In this study, we investigated the immunological phenotype of blastcells from patients BC disease, determined the subpopulations of blast cells and its modification duringtherapy, estimated the significance of immunophenotype blast cells for prognosis in CML.

Materials and methods

Patients. 54 adult patients (30 males and 24 females) hospitalized for CML BC were studied. Their ages ranged from 21 to 68 years. Diagnosis of CML BC was made based on the presence of more than 30 % blasts in blood, typical clinical and hematological features. 48 patients had the Ph'-chromosome in bone marrow cells; there were not occured mitosises in another 6

NATO ASI Series, Vol. H 94
Gene Technology
Edited by A. R. Zander, W. Ostertag,
B. V. Afanasiev, F. Grosveld

cases. The estimation of efficiency of chemotherapy in BC is based on following criteria (P.M. Derderian, 1993):

1.Complete remission (CR): hemograms normalization with granulocytes >1000/ml, trombocytes >100000/ml and disappereance of blast cells from blood till 4 weeks, blast cells in the bone marrow < 5 %.

2. Partial remission (PR): Absence of the blast cells in peripherial blood , normalization of hemoglobin level, quantity of granulocytes and trombocytes as in CR, however blast cells in the bone marrow were 6 - 25 %.

3. Resistant (R): This group included patients which did not refer to the first or second group. Monoclonal antibodies (Mabs). In this study we used Mabs against human leukocyte differentiation antigens (Table 1). Each marker was diagnostic valuable if it detected > 20 %cells.

Immunofluorescence analysis. Leukocytes were isolated from 10 ml of peripherial blood by sedimentation in 1 % gelatin solution for 45 min, at 37°C. For the total lysis of erythrocytes we added 0,5 ml of lysing solution (Becton Dickinson) to precipitate the cells and incubated for 4 min. Then the cells were washed twice in PBS and used in immunofluorescence analysis. $5x10^6$ cells were incubated with 20 µl of Mabs for 30 min. at 4°C and washed twice in PBS. Then cells were incubated for 30 min at 4°C with 20 µl $F(ab')_2$ fragments rabbit anti-mouse serum conjugated with FITC or the isotype specific antiserum conjugated with FITC or phycoerythrin (PE). At the next stage cells were washed twice in PBS and resuspended in PBS, containing 1 % formaldehyde. An expression of antigens on the blast cells was detected by means of flow cytometer FACScan (Becton Dickinson) using two flow cytometric parameters as light scatter and size of cells.

Two-colour analysis. $1x10^7$ cells were incubated with 20 µl of the Mabs for 30 min. at 4°C and washed in PBS. Then the cells were incubated for 30 min. at 4°C with 20 µl of the isotypespecific antiserum conjugated with FITC or phycoerythrin (PE). The cells were washed in PBS and incubated with the second Mabs conjugated with FITC or PE for 30 min at 4°C. At the next stage the cells were washed twice in PBS medium and resuspended in PBS, containing 1 % formaldehyde. Mouse immunoglobulins of the same isotype as Mabs and directly conjugated with FITC or PE were used as controls. Sorting windows were established for four separate parametrs: forward and light scatter, FITC and PE fluorescence.

Table 1
Monoclonal antibodies against human leukocyte differentiation antigens using in the work

Antibody	Antigen	Main distribution
Anti-CALLA*	CD10	Lymphoblastic cells
ICO-GM1**	CD11b	Granulocyte, monocyte,
Leu-M3*	CD14	Monocyte
My1***	CD15	Granulocyte
HD37****	CD19	B-cells
HD39****	CD22	B-cells
My10***	CD34	Progenitor cells
ICO-1**	HLA-DR	Progenitor-, B-cells, monocyte
HAE-9**		Erythroblast cells

* - Becton Dickinson
** - Cancer Research Center, Moscow, Russia
*** - Mabs kindly provided by Dr. T.Trishman (Oncology Center of Jhon Hopkins, Baltimor, USA)
**** - Mabs kindly provided by Dr. G.Moldenhauer (Oncology Research Center, Heidelberg, Germany)

Results

We studied the expression of the following antigens on the blast cells from 54 patients with BC of CML: CD34 of hematopoietic stem cells, CD10 of lymphoblasts, CD19 and CD22 of B cells, CD15 of myelocytes, CD11b and CD13 of myelomonocytes, CD14 of monocytes, antigen HAE-9 of erythroblasts and HLA-DR antigens. In relation to the blast cells immunophenotype the following 5 immunological variants of CML BC were identified (Table 2).

Table 2
Immunological variants of chronic myelogenous leukemia in blastic crisis

Immunological variants	No (%) of Patients	Immunological phenotype of blast cells
Primitive	5 (9,2)	CD34
Lymphoid	13 (24,1)	CD34, CD10, CD19, CD22
Myeloid	15 (27,8)	CD34, CD15,CD11b,CD13,CD14,HAE-9
Mixed	12 (22,2)	CD34,CD10,CD19,CD15,CD11b,CD13,CD14
Undifferentiated	9 (16,7)	

The primitive variant of CML BC was described as a result of CD34 antigen expression in the absence of other markers of the later differentiation stages. It was found in 5 patients (9.3 %). 13 (24.1 %) patients had a lymphoid variant because their blast cells expressed antigens CD10, CD19 and CD22. Blast cells of 15 patients (27.8 %) expressed antigens of myelocytes, monocytes and erythrocytes: CD11b, CD13, CD14, CD15 and HAE-9. Thus the diagnosis of myeloid variant of CML BC was made in these cases. The mixed variant of BC was diagnosed in 12 (22.2 %) patients because we detected on their blast cells the antigens of two cell lines: myeloid and lymphoid ones. Undifferentiated variant was documented in 9(16.7 %) patients because of lack of the expression of differentiation antigens.

However the majority of patients had none of the markers on the 100 % of blast cells. It was supposed about the existence of several blast cells subpopulations in BC CML. In order to elucidate it we studied the structure of blast cell pool of 10 patients by two-colour immunofluorescence. As a result the subclones of blast cells were found in each of these 10 patients. For instance the distribution of the following pairs of differentiation antigens on the patient`s D. blast cells were analysed: CD34 and CD10, CD34 and CD19, CD34 and HLA-DR, CD10 and CD19, CD19 and HLA-DR, CD10 and HLA-DR (Fig.1). It was found that the blast cells of the patient D. had the 3 subpopulations of blast, expressing only CD34 antigen, CD34+CD10+ antigens and CD10 antigen (Fig.1c). CD34 and CD19 or CD34 and HLA-DR antigens were also expressed on 3 subpopulations of the blast cells (Fig.1b). CD10 and CD19 antigen were expressed on the 2 subpopulations of blast cells (Fig.1d). CD19 and HLA -DR antigens were also expressed on the 2 subpopulations of blast cells (Fig.1a). Two-colour analysis permited us to suggest that the cells of patient D. were on the successive stages of differentiation of lymphoid lineage. The analogical results were obtained by analysis of blast cells from another 9 patients.

On the next stage of our work the stability of immunological phenotype of blast cells was evaluated. We studied the immunological phenotype of blast cells of 16 patients with BC CML repeatedly in the intervals between the chemotherapy courses from 2 to 5 times. It was found that in 10 (62.5 %) cases the immunological phenotype of cells indicated the conservation. In 6 (37.5 %) cases the marked change of immunological phenotype was found, indicating the appereance of new cell subpopulation. Three patients withmixed variant of BC CML had changes in their phenotype. In one of these patients the mixed variant turned into the primitive one, the cell population "rejuvenated" to the level of stem cell. In the next patient there was turning of mixed variant into the preferentially lymphoid one. The third case was characterized by changing of mixed variant into myeloid one. The primitive variant of BC in patient No. 4 changed into myeloid one. In two cases the lymphoid variant of BC was substituted by mixed and myeloid ones accordingly. Like that the change of immunological phenotype was characterized by appearence of new leukemic population both intralinear and interlinear. Fas/APO-1 antigen expression was determined in 8 patients with CML BC

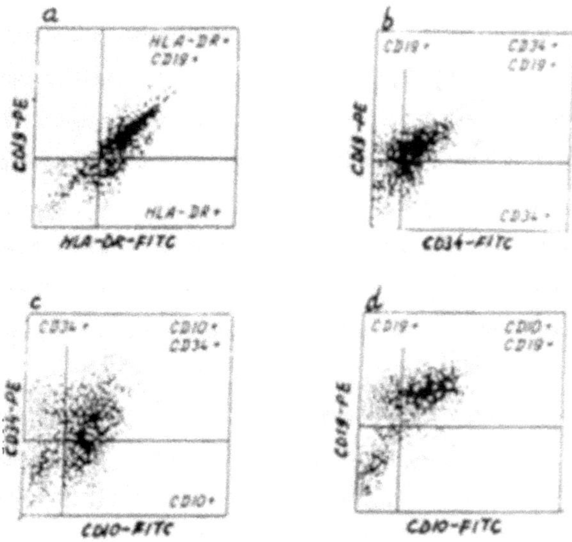

Fig.1 Two-colour immunofluorescence analysis of blast cells from the patient with lymphoid variant of CML BC.

Patients in blast crisis were examined 1 to 3 times during each aggravation and detection of blast cells in the peripheral blood (15 examinations were performed).

Fas/ APO-1 antigen was expressed on blast cells in 11 out of 15 cases tested. The percentage of antigen-positive cases varied from 4 to 88 %. High correlation was noted (r=0.89) between the expression of the CD34 and Fas antigens. This finding indicates that these antigens are expressed on the same cell.

Clinical and hematological features of the patients with different variants of CML BC were analysed. We did not find any marked differences in sex, age, liver size, the quantity of leukocytes, blast cells and trombocytes of peripherial blood and duration of chronic and blastic phases in different variants of BC.

The special features of clinical and hematological picture of the patients with lymphoid BC were the following: the higher level of hemoglobin at the moment of diagnosis of BC, absence of marked splenomegaly. In comparison with another variants of BC the more frequent complication was the development of neuroleukemia (3 from 13 patients). At the same time the patients with primitive variant had the hemoglobin level lower than the lymphoid variant and also the larger size of spleen and longer acceleration phase. The likeness of clinical and hematological features of the patients with lymphoid and mixed BC CML was noted. Using multifactorial correlation analysis a negative correlation ($p<0.05$) between the expression of CD10, CD19, HLA-DR antigens and the size of liver, spleen and duration of PA was obtained. The positive connection ($p<0.005$) was discovered between these antigens and the level of hemoglobin and duration of BC CML. In the group of patients with blast cells expressing myeloid, myelomonocytic and erythroid antigens the same correlation was not found.

The analysis of treatment of patients with different variants of CML BC showed that the best results of treatment were obtained in the group of patients with the lymphoid variant: 50 % of patients responded to the treatment. After treatment of the mixed variant CP or PR was recieved in 33% of the patients.

Thus the best response to chemotherapy was obtained in the group of patients having mainly lymphoid antigens though it was worse with the expression of some myeloid antigens. The shortest period of blast crisis was in patients with primitive and undifferetiated variants. None of these patients had remission.

The treatment courses of patients with lymphoid and mixed variants of BC CML involved as a rule prednisolone and vincristine. It was found that remission (CP or PR) was reached without the period of profound cytopenia. At the same time, the treatment by vincristine and prednisolone of the patients with myeloid, primitive and undifferentiated variants was not effective.

By using more "aggressive" courses of treatment as Ara-C+Rubomycin or Novantron the profound cytopenia occured, but there was not a remission. The duration of BC CML in patients, receiving no aggressive treatment, was longer.

As we can see from the Fig.2 the median range among patients with different variants of BC CML does not reveal anysignificant differences although there are tendencies to thelonger survival of the patients with lymphoid variant BC withcompare of the patients having the other variants BC.

Using the multifactorial correlation analysis we discovered that those patients whose blast cells expressed lymphoid antigens had longer median range than those patients with myeloid antigens ($p<0.05$). It can be noticed that the patients with lymphoid blastic crisis disease had a shorter survival than those with undifferentiated BC, although the differences were not statistically significant

Discussion.

Thus the investigation of immunologic phenotype of blast cells from the patients with CML BC showed the heterogeneity of leukemia population. By their immunologic phenotype blast cells represented all differentiation lineages. They had markers of myeloid, monocytic, erythrocytic and lymphoid cells. And even more: In some cases one and the same patient had a pool of blast cells, involving few subclones that represented either different stages of differentiation into one lineage or differentiation into two different lineages, such as the myeloid and lymphoid one. All this supportes the conception about the development of CML as a result of malignant transformation of pluripotent stem cell, capable of differentiation both in myeloid and lymphoid subpopulations (Bettelheim P. et.al., 1985; Hermandez P. et al., 1982; Bakhshi A. at al, 1983; Kitc K et al, 1986; Bernstein ID et al., 1992; Mohamed AN et al., 1992; Verfaillie CM et al., 1992).

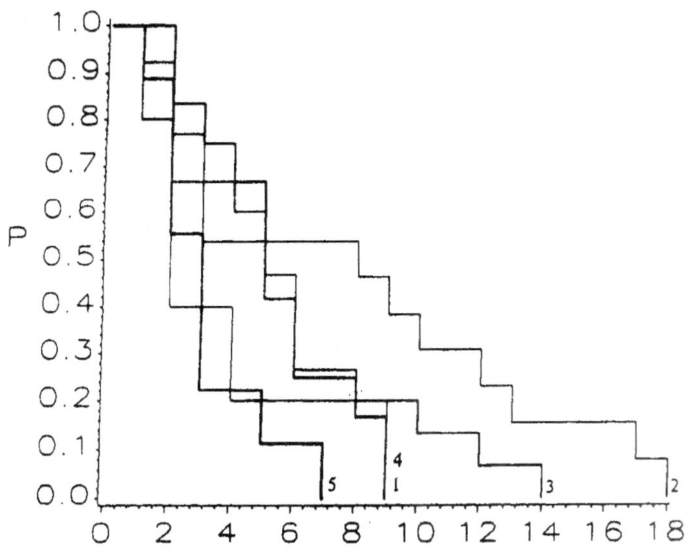

Fig.2 Survival of the patients with different variants CML BC

1. primitive 2.lymphoid 3.myeloid 4.mixed 5.undifferentiated

On the basis of our investigations we could describe 5 variants of CML BC. We consider that it is expediently to describe a primitive variant of BC in which blast cells have their mmunological phenotype analogous to stem cells, which express only the CD34 antigen. The clinical and hematological picture of this variant is characterized more extensively by comparison with others, anemia, splenomegalia, long phase of acceleration, the total resistance to polychemotherapy and the shorter survival.

Blast cells of undifferentiated variant of BC during theCML according to our investigations had no immunologic markers. This could be the consequence of a low level of antigen expression or lack of Mabs against megacaryoblasts and immature cells of basophilic and eosinophilic lineage in our panel. These cells can appear as a result of malignant transformation or they represent an unknown subpopulation of normal progenitor cells whose very low number does not permit an identification in normal blood.

The study of clinical-hematological characteristics in patients with these variants showed that only lymphoid variant differed from other ones (Bettelheim P et al, 1986; Janossy G et al,1979). P.M. Derderian et al.(1993) also reported about particular characteristics of the lymphoid variant: the patients are younger, they often develop BC without an antecedent accelerated phase, they present with lower leukocytes and blasts in peripherial blood, in spite of heavy infiltration of bone marrow, higher level of hemoglobin. From the clinical position it is probably advisable to distinguish only lymphoid and unlymphoid variants of BC during the CML. These and other authors prefer to divide BC of CML into two variants: lymphoid and unlymphoid (Griffin JD et al, 1983).

Our investigations on the estimation of stability of immunologic phenotype of blast cells during the disease led us to the same conclusion. The majority of patients with the relapse of the disease had the immunologic phenotype of blast cells distinguished from the primary one. By their phenotype the blast cells became younger or vice verca they became more mature. In 30 % of cases the cells lineage has changed. This can be the consequence of malignant transformation of stem cell. Analogical results were received by T.Ikeda et al.(1990).

The potentials to new level of differentiation which is ,defined by immunophenotyping, can originate from the own nature of tumor cell under the influence of chemotherapy or they can be the concequence of microenvironment (stroma), which also is exposed to the agents used for treatment of disease. The least changes in restorational stroma can produce the signals for the changing of blast cell differentiation.

It is very important from our point of view that the blast cells have also CD95 (Fas/APO-1) antigen mediated apoptosis and what is more a correlation between the expression of this antigen and CD34, which is the marker of stem cell. Firstly, this points out that the apoptosis can be the regulator of proliferation and self-support of stem cell. Secondly, apoptosis can be the mechanism of changing of block differentiation level of blast cells. The potential endogenic ligand of the molecule defined the apoptosis (CD95 antigen) can be also sensitive to the used chemotherapy.

Thus on the basis of our investigations the conception about the heterogeneity of blast cells population at the CML BC and its "interlinear" and "intralinear" unstability during the disease was confirmed.

References

Bakhshi A, Minowada J, Arnold A. et al. (1983) Lymphoid blast crises of chronic myelogenous leukemia represent stages in the development of B-cell precursors. New Engl J Med 309: 826-831

Baryshnikov AYu, Morozova LF, Frinovskaya IV et al. (1984) Immunological phenotype of blast cells from the patients with chronic myelogeneous leukemia and ist importance. Therapheutical archive 6:50-52

Bernstein ID, Singer JW, Smith FO et al.(1992) Differences in the frequency of normal and clonal precursors of colony-forming cells in chronic myelogenous leukemia and acute myelogenous leukemia. Blood 79: 1811-1816

Bettelheim P, Lutz D, Megdic O. (1985) Cell lineage heterogeneity in blast crisis of chronic myeloid leukemia. Brit J Hematol 59:395-409

Bettelheim P, Knapp W (1986) Philadelphia-chromosome-negative chronic myelogeneous leukemia with lymphoid stem cell blastic transformation. Amer J Hematol 21 115-116

Derderian PM, Kantarijan HM, Tapaz M (1993) Chronic myelogeneous leukemia in the lymphoid blastic phase: characteristics, treatment responce and prognosis. Amer J Med 94:60-74.

Fialkov PG, Martin PJ, Najfeld V, Fenfold GK (1981) Evidence for multi-step pathogenesis of chronic granulocytic leukemia. Blood 58:158-163

Gale RP (1991) Chronic myelogeneous leukemia: molecule to man. Henry Ford Hosp Med J 39:108-111

Griffin JD, Tantravahi R, Canellos GF (1987) T-cell surface antigens in patient with blast crisis of chronic myeloid leukemia. Blood 61: 640-644.

Hermandez P, Carnot J, Orum C (1982) Cronic myeloid leukemia blast crisis with T-cell features. Brit J Heamat 61:176-177

Ikeda T, Kawakami K, Anazawa H et al. (1990) A case of chronic myelocytic leukemia in blast crisis: the myeloblasts became over from lymphoid and mieloid population of the blast crisis cell after chemotherapy. Acta Haematol JPN 53: 62 -69

Janossy G, Greaves MF, Revesz T (1976) Cell surfacemarker analysis of lymphoid and myeloid cases. Brit J Hematol34:179-192

Janossy G, Woodruff RK, Paxton A (1979) Relation of lymphoid phenotype and response to chemotherapy incorporeiting vincristine-prednisolone in the acute phase of Ph'-positive leukemia. Cancer 49:426-434

Kantarjian HM, Talpaz M, Gutterman JU (1989) Chronic myelogeneous leukemia - past, present and future. Hematol Patology 2:91-120

Kitc K, Tezuka H, Komesati H (1986) Blast crisis in chronic myelocytic leukemia with blasts expressing both lymphoid and myeloid antigens. Acta Haematol 49:885-892

Mohamed AN, Smith MR, Al-Katib A et al. (1992) Emergence of myeloid stem cell line from T-lymphoid blastic phase of chronic myeloid leukemia in culture. Leuk Res 16: 521-527

Valiron O, Clemancey-Maroille G, Troesch A (1989) Immunophenotype of blast cells in chronic myeloid leukemia. Leukemia Res 12:861-872

Verfaillie CM, Miller WJ, Boylan K, McGlave PB (1992) Selection of benign primitive hematopoietic progenitors in chronic myelogenous leukemia on the basis of HLA-DR antigen expression. Blood 79: 1003-1010

The treatment of acute lymphoblastic leukemia in children according to BFM protocols.

E. Samochatova, A. Karachunsky, O. Kryzhanovsky,
A. Maschan, N. Zhestkova, L. Baydun, A. Rumyantsev.
Institute of Children Hematology
Leninsky prospect 117,
Moscow, Russia.

For the past 7 years more than 300 previously untreated patients with ALL admitted to the clinic of our Institute. During the first 3 years 69 patients were treated according to a guideline accepted at that time in all hematological clinics in the USSR. This kind of treatment included standard induction with 2-3 drugs given over 4-6 weeks, then (in some cases) cranial irradiation and prolonged maintenance therapy with 6-MP, Mtx and Cph and reinduction courses every 8-12 weeks during the first 3 years of remission. During induction children received 3-5 intrathecal injections of Mtx or Mtx + Ara-C; doses were counted according to body surface. It was hardly a programmed therapy because it was not strictly identical for specific risk groups, the doses and the time schedule were varied arbitrary and supportive care wasn't adequate. The results of such kind of treatment were by no means satisfactory with high frequency of relapses; the rate of continuous complete remissions was not more than 40% (Rumjantsev, 1980). So the results were comparable with those major centers abroad had achieved at the beginning of the 70's (Pinkel et al, 1971).

In the meantime treatment intensification as used for the last 15 years has had major positive impact on outcome for children with acute lymphoblastic leukemia especially for patients with medium and high risk ALL (table 1) (Rivera et al., 1994). That was the reason why we decided to find an effective programmed ALL treatment and to use it in our clinic. We chose one of the most successful leukemia treatment regimens developed by German specialists, well-known as the BFM programs (Ritter et al, 1990)

NATO ASI Series, Vol. H 94
Gene Technology
Edited by A. R. Zander, W. Ostertag,
B. V. Afanasiev, F. Grosveld

Table 1

Current Results of International Trials for Treatment of Pediatric/Adolescent Acute Lymphoblastic Leukemia (protocols opened from 1981-1986)

Protocol	Year	No. Pts.	Patient Eligibility	EFS rates
DFCI 81-01[16]	1981-1985	289	All patients up to age 18 years	7 yr, 72%
AEIOP[67]	1982-1987	902	All patients 1-18 years	7 yr, 54%
CCG-105[3]	1983-1989	1675	Predominantly B-progenitor ALL, 1-18 years, WBC<50×10^9/L	5 yr, 67%
ALL-BFM 83[2]	1983-1986	653	All patients up to age 18 years	5 yr, 68%
CCG-123[18]	1983-1989	707	Predominantly T-ALL, 1-18 years, lymphoma syndrome and/or WBC> 100×10^9/L	7 yr, 60%
FRALLE 83-87[19]	1983-1987	559	All patients up to age 18 years	7 yr, 57%
St. Jude XI[20]	1984-1988	358	All patients up to age 18 years	5 yr, 72%
UKALL X[21]	1985-1990	1614	All patients up to age 15 years	4 yr, 66%
ALL-BFM 86[22]	1986-1989	998	All patients up to age 18 years	5 yr, 72%
ALinc 14[6]	1986-1991	1854	B-progenitor ALL, 1-21 years, all levels of WBC	4 yr, 72%

We have been using this kind of therapy since 1990 and up to May 1994 159 consecutive previously untreated patients were treated according to BFM-ALL 90 protocol. There were some modifications in the original protocol during the 2 first years of our work: we gave 1 gramm of Mtx per square meter for 36 hours in protocol M and in blocks of high-risk protocol. Dose of cranial irradiation was 18 Gy for all patients in spite of the risk group (Rumjantsev et al, 1994).

Patients at the age of 15 and younger who met diagnostic criteria of ALL are included in this study. Patients who had been treated with BFM-protocol at first but then for some reasons with a different treatment were included in the nonprogram group (table 2).

The nonprogram group includes 91 patients (88 patients with more then 2 years of follow-up). The BFM-group includes 159 patients (86 patients with more than 2 years of follow-up).

The number of early deaths, nonresponses and complete remission achievements were determined for all patients. The rate of treatment failure because of deaths in remission, relapses and rate of continuous complete remissions were counted only for patients with more than 24 months follow-up duration. Demographic data for patients with ALL admitted to the clinic during 7 years are presented in table 3.

Table 2

Total group of ALL pts (01.87 - 05.94)	- 301 pts
Group of nonprogram treatment (1987 - 1994)	- 91 pts
Group of BFM-treated patients	- 159 pts
Other protocols	- 51 pts

Table 3
Demographic Characteristic of Patients with ALL

	Total	Nonprogram group		BFM-treated group	
		Total	24 mts Follow-up	Total	24 mts Follow-up
No of pts	301	91	87	159	86
Male	167	45	42	97	54
Female	134	46	45	62	32
M : F	1.25 : 1	1 : 1.02	1 : 1.07	1.6 : 1	1.7 : 1
Median Age Range	6.6	6.2	6.2 1 - 15	6.7	6.6
Study period	01.87-05.94	01.87-05.94	01.87-05.92	01.90-05.94	01.90-05.92

Clinical investigations were traditional and according to its results patients were enrolled in one of BFM-postulated risk group (table 4).

Table 4
Risk Groups

	Standart	Medium	H i g h
		No / %	
Nonprogram group	19/ 20.9	63/69.2	9/9.9
BFM-treated group			
Total	39/24.5	104/65.4	16/10.1
≥ 24 mts up	17/19.8	60/69.8	9/10.5

ALL was diagnosed according to classical criterias (Bennet et al, 1981).

Morphological and cytochemical identification was performed for all patients, immunophenotyping for 27 patients of nonprogram group and for 71 patients from BFM-treated group. Leukemic cells were analyzed for T- and B-lineage-associated antigen expression using various antibodies without uniform methods. Perhaps this in part explains why our results differ from published data. Immunological data are presented in the table 5.

Table 5
Immunophenotype

	O	T	pre-pre B CALAA+	pre-B	B -	Total
Nonprogram group	3/11.1%	9/33%	11/40.7%	4/14.8%	-	27
BFM-treated group	1/1.4%	23/32.9%	36/50%	10/14.3%	1/1.4%	71
Totally Made	8/6%	39/29.1%	70/52.2%	16/12%	1/0.75%	134

Results of the treatment are presented in the tables 6 and 7.

Table 6. Remission achievements for all patients.

No of pts	Nonprogram group (91 pts)	BFM-treated group (159 pts)
Early deaths	1	3
Nonresponses	2	2
CR achieved	88 / 96.7%	154 / 96.85%

The early deaths and nonresponse rates were the same for both groups. The complete remission achievement was then 97% for each group.

Table 7.
Treatment results for patients with >24 months of follow up

	Nonprogram group	BFM-treated group
No of pts.	87	86
Death in remission	8	6
Relapses	30 (45.45%)	6 (8%)
1st CCR > 24 months	33 (44.6%)	59 (78.7%)
LFU	13	11
Alive after relapse	4 (2 after allo-BMT)	0

According to these figures the rate of remission achievement and deaths in remission are the same in both groups but the rate of relapses and continuous complete remission more than 24 months improved significantly with the BFM regimen. Life table analyzes of actuarial event free survival were performed using the Kaplan-Mejer method. Failure to achieve complete remission (nonresponse, early deaths) or termination of the first remission (relapse, death in remission) were evaluated as events. The results for both groups are presented in figure 1.

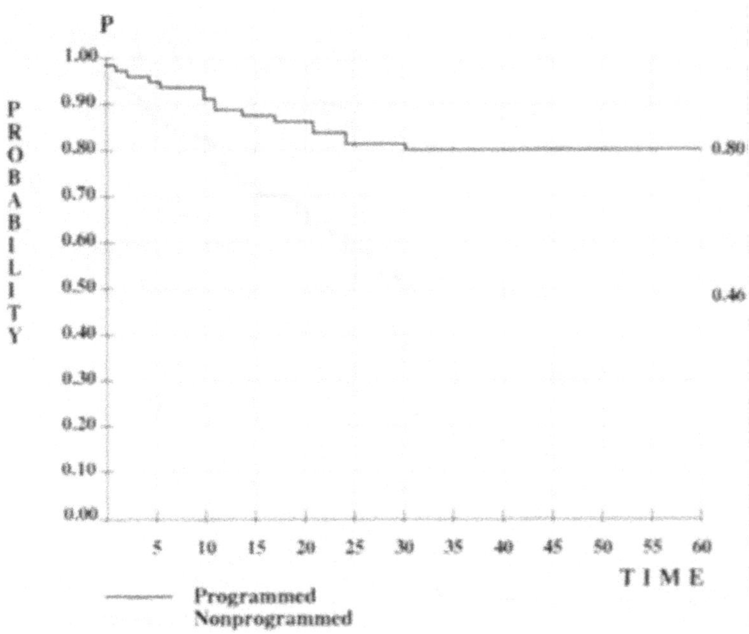

Fig. 1

Although this data are preliminary, it appears that the BFM-group will have much better treatment outcome than the nonprogram group. This is suggested by the fact that the rates of late relapses and deaths in remission are low for patients who are followed up more then 4 years.

The treatment outcome for patients into BFM-treated group whose 2-year treatment course had been finished we analyzed separately according to the year of therapy beginning, so it were subgroups 1990-1992, 1991-1993 and 1992-1994. Actuarial life tables for those groups are shown in figure 2.

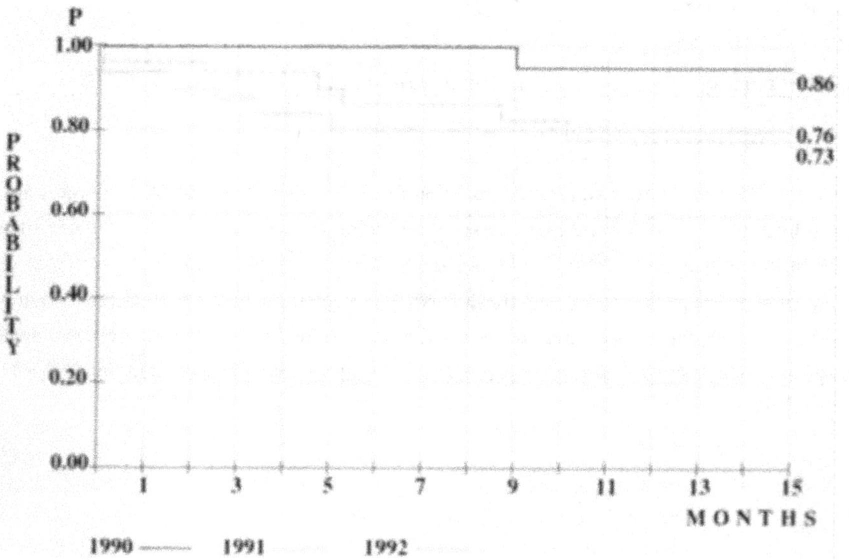

Fig. 2

Event - free survival for 86 patients treated according BFM protocols in periods 1990-1992, 1991-1993, 1992-1994 years

There are no significant differences in the figures of EFS, but it is possible to note a certain sign of improvement. The causes of death also have changed: while Mtx toxicity and hepatites B were the causes of 5 deaths during the first 1 and a half year of using BFM protocols, for the next 3 years we lost 3 patients because of bacterial and fungal infections.

The only changes we made during this period were principal improvement of supportive care, improvement of organizational know-how of nursing staff and paying special attention to setting up out-patient principles of chemotherapy.

In conclusion based on our experience we can conclude:

1. Intensive treatment of ALL can be used in clinics of Russia with high effectiveness if organizing principles of work of oncohematological departments are changed.

2. Adequate supportive care is one of the essential conditions for the successful intensive treatment.

Acknowledgments: to all doctors, nurses and lab.workers who took part in the treatment process. To collegues and teachers in the foreign clinics: Prof. G. Shellong, University Clinic of Munster, Prof. G. Henze, Free University of Berlin, Prof. H. Gadner, St. Anna Kinderspital, Viena, Dr. D. Sandlund, St. Jude Children Research Hospital, Memphis.

REFERENCES

Bennet JM, Catovsky D, Daniel MT et al. (1981) The morphological classification of acute lymphoblastic leukemia: concordance among observers and clinical correlations. Br. J Haematology 47: 553-561

Pinkel D and others (1971) Drug dosage and remission duration in childhood lymphocytic leukemia. Cancer 27: 247

Rivera GK, Crist WM, Sallan SE (1994) Biology and therapy of childhood acute lymphoblastic leukemia. XXV Congress of the International Society of Hematology, 26-33

Rumjantsev AG, Samochatova EV, Zhestkova NM, et al. (1994) The results of treatment of childhood ALL with the programs of intensive therapy. Hematology and transfusiology, N 2: 2 - 10.

Rumjantsev AG (1980) The modern treatment of acute leukemias in children, Pediatria, N 5: 57 - 62.

Ritter J., Creutzig U., Reiter A., Riehm., Shellong G. (1990) Childhood leukemia: cooperative Berlin-Frankfurt-Munster trials in the Federal Republic of Germany. J.Cancer Res Clin Oncol, 116, 100-103.

Part 9

Bone Marrow Transplantation

Carcinoembryonic antigen (CEA) in blood serum of patients with bone marrow transplantation

Uvarov M, Okulov V, Afanasiev, Karetkina E

Clinical Center for Advanced Medical Technologies and Petrov Research Institute of Oncology, St. Petersburg, Russia

Introduction

Bone marrow transplantation (BMT) is a potentially life-saving procedure that is used to treat a variety of malignant diseases. The most severe life-threatening complication of BMT is graft-versus-host disease (GVHD). It's main target organs are skin, liver and gastrointestinal tract. Usually the diagnosis of GVHD is based on results of clinical, biochemical, and histological data. The latter are the most objective. However, biopsies are invasive and possibly endangering immunodeficient and pancytopenic patients. Therefore the search for new, non-invasive methods to monitor patients after BMT is of great importance.

We assumed that the estimation of blood CEA levels might be an additional method to monitor intestinal complications in patients after BMT. It is known that CEA levels are influenced by damage of intestinal epithelium and subsequent proliferation. In patients undergoing BMT this might be due to conditioning regimen's (CR) toxicity and GVHD. Here we present preliminary data on CEA levels in patients undergoing BMT as a therapy of various malignancies.

Materials and Methods

30 Patients (18 female, 12 male) with hematological diseases were included into this study. Table 1 shows the distribution of patients according to diagnosis.

Table 1. Distribution of patients according to diagnosis.

Diagnosis	AML	ALL	CML	NB	MDS	SAA	HD	NHL	MM	SySa
No. of Patients	8	4	4	2	1	1	3	1	5	1

NATO ASI Series, Vol. H 94
Gene Technology
Edited by A. R. Zander, W. Ostertag,
B. V. Afanasiev, F. Grosveld

16 patients underwent allogeneic BMT (group 1) and 14 patients autologous BMT (group 2). The median age of the patients at the time of BMT was 20 yrs. (6 -52 yrs.). CR was carried out according to standard protocols (Bu, Cy, CBV, Melphalan, FBT etc.). All patients underwent gastrointestinal tract decontamination during conditioning.

CEA levels were measured by radioimmunoassay. Commercial available kits (Institute of Bioorganic Chemistry, Minsk) were used with a normal upper limit of 10 ng/ml in blood serum of healthy donors. The measurement was performed prior to conditioning (usually day -7), on the day of BMT (day 0) and once a week until patients were discharged. Engraftment was monitored by peripheral blood counts and bone marrow cytology.

Student's t-test was used to estimate the statistical significance of the obtained results.

Results and Discussion

All 30 patients became leukopenic after BMT ($<0.2*10^9/l$). There was no significant difference in severity of leucopenia in patients of either group. The median time of recovering from neutropenia (to $> 0.5*10^9/l$) was 28 days (14 - 56 days).

15 of 16 patients of group 1 developed acute GVHD with involvement of skin (15 patients), liver (8 patients) and intestinal tract (4 patients); group 2 included only patients, undergoing autologous BMT. CEA-levels are shown in tables 2 to 5. Two patients in group 1 and 3 patients in group 2 had increased serum-levels of CEA (14 - 25 ng/ml) at day -7 before conditioning was started. The average CEA level in patients of both groups before or at the day of engraftment (days 21, 28 and 35) was higher than on days -7, 0 and 7 (table 2).

Tabelle 2. Serum CEA levels of patients of group 1 and 2 before and after BMT (n=30), data are shown ± SD

Day	-7	0	7	14	21	28	35
CEA (ng/ml)	7.0±1.1	4.5±0.7	5.0±0.9	10.0±1.9	16.6±1.8#	21.0±2.0#	12.8±1.7#

# $p<0.02$ as compared to days -7, 0 and 7

While only 4 of 30 patients (13%) had elevated levels of serum-CEA (>10 ng/ml) on the day of BMT, 20 (67%), 21 (70%) and 15 (50%) of these patients had elevated

CEA-level on days 21, 28 and 35, respectively. Of 16 patients undergoing allogeneic BMT 12 (75%), 13 (81%) and 8 patients (50%) had increased CEA-levels on days 21, 28, 35 in comparison with 8 (57%), 8 (57%) and 7 (50%) patients undergoing autologous BMT (table 3).

Table 3. Serum CEA levels of patients of group 1 (n=16) and group 2 (n=14) before and after BMT, data are shown ± SD

Day	-7	0	7	14	21	28	35
GROUP 1							
CEA (ng/ml)	5.7±1.6	4.2±1.1	5.7±1.4	10.8±3.1	19.4±2.8*	24.6±3.1*	12,9±2.6#
GROUP 2							
CEA (ng/ml)	8.1±1.8	4.8±1.1	4.3±0.9	9.1±1.8	13.3±2.7#	17.6±3.3*	11.1±2.8

* $p<0.02$; # $p<0.05$ as compared to days -7, 0 and 7

11 patients (87%) in group 1 had GVHD and increased levels of CEA on day 28 after allogeneic BMT. 4 patients in group 1 with intestinal involvement of GVHD had normal CEA concentrations (table 4).

Tabelle 4. Serum CEA levels of patients of group 1 (n=15) with GVHD and without (upper line) and with (bottom line) involvement of the intestine, data in the upper line are shown ± SD

Day	-7	0	7	14	21	28	35
CEA (ng/ml)							
wo. Int. inv.	6.3±1.7	4.2±1.2	5.8±1.5	11.4±3.2	20.5±3.0*	26.1±3.3*	13.7±2.6#
w. Int. inv.§	3.0	2.0	3.8	8.8	21.5	18.0	17.2

* $p<0.02$; # $p<0.05$ as compared to days -7, 0 and 7 of the same line

§ SD not evaluated

Independent of GVHD, increases of serum CEA concentration were associated with engraftment. Among other possible explanations, we are inclined to assume that this increase of CEA levels is closely linked to the recovery of subsets of intraepithelial lymphocytes (IEL). On one hand , CEA family molecules are synthesized by lymphoid cells and posses immunomodulatory functions (Medoff et al. 1984; Hakim 1984; Kuroki et al 1990). On the other hand, IEL contribute to proliferation of enterocytes. The number of the former may be dramatically decreased in intestinal disorders (Corazza et al. 1984; Sydora et al. 1993). This may be the case in patients undergoing BMT during conditioning and subsequent engraftment of hematopoietic cells. A simultaneous increase of IEL may result in reconstitution of CEA-synthesizing cells (both lymphocytes and enterocytes) and recovery of the intestinal tract as a digestive and also immunocompetent organ.

References

Corazza GR, Frazzoni M, Gasbarrini G (1984) Jejunal intraepithelial lymphocytes in coeliac disease: are they increased or decreased? Gut 25: 158 - 162

Hakim AA (1984) Carcinoembryonic antigen, a tumor-associated glycoprotein induces defective lymphocyte function. Neoplasma 31: 385 - 397

Kuroki M, Matsuo Y, Ohtani T et al. (1990) A novel CEA-cross-reacting antigen of molecular weight 140,000 expressed on human lymphoid cell lines. Mol Immunol 27: 689 - 896

Medoff JR, Jegasothy BV, Roche JK (1984) CEA induced release of suppressor factor from normal human lymphocytes in vitro. Cancer Res 44: 5822 - 5827

Sydora BC, Habu S, Taniguchi M (1993) Intestinal intraepithelial lymphocytes preferentially repopulate the intestinal epithelium. Int Immunol 5:743 - 751

CYTOGENETIC MONITORING IN THE POST BMT PERIOD

MARINETS O.V., OGORODNICOVA J.S.
DEPARTMENT OF MOLECULAR GENETICS
CLINICAL CENTER OF ADVANCE MEDICAL TECHNOLOGIES
DYNAMO 3
197042 ST-PETERSBURG
RUSSIA

INTRODUCTION

The evaluation of the bone marrow (BM) cells after bone marrow transplantation (BMT) is important. Post BMT cells may be of donor or recipient origin, and the latter may belong to the normal or the leukemic population. The existance of recipient cells as well as donor cells in the peripheral blood (PB) or BM is defined as hematopoetic chimerism. Detailed study of such hematopoetic chimerism is necessary to improve the understanding of the interaction between the donor and the recipient BM cells and to study the therapeutic effect of BMT (Gale and Champlin, 1986; Gilliland et al., 1990).

Another important problem in transplantation is relapse. Recurrent leukemia may derive from the regrowth of the original leukemic cells, from development of secondary leukemia in the recipient and leukemic transformation of donor cells. In most cases post BMT relapse is of recipient origin (Lawler et al., 1990). However, examples of relapse in donor cells have been documented (Schitz et al., 1987).

To address these questions, genetic markers are used in cytogenetic studies. In these studies, the sex difference between the donor and the recipient can be used. Moreover, there are specific chromosomal aberrations in patients with acute leukemia (AL) and in chronic myelogenous leukemia (CML). Thirdly, secondary chromosomal abnormalities have been described. They could be the result of clonal evolution in the patients at the time of relapse and disease progression or the results of high-dose chemotherapy/TBI for BMT.

NATO ASI Series, Vol. H 94
Gene Technology
Edited by A. R. Zander, W. Ostertag,
B. V. Afanasiev, F. Grosveld

MATERIALS AND METHODS

Patients

The study included 9 patients who were examined between 1992 and 1994. There were 4 patients with CML, 2 with acute nonlymphoblastic leukemia (ANLL), 2 with acute lymphoblastic leukemia (ALL), and 1 with myelodysplastic syndrome (MDS).

Procedure

Bone marrow samples obtained at diagnosis, before and after BMT were processed according to the short-term unstimulated culture and stained by Giemsa banding by a modification of the trypsin method of Seabright (Seabright et al., 1977; Verma, Babu, 1989). Chromosomal abnormalities were identified and classified according to the International System of Human Cytogenetic Nomenclature (ISCN, 1981).

Karyotypes of donors were studied by use of 72 hour culture with additional PHA and G-banding staining.

RESULTS AND DISCUSSION

Patient characteristics are summarized in Tables 1 and 2. All patients were treated with standards protocols for the initial disease. Conditioning and GvHD prophylaxes for BMT was done according to Hamburg protocols. 8 patients received allogeneic and 1 syngenic BMT. In one patient with ALL (V.A.) BMT was performed twice.

2 patients with AL (M.J., V.A.) were sex-different from their respective donors, 2 patients with AL (N.) and one with MDS (R.J.) had the abnormal karyotype in the initial phase of the disease. Patients with CML had donors with another sex.
Cytogenetic studies performed before and after BMT allow the definition of origin of cells after allogenic bone marrow transplantation. In our investigation we used such cytogenetic markers as sex difference between donor and recipient and specific chromosomal aberration in patients with leukemia.

Table 1. Clinical and cytogenetic findings of patients with AL.

Patients age sex	D-s	Karyotype at D-s	Type BMT	Donor	Status at BMT	Post BMT Wks	Post BMT Blasts BM (%)	Post BMT Karyotype	Survival post BMT (wks)
L.P.	ALL	46,XY,Ph+	Allo	Broth.	3 Rel	7	0.2		133.0+
5/M		t(9;22)		46,XY		8	0.5		
CR		(q34;q11)				10	0.8	Ph+-8%	
						13	0.2	46,XY	
						58	1.4	46,XY	
						63	1.8	46,XY	
						68	1.0	46,XY	
V.A.	ALL	46,XY	Allo	Sister	2 CR	3		46,X 97%	84.0+
10/M				46,XX		8		46,XX	
CR						17	4.2	46,XX	
						25	3.3	46,XX-93%	
						30	1.2	46,XX	
						41	4.4	46,XX-94%	
						54	26.4	46,XX-87%	2 Rel
						59	28.0	46,XX-91%	
						64	0.4	46,XX-92%	
						69	6.8	46,XX-35%	
			2BMT			79	1.8	46,XX-25%	
						75		46,XX-40%	
						83	0.4	46,XX-50%	
M.J.	ANLL	46,XY	Allo	Sister	2 CR	2		46,XX-90%	54.0+
12/M				46,XX		10	0.4	46,XX-80%	
R						25	0.4	46,XX-93%	
B.J.	ANLL	46,XY,	Syn	Broth.	1 CR	3	1.2	46,XY	46.0+
7/M		t(8;21)		46,XY		12	0.6		
CR		q(22;q22)				22	0.4	46,XY	
						31	1.2	46,XY	
R.J.	MDS	46,XX,	Allo	Sister	RAEB	2	0.6	7q- 93%	153.0+
2/F		del(7)		46,XX		4	1.5	46,XX	
		(q33)				6	2.5	46,XX	
						8	0.8	46,XX	
						30	1.0	46,XX	
						35	2.0	46,XX	
						86	2.4	46,XX	
						96	2.6	46,XX	
						104	0.6	46,XX	

Table 2. Clinical and cytogenetic findings of patients with CML.

Patients age/ sex/ clin. outcome	Karyotype at D-s	Type BMT	Donor	Status at BMT	Post BMT Wks	Blasts+ promyeloc. BM (%)	Karyo- type	Sur- vival post BMT (wks)
F.S.	46,XY,Ph+	Allo	Sister	CP	2	0.4	46,XX	9.0
14/M	t(9;22)		46,XX		3	0.4	Ph+ -15%	
Death	(q34;q11)				4	1.6	46,XX	
GvHD IV					5	0.8	46,XX	
F.G.	46,XX,Ph+	Allo	Sister	CP	3		Ph+ -22%	56.0+
40/F	t(9;22)		46,XX		5	0.2	Ph+ -10%	
CR	(q34;q11)				7		46,XX	
					9	1.0	46,XX	
					15	1.5	Ph+ -15%	
					20	0.4	46,XX	
					27	0.2	46,XX	
					46	1.0	46,XX	
V.J.	46,XY,Ph+	Allo	Broth.	BC	5	1.0	Ph+,+t(2;7) - 81%	12.0
16/M	t(9;22)		46,XY					
Death								
BCCML	(q34;q11)-CP				8	9.0		
	46,XY,Ph+				9	3.0		
	t(9;22)(q34q11)				11	14.0		
	t(2;7)(p11;p22)-BC				12	30.0		
B.V.	46,XX,Ph+	Allo	Broth.	CP	2		Ph+ -40%	54.0
36/F	t(9;22)		46,XY		7	0.5	46,XY	
BC	(q34;q11)				18	6.0	Ph+ -91%	
					22	16.6	Ph+ -70%	
					24	18.4	Ph+ -97%	
					30	25.4	Ph+ -100%	
					41	12.8	Ph+ -95%	

For example, in two patients with AL we observed the ratio of cells with the male and female karyotypes. In one of them (V.A.) the percentage of recipient cells increased during development of relapse, and vice versa. Another patient (M.J.) demonstrated rather high percentage of donor metaphases (from 80.0% to 95.4%) during remission.

Ph-chromosome is an informative cytogenetic marker used for the definition of hematopoietic chimerism after BMT. In CML, Ph-positive cells may be present in clinical remission induced by chemotherapy. The persistence of Ph-positive cells after BMT is a feature in some patients with CML despites successful treatment by BMT. Two of our patients with CML, at 3 (F.C.). 5 and 15 (F.G.) weeks after BMT showed transitory Ph-positive cells in 15% metaphases which later disappeared at 4 weeks (F.S.) and 20 weeks (F.G.) after BMT.

In other cases Ph-positive cells after BMT are indicating recurrence of disease. Percentage of Ph-positive cells increased in one of our patients with CML (B.V.) from 40% (on two weeks) to 100% (30 weeks post BMT). These changes were accompanied by progression of the disease to blast crisis (BC).

In another patient (V.J.) cytogenetics was performed in BC CML. Secondary chromosomal changes, t(2;7)(p11;p22), were observed in 86% BM cells during BC before BMT. After BMT the same t(2;7) translocation was found in addition to the t(9;22) translocation. The patient died of disease progression.

Cytogenetic investigation was also used to confirm the disease remission in those patients with AL and MDS, in whom specific cytogenetic markers were defined at diagnosis. Translocation t(8;21)(q22;q22) in one patient with ANLL (B.J.), translocation t(9;22)(q34;q11) in one patient with ALL (L.P.) and deletion of chromosome 7 in one MDS patient (R.J.) were absent during remission and after BMT.

REFERENCES

Gale R., Champlin R. (1986) Bone marrow transplantation in acute leukemia. Clinics in Haematology, 15

Gilliland G., Patterson E. (eds) (1990) Multiple clones with(3;21) translocation in a case of Ph-positive chronic myelogenous leukemia during relapse after allogeneic BMT. Cancer Genet.Cytogenet. 7:55-60

ISCN(1985) International System for Human Cytogenetic Nomenclature

Lawler S., Khokhar M., (eds) (1990) Cytogenetic studies of leukemic recurrence in recipients of bone marrow allograft. Cancer Genet. Cytogenet. 47:249-263.

Schmitz N., Iohanson W., (eds) (1987) Recurrence of acute lymphoblastic leukemia in donor cells after allogenic marrow transplantation assotiated with a deletion of the long arm of chromosome 6. Blood, 70:1099-1104.

Seabright M.(1971) A rapid banding technique for human chromosomes. Lancet 2: 971.

Verma R., Babu A. (1989) Human chromosomes: Manual of basic techniques. Pergamon Press. Inc. New-York.

FIRST EXPERIENCE IN THE FIELD OF BONE MARROW TRANSPLANTATION AT THE ST.-PETERSBURG BMT CENTER

B.V.Afanasiev

for St.-Petersburg BMT Center Team,
189646 St.-Petersburg
Russia.

INTRODUCTION.

A significant proportion of patients, who did develop acute leukemia (AL), chronic myelogenous leukemia (CML), non-Hodgkin`s (NHL), Hodgkin`s disease (HD), multiple myeloma (MM), severe aplastic anemia (SAA), myelodysplastic syndrom (MDS) and other malignancies may be cured with allogeneic (allo), autologous (auto) bone marrow and/or peripheral blood stem cell transplantation (PBSCT). This article deals with first results of BMT and PBSCT performed at Petrov Research Institute of Oncology

Table 1. GENERAL DATA ABOUT BMT TRANSPLANTATION AT PETROV RESEARCH INSTITUTE OF ONCOLOGY

YEARS	ALLO-BMT		AUTO-BMT		SING-BMT		PBSC	PBSC+BM	TOTAL
	ADL	CHILD	ADL	CHILD	ADL	CHILD			
1990			5			1			6
1991	6	3	13		1	1			23
1992	5	5	21			SECOND 1			32
1993	3	6	12	1		1			23
1994		SECOND 1	6	7			9	6	28

and St.-Petersburg City BMT Center between February 1989 and December 1994 (see table 1).

NATO ASI Series, Vol. H 94
Gene Technology
Edited by A. R. Zander, W. Ostertag,
B. V. Afanasiev, F. Grosveld

MATERIALS AND METHODS.

Thirty-two alloBMT (16 children and 16 adults), two second alloBMT (both in children), 3 syngeneic BMT, sixty-one autoBMT (9 children and 52 adults), nine auto-PBSC and 6 auto-PBSC plus autoBMT transplantations were performed during the abovementioned time period. Patient characteristics of allo- and auto-BMT patients are shown in Tables 2 and 3.

Table 2. PATIENT CHARACTERISTICS FOR ALLOGENEIC BONE MARROW TRANSPLANTATION

DIAGNOSIS	STAGE	QT.	SEX	QT.	AGE
ACUTE MYELOBLAS-	CR	8	M	8	
TIC LEUKEMIA	RELAPSE	2	F	2	5-54
ACUTE LYMPHOBLAS-	CR	5	M	8	
TIC LEUKEMIA	RELAPSE	5	F	2	7-36
CHRONIC MYELOGE-	CPH	3	M	3	
NOUS LEUKEMIA	BLCR	3	F	3	14-42
RAEB	CR	1	F	1	2
SAA			M	1	12
		2	F	1	5
NEUROBLASTOMA	IV	1	F	1	8
NHL	IV	1	M	1	33
ESSENT.THROMBOCYTHEMIA		1	F	1	21
SECOND BMT	CR	2	M	2	9-10

Conditioning.regimens for allo- and syngeneic BMT included busulfan (Bu) plus cyclophosphamide (Cy) (Tutchka protocol), Bu plus Cy plus VP-16 (Hamburg University protocol). In SAA, we used Cy plus antithymocyte globulin (ATG).

Table 3. PATIENT CHARACTERISTICS FOR AUTOLOGOUS BONE MARROW TRANSPLANTATION

DIAGNOSIS	STAGE	QT.	SEX	QT.	AGE
ACUTE MYELOBLAS-	CR	7	M	4	
TIC LEUKEMIA	RELAPSE	0	F	3	9 - 57

(To be continued)

ACUTE LYMPHOBLAS	CR	4	M	3	
TIC LEUKEMIA	RELAPSE	0	F	1	7 - 34
HODGKIN`S	III	9	M	9	
DISEASE	IV	8	F	8	12 - 42
(Is continued)					
NON - HODGKIN`S	III	2	M	4	
LYMPHOMA	IV	7	F	5	15.- 39
MULTIPLE	AGRESSIVE		M	5	
MYELOMA	COURSE	14	F	9	42 - 62
SOLID TUMORS	III		M	8	
	IV	19	F	11	4 - 54
HYSTIOCYTOSIS	RESIST.FORM	1	M	1	28

For autoBM and PBSC transplantations, various conditioning regimens were applied, including CBV, or BEAM, or TACC, or high doses of Melphalan for MM, TBI plus Melphalan plus VP-16 for solid tumors.

For PBSC mobilization and priming, G-CSF or GM-CSF have been used (NEUPOGEN "Roche" and LEUCOMAX "Sandoz").

Treatment efficiency of haematological malignancies can be predicted using some common prognostic signs which are assessed during primary examination, or prior to BMT. Therefore, we have analyzed the possible prognostic significance of such features as age, stage of disease, Karnovsky performance scores.

RESULTS AND CONCLUSION.

As seen from Table 1, an average of twenty-eight BMTs per year were performed at the St.- Petersburg BMT Center. Fig. 1 and 2 show survival rates in the patients with AL and in patients with AML treated with BMT. These results are in accordance with recent publications.

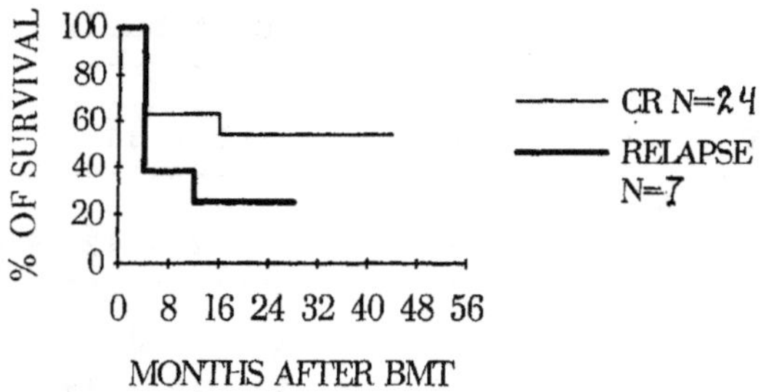

Fig.1 Survival rates in BMT patients with AL

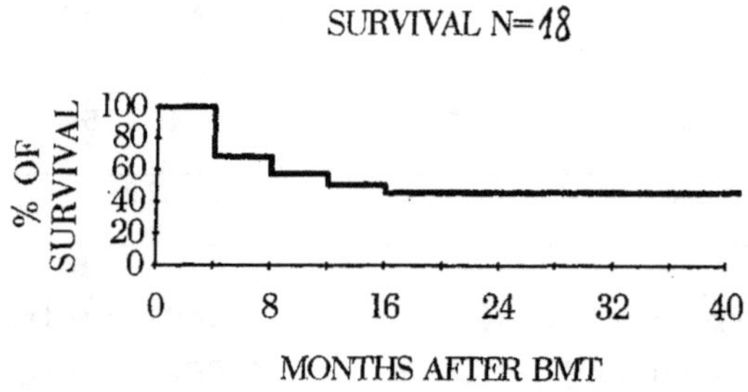

Fig.2. Survival rates in BMT patients with AML

During the initial phase of transplantation program, we could not strictly stick to the common eligibility criteria, and included the patients with rather poor performance scores as well as with other contraindications for BMT. Hence, initial results are to be interpreted by considering an increased procedure-related mortality (Fig.3).

The significant relapse rates are the main problem at present time, due to changed inclusion criteria of patients.

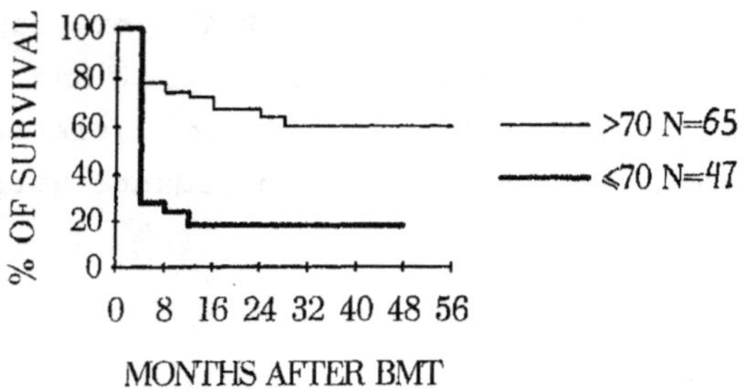

Fig.3. Survival of BMT patients with different Karnovsky score

Reasons of mortality after allo- and auto-BMT included clinical relapses - 30,3%, regimen-related toxicity - 15,6%, acute GVHD - 12,1%, fatal septic infections -12,1%.

The most important prognostic factors for clinical outcome of BMT were: patient age, Karnovsky score, and stage of the disease at the time of transplantation (Fig.3,4,5).

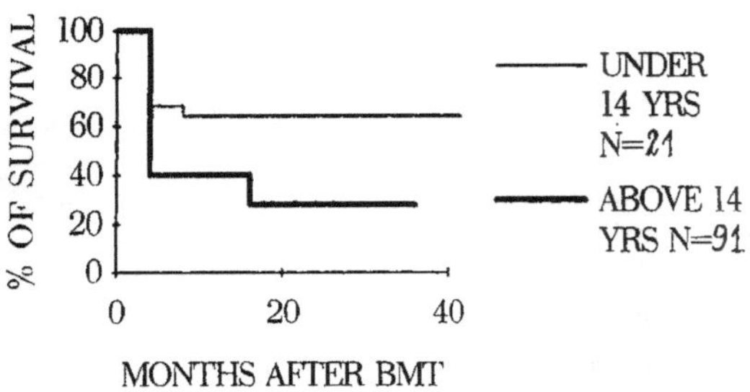

Fig.4. Survival of BMT patients from different age groups

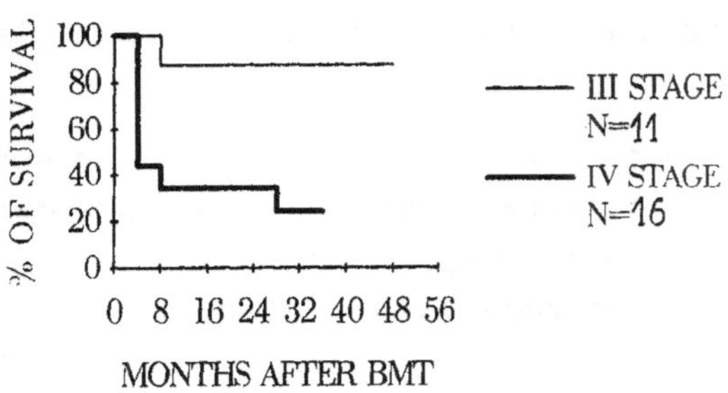

Fig.5 Survival of BMT patients with lymphomas (stage III vs. IV)

Overall survival of patients with a pre-treatment Karnovsky score of more than 70% were 62% (56 months follow-up), in children 65% (48 months follow up), in patients with AL (CR state) 57 % (52 months follow-up) and and in patients with an early stage of malignancy 85 % (50 months follow-up), thus being much higher compared with other groups($p< 0.05$). High-dosage chemotherapy followed by auto-BMT increased the overall survival rate in patients with aggressive multiple myeloma (MM), compared to conventional therapy,(62% during 32 months and 15% during 96 months, respectively, $p < 0,05$), as shown in Fig.6.

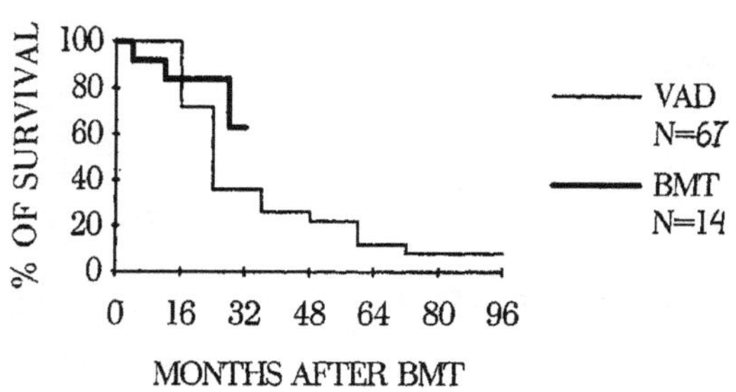

Fig.6 Survival of BMT patients with aggressive multiple myeloma (BMT vs.VAD protocol)

The preliminary experience with PBSC transplantation showed that this procedure is highly efficient, and in some instances it seems to be superior to BMT with respect to an accelerated recovery and decreased transfusion requirements (Tab.4).

Table 4. BLOOD RECOVERY (days) AND REQUIREMENT FOR BLOOD PRODUCT TRANSFUSION (number of transfusion days) AFTER BM, BM+PBSC AND PBSC TRANSPLANTATIONS

	RECOVERY L > 0,5	RECOVERY TR > 20	NECESSITY RBC	NECESSITY Platelets
BM	15,6 + - 34	24,3 + 59 -	8 + - 2,9	14 + - 3,1
BM + PBSC	12,6 + -1,4	14,8 + - 51	3,0 + - 1,7	9,7 + - 4,9
PBSC	15,25+ -4,75	15,75+ -4,9	5,7 + - 3,7	10 + - 8

When summarizing the aforementioned data, a conclusion may be drawn, that the latest experience of St.-Petersburg BMT Center yielded sufficient clinical results which are in good agreement with results of multicentric studies. One of the most promising approaches will be the wider use of intensive chemo/radiation therapy employing hematopoietic stem cells transplantation. Treatment should be started at the earliest stage of malignancy, especially in patients with poor prognosis.

ST.PETERSBURG'S BMT CENTER TEAM.

Boris Afanasiev - Chief

Julia Alexandrova
Natalia Anikina
Vladimir Andreev
Dmitri Alexeev
Irina Baldueva
Margarita Belogurova
Tatiana Bykova
Alexey Chukhlovin
Elena Darskaya
Konstantin Efimov
Nikolay Fedtchenko
Alexander Figursky
Lubov Fregatova
Olga Frolova
Ludmila Gavrilova
Galina Goroshko
Andrey Katyshev
Marina Kirichenko
Anna Klimovich
Sergey Konoplev
Marina Konopleva
Irina Koshkarova
Anton Lyzhov
Olga Marinetz
Nadezhda Medvedeva
Elena Leenman
Natalia Myhailova
Elena Morosova
Dmitri Nikitin
Julia Ogorodnikova
Valery Okulov
Marina Osipova
Vladimir Osipov
Michail Ouvarov
Galina Platonova
Eleonora Podoltseva
Nikolay Podoltsev
Galina Radulesku
Alexey Ryskin
Xenia Sytskaya
Stanislav Shavva
Alexander Shmidt
Irina Stuff
Tatiana Viktorovich
Olga Volkova
Svetlana Vydayko
Tatiana Zabelina
Michail Zaraisky
Andrey Zaritsky
Ludmila Zubarovskaya

ACKNOWLEDGEMENTS

We would like to express our deep gratitude to Prof. Axel Zander and to all his colleagues at BMT Center, University Hospital Eppendorf (Hamburg), for valuable aid, clinical consulting and stimulating discussions.

IMMUNOGLOBULIN E LEVELS IN ALLOGENIC BMT

Anna Klimovich*, Maria Teresa van Lint**, Marina Samoilovich***, Vladimir Klimovich***

* Petersburg BMT-Centre, Centre for Advanced Medical Technologies, St.Petersburg, Russia
** Ospedale San Martino, Genova, Italy
***Research Roentgen Radiological Institute, St.Petersburg, Russia

INTRODUCTION

It is clear now that different events, such as neoplastic haematological diseases, chemotherapy, total body irradiation (TBI) and bone marrow transplantation may disturb T-cell regulation and may lead to increase of immunoglobulin E (IgE) levels (Bengtson M,1989; Korver K, 1987, 1988). Also liver damage has been reported to be a cause of serum IgE elevation (Joske RA, 1976; Van Epps E, 1976). Increase of IgE levels has been reported after allogenic bone marrow transplantation (allo-BMT) in 80-95% of the patients (Akasaki M, 1987; Geha RS,1980; Ringden O, 1983). Reasons for this increase include: infectious episodes, allergic drug reactions (to cyclophosphamide and others) (Bengtson M, 1989; Chiorazzi N, 1976; Korver K, 1988) or TBI (Korver K, 1988; Heyd J, 1988) and graft-versus-host disease (GvHD) (Korver K, 1988; Saarinen UM, 1987), especially with liver involvement (Heyd J, 1988). These data were obtained during the period of 1979-1989. Since trends in therapy and conditioning regimens have changed, it is interesting to re-evaluate the role of IgE-levels in BMT-patients.

MATERIALS AND METHODS

We have analyzed 87 sera samples of 17 patients, treated at the Ospedale San Martino Allo-BMT Department, Genova, between day 29 to day 89 after BMT. There was 1 patient with NHL, 2 patients with ALL, 1 patient with MDS, 2 patients with SAA, 5 patients with CML, and 6 patients with AML.

The measurement of total serum IgE levels was carried out by the immunometric assay "Polygnost" using monoclonal antibodies developed by the Hybridoma Technology Laboratory, Roentgen- Radiological Institute, St-Petersburg, "Polygnost". (Samoilovich M,1992). The assay was validated by comparison with analogous systems manufactured by "Bio-Merieux" and CMG (Switzerland). Results are presented in international units (IU).

Due to wide interindividual variation in IgE level (Saarinen UM,1987) (genetic factors, latent atopic disease, age), baseline levels of IgE were determined individually and the estimation of IgE variation was performed in each patient (2-8 times) separately. The pretransplant or early posttransplant data (day 1 to 5) were considered basal level of IgE.

NATO ASI Series, Vol. H 94
Gene Technology
Edited by A. R. Zander, W. Ostertag, B. V. Afanasiev, F. Grosveld

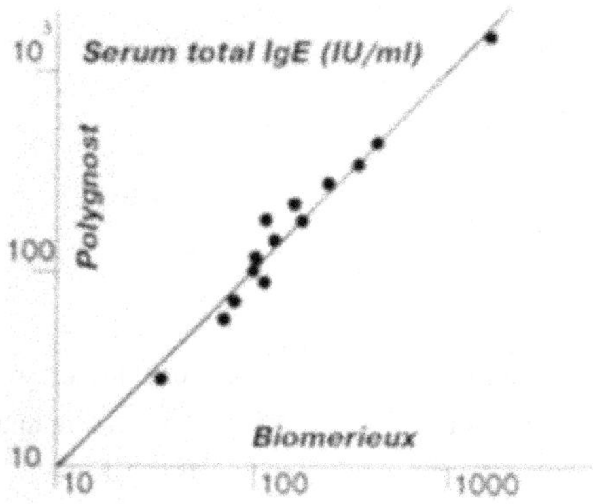

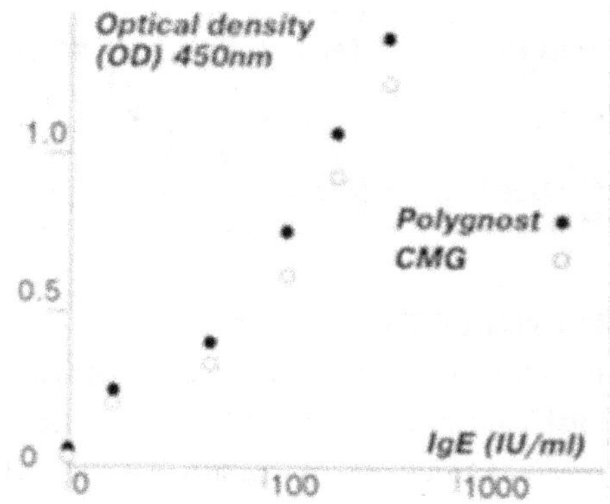

Fig.1A Serum total IgE, measured by Bio-Merieux and Polygnost assays and CMG-systems

Fig.1B. IgE calibrators determination in Polygnost

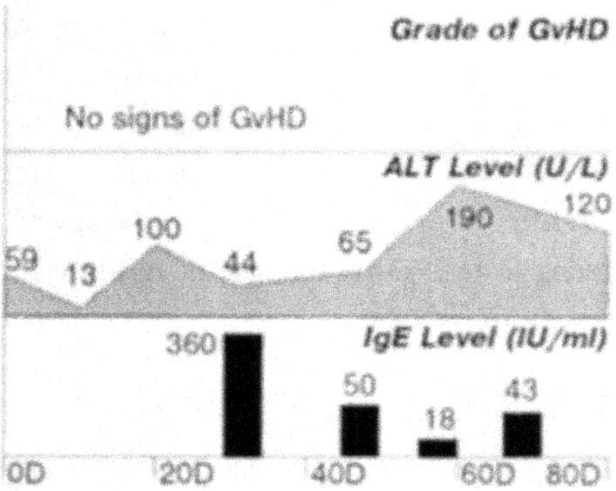

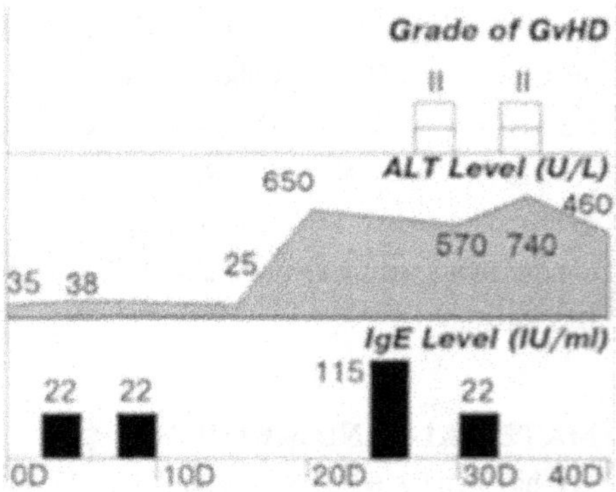

Fig.2 Patient A

Fig.3. Patient B

RESULTS AND DISCUSSION

The sensitivity of "Polygnost"-IgE assay is comparable with analogous assays, manufactured by "Bio-Merieux" and CMG (Fig.1A and B).

In the majority of samples, IgE levels were much lower than normal as defined by WHO (100 IU/mL): in 60 samples IgE level did not exceed 20 IU/L. 9 of 17 patients showed spikes of IgE level.Peak levels ranged from 42 to 3000 IU. One spike (day 0) was due to.a severe drug-related allergic reaction.

Increase of IgE levels after BMT was seen in 8 of 17 cases. This is slightly less than previously reported.

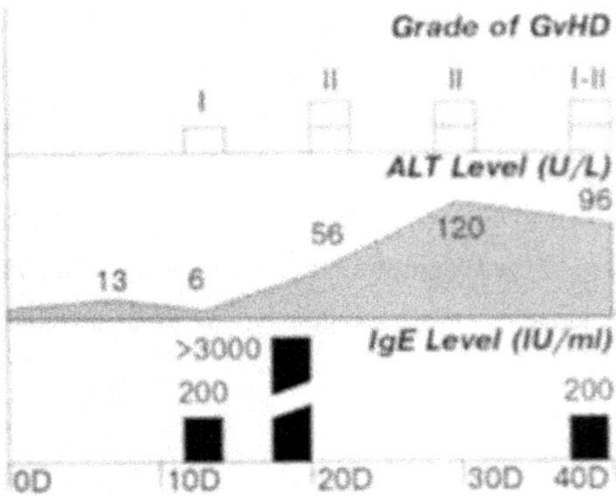

Fig.4. Patient C

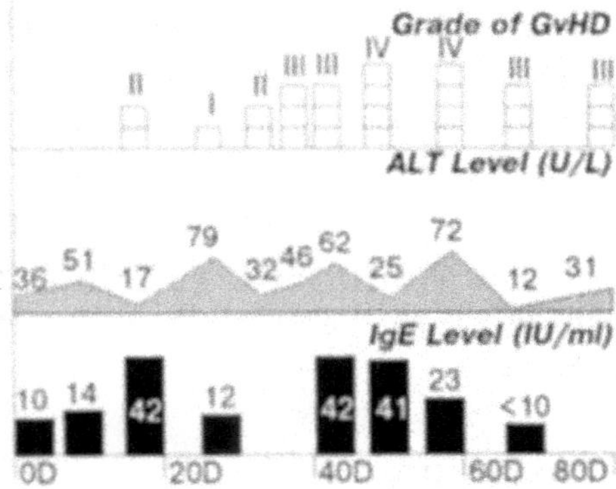

Fig 5. Patient F

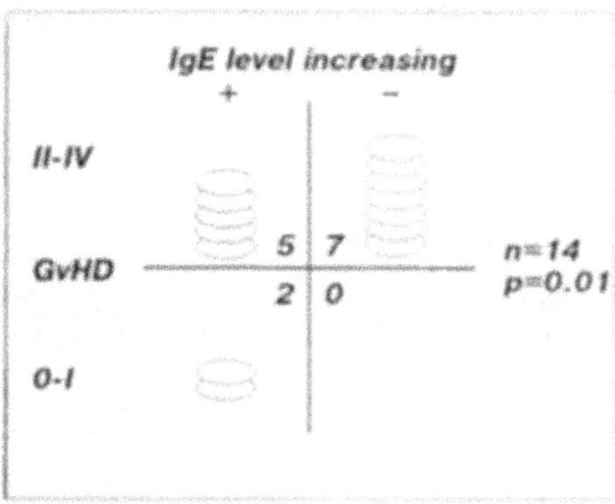

Fig.6 Grade of GvHD and IgE-level

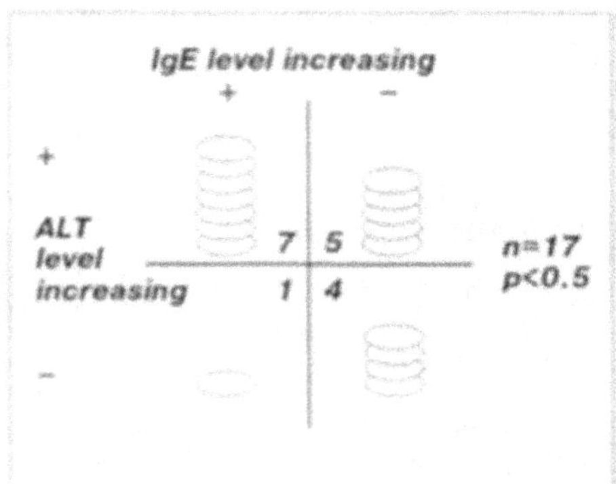

Fig.7 IgE level and liver damage.

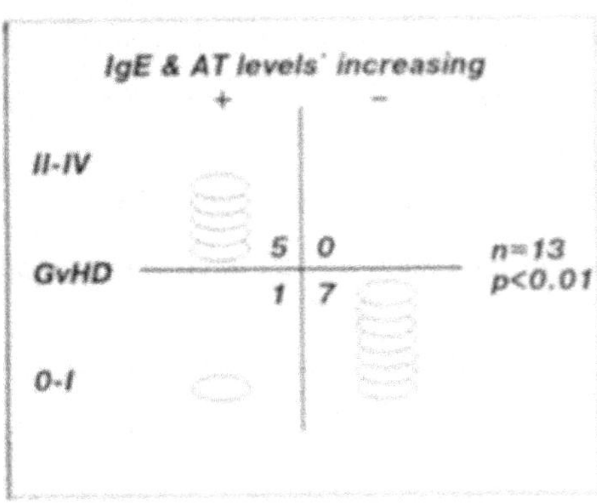

Fig.8. Grade of GvHD and IgE associated with liver damage

The grade of GvHD and IgE elevation was associated with liver damage. In 1 patient (Fig.2), the IgE-peak was noticed during an episode of acute viral hepatitis (high aminotransferases' levels but without any sign of GvHD). In 5 patients suffering from GvHD, significant IgE elevations were measured from day 7- day 65 (Fig.3,4,5). A rise of IgE levels was noticed only in cases with severe GvHD (grade II-IV) as shown in Fig.6. The majority of IgE elevations was accompanied by signs of liver damage with increased aminotransferases in serum (Fig.7). But only in GvHD, grade II-IV, signs of liver damage were accompanied by IgE elevation (Fig.8).

CONCLUSION

1.The increases of IgE levels appear only in severe GvHD (grade II-IV).

2.Further studies are necessary to determine the role of GvHD, liver damage and other medical procedures (TBI, CY-therapy, bone marrow transplantation) in regulating IgE production.

REFERENCES

Abedi MR (1989) Serum IgE after BMT. Bone Marrow Transplantation 4: 225

Akasaki M, Ishizaka K (1987) Effects of cyclophophamide treatment and gamma irradiation of SJL mice on the IgE antibody response and the nature of IgE-binding factors.Int Arch Allergy Appl Immunol 82: 417

Bengtsson M (1989) B-cell reconstruction after auto BMT: increase in serum CD23 precedes IgE and B-cell regeneration. Blood 73: 2139

Chiorazzi N, Fox DA, Katz DH (1976) Hapten-specific IgE-antibody response in mice. VI. Selective enhancement of IgE-antibody production by low X-irradiation and cyclophosphamide. J Immunol 117: 1629

Geha RS (1980) Increased serum IgE level following allo-BMT.J Allergy Clin Immunol 66: 78

Heyd J (1988) IgE levels following auto-, allo- and syngenic BMT: An indirect association between hyperproduction and acute GvHD in allo-BMT. Blood 72: 442

Joske RA (1976) Serum IgE levels in patients with liver disease. Med. J Austral 2: 555

Korver K (1987) Transient elevation of serum IgE after alloBMT. Int Arch Allergy Appl Immunol 84: 45

Korver K (1988) Transient elevation of serum IgD after allo-BMT. Clin Exp Immunol 72: 337

Ringden O (1983) Are increased IgE levels a signal of an acute GvH reaction? Immunol Rev 71: 57

Ringden O (1983) Markedly elevated serum IgE levels following allo and syngenic BMT. Blood 61: 1190

Saarinen UM (1987) Differentiation of presumed sepsis from acute GvHD by C-reactive protein and serum total IgE in BMT recipients. Transplantation 44: 540

jSamoilovich M, Klimovich V (1992) A new family of monoclonal antibodies against human IgE- immunochemical characteristics. Allergy Clin Immunol News 4: N1

Saryan JA (1983) Regulation of human IgE synthesis in acute GvHD.J Clin Invest 71: 556

Van Epps E (1976) Liver disease - a prominent cause of serum IgE elevation. Clin Exper Immunology 23: 444

ACKNOWLEDGEMENTS
We would like to thank Prof.Boris Afanassiev for the kind attention to our work and to Dmitry Kozunov for word and layout processing.

PROGNOSTIC FACTORS AND INDICATIONS FOR HAEMOPOIETIC STEM CELLS TRANSPLANTATION IN PATIENTS WITH MULTIPLE MYELOMA.

E.I. Podoltseva, M.I. Zaraiski

Clinical Center for Advanced Medical Technologies , 197042 , St-Petersburg , Russia.

INTRODUCTION

Multiple myeloma (MM) is a malignant clonal B-lymphoproliferative disease. The survival rate of MM patients varies from few months to many years (Kyle , 1988). Determination of the prognostic factors in MM patients is important for choosing the treatment. Some patients don't need specific therapy for a long time, while the management of others may demand intensive chemotherapy with subsequent bone marrow transplantation (Barlogie, Jagannath, 1994). Since the pioneering report by Carbone and coworkers in 1967, a large number of studies on prognostic factors for MM have been carried out. The most important factors influencing survival in MM patients are: renal function, haemoglobin level, performance status, serum calcium, extent of skeletal lesions and percentage of bone marrow plasma cells (Galton and Peto, 1973; Alexanian et al., 1975; Kyle and Elvebach, 1976;Matzner et al., 1978; Merlini et al., 1980). Recently, other prognostic factors such as plasma cells morphology (Carter et al., 1987), serum B2-microglobulin, LDH, and DNA-RNA content in plasma cells have been identified (Barlogie, Salmon 1990). Moreover, efforts have been made to use prognostic factors for separation of patients into different risk groups in order to improve the therapeutic approach. There are some classifications of MM: Dure and Salmon (1975), Merlini-Waldenstrom-Jayakar (1980), Medical Research Council (1980), and others. But prognosis in an individual MM patient is still hard to evaluate.

The primary aim of the present study was to analyze different features in our MM patients, as well as to derive a prognostic system from the results of this analysis.

MATERIALS AND METHODS

MM was diagnosed in 250 patients of the Haematological Department of St-Petersburg Medical University and St-

NATO ASI Series, Vol. H 94
Gene Technology
Edited by A. R. Zander, W. Ostertag, B. V. Afanasiev F. Grosveld

Petersburg's Clinical Center for Advanced Medical Technology during the period from January 1976 to December 1992. The diagnosis of MM was made according to criteria published previously: Chronic Leukemia - Myeloma Task Force, 1973.

Maturation of plasma cells was determined according to the criteria of the morphological classification for MM (Carter et al., 1987).

HLA-typing was performed in a routine 2-step lymphotoxic test with complement (Friemel 1984) using the national full-set anti-HLA sera. Blood donors of St.-Petersburg were used as regional control (Shabalin et al., 1988).

Data were analyzed by means of BMDP Statistical Software (University of California, Los-Angeles, USA, 1981): Kaplan-Meyer estimates of overal survival curves (Kaplan and Meier 1958); log-rank test (Mantel 1966) and covariance analysis (Breslow 1974), using a program for comparison of curves. Multivariate analysis was performed with discriminant models. Statistical analysis of HLA data (X^2 and relative risk- RR) was performed according to Lefkovits (1979).

RESULTS

We have determined 8 features from 21 basic clinical signs, which have the strongest prognostic value for survival of MM patients: performance status, bone marrow plasmacytosis, morphological type of myeloma cells, severity of bone lesions, serum creatinine concentration, haemoglobin level, "shift to the left" in peripheral blood leukocytes count and concentration of serum B2-microglobulin (P=0.0000 - 0.0017). The best overall survival was in MM patients with performance status "0" and the worst in patients with performance status III and IV (Fig. 1) Median survival in patients with bone marrow plasmacytosis less than 10% was 110 months; in patients with bone marrow plasmacytosis of 10%-30%, 60 months, and in patients with bone marrow plasmacytosis of more than 30%, 36 months (Fig. 2)

Median survival in patients with mature MM was 120 months; in patients with immature MM, 35.2 months, and in patients with plasmablastic MM, 11.2 months (Fig. 3).

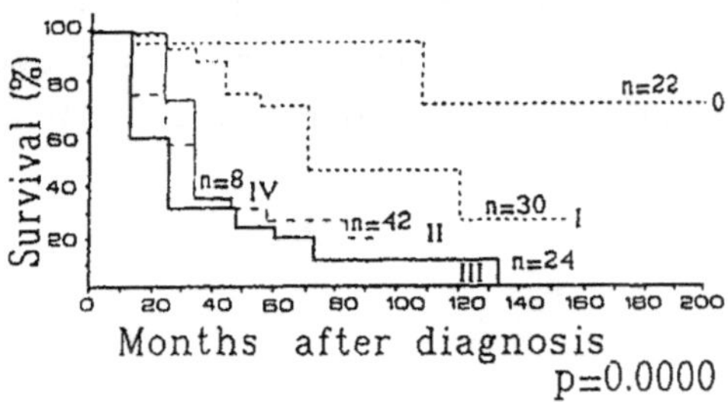

Fig. 1. Overall survival in multiple myeloma patients with different performance status.

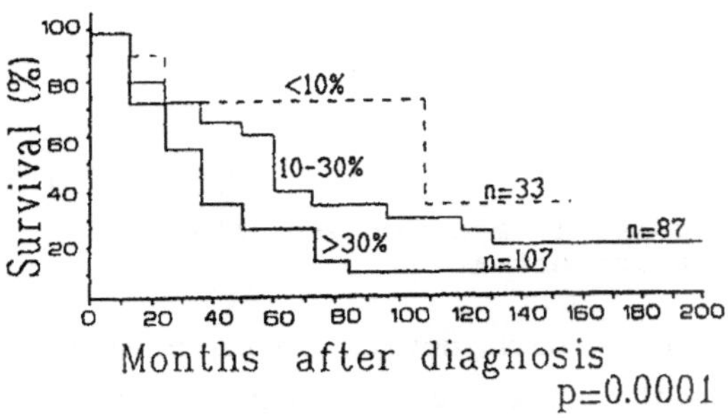

Fig. 2. Overall survival in multiple myeloma patients with different degree of bone marrow plasmacytosis.

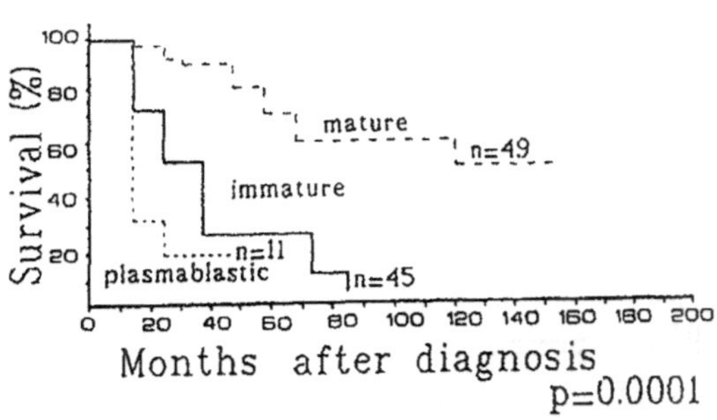

Fig. 3. Overall survival in multiple myeloma patients with different morphological types.

Median survival was highest in patients with blood haemoglobin concentration of more than 110 g/l- 72 months, and lowest in the patients with haemoglobin levels of less than 85 g/l- 24 months (Fig. 4).

Overall survival in MM patients with normal bones on X-ray was significantly higher than in patients with osteoporosis and especially with moderate and extensive bone lesions (Fig. 5)

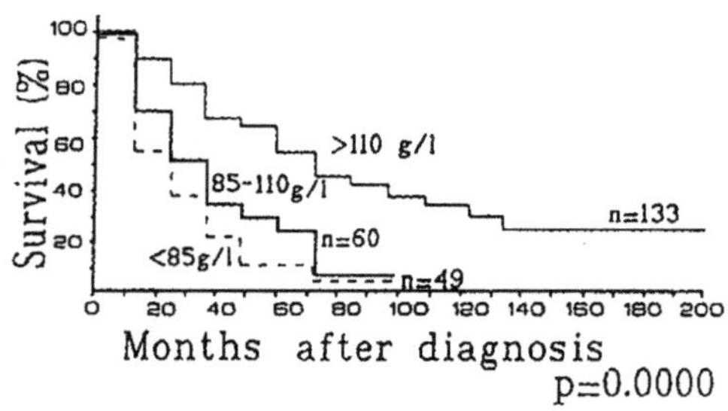

Fig. 4. Overall survival in multiple myeloma patients with different haemoglobin concentration.

Median survival of MM patients with normal serum creatinine level was 60 months, in patients with elevated serum creatinine level, 10 months (Fig. 6).

Median survival of MM patients with "a shift to the left" in the differential leukocytes counts was significantly lower (22 months) than in patients without such a shift (58 months) (Fig. 7).

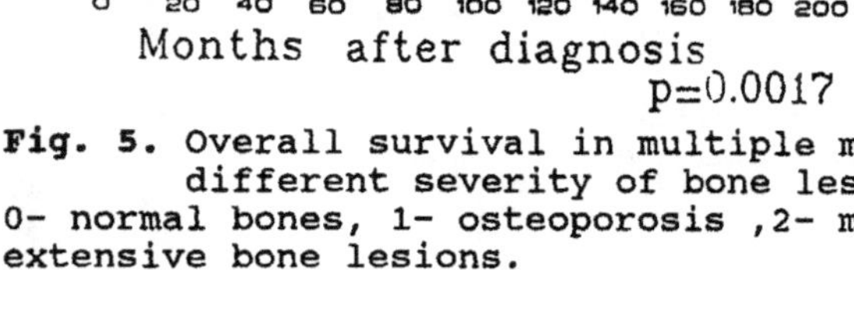

Fig. 5. Overall survival in multiple myeloma patients with different severity of bone lesions
0- normal bones, 1- osteoporosis ,2- moderate bone lesions,3- extensive bone lesions.

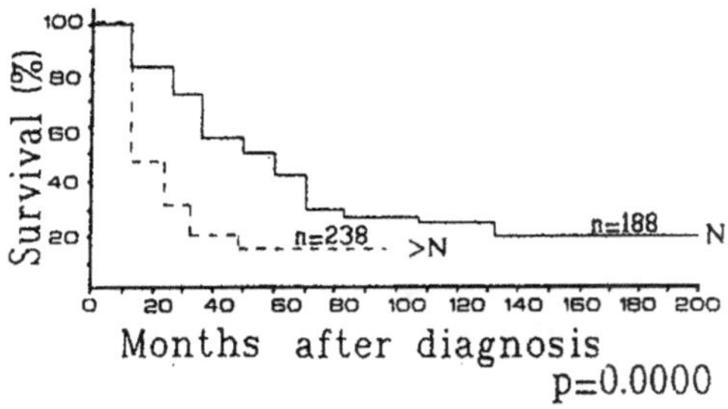

Fig. 6. Overall survival in multiple myeloma patients with normal and elevated serum creatinine concentration.

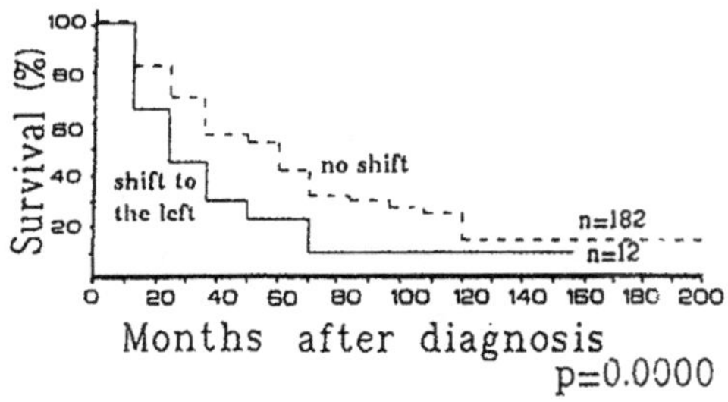

Fig. 7. Overall survival in multiple myeloma patients with or without "shift to the left" in leukocytes count.

Overall survival in MM patients with serum B2-microglobulin concentration of more than 6 g/l was significantly worse than in patients with serum B2-microglobulin concentration of less than 6 g/l (Fig.8).

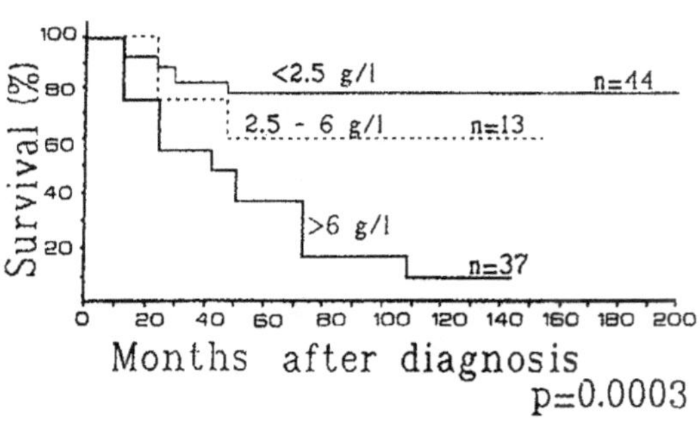

Fig. 8. Overall survival in multiple myeloma patients with different concentration of serum B2-microglobulin.

It was, however, impossible to predict the individual prognosis of MM patients according to the abovementioned factors of disease's presentation. That is why we classified our patients by their life expectancy, which depends on the activity of the disease. This classification is the result of prospective investigation. We have detected three groups : first - indolent myeloma , second - active myeloma , third - aggressive myeloma. The diagnosis was based on the criteria which are presented in Table 1. Survival curves in patients with different disease courses are shown in Fig. 9.

Multivariate discriminative analysis has shown that the main independent prognostic factors for survival (bone marrow plasmacytosis, plasma cells morphology, serum creatinine level, severity of bone lesions, "shift to the left" in differential leukocyte counts, serum B2-microglobulin concentration and also sex and dishaemopoietic signs in granulocytic compartment of bone marrow) provides accurate prognosis of MM course (indolent, active, aggressive) in 83.3% of patients at diagnosis (Tab. 2,3).

Table 1. Clinical staging system of multiple myeloma based on the disease activity.

Type of course	Criteria	Median survival (months)
Indolent	Patients with smoldering MM according to criteria of Kyle and Greipp (1980); patients who had been in remission for more than 5 years without chemotherapy.	---
Active	Patients who responded to first line chemotherapy.	72
Aggressive	Patients who showed primary or secondary resistance to che-chemotherapy.	24

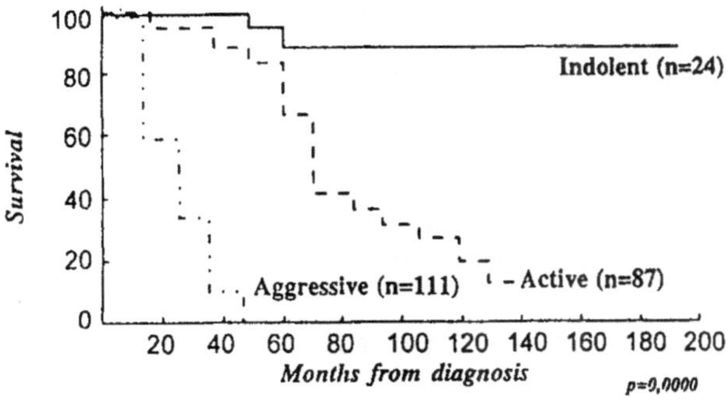

Fig. 9. Overall survival in multiple myeloma patients with different disease courses.

50% of MM patients presented with aggressive course, 39.5% with active course.

Study of the distribution of HLA-antigens (class I and II) revealed a significantly ($p<0.05$) increased frequency of HLA-B13 antigen (20.37%) in MM patients compared to controls (6.9%). The frequency of this antigen in patients with aggressive MM was 35,3%.

Table 2. Classification functions for determination type of MM course.

Coefficients for different forms of MM	Indolent	Active	Aggresive
Sex(1-Male; 2-Female)	17.32957	13.50706	11,37061
(Continued) % of plasma cells in bone marrow (1- 10; 2 - 10-30 ; 3 > 30)	-0.66950	2.15809	1.81066
Plasma cells morphology (1 - mature ; 2 - immature ;3 - plasmablastic)	6.77402	8.98926	10.27213
Serum creatinine level (mmol/l) 1 < 0,133 ;2 > 0,133	9.05185	7.90490	9.26515
Bone lesions 1-normal;2-osteoporosis; 3- moderate;4- extensive	5.36817	6.55825	7.06294
"Shift to the left" in the differential leucocyte count 1-Yes ; 2-No	20.31238	20.16097	15.66659
Serum B2-micro - globulin level (g/l) 1<2,5; 2 -2,5-6; 3>6	1.27943	1.15167	2.96559
Dishaematopoietic signs in granulocytic compartment of bone marrow 1- Yes;2- No	9.21029	12.63615	14.99142
Constant	-58.43399	-67.85653	-69.38179

Table 3. Classification matrix.

Type of course	Percent correct
Indolent	90.9
Active	85.7
Aggressive	77.3
Total	83.3

(p<0.05), in patients with active MM 12.0% , and in patients with indolent MM - 16,67 % (Tab. 4).

Table. 4. Frequency of HLA-B13 antigen in patients with different type of MM.

Type of MM course	Number of patients	Frequency of HLA-B13 (%)	X2	RR**
Aggressive	17	35.30*	19.62	7.41
Active	25	12.0	0.99	1.85
Indolent	12	16.67	1.75	2.71
Regional control	845	6.9	---	--

* - p< 0.05.
** - reliable when RR > 2

CONCLUSIONS

1. The use of the independent prognostic factors for survival (bone marrow plasmacytosis, plasma cells morphology, serum creatinine level, severity of bone lesions, "shift to the left" in differential leukocyte count, serum B2-microglobulin concentration and also sex and dishaemopoietic signs in granulocytic compartment) provide accurate prognosis of the course of myeloma in 83.3% of patients at diagnosis.

2. 50% of MM patients present with an aggressive course, 39.5% with an active course.

3. The frequency of the HLA-B13 antigen in aggressive MM is significantly higher compared to indolent, active MM, and controls.

4. HLA B-13 may be used as a reliable marker of the aggressive course and may serve as an additional criteria for

selection of patients for haemopoietic stem cells transplantation.

REFERENCES

Alexanian R, Balcerzak S, Bonnet J, Gehan E, Haut A, Hewlett J, Monto R (1975) Prognostic factors in multiple myeloma. Cancer 35 : 1192-1201.

Barlogie B, Jagannath S (1994) Autologous bone marrow transplantation: in Forman S, Blume K., Thomas E. (eds). Bone Marrow Transplantation pp 754.

Barlogie B, Salmon S, (1990) New treatment approaches for multiple myeloma. American Society of Clinical Oncology Educational Symposia pp 171-176

Carbone P, Kellerhouse L, Gehan E, (1967) Plasmacytic myeloma: a study of the relationship of survival to various clinical manifestations and anomalous protein type in 112 patients. Am. J. of Med. 42: 937-948.

Carter A, Hocherman J, Linn S, Cohen J, Tatarsky J, (1987) Prognostic significance of plasma cell morphology in multiple myeloma. Cancer 60: 1060-1065.

Chronic Leukemia-Myeloma Task Force (1973) National Cancer Institute: Proposed guidelines for protocol studies - II. Plasma cells myeloma. Cancer Chemother. Reports 4: 145-158.

Dure B, (1994) Beta-2 Microglobulin as a Prognostic Factor in mul-tiple myeloma, International Conference "Multiple Myeloma from biology to therapy. Current concepts", Mulhouse, France, Abstr.

Dure B, Salmon S. (1975) A clinical staging system of multiple myeloma. Correlation of measured myeloma cell mass with presenting clinical features, response to treatment and survival. Cancer 38: 842-854.

Friemel H, (1984) Immunologische Arbeitsmethoden, pp 176- 186.

Galton D, Peto R, (1973) Report on the first myelomatosis trial. Analysis of presenting features of prognostic importance.Brit. J. of Haematol. 24: 123- 139.

Kaplan E, Meier P, (1958) Nonparametric estimation from incomplete observations. J. Am. Stat. Assoc. 53 : 457-481.

Kyle R, (1988) Prognostic factors in multiple myeloma. Hematol. Oncol. 6 : 125-130.

Kyle R, Elveback L, (1976) Management and prognosis of multiple myeloma. Mayo Clinic Proceeding 51: 751- 760.

Kyle R, Grepp P. (1980) Smoldering multiple myeloma. New. Engl. J. Med. 302 : 1347-1349.

Lefkovits I, Pernis B, (1979) Immunological Methods pp 428-467.

Mantel N, (1966) Evaluation of survival data and two new, rank order statistics arising in its consideration. Cancer Chemother. Reports 50: 163.

Matzner J, Benbassat J, Polliack A, (1978) Prognostic factors in multiple myeloma. A retrospective study using conventional methods and a computer programme. Acta Haematol. 60: 257-268.

Medical Research Council's Working Party on Leukemia in Adults (1980) Prognostic features on the third MRC myelomatosis trial. Brit. J. Cancer 42 831-840.

Merlini G, Waldenstrom I., Jayakar S., (1980) a new improved clinical staging system for multiple myeloma based on the analysis of 123 treated patients. Blood 55 : 1011-1019.

Shabalin V, Serova L, (1988) The Clinical Immunohaematology pp 75-80.

Mobilization of Progenitor Cells for Autologous and Allogeneic Transplantation

A. Zander[1], W. Zeller[1], M. Stockschläder[1], C. Berger[1], W. Krüger[1], J. Grimm[1], K. Hummel[2], C. Löliger[2], P. Kühnl[2], H. Kabisch[3]
[1] Dept. Oncology and Hematology, BMT-Unit, [2] Dept. of Transfusion Medicine, [3] Dept. Pediactric Hematology, University Hospital Eppendorf, Hamburg, Germany

Introduction

The hematopoietic growth factors G-CSF and GM-CSF are able to stimulate granulopoiesis after autologous transplantation of bone marrow or blood stem cells (PBSC) and can also mobilize peripheral blood progenitor cells [1, 2, 3, 4]. Rapid recovery of platelets has been observed following addition of G-CSF mobilized PBSC to autologous bone marrow transplants [3]. The kinetics of engraftment are mainly dependent on the number and qualitiy of progenitor/stem cells in the graft [5, 6, 7, 8]. We have studied the impact of higher doses of G-CSF (Filgrastim) on the yield of CD34 positive cells, CFU-GM and mononucelar cells in leukapheresis products in patients with Non Hodgkin Lymphomas, Hodgkins disease, testicular cancer, AML and neuroblastoma as well as in normal donors for allogeneic progenitor cell transplantation. We report our experience with mobilization of G-CSF in two different doses: 10 µg/kg per day versus 24 (2 x 12) µg/kg per day for stem cell collection, separation and transplantation.

Materials and methods

Mobilization: G-CSF (Filgrastim) was administered subcutaneously for 5 - 6 days at a daily dose of 10 µg/kg (group A) or 2x12 µg/kg (group B) to 45 patients. Beyond this a total of 18 mobilizations followed by 1 to 4 leukapheresis procedures, was carried out in normal donors. 15 transplants, involving mobilized progenitor cells either alone or in conjunction with bone marrow, were carried out. The characteristics of the autologous patients are shown in table 1.

NATO ASI Series, Vol. H 94
Gene Technology
Edited by A. R. Zander, W. Ostertag,
B. V. Afanasiev, F. Grosveld

	group A (10μg) n = 25	group B (2x12μg) n = 20
Age (mean, Range)	35.6 (19-56)	35,8 (19-55)
Sex	9f, 16m	6f, 14m
NHL, high grade	12	8
NHL, low grade	6	5
Hodgkins disease	5	3
Testicular Cancer	1	2
AML	0	2
Neuroblastoma	1	0

Table 1. Patient characteristics

15 patients with the following diagnosis: AML: 4, ALL: 5, Multiple Myeloma: 1, Severe Aplastic Anaemia: 3, CML blastic crisis: 1, Myeloproliferative Syndromes 1 received an allogeneic stem cell transplant. The donor-recipient relationships were the following: matched siblings: 8, one mismatch: 2, parent to child: 3, child to parent: 1, matched unrelated donor: 1. Conditioning regimens: without conditioning as a rescue graft for rescue from insufficient bone marrow function: 5; TBI-Cyclophosphamide-VP16: 5; TBI-Cyclophosphamide: 1, Busulfan, Cyclophosphamid ± Etoposid: 4.

Colony forming assay: Colony forming units granulocyte-monocyte (CFU-GM) were assayed in Iscove's methylcellulose. Colonies were enumerated after 14 days of culture in 37° C, 5 % CO_2 and 100 % humidity by using an inverted phase microscope (9).

Flow cytometric analysis: Surface antigen expression was evaluated using the monoclonal antibody HPCA-2, directly conjugated to FITC (Becton-Dickinson). Analysis was performed on a FACScan (Becton-Dickinson) using Lysis II research software [5, 10, 11].

Statistics: Data are shown as median (range), statistical significance was determined by the Student's t-test.

Results

Autologous: The percentage of CD34+ mononuclear cells (MNC) in the first leukapheresis product increased significantly ($p=0.024$) from 0.68 (0.03-1.96) (group A) to 1.23 (0.4-4.28) (group B) (Fig. 1).

Fig.1 Leukapheresis products
10µg vs. 2x12µg G-CSF

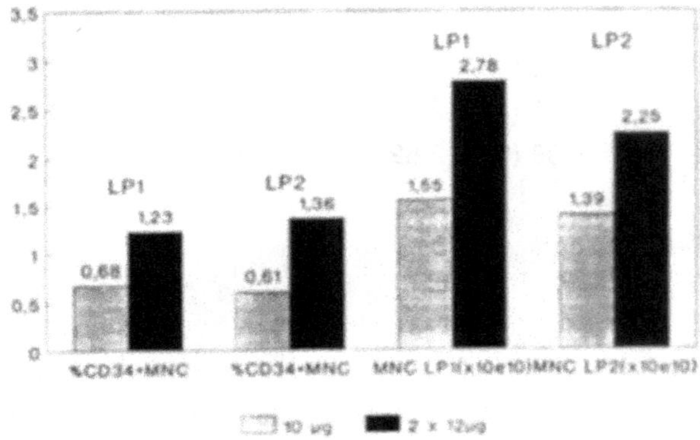

Statistically significant differences were also found with respect to total CD34+ MNC per leukapheresis product (LP1): 10.08×10^7 (0.4-44.0) (group A) versus 42.51×10^7 (3.46-86.3) (group B) ($p=0.012$) (Fig. 2, Tabl. 2).

Fig. 2 Leukapheresis products
10µg vs. 2x12µg G-CSF

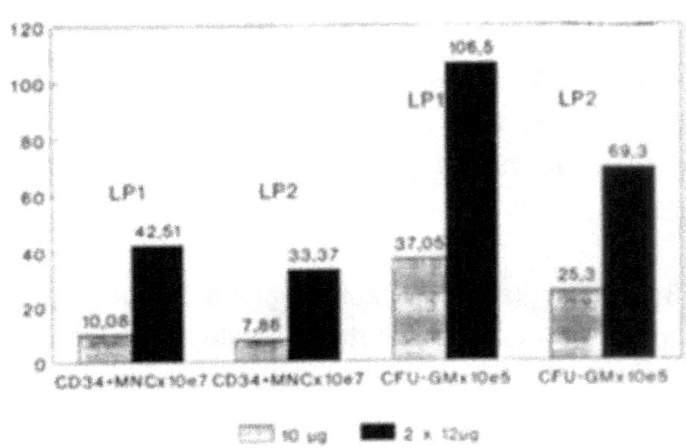

1st Leukapheresis	group A (n=25)	group B (n=20)	p
% CD34+ MNC	0,68 (0,03-1,96)	1,23 (0,4-4,28)	0,024
CD34+ MNC $x10^7$(total)	10,08 (0,4-44,0)	42,51 (3,46-86,3)	0,012
CFU-GM $x10^5$(total)	37,05 (1,6-95,1)	106,5 (8,7-390,5)	0,004
MNC $x10^{10}$(total)	1,55 (0,39-3,09)	2,78 (0,66-4,88)	0,02
2nd Leukapheresis			
%CD34+ MNC	0,61 (0,04-2,12)	1,36 (0,32-4,28)	0,011
CD34+ MNC $x10^7$ (total)	7,86 (0,2-30,1)	33,37 (5,5-91,6)	<0,001
CFU-GM $x10^5$ (total)	25,3 (1,64-267,2)	69,3 (3,9-199,2)	0,047
MNC $x10^{10}$ (total)	1,39 (0,35-2,62)	2,25 (0,95-4,82)	0,003

Table 2. Graft Data

CFU-GM increased from 37.05 (1.6 - 95.1) x 10^5 (group A) to 106.5 (8.7 - 390.5) (group B) (p = 0.004). These results confirm, that higher doses of G-CSF can significantly increase the number of CD34 positive cells and CFU-GM in leukapheresis products.

Allogeneic: An average of 3 leukapheresis (range 1 to 4) were carried out, which yielded 13 x 10^8 (range 2.9 - 29) mononuclear cells/kg BW of the recipient, 5 x 10^6 CD34 positive cells/kg BW of recipient (range 1.2 - 39), CFC 3 x 10^5 /kg BW (0,88 - 60,4). The mobilization with G-CSF was generally well tolerated with the exception of moderate bone pain in 2/3 of the donors.

Engraftment in 9 evaluable patients revealed neutrophil engraftment of more than 0,5/nl by day 14 (range 9 - 22), neutrophils more than 1.0/nl day 15 (range 10 - 24), thrombocytes more than 20/nl day 16 (range 8 - 32), more than 50/nl 24 (range 13 to more than 100 days).

GvH grade 0 - 1: 3 patients, grade 2: 4 patients, grade 4: 2 patients and moderate chronic GvH in 1 patient. Total of 12 of the 15 patients are alive in continuous complete remission, 3 have died of GvH and infection (aspergillus 1, candida crusei 1).

Discussion

We could show that the mobilization with a higher than usual cosis of G-CSF is well tolerated in normal donors as well as in patients and yields grafts which allow engraftment after myeloablative therapy. GvH occured in 2/3 of evaluable patients and led in conjunction with infection to death in 3 patients. 1 patient received an allogeneic graft from a matched unrelated donor, showed engraftment but unfortunately died of a preexisting fungus disease.

Based on our experience in the autologous setting, where we found statistically significant differences in the number of CD34+ cells and CFU-GM in patients mobilized with 10 µg/kg G-CSF compared with 24 µg/kg and based on the observation that the side effects in both dose levels are comparable, we would recommend 24 µg/kg BW G-CSF per day, given in 2 divided doses or continuously intravenous infusion for mobilization of normal donors.

References

1. Socinski MA, Cannistra SA, Elias A, Antman KH, Schnipper L, Griffin JD: Granulocyte-macrophage colony stimulating factor expands the circulating haemopoietic progenitor cell compartment in man. Lancet 1: 1194, 1988

2. Dührsen U, Villeval JL, Boyd J, Kannourakis G, Morstyn G, Metcalf D: Effects of recombinant human granulocyte colony-stimulating factor on hematopoietic progenitor cells in cancer patients. Blood 72: 2074, 1988

3. Sheridan WP, Begley CG, Jutner CA, Szer J, To LB, Maher D, McGrath KM, et al.: Effect of peripheral-blood progenitor cells mobilised by filgrastim (G-CSF) on platelet recovery after high-dose chemotherapy. Lancet 339: 640 1992

4. Gianni AM, Siena S, Bregni M, Tarella C, Stern AC, Piler A, Bonnadonna G: Granulocyte-macrophage colony-stimulating factor to harvest circulating haemopoietic stem cells for autotransplantation. Lancet 2: 580, 1989

5. Siena S, Bregni M, Brando B, Belli N, Ravagnani F, Gandola L, et al.: Flow cytometry for clinical estimation of circulating hematopoietic progenitors for autologous transplantation in cancer patients. Blood 77: 400, 1991

6. Weaver CH, Buckner CD, Longin F, Appelbaum FR, Rowley S, Lilleby K, et al.: Syngenic transplantation with peripheral blood mononucelar cells collected after administration of recombinant human granulocyte colony-stimulating factor. Blood 82: 1981, 1993

7. Chao NJ, Schriber JR, Grimes K, Long GD, Negrin RS, Raimondi CM, et al.: Granulocyte colony-stimulating factor "mobilized" peripheral blood progenitor cells accelerate granulocyte and platelet recovery after high-dose chemotherapy. Blood 81: 2031, 1993

8. Bensinger W, Singer J, Appelbaum F, Lilleby K, Longin K, Rowley S, et al.: Autologous transplantation with peripheral blood mononuclear cells collected after administration of recombinant granulocyte stimulating factor. Blood 81: 3158, 1993

9. Messner HA, Curtis JE, Minden MD, Tritchler D, Lockwood G, Takahashi T, Lepine J, Jamal N, Tweeddale M, Wandl U: Clonogenic hemopoietic precursors in bone marrow transplantation. Blood 70: 1425, 1987

10. Berenson RJ, Andrews RG, Bensinger WI, et al.: Selection of CD34+ marrow cells for autologous marrow transplantation. Autologous Bone Marrow Transplantation: Proceedings of the 4th International Symposium, Dicke KA, Spitzer G, Jagannath S, et al. (eds.). Houston, Univ. of Texas, 1989, 55

11. Shpall EJ, Jones RB, Bearman SI, Franklin WA, Archer PG, Curiel T, Bitter M, et. al.: Transplantation of enriched CD34-positive autologous marrow into breast cancer patients following high-dose chemotherapy: influence of CD34-positive peripheral-blood progenitors and growth factors on engraftment. Journal of Clinical Oncology 12: 1994, 28

Transfusion problems of allogenic bone marrow transplantation (BMT).

L.Fregatova, G.Platonova, O.Volkova, S.Shavva, B.Afanasiev.

Department of BMT, N.N.Petrov
Research Institute of Oncology,
Leningradskaya 68,
198646 St-Petersburg, Russia

It is well known that BMT is associated with certain immunohematological and transfusion problems. Such problems are related to the fact that inheritance of blood group antigens is independent of the HLA gene complex so that even if the donor and the recipient are HLA-matched, their red blood cells (RBC) may be of different genotypes. There may be a minor or a major incompatibility, which is important for the ABO, Rh, and Kell antigens systems.

In our BMT Center 35 allogeneic bone marrow transplantations (allo-BMT) have been performed in patients with hematological diseases from HLA-identical siblings (16 were RBC-matched and 19-mismatched: 11 major, 6 minor, 2 major minor). This number of cases of RBC-mismatched BMT is rather high when compared to the literature (Petz L.D. 1987), but might be explained by ethnic and geographical peculiarities of our country.

Since RBC incompatibilities between the donor and recipient pairs were present, a number of immunohematological complications and transfusion problems after BMT can be anticipated.

The first problem is the possibility of acute hemolysis at the time of donor bone marrow (BM) infusion, especially in cases with major ABO-mismatching. Several pretransplant measurements have been undertaken in an attempt of preventing the development of hemolysis at the time of marrow graft infusion. This is necessary because the marrow contains about the same concentration of RBC as whole blood, and may induce severe hemolytic transfusion reactions.

Prophylaxis of acute hemolysis involved attempts of reducing anti-A and anti-B titers in recipient before BMT (Buchner C.D.et al, 1978). All these methods are, however, time-consuming, and should be carried out before transplantation, at a time, when the patient is undergoing conditioning regimen. In some patients, the reduction of titers is not efficient. There are alternative, commonly used methods of RBC-depletion in donor marrow products. Several teams have reported the results on gravity sedimentation of RBC with different agents and the use of cell separation equipment (Miller A. et al. 1989, Ratajczak. M. et al. 1989, Dragani A. et al. 1990, Rodriger J.M., 1992).

We compared the results of RBC sedimentation with 6% HES and our modified method (combination of the "buffy coat" method with consequent removal of RBC by sedimentation with low weight 10% dextran - "Polyglucin"). "Polyglucin" was used at three different ratios to BM (2:3; 1:1; 2:1), 6% HES at two ratios (1:8; 1:4). We found, that all types of sedimentation with "Polyglucin" provided the same recovery of nucleated cell (78.0 $\pm$ 3.0%; 81 $\pm$ 2.78% and

NATO ASI Series, Vol. H 94
Gene Technology
Edited by A. R. Zander, W. Ostertag,
B. V. Afanasiev, F. Grosveld

84.3 ± 2.98%, respectively) with moderate admixture of RBC (5.7 ± 1.14%, 6.6 ± 0.9% and 6.0 ± 0.8%, respectively).

In general, the results of RBC sedimentation with 6% HES provided a higher recovery of nucleated cells (89.7 ± 3.33% and 88.0 ± 4.0%) with RBC admixture of 11.1 ± 1.8% and 6.8 ± 0.8%, respectively.

Based on these results, we decided to use sedimentation of RBC with 6% HES in 1:4 ratio with BM for separation in cases of RBC-mismatching. BM separated in such a way was well-tolerated and there was no hemolytic complication at the time of infusions.

In four cases with extremely small volumes of bone marrow (less than 100 ml) in young patients, we used a simple and reliable technique for RBC removal as described by Chan (1991). The first centrifugation was followed by substitution of 50% of donor-incompatible RBC with RBC-matched with both donor and recipient (usually 0-group). The RBC were washed and irradiated. They created a "shelf", in order to facilitate cell separation. Hence, in the resulting marrow product only 50% of RBC (1 - 1.5%) were still incompatible with the recipient.

The second problem was delayed onset of immune hemolysis after BMT. In cases of major ABO incompatibility the development of immune hemolysis in the post-BMT period is due to erytropoiesis of BM graft (Hows J. et al, 1986). The management of this condition included the transfusion of fluids, 0-group RBC, corticosteroids, and hemodialysis in cases with renal failure.

We may illustrate this with a case report (#5163). A young patient, 9 years old, with ALL (0-blood group, Rh-positive) was transplanted from his sibling, 11 years old (A-blood group, Rh-positive). Marrow infusion of the graft separated with "Polyglucin" was well tolerated, without evidence of transfusion reaction. The patient's pretransplant serum contained anti-A (IgM/IgG titers of 8/8). The increase of anti-A titers occured on day 37th after BMT (up to IgM/IgG titers of 16/128). The hematocrit dropped to 22%, the direct antiglobulin test became positive for IgG, and a delayed onset of erythropoiesis until 206 days was observed as well. The patient's condition improved following therapy with i.v. fluids, corticosteroids and transfusion of 20 units of 0-group RBC.

In cases of minor compatibility the most probable mechanism of delayed hemolysis is the active production of antibodies by the donors lymphocytes, rather than passive transfer of antibodies with marrow grafts (Branch D.R. et al, 1984). We did not observe such complications.

The third problem, which is closely related with the previous one, is the delayed onset of hematopoiesis, especialy erythropoiesis, in some patients with AB0-mismatched BMT. Bone marrow examination revealed that delay of erythroid maturation may be advantageous for the patient as a protective factor against the consequences of major hemolytic anemia, until the circulating hemagglutinins will be reduced to a clinicaly insignificant level (Petz L.D., 1987). In our study, onset of erythropoies differed between RBC-matched and mismatched transplants. A delay of reticulocyte recovery (<1%) was seen in the group with major RBC incompatibility, compared to groups with minor incompatibility and compatible transplants (on the average, 40, 34, and 25 days, respectively). In two cases with immunohematologic problems this delay was significant (118 and 206 days).

As a result, there was increased posttransplant requirement for RBC transfusions. This is the fourth problem of AB0-mismatched allogeneic BMT. On the average, 12 units of RBC were transfused during the posttransplant period in the group of AB0-mismatched patients as compared to an average of 6 units in AB0-matched transplants. Two of 19 patients had extreme tranusfusion requirement (20 and 22 units of RBC).

CONCLUSIONS

This study showed that manual methods of RBC depletion of BM (with "Polyglucin" and 6% HES) provide satisfactory results. We also revealed the necessity of a constant antibody screenig and control of the direct antiglobulin test in the post-BMT period for early detection of immune hemolysis and for timely treatment. We now suggest a greater requirement for transfusions in cases with immunohematological problems in post-BMT period.

REFERENCES

Branch D.R., Sniecinski I., Spruce W., Krance R.A., Petz L.D., St.Jean J. (1984) Delayed immune hemolytic anemia following minor ABO - mismatched bone marrow transplantation. Blood 64; suppl.1:755A.

Bucher C.D., Ceift K.A., Sanders J.E., Williams B., Gray M., Storb R., Thomas E.D. (1978) ABO-incompatible marrow transplants. Transplantation 26:233-238

Chan K.W., Herd S.L., Stanley C.E., Wadsworth L.D. (1991) ABO-incompatible pediatric bone marrow transplantations: A simple method to remove red blood cells from small volume marrow
grafts. Bone marrow transplantation 8:473-475.

Dragani A., Angelini A., Jacone A., D'Antonio D.,Forlontano G. (1990) Comparison of five methods for concentrating progenitor cells in human marrow transplantation. Blut 60,5:278-281

Hows J.,Beddow K., Gordon-Smith E., Branch D.R., Spruce W., Sniecinski I., Krance R.I.,Petz L.D. (1986) Donor derived red cell antibodies and immune hemolysis after allogenic bone marrow transplantation. Blood 67,7:177-181

Miller A., Hermann J., Fuchs D., Prager V., Burger E., Kob D., Goetz I., Zintl F. (1989) Transplantation of major ABO-incompatible bone marrow: Removal of red cells by dextran sedimentation. Folia Hematology 3 - 4:475-479.

Petz L.D. (1987) Immunohematologic problems associated with bone marrow transplantation. Transfusion Medicine Reviews 1,2:85-100.

Ratajczak M.Z., Szczylik C., Wiktor-Jedrzejezak W. (1989) Elimination of red blood cells from marrow suspension prior to transplantation using polysaccharide-ditrizoate gradient centrifugation. Folia Hematology 3-4:469-473

Rodriguez J.M., Carmona M., Noguerol P., Parody R., Vidal F., Pevez - Hurtato J.M., Espigado I. (1992) A fully automated method for mononuclear bone marrow cell concentration. Jornal of Clinical Apheresis 7:101-109.

NATO ASI Series H

Vol. 1: **Biology and Molecular Biology of Plant-Pathogen Interactions.** Edited by J.A. Bailey. 415 pages. 1986.

Vol. 2: **Glial-Neuronal Communication in Development and Regeneration.** Edited by H.H. Althaus and W. Seifert. 865 pages. 1987.

Vol. 3: **Nicotinic Acetylcholine Receptor: Structure and Function.** Edited by A. Maelicke. 489 pages. 1986.

Vol. 4: **Recognition in Microbe-Plant Symbiotic and Pathogenic Interactions.** Edited by B. Lugtenberg. 449 pages. 1986.

Vol. 5: **Mesenchymal-Epithelial Interactions in Neural Development.** Edited by J. R. Wolff, J. Sievers, and M. Berry. 428 pages. 1987.

Vol. 6: **Molecular Mechanisms of Desensitization to Signal Molecules.** Edited by T M. Konijn, P J. M. Van Haastert, H. Van der Starre, H. Van der Wel, and M.D. Houslay. 336 pages. 1987.

Vol. 7: **Gangliosides and Modulation of Neuronal Functions.** Edited by H. Rahmann. 647 pages. 1987.

Vol. 8: **Molecular and Cellular Aspects of Erythropoietin and Erythropoiesis.** Edited by I.N. Rich. 460 pages. 1987.

Vol. 9: **Modification of Cell to Cell Signals During Normal and Pathological Aging.** Edited by S. Govoni and F. Battaini. 297 pages. 1987.

Vol. 10: **Plant Hormone Receptors.** Edited by D. Klämbt. 319 pages. 1987.

Vol. 11: **Host-Parasite Cellular and Molecular Interactions in Protozoal Infections.** Edited by K.-P. Chang and D. Snary. 425 pages. 1987.

Vol. 12: **The Cell Surface in Signal Transduction.** Edited by E. Wagner, H. Greppin, and B. Millet. 243 pages. 1987.

Vol. 13: **Toxicology of Pesticides: Experimental, Clinical and Regulatory Perspectives.** Edited by L.G. Costa, C.L. Galli, and S.D. Murphy. 320 pages. 1987.

Vol. 14: **Genetics of Translation. New Approaches.** Edited by M.F. Tuite, M. Picard, and M. Bolotin-Fukuhara. 524 pages. 1988.

Vol. 15: **Photosensitisation. Molecular, Cellular and Medical Aspects.** Edited by G. Moreno, R. H. Pottier, and T. G. Truscott. 521 pages. 1988.

Vol. 16: **Membrane Biogenesis.** Edited by J.A.F Op den Kamp. 477 pages. 1988.

Vol. 17: **Cell to Cell Signals in Plant, Animal and Microbial Symbiosis.** Edited by S. Scannerini, D. Smith, P. Bonfante-Fasolo, and V. Gianinazzi-Pearson. 414 pages. 1988.

Vol. 18: **Plant Cell Biotechnology.** Edited by M.S.S. Pais, F. Mavituna, and J. M. Novais. 500 pages. 1988.

Vol. 19: **Modulation of Synaptic Transmission and Plasticity in Nervous Systems.** Edited by G. Hertting and H.-C. Spatz. 457 pages. 1988.

Vol. 20: **Amino Acid Availability and Brain Function in Health and Disease.** Edited by G. Huether. 487 pages. 1988.

NATO ASI Series H

Vol. 21: **Cellular and Molecular Basis of Synaptic Transmission.** Edited by H. Zimmermann. 547 pages. 1988.

Vol. 22: **Neural Development and Regeneration. Cellular and Molecular Aspects.** Edited by A. Gorio, J. R. Perez-Polo, J. de Vellis, and B. Haber. 711 pages. 1988.

Vol. 23: **The Semiotics of Cellular Communication in the Immune System.** Edited by E.E. Sercarz, F. Celada, N.A. Mitchison, and T. Tada. 326 pages. 1988.

Vol. 24: **Bacteria, Complement and the Phagocytic Cell.** Edited by F. C. Cabello und C. Pruzzo. 372 pages. 1988.

Vol. 25: **Nicotinic Acetylcholine Receptors in the Nervous System.** Edited by F. Clementi, C. Gotti, and E. Sher. 424 pages. 1988.

Vol. 26: **Cell to Cell Signals in Mammalian Development.** Edited by S.W. de Laat, J.G. Bluemink, and C.L. Mummery. 322 pages. 1989.

Vol. 27: **Phytotoxins and Plant Pathogenesis.** Edited by A. Graniti, R. D. Durbin, and A. Ballio. 508 pages. 1989.

Vol. 28: **Vascular Wilt Diseases of Plants. Basic Studies and Control.** Edited by E. C. Tjamos and C. H. Beckman. 590 pages. 1989.

Vol. 29: **Receptors, Membrane Transport and Signal Transduction.** Edited by A. E. Evangelopoulos, J. P. Changeux, L. Packer, T. G. Sotiroudis, and K.W.A. Wirtz. 387 pages. 1989.

Vol. 30: **Effects of Mineral Dusts on Cells.** Edited by B.T. Mossman and R.O. Begin. 470 pages. 1989.

Vol. 31: **Neurobiology of the Inner Retina.** Edited by R. Weiler and N.N. Osborne. 529 pages. 1989.

Vol. 32: **Molecular Biology of Neuroreceptors and Ion Channels.** Edited by A. Maelicke. 675 pages. 1989.

Vol. 33: **Regulatory Mechanisms of Neuron to Vessel Communication in Brain.** Edited by F. Battaini, S. Govoni, M.S. Magnoni, and M. Trabucchi. 416 pages. 1989.

Vol. 34: **Vectors asTools for the Study of Normal and Abnormal Growth and Differentiation.** Edited by H. Lother, R. Dernick, and W. Ostertag. 477 pages. 1989.

Vol. 35: **Cell Separation in Plants: Physiology, Biochemistry and Molecular Biology.** Edited by D. J. Osborne and M. B. Jackson. 449 pages. 1989.

Vol. 36: **Signal Molecules in Plants and Plant-Microbe Interactions.** Edited by B.J.J. Lugtenberg. 425 pages. 1989.

Vol. 37: **Tin-Based Antitumour Drugs.** Edited by M. Gielen. 226 pages. 1990.

Vol. 38: **The Molecular Biology of Autoimmune Disease.** Edited by A.G. Demaine, J-P. Banga, and A.M. McGregor. 404 pages. 1990.

NATO ASI Series H

Vol. 39: **Chemosensory Information Processing.** Edited by D. Schild. 403 pages. 1990.

Vol. 40: **Dynamics and Biogenesis of Membranes.** Edited by J. A. F. Op den Kamp. 367 pages. 1990.

Vol. 41: **Recognition and Response in Plant-Virus Interactions.** Edited by R. S. S. Fraser. 467 pages. 1990.

Vol. 42: **Biomechanics of Active Movement and Deformation of Ce ls.** Edited by N. Akkas. 524 pages. 1990.

Vol. 43: **Cellular and Molecular Biology of Myelination.** Edited by G. Jeserich, H. H. Althaus, and T. V. Waehneldt. 565 pages. 1990.

Vol. 44: **Activation and Desensitization of Transducing Pathways.** Edited by T. M. Konijn, M. D. Houslay, and P. J. M. Van Haastert. 336 pages. 1990.

Vol. 45: **Mechanism of Fertilization: Plants to Humans.** Edited by B. Dale. 710 pages. 1990.

Vol .46: **Parallels in Cell to Cell Junctions in Plants and Animals.** Edited by A. W Robards, W. J . Lucas, J . D. Pitts, H . J . Jongsma, and D. C. Spray. 296 pages. 1990.

Vol. 47: **Signal Perception and Transduction in Higher Plants.** Edited by R. Ranjeva and A. M. Boudet. 357 pages. 1990.

Vol. 48: **Calcium Transport and Intracellular Calcium Homeostasis** Edited by D. Pansu and F. Bronner. 456 pages. 1990.

Vol. 49: **Post-Transcriptional Control of Gene Expression.** Edited by J. E. G. McCarthy and M. F. Tuite. 671 pages. 1990.

Vol. 50: **Phytochrome Properties and Biological Action.** Edited by B. Thomas and C. B. Johnson. 337 pages. 1991.

Vol. 51: **Cell to Cell Signals in Plants and Animals.** Edited by V. Neuhoff and J. Friend. 404 pages. 1991.

Vol. 52: **Biological Signal Transduction.** Edited by E. M . Ross and K . W. A. Wirtz. 560 pages. 1991.

Vol. 53: **Fungal Cell Wall and Immune Response.** Edited by J. P. Latge and D. Boucias. 472 pages. 1991.

Vol. 54: **The Early Effects of Radiation on DNA.** Edited by E. M. Fielden and P. O'Neill. 448 pages. 1991.

Vol. 55: **The Translational Apparatus of Photosynthetic Organelles.** Edited by R. Mache, E. Stutz, and A. R. Subramanian. 260 pages. 1991.

Vol. 56: **Cellular Regulation by Protein Phosphorylation.** Edited by L. M. G. Heilmeyer, Jr. 520 pages. 1991.

NATO ASI Series H

Vol. 57: **Molecular Techniques in Taxonomy.**
Edited by G . M . Hewitt, A. W. B. Johnston, and J. P. W. Young . 420 pages. 1991.

Vol. 58: **Neurocytochemical Methods.**
Edited by A. Calas and D. Eugene. 352 pages. 1991.

Vol. 59: **Molecular Evolution of the Major Histocompatibility Complex.**
Edited by J. Klein and D. Klein. 522 pages. 1991.

Vol. 60: **Intracellular Regulation of Ion Channels.**
Edited by M. Morad and Z. Agus. 261 pages. 1992.

Vol. 61: **Prader-Willi Syndrome and Other Chromosome 15q Deletion Disorders.**
Edited by S. B. Cassidy. 277 pages. 1992.

Vol. 62: **Endocytosis. From Cell Biology to Health, Disease and Therapie.**
Edited by P. J. Courtoy. 547 pages. 1992.

Vol. 63: **Dynamics of Membrane Assembly.**
Edited by J. A. F. Op den Kamp. 402 pages. 1992.

Vol. 64: **Mechanics of Swelling. From Clays to Living Cells and Tissues.**
Edited by T. K. Karalis. 802 pages. 1992.

Vol. 65: **Bacteriocins, Microcins and Lantibiotics.**
Edited by R. James, C. Lazdunski, and F. Pattus. 530 pages. 1992.

Vol. 66: **Theoretical and Experimental Insights into Immunology.**
Edited by A. S. Perelson and G. Weisbuch. 497 pages. 1992.

Vol. 67: **Flow Cytometry. New Developments.**
Edited by A. Jacquemin-Sablon. 1993.

Vol. 68: **Biomarkers. Research and Application in the Assessment of Environmental Health.** Edited by D. B. Peakall and L. R. Shugart. 138 pages. 1993.

Vol. 69: **Molecular Biology and its Application to Medical Mycology.**
Edited by B. Maresca, G. S. Kobayashi, and H. Yamaguchi. 271 pages. 1993.

Vol. 70: **Phospholipids and Signal Transmission.**
Edited by R. Massarelli, L. A. Horrocks, J. N. Kanfer, and K. Löffelholz. 448 pages. 1993.

Vol. 71: **Protein Synthesis and Targeting in Yeast.**
Edited by A. J. P. Brown, M. F. Tuite, and J. E. G. McCarthy. 425 pages. 1993.

Vol. 72: **Chromosome Segregation and Aneuploidy.**
Edited by B. K. Vig. 425 pages. 1993.

Vol. 73: **Human Apolipoprotein Mutants III. In Diagnosis and Treatment.**
Edited by C. R. Sirtori, G. Franceschini, B. H. Brewer Jr. 302 pages. 1993.

NATO ASI Series H

Vol. 74: **Molecular Mechanisms of Membrane Traffic.**
Edited by D. J. Morré, K. E. Howell, and J. J. M. Bergeron. 429 pages. 1993.

Vol. 75: **Cancer Therapy. Differentiation, Immunomodulation and Angiogenesis.**
Edited by N. D'Alessandro, E. Mihich, L. Rausa, H. Tapiero, and T. R. Tritton. 299 pages. 1993.

Vol. 76: **Tyrosine Phosphorylation/Dephosphorylation and Downstream Signalling.**
Edited by L. M. G. Heilmeyer Jr. 388 pages. 1993.

Vol. 77: **Ataxia-Telangiectasia.** Edited by R. A. Gatti, R. B. Painter. 306 pages. 1993.

Vol. 78: **Toxoplasmosis.** Edited by J. E. Smith. 272 pages. 1993.

Vol. 79: **Cellular Mechanisms of Sensory Processing. The Somatosensory System.**
Edited by L. Urban. 514 pages. 1994.

Vol. 80: **Autoimmunity: Experimental Aspects.**
Edited by M. Zouali. 318 pages. 1994.

Vol. 81: **Plant Molecular Biology. Molecular Genetic Analysis of Plant Development and Metabolism.**
Edited by G. Coruzzi, P. Puigdomènech. 579 pages. 1994.

Vol. 82: **Biological Membranes: Structure, Biogenesis and Dynamics.**
Edited by Jos A. F. Op den Kamp. 367 pages. 1994.

Vol. 83: **Molecular Biology of Mitochondrial Transport Systems.**
Edited by M. Forte, M. Colombini. 420 pages. 1994.

Vol. 84: **Biomechanics of Active Movement and Division of Cells.**
Edited by N. Akkaş. 587 pages. 1994.

Vol. 85: **Cellular and Molecular Effects of Mineral and Synthetic Dusts and Fibres.**
Edited by J. M. G. Davis, M.-C. Jaurand. 448 pages. 1994.

Vol. 86: **Biochemical and Cellular Mechanisms of Stress Tolerance in Plants.**
Edited by J. H. Cherry. 616 pages. 1994.

Vol. 87: **NMR of Biological Macromolecules.**
Edited by C. I. Stassinopoulou. 616 pages. 1994.

Vol. 88: **Liver Carcinogenesis. The Molecular Pathways.**
Edited by G. G. Skouteris. 502 pages. 1994.

Vol. 89: **Molecular and Cellular Mechanisms of H^+ Transport.**
Edited by B. H. Hirst. 504 pages. 1994.

Vol. 90: **Molecular Aspects of Oxidative Drug Metabolizing Enzymes: Their Significance in Environmental Toxicology, Chemical Carcinogenesis and Health.**
Edited by E. Arınç, J. B. Schenkman, E. Hodgson. 623 pages. 1994.

NATO ASI Series H

Springer-Verlag and the Environment

We at Springer-Verlag firmly believe that an international science publisher has a special obligation to the environment, and our corporate policies consistently reflect this conviction.

We also expect our business partners – paper mills, printers, packaging manufacturers, etc. – to commit themselves to using environmentally friendly materials and production processes.

The paper in this book is made from low- or no-chlorine pulp and is acid free, in conformance with international standards for paper permanency.

GPSR Compliance
The European Union's (EU) General Product Safety Regulation (GPSR) is a set of rules that requires consumer products to be safe and our obligations to ensure this.

If you have any concerns about our products, you can contact us on

ProductSafety@springernature.com

In case Publisher is established outside the EU, the EU authorized representative is:

Springer Nature Customer Service Center GmbH
Europaplatz 3
69115 Heidelberg, Germany

www.ingramcontent.com/pod-product-compliance
Lightning Source LLC
Chambersburg PA
CBHW071407230926
51846CB00001B/30
* 9 7 8 3 6 4 2 6 4 7 0 4 8 *